ELEMENTS OF ENGINEERING DESIGN: An Integrated Approach

Martyn S. Ray

*Department of Chemical Engineering,
Western Australian Institute of Technology*

PRENTICE HALL

New York   London   Toronto   Sydney   Tokyo

First published 1985 by
Prentice Hall International (UK) Ltd,
66 Wood Lane End, Hemel Hempstead,
Hertfordshire, HP2 4RG
A division of
Simon & Schuster International Group

Printed and bound in Great Britain at
the University Press, Cambridge

Library of Congress Cataloging-in-Publication Data

Ray, Martyn S., 1949–
  Elements of engineering design.
  Includes bibliographies and index.
  1. Engineering design
  I. Title
  TA174.R37 1985 620'.00425 84-18052

  ISBN 0-13-264185-2

British Library Cataloguing in Publication Data

Ray, Martyn S.
  Elements of engineering design: an integrated
    approach.
  1. Engineering design
  I. Title
  620'.00425   TA174

  ISBN 0-13-264185-2
  ISBN 0-13-264177-1 Pbk

---

2 3 4 5   92 91 90 89 88

Contents

Preface

Our whole life is affected by engineering — home, work and leisure, in fact everything we do or touch. The engineers who have such an influence upon our lives spend a large proportion of their time involved in design activities. However, the meaning of this term seems shrouded in mystery.

What is design?

What does a designer actually do?

The student's question is often:

'But how is design involved in (the topics I'm studying in) my course?'

Just as engineering influences all aspects of our life, so design is inherent in all aspects of engineering. Engineering is concerned with using available resources to satisfy the needs of society, and design is the definition of that need and the implementation of the plan required to produce a solution (see Chapter 1 for a more detailed discussion).

Engineering courses are divided into units, each of which is concerned with teaching particular principles and applications. The student should (eventually) appreciate how all this knowledge can be applied or utilised in a design problem, and also how the many aspects of a problem are interrelated. The aim of this book is to provide an integrated approach to the many elements involved in a design problem.

Chapter 1 contains some questions and ideas which should help answer such questions as: 'What does an engineer do?' and 'What is design?'

Chapter 2 describes the design process — the concept of design, the stages that are involved and the way in which a design study should evolve.

Chapters 3 to 10 each concentrate on a particular topic of importance in engineering design, e.g. material selection, CAD, economics. These topics are discussed in relation to engineering situations: the aim is to show their interrelationship in a design study.

Chapter 11 discusses the importance and the role of engineering in society, and the responsibilities of engineers in solving world problems.

The problems in this book do not have definite 'correct' solutions. The answers depend upon how the problems are defined and interpreted, and the data that are used. The chapter material is deliberately non-mathematical and non-specialist; it is intended as an introduction to the topic, not as an in-depth study.

Problems are provided at the beginning of each chapter, with suggestions and ideas in Chapters 2 and 3. The reader should consider these problems initially, and then reassess the answers after more detailed study and discussion. A list of keywords is included at the end of most chapters as a revision and reference aid. Each chapter also has an annotated bibliography.

My hope is that the reader will view engineering as an interesting and challenging activity with far-reaching possibilities and consequences. The successful and happy engineer will see his or her work as more than just a series of technical problems, but rather as an overall design activity.

I wish to acknowledge the support, help and time contributed by my wife Cherry.

Martyn Ray

Objectives

The first objective of this book is to ask questions.

The second objective is to make the reader (student *and* teacher) ask questions.

Most of the answers (and ideas) should come from the student, but the enthusiasm and interest must come from *both* student and teacher.

Education should be concerned with thinking up relevant questions when we do not know what to do, and then attempting to find the answers. Knowledge and memory should be the tools we use — not the end-product!

Note

The male gender is used throughout this book for reasons (or excuses) of readability (i.e. avoiding his/her), consistency and convention. The latter is a poor reason in view of the appeals for innovation made in the text. I apologise for this and look forward to the appearance of more textbooks written by women engineers.

Where the male gender appears it should be taken to include both sexes.

ELEMENTS OF
ENGINEERING DESIGN:
An Integrated Approach

1

Where to Begin?

Chapter Objectives

1 To ask some questions about engineering and design. Some of these questions may appear so obvious that they may have been overlooked — by students and tutors!

2 To introduce the concept of design.

3 To stimulate the reader's thoughts towards new ideas and new ways of looking at established concepts.

4 To produce a new horizon to the student's abilities and achievements beyond the limits normally defined by traditional teaching, and the constraints of the syllabus.

* * *

Exercises

1 Examine a chair. List its main functional features. List any other features. List any other requirements. Prepare a design specification for a chair.

2 Present freehand sketches, not detailed drawings, of a supermarket shopping trolley, clearly showing the main features.

3 Perform simple calculations required for the manufacture of a car jack.

4 Design a pressure cooker.

5 Design a paper clip, a drawing pin or a safety pin.

6 Recommend a biological deterrent as an alternative to conventional weapons.

7 Assess the effectiveness of domestic underground shelters in the event of a nuclear explosion.

8 Investigate ways of providing water in desert areas.

9 Consider methods of reducing deaths due to car accidents.

10 Determine the feasibility of low-cost house construction in the Third World.

* * *

INSTRUCTIONS

Read through all the exercises 1 to 10, and then think awhile.

Select one exercise from 1 to 5.

Alternative problems can be considered — preferably after discussion with the tutor.

Discuss the problem in a group ⎫
Decide what is required   ⎬ Make notes.
How is it to be achieved?   ⎭

Spend 30 to 60 minutes (for example) on the above activities, then read the following instructions.

The easiest way to start this type of problem is to examine an existing model. If this is not available then try to obtain published information relevant to the problem.

Make freehand sketches of the item to be designed. These should be presented from different angles and give details of all relevant features. Students often seem reluctant to prepare sketches and diagrams, but without them the work will proceed slowly and the end result will not be satisfactory. Anyone who can perform this type of study without diagrams will already be able to earn a small fortune in industry, as this skill will eliminate many hours of work!

Decide the extent of the work.

Decide with the tutor what is actually required. Ideally this decision should be left to the students, but this often raises problems of confidence associated with a lack of previous experience.

Decide a plan of action. Decide how the objectives are to be achieved.

What features and properties are required by the item?

How are their values calculated?

What secondary features are to be incorporated?

What are the constraints or limitations of the design?

Calculate, draw, revise, discuss.

Continue these operations until you are satisfied with the outcome.

Time allowed: to be determined by the tutor. Probably 1 to 3 hours without any previous experience. This could be extended to several hours depending upon assistance and time available, and the amount of enthusiasm generated!

When this exercise is completed to the satisfaction of the student and the tutor, select a second exercise from 6 to 10 in the list at the beginning of this chapter.

Consider how the approach to this type of problem differs from the first exercise, and the role required of the engineer.

To the student: *What were your reactions upon being presented with these exercises, and after initial consideration of the problem?*

Various common reactions and comments are likely at this stage. They will probably include most, if not all, of the following:

This problem is too trivial.
This problem is too difficult.
This problem is too vague.
What has this got to do with my particular engineering discipline?
The answer to this problem has already been found, it's obvious.
When can we start on the essential parts of the course?
I need more information.
I've got too many ideas/variables.
I'm stuck.
I don't know where to start.
I don't understand.
I can't do this.
What does it mean by 'design'?
I think this is boring.
What do you really want?
How far should I go?
How many marks are allocated for this work?
How much time should I spend?
When is it to be handed in?

Discuss with the teacher why these questions are asked (or comments made) by undergraduates who are already well qualified. What are the teacher's reactions and answers to these comments? On reflection, what are the students' reactions?

COMMENTS

Refer to the objectives at the beginning of the book.

Why are design problems presented at the beginning of the chapter and not at the end?

I believe that an appreciation of design can, and should, be obtained through a series of problem-solving exercises. Some courses are taught by first presenting a foundation of knowledge which is used later for solving project-type problems. It is hoped that the student's interest can be stored for this length of time and then be released like a coiled spring!

Many courses 'reserve' design for the last year of an undergraduate course. I believe that the students entering a degree course already have sufficient knowledge to tackle certain design problems. They may need more specialised knowledge to improve the calculations and technical aspects of their ideas, but what they need most are *enthusiasm, motivation and ideas* from the teacher.

Engineering and design are activities utilising new applications, new methods and new ideas. This will also be the approach adopted by the successful teacher when considering the methods of teaching design.

QUESTIONS — 1

What is engineering?

What are the main branches of engineering?

Who is involved in engineering?

Who is affected by engineering?

Suggestions for these questions are now presented, but the reader should try to obtain answers *before* reading the following pages.

What is engineering?

Did you arrive at a formal definition? If so, compare it with answers produced by other students in your class.

Obtain answers from students of engineering, science, social science, law, etc. Compare your answer with those of the other students.

Obtain definitions from textbooks.

Did your definition, and the others, include the following?
- (a) Producing, making or building something.
- (b) An efficient process.
- (c) The best utilisation of resources.
- (d) Optimum overall cost.
- (e) Reliability.
- (f) Safety.

Any other factors?

Did it include the following, less tangible, factors?
- (a) Satisfaction
- (b) Need
- (c) Value

Did these factors lead to discussion or thoughts about the following?
- (a) *Satisfaction for whom?* The producer, the designer, the company, the shareholders, the user, society in general, etc.?
- (b) *Who expressed the need for the product?* Society, government, other engineers, a small group, a large population, the company?

Can needs be artificial as well as real? What is the difference? Give examples of each type.
- (c) *Of value to whom?* The company, the country, society?

What are the main branches of engineering?
Chemical, civil, electrical, mechanical.

What are the differences in training and the work performed by engineers qualified in these disciplines?

What areas of their training and work overlap?

Consider the training and work of the following:

  (a) Production engineers
  (b) Industrial engineers
  (c) Systems engineers
  (d) Petroleum engineers
  (e) Materials engineers
  (f) Pollution engineers

Consider the feasibility of a general engineering degree programme. What syllabus, i.e. course units, should be included?

  (a) How many years to qualify?
  (b) Advantages?
  (c) Objections?

Compare what you think the course units for your degree programme should be with those actually scheduled.

Who is involved in engineering?
Did the list include all or some of the following?

  (a) Engineers (various disciplines)
  (b) Scientists
  (c) Computer personnel
  (d) Draughtsmen
  (e) Technicians
  (f) Craftsmen
  (g) Specialist consultants: patent officers, lawyers, economists, accountants, etc.
  (h) Supporting administrative staff: personnel department (employee relations, training, recruitment), secretaries, typists, printing, photocopying, cashiers, etc.
  (i) Unions: professional, administrative personnel, manual workers
  (j) The Board of Directors (or Governors):
       What are their functions?
       How are they appointed?
       What are their qualifications and expertise?
  (k) Shareholders:
       What is their power?
       How are they protected?
  (l) The public:
       Is the public adequately considered and protected?
       Is public opinion (and pressure) important?

(m) Others:
   The police
   The government
   The civil service
   Water and sewage authorities
   Dept. of Trade and Industry
   Organisations such as the European Economic Community (EEC), oil producing and exporting countries (OPEC), etc.
   International organisations, e.g. United Nations (UN), World Health Organisation (WHO)

Who is affected by engineering?
Everyone!
   (a) The producers, i.e. the company employees.
   (b) Suppliers of raw materials and components.
   (c) People who buy or use the product.
   (d) The government (by the effects of public opinion, the number of people employed, taxes collected, exports, etc.).
   (e) Society in general — both directly and indirectly.
Discuss this last statement.

QUESTIONS — 2

What is design?

What is a 'good' design?

What is the difference between engineering design and industrial design?

Identify different types of design projects

Identify different types of design work

Suggestions for these questions are now presented, but the reader should try to obtain answers *before* reading the following pages.

What is design?

Design is the formulation of an enquiry or a plan or a scheme in order to arrive at the required end-product.

This will involve a systematic and detailed evaluation of the problems, the alternatives and the solutions.

A detailed study of the stages involved in the design process is presented in Chapter 2.

What is a 'good' design?

If engineering was an occupation where the unknown and the unpredictable had been eliminated, then a 'good' design would provide a foolproof solution to a problem, obtained at an economic cost and comprising the minimum number of parts.

However, in the real world these aims must be modified and a 'good' design probably has the following features:

(a) It fulfils its functional purpose.

(b) It is economical with respect to resources, for both the producer and the user.

(c) It embodies satisfactory properties, e.g. strength, durability, size, etc.

(d) It has lasting aesthetic qualities, appropriate to the users and other observers.

Criticise and extend, or modify, this list.

Can you think of projects that do not satisfy (b) above?

What is the difference between engineering design and industrial design?

The reader should by now have established a personal idea of the activities that make up engineering, and the factors to be incorporated in the design process.

Broadly speaking, the design engineer is concerned with a *functional* solution to a problem.

Therefore, an important question is: *'What is the role of the industrial designer?'*

Briefly, to improve the aesthetics of the product, to combine engineering ability with artistic sensitivity. In particular, if the end-product is to be sold then he creates a 'style' that provides appeal and is attractive to the consumer.

The industrial designer must possess an aesthetic appreciation and have undertaken some formal art training. However, this alone is not a sufficient qualification for success. These attributes are what sets him apart from the design engineer, but he will also need specialised engineering knowledge. This will include practical experience on machine tools, some draughting experience and a wide knowledge of manufacturing processes and techniques.

The aim of the industrial designer is to make the product easier to sell, but he should also try to make it easier to produce, or at least not make that task any more difficult. To fulfil this role he must understand and appreciate many of the engineering operations.

* * *

Exercise

Select a common consumer item and outline those aspects attributable to the design engineer and to the industrial designer. Which features are considered by both the design engineer and the industrial designer?

* * *

Identify different types of design projects

Design projects can be broadly classified into two groups, although there will often be overlap between them.

Firstly, there are projects that start from the desire for a particular 'solution' concept. The project specification is thus primarily concerned with a physical product, e.g. design a car steering wheel, an electronic calculator, a bridge. This could be called 'component design'.

Secondly, there are projects that are defined in terms of a 'problem' situation, e.g. absenteeism is high, public transport is under-utilised. The design study will then be concerned with physical and non-physical aspects, e.g. personal comfort, access, price, taxation, public image. This could be called 'policy design'.

The first type of project will be concerned with technical plans and calculations. Most of the work will be performed by qualified engineers, at least in the pre-promotional stages. The need for the product has previously been decided.

The second type of project is concerned with a design policy, organising an approach or an enquiry. This is the establishment of a policy that may need to be implemented, or strategies to be tested. It will involve aspects of planning, management, research, politics, social programmes. For this type of project or this stage in a design project, the engineers will be involved mainly as specialist consultants.

* * *

Exercise

Identify a major engineering project such as the need for a large bridge or a high speed railway. List various features required in the 'solution', and the personnel involved.

* * *

Identify different types of design work

For both the university student and the graduate engineer in industry, there are different types of design work which can be performed. (This is different from identifying component design and policy design problems.)

As an example, consider the possible alternatives for an engineer required to produce a design for a chair.

Take 5 minutes.

No calculations or drawings are required.

For a common item, such as a chair, the engineer has three alternatives. These are:

(a) Find an existing chair that satisfies *all* the requirements of the chair to be designed and then *copy* this design.

(b) Study the designs of existing chairs and some manufactured products. Determine the requirements of the new chair. Produce a new design based on existing designs, incorporate the 'good' features of each existing design and adapt, modify and improve certain features to obtain the required end-product.

(c) Produce a completely new, innovative design.

These activities will be considered in more detail in Chapter 2, and especially in Section 2.3, dealing with creativity. However, it should be noted that work of type (a) can usually be performed by a competent technical draughtsman and requires little in the way of particular engineering expertise.

The second type (b) is the work most often carried out by graduate engineers.

* * *

Exercise

Consider a few examples of completely new, innovative designs. Identify the features that make them original. Discuss these suggestions with other students and your tutor.

* * *

QUESTIONS — 3

> **Where should design be taught in an engineering degree programme?**
>
> **If design is fully integrated within the syllabus, does it also need to be taught as a separate unit?**
>
> **Can design be taught, or is it an 'art' only appreciated from experience?**
>
> Suggestions for these questions are now presented, but the reader should try to obtain answers *before* reading the following pages.

Where should design be taught in an engineering degree programme?

Everywhere!

Design is the primary function of engineering, and is concerned with problems requiring definite solutions and unsatisfactory situations needing remedial action. It makes no sense to isolate design work in a particular course unit, separate from the topics taught in the rest of the course. The knowledge to be acquired in a degree course is divided into convenient units, each one concerned with underlying principles, theory, calculations, etc. However, these ideas should be brought together and given relevance by requiring the students to think for themselves and make their own decisions for design-type problems.

Some people may argue that topics such as mathematics are concerned with methods of analysis to be applied in engineering situations, and are not open to teaching by project work. However, a diet of abstract ideas is a certain way to minimise student enthusiasm, and any teacher of 'methods' ought to be able to provide practical situations to use those ideas. Ultimately, students should attempt problems that require them to choose and test the method of analysis.

If design is fully integrated within the syllabus, does it also need to be taught as a separate unit?

I believe the answer is yes. Mathematics should be an integrated subject, i.e. used in other areas of a course, but it can still be taught as an entity. The inclusion of a separate design unit has two main advantages. First, it can bring together the teaching and ideas in many different parts of an engineering course. Second, it can provide an opportunity for the student to obtain a wider view of engineering, not only in industry but within the whole structure of society.

Economic, political and social pressures that are relevant in the real world can then be discussed.

In this context, I would make two recommendations to the educators, and to the students who should be providing feedback to the educational system. First, that a design unit is incorporated as early as possible into the degree scheme, and not kept back until 'sufficient' knowledge has been covered. Second, that design should be 'connected' in some way with humanities, social sciences or general studies courses within the syllabus.

I hope these points provide food for thought and discussion!

Can design be taught, or is it an 'art' only appreciated from experience?

If design or at least aspects of design cannot be taught, then this book and many others will serve no useful purpose. It is true that successful designs will only emerge from a mix of adequate study and sufficient experience. However, there are aspects of the design process that can be identified and taught (as detailed in Chapter 2), and other factors to be considered for the successful execution of a design.

The graduate engineer will not be immediately set loose with a free rein on industrial design projects. Responsibility will come with experience.

BIBLIOGRAPHY

General Comments

I have a feeling that this section of a textbook is often ignored by students. This could be for several reasons. The tutor selects a recommended course text because it covers the majority of the course material. The tutor also provides references for other topics and background reading. Students often consider background or further reading a secondary claim on valuable working time. Locating additional references requires effort, especially if they are not in your library. Finally, the student may prefer to survey the particular subject group in his own library, and select appropriate books written in a style that he personally prefers.

This latter action will provide experience in an important aspect of problem-solving, namely information retrieval.

Bibliographies are included at the end of each chapter to provide some direction in the choice of supplementary and complementary material. Also this book would be incomplete, and less useful, if references were omitted entirely.

My aim is to provide only a few references for each chapter, including some journals which can provide information about new developments and ideas. Reference is made to the appropriate bibliography for sources quoted in the chapter.

I have also included my personal comments.

The majority of these references should be stimulating, interesting, thought-provoking and visual.

Armytage, W.H.G., *A Social History of Engineering*, 4th Ed. Faber and Faber, London (1976).
  As title.

Baldwin, A.J., and Hess, K.M., *A Programmed Review of Engineering Fundamentals*, Van Nostrand Reinhold Co., New York (1978).
  Basic material; applications and problems to be mastered if you intend to be a successful engineering undergraduate.
  Note 'programmed' and 'review' in title.

Baldwin, J., and Brand, S. (Eds.), *Soft-Tech*, The Whole Earth Catalog, Penguin Books, New York (1978).
  If you do not know what is meant by 'alternative technology' read this book.

Clarke, D. (Ed.), *The Encyclopedia of How it Works*, Marshall Cavendish Books Ltd, London (1977).
  Not a children's book. So you think you know it all? If so, then you do not need an engineering degree! Excellent diagrams and explanations. Visual presentation.

de Moll, L., and Coe, G. (Eds.), *Stepping Stones: Appropriate Technology and Beyond*, Marion Boyars Publishers Ltd, London (1979).
  It may open your eyes to the alternatives, although *Soft-Tech* in this list is more readable.

Gregory, M.S., *History and Development of Engineering*, Longman Group Ltd, London (1971).
  As title, not too lengthy, interesting and readable.

Kemper, J.D., *Engineers and their Profession*, 3rd Ed. Holt, Rinehart and Winston, New York (1982).
  Plenty of discussion material. Keyword is 'profession'. Not visual.
  Read the sections of personal interest. Describes the 'profession' rather than the 'nuts and bolts' work.

Kingery, R.A., Berg, R.D. and Schillinger, E.H., *Men and Ideas in Engineering*, University of Illinois Press, Urbana, Illinois (1967).
  Background reading and hopefully some personal interest. Twelve case histories.

Kranzberg, M., and Davenport, W.H. (Eds.), *Technology and Culture: An Anthology*, Meridian Books, New American Library Inc., New York (1972).
  If you dont know what an anthology is — find out!

Laithwaite, E.R., *All Things are Possible — An Engineer Looks at Research and Development*, IPC Electrical Electronic Press, London (1976).
  As per title. If this doesn't generate some interest or enthusiasm, why are you studying engineering?

Marsh, K., *The Way the New Technology Works*, Century Publishing Co. Ltd, London (1982).
  Essential — easy to read and visual.

Presence, P. (Ed.), *Purnell's Encyclopedia of Inventions*, Purnell Books, Maidenhead (1975).
See comments for Clarke.

Roadstrum, W.H., *Excellence in Engineering*, John Wiley and Sons Inc., New York (1967).
Not visual, but well written and some good ideas.

Sandstrom, G.E., *Man the Builder*, McGraw-Hill Book Co., New York (1970).
A history book with some good illustrations. Worth a browse.

Sempler, E.G. (Ed.), *Engineering Heritage*, Volumes 1 and 2, Institution of Mechanical Engineers, London (1972).
A good reference source.

Stephens, J.H., *Towers, Bridges and Other Structures*, Sterling Publishing Co. Inc., New York (1976).
Technical and visual.

Thring, M.W., *The Engineer's Conscience*, Northgate Publishing Co. Ltd, London (1980).
Interesting!

Twentieth Century Engineering, from an exhibition at the Museum of Modern Art, New York (1964).
As per title, completely visual. Stunning!

Vesper, K.H., *Engineers at Work: A Casebook*, Houghton Mifflin Co., Boston, Massachusetts (1975).
Fascinating case studies, should be essential reading!

Whyte, R.R. (Ed.), *Engineering Progress Through Trouble*, Institution of Mechanical Engineers, London (1975).
See comments for Laithwaite.

Journals

The following journals provide a wide coverage of topics in engineering and technology. Most have some specialisation but still publish articles in many broad areas. Students should be encouraged to 'browse' through the current issues of a range of journals, rather than be misled into thinking that all knowledge is found in textbooks.

Atmospheric Environment, Pergamon Press, Oxford.

Automotive Engineering, Society of Automotive Engineers, Dallas, Texas.

Design, The Design Council, London.

Designer's Journal, Architectural Press Ltd, London.

Energy Policy, Butterworth Scientific Ltd, Guildford, Surrey.

Energy World, Institute of Energy, London.

The Engineer, Morgan-Grampian (Publishers) Ltd, London.

Engineer's Digest, United Trade Press Ltd, London.

Engineering, Design Council, London.

Environmental Pollution. Series A: Ecological and Biological, Series B: Chemical and Physical, Elsevier Applied Science Publishers, Barking, Essex.

Impact of Science on Society, UNESCO.

Industrial Engineering, Institute of Industrial Engineers, Atlanta, Georgia.

The International Journal of Environmental Studies, Gordon and Breach Science Publishers, London.

Issues in Engineering — Journal of Professional Activities, Proceedings of the American Society of Civil Engineers, New York.

New Scientist, New Science Publications, London.

Nuclear Engineering International, Quadrant House, Sutton, Surrey.

The Plant Engineer, Journal of the Institution of Plant Engineers, London.

Process Engineering, Morgan-Grampian (Publishers) Ltd, London.

Resources and Conservation, Elsevier, Amsterdam, Netherlands.

Science, American Association for the Advancement of Science, Washington, DC.

Science of the Total Environment, Elsevier, Amsterdam, Netherlands.

Scientific American, Scientific American Inc., New York.

Technology Review, Massachusetts Institute of Technology, Boston, Massachusetts.

The Wheel Extended, Toyota Motor Sales Co., Japan.

2

The Design Process

Chapter Objectives

To obtain an understanding of:

1 the work performed by a design engineer;

2 the activities to be carried out in the design process;

3 the role of the project engineer;

4 the factors influencing the success of a project;

5 the life cycle and stages of a project from inception to obsolescence.

---

* * *

Exercises

Read through the following problems:

1 You work for a company whose main activities and products are related to energy sources, new forms of energy storage and transmission (originally a battery manufacturer) and a range of electronic components. You are an engineer in the Product Design and Development Department. It has been disclosed that the Research Section have produced a new type of small, long-life, renewable energy source. They have also made a disc of material that can produce a high-intensity beam of collimated light when connected to this new power source.

Your brief: appointed as the project engineer responsible for the design and development of these combined products for use as a hand torch.

2 You have recently joined the Engineering Section of a large multi-national company that specialises in the bulk production and sales of a wide variety of chemicals and gases, and associated equipment. Technical Gases Ltd is a subsidiary company responsible for high-purity, cylinder gas sales. You are seconded to work on a project for this company and the Marketing Manager outlines the following brief.

The company does not make, buy or sell carbon monoxide gas, either pure or in gas mixtures. The present market demand is satisfied by a competitor (a smaller company) which specialises in medical gases. Their carbon monoxide is produced in a plant that is now 20 years old and becoming increasingly unreliable. The only other source of supply is from a foreign country, which involves shipment in large pressurised cylinders and a sea journey of 160 km. The price of the home-produced gas is therefore fixed at the cost price of importing the gas.

Should Technical Gases Ltd. become a supplier of carbon monoxide?

3 Since graduating as an engineer you have worked in the Research and Development Department of a small state-owned corporation. All the current projects are financed by the Ministry of Defence for military use, and the main work is the development of pre-programmed and remote-controlled robots for bomb disposal and handling of radioactive materials.

   You have recently been assigned to a small project team. The project specification is to develop and produce an artificial human hand that can be coordinated by a small control panel strip fixed behind the front teeth, and operated by tongue touch. The product must be commercially viable.

*   *   *

INSTRUCTIONS

Read through all three exercises and then select the one that most interests you. Without interest the project will not be a success—either from the technical viewpoint or in terms of personal satisfaction.

Outline the progress of the project at this time, the work performed, decisions made, etc. Prepare a schedule (work and time flowchart) for the work still to be performed through to commercial success (hopefully!).

Identify particular areas or stages of the project, the work involved in each stage, the decisions to be made and the personnel involved.

Prepare in more depth the activities and work performed during the research, design and development stages. Detailed calculations are not required but you are expected to prepare some basic technical data, and a set of specification drawings.

You are *not* to write essays; all aspects of this work should be in brief note form—but still understandable to someone with experience in design work.

It is preferable to perform this work in a small group, say three or four students. At each stage, and as problems arise, discuss the work with your tutor and with other students.

Record your feelings and reactions (or those of the group) as the work proceeds. Keep a brief record of the progress and setbacks which you encounter. Record the time spent on various aspects and tasks, e.g. primary evaluation, data collection, discussion, arguments, and so on. Try to identify how the work began, perhaps with a feeling of lost hopelessness or as a rush of ideas! Write down how major obstacles and problems were overcome. It will be useful and illuminating later in the chapter to compare your experiences and problem-solving methods with the typical design phases described.

There are ideas, suggestions and recommendations for further work for each exercise within this chapter. *Do not* use these notes unless other help is not available, or you are *totally* unable to proceed, or it is recommended by your tutor. Carry out the exercise as far as possible *before* referring to these pages.

Exercise 1: page 27.
Exercise 2: page 28.
Exercise 3: page 29.

2.1 WHAT IS DESIGN?

(Revision from Chapter 1.) The design process involves applications of technology for the transformation of resources, to create a product that will satisfy a need in society. The product must perform its function in the most efficient and economic manner within the various constraints that may be imposed. The major restraint is cost, although other factors such as safety, pollution and legal requirements will have to be considered. It is the ability to design an efficient product that sets apart the successful designer.

It is important that the designer is not referred to as a draughtsman or planner or any other description. These are specialised functions which are required in the overall process, and they are often performed by the designer. However, the role of the designer goes beyond these particular functions. The designer should be involved in all stages of the design process, and it is his intellectual ability and conceptualisation of the problem and its solution that determines the final product.

The designer is concerned with solving a problem or producing a product. This work has resulted from a need within society and the decision to proceed to a solution. A designer will often have been trained as an engineer; however, he and other members of a team must remember that the product is meant to satisfy a need in society.

If society decides to reject a product, whether for functional reasons or merely because of habit, then the design effort is wasted.

The sociological view of the product's place in society must also be considered. This will require an understanding of the structure and needs of society, and any changes that may occur, e.g. occupational changes or wealth, during the lifetime of the product. Statistics and statistical analysis are often useful in this context, although they should not be considered to the exclusion of other considerations. The designer is not expected to be an expert on sociology or statistics but he should be aware of their importance to his work.

The project engineer is not necessarily a designer. He may perform design tasks but his main function is to oversee and organise a project or a particular aspect of the work. Whatever his professional qualifications he will become involved in all aspects of engineering, electrical, mechanical, etc. If he lacks the necessary expertise he will consult the appropriate people. For small projects or one-off items, the project engineer will often be responsible from inception of the product to the sales and commissioning stages. For larger projects or products to be mass-produced, part of the responsibility may be shared at a later stage, e.g. with a production engineer.

2.2 TYPES OF DESIGN WORK

There are three main types of design work, each of which requires a different level of intellectual ability and creativity.

Primarily, there is the adaptation of existing designs which require only minor modifications, often in the dimensions of the product. This type of work represents a large proportion of the total design work undertaken. The designer engaged entirely in this type of work needs only the basic technical skills and will not appreciate the nature of the design process unless other work is undertaken.

Secondly, there is development design, which uses an existing design only as a basis. The technical work involved can be considerable and the final product may be very different from the original item. In certain cases the use of an existing design may make the solution more difficult to obtain, and the level of ability and time spent by the designer may be considerable. This is sometimes called *adaptive design*.

Finally, the most demanding work concerns the design of a new product that has no precedent. This needs a high level of ability and relatively few designers will be engaged in this type of activity. It is sometimes referred to as *creative design*.

* * *

Exercises

1 Consider particular examples and discuss the statement: 'In certain cases the use of an existing design may make the solution more difficult to obtain'.

2 How important is 'creativity' or 'intuition' or 'invention' in the design process?

3 What is the relationship between IQ and design ability?

4 Are the 'best' qualified engineers the most successful designers?

5 What abilities and personal qualities would you expect to characterise a successful design engineer?
Consider this last question and then compare your thoughts with the ideas on the following pages.

* * *

2.3 THE ROLE OF THE DESIGNER

Before considering the role of the designer in an organisation or a team, it will be useful to assess some of the qualities that a designer would be expected to possess. Obviously any discussion or list of this type should not be considered exhaustive, but the following is intended to emphasise the main qualities and capabilities of a successful designer.

(a) Conceptual ability

The designer should be able to understand and visualise the whole concept of the design project. This requires imagination of simple shapes and their combinations and interrelationships, before the detailed design stage begins.

(b) Logical thought

The ability to think logically is an important skill which is acquired early in life, mainly in the school years. A logical approach to problem-solving is necessary at each stage of the design process, and also in the preparation of the overall design strategy which will be considered later.

(c) Perseverance, concentration and memory

These are personal qualities which are also developed mainly in the school years. A person who lacks these attributes is unlikely to achieve the qualifications required to begin a career in design.

(d) Responsibility, integrity, will-power, temperament

These are personal or moral qualities which determine not only the success of the designer's work, but also the direction and success of the designer's career. A person who is responsible for other people's actions must possess more than technical skills.

(e) Invention or creativity

Problem solutions are obtained by hard work and logical thinking, combined with a certain amount of inventiveness, intuition or creative ability. The idea of a flash of inspiration or the existence of a creative genius is misplaced. The 'novel' solution is obtained by a clear understanding of the problem followed by logical analysis and deduction, even though these stages may not be remembered or recognised as such when the solution is obtained.

(f) Communication

The most superior design will be ineffective if the designer cannot communicate essential information to other members of an organisation or to a customer. This will require the presentation of information in written, visual and spoken form.

(g) Scientific knowledge

Throughout his training the designer will be acquiring basic scientific knowledge which will be used to produce satisfactory designs. However, this knowledge is only useful when it is applied with the skills outlined above. The knowledge that will be required depends upon the areas in which the designer specialises. Subjects of basic importance include:

(i)   mathematics—elementary and advanced levels, geometry and technical drawing;
(ii) principles of physics;
(iii) chemistry;
(iv) engineering and technology—materials, manufacture, machines, structures, production, etc;
(v)   management studies;
(vi) economics.

The role of a design engineer within an engineering department or design section of an organisation can be appreciated by reference to Fig. 2.1. In a small department all staff may be actively engaged in all aspects of design work. In larger departments senior personnel will spend more of their time concerned with management tasks. Also, within larger companies staff may specialise in particular design problems and sections may exist concerned primarily with these aspects. Although not always adhered to, it is preferable if the names 'designer' and 'design engineer' are reserved for persons capable of working independently and creatively towards the solution of problems. Less-qualified staff may be designated as assistants, draughtsmen, technicians, etc.

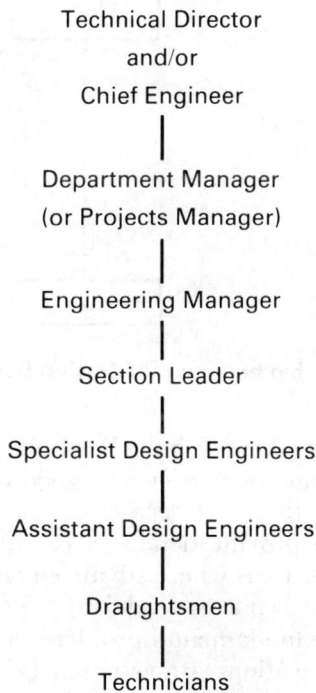

Technical Director

and/or

Chief Engineer

|

Department Manager

(or Projects Manager)

|

Engineering Manager

|

Section Leader

|

Specialist Design Engineers

|

Assistant Design Engineers

|

Draughtsmen

|

Technicians

Figure 2.1   Hierarchy of Staff within an Engineering Design Department

For small projects the entire process of design, production and sales may be performed by the design engineer. For larger projects, and within some companies, the designer acts as a member of a team. In larger companies, departments or sections exist which deal primarily with different aspects of the production process, e.g. sales, commissioning and legal aspects. Each department will need to co-operate with several others and none will, or should, act with complete independence. The relationship that may exist between the design department and other departments within a company is shown in Fig. 2.2. The relationship between each department is not shown.

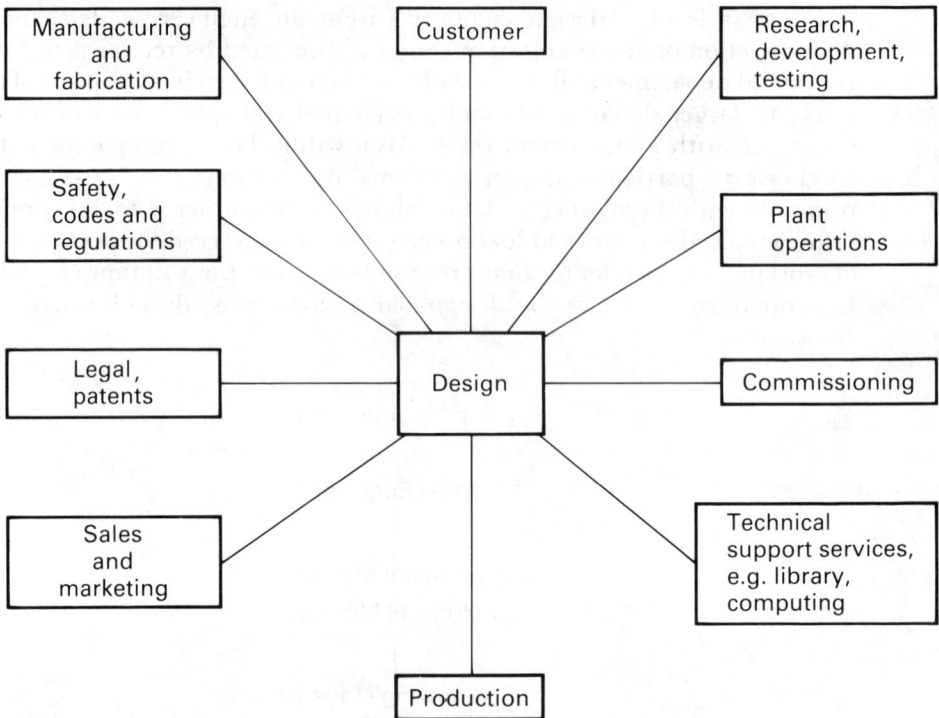

Figure 2.2   Relationship between the Design Department and other Departments

The main areas where technical creativity occurs are design and production, and liaison should be particularly good between these departments. It is also important that there are reciprocal communications between scientific departments, which provide details of recent advances, and design departments, which outline areas where advancement is required. If this is not done, new developments will not be exploited. The product will then become less competitive because fundamental problems are not being investigated. However, good communications are essential between all departments, such as customer requirements from the sales office to the designer, if the product and the company are to succeed.

* * *

Exercises

1 Try to describe what is meant by creativity or inventiveness.

2 Discuss the importance of IQ (intelligence quotient) for creative thought.

3 What correlation would be expected between IQ and creative ability?

4 What particular traits would you expect a creative person to possess?

5 List the factors that are likely to help or hinder the search for a creative solution to a problem.

6 Can creativity or creative thinking be taught, and hence improved?

Prepare your own answers and then compare them with the ideas put forward on the following pages.

* * *

2.3.1 Creativity

Engineers and designers frequently strive to achieve creativity or inventiveness in their work by searching for a solution to a problem which is the 'best' solution in as many ways as possible. Thus, the engineer attempts to produce a solution which will satisfactorily perform the functions required of it, requiring the minimum of materials, to be produced at the least cost, to be aesthetically pleasing, etc. In other words, an attempt is made to achieve the impossible and produce the perfect design. In practice, an optimum solution is the best which can be obtained and this balances ideal aims and practical limitations.

Creativity and inventiveness are not, in this context, the same as innovation. All engineers would like to produce solutions that display innovative ideas, but only a few possess the skill to achieve this aim. Innovation involves the use of new ideas or technology, or old ideas that have not previously been adapted in a particular situation, to produce an improved product.

The design problem gives the design engineer the opportunity to utilise and develop his creative skills to their full extent. In order to select a technique or idea that will provide a solution to the problem, the engineer requires both practical experience and a depth of knowledge. He must then be able to relate widely different elements and recognise their mutual adaptability. Ultimately, a truly creative solution will only emerge if the designer can apply open-ended thinking and can also appreciate alternative solutions and strategies.

The alternative type of investigation is the analytical process. This requires particular specialised knowledge, the ability to remember and recognise specific facts, mathematical skill and the determination to continue with a problem until a single solution is obtained.

It seems strange that the skills we most value in our engineers and designers are the ones that are given least attention in the teaching of undergraduate students. Undergraduate courses are often composed largely of a diet of analysis, combined with the task of predicting the performance of something which already exists. Problems are often no more than mathematical manipulation in order to obtain a 'standard' solution. How many problems leave the tasks of identifying and subsequently retrieving the data required to the students? How many projects allow the students to participate in the formulation and feasibility stages?

Students need training and experience in solving problems analytically, but of equal importance should be the development of creative thinking. The opportunity to learn by a 'design–build–test' process should be a vital part of every syllabus.

2.3.2 IQ

Before deciding the importance of IQ it will be necessary to clarify our thoughts about the meaning of IQ, and what it is meant to measure. It is often assumed, not only by the general public but by graduates, that IQ quantifies a person's intelligence or ability to think logically. Alternatively, that it is a measure of how 'clever' a person is or the level to which someone can be educated. These are all very imprecise and ill-defined ideas.

Assume for a moment that IQ is a measure of intelligence.

Now consider the following:

Should a graduate's IQ increase at each stage of his university training?

Should a Ph.D. graduate naturally possess a higher IQ than a B.Sc. graduate?

If you have time, try to test these ideas.

Many books are published entitled 'Improve your IQ', or similar. This improvement is usually brought about by becoming aware of alternative ways of considering a problem, i.e. by using divergent rather than convergent thinking, and by familiarisation with IQ tests.

Surely, if IQ tests really measure our intelligence, we do not become more intelligent merely because of practice at answering IQ test questions?

* * *

Exercise

Discuss in a group the possibility that IQ tests measure our ability to learn how to do IQ tests! Such a discussion should influence your opinion of the importance of IQ for creative thought.

* * *

The discussion of this exercise should help you to decide what correlation would be expected between IQ and creative thought. The general 'feeling' is that there is no correlation for an IQ of between 110 and 150. Outside these limits there is little information available. An IQ score of 60 to 80 is usually interpreted to mean that the person is only capable of acquiring very basic knowledge.

Wait just a minute!

How do we explain the anomalies of those schoolchildren who are dismissed as ineducable and yet possess superior skills in craft, drawing, motor mechanics, etc.?

What of the very young child performing creative acts with construction toys?

What of primitive races untouched by technology, but with the ability to survive in extremely adverse conditions?

IQ tests have been developed for different races in attempts to avoid cultural factors affecting the results. Tests have also been developed that are (supposedly) not dependent upon verbal skills, i.e. they are visual or pictorial tests.

How much time is required to ascertain what the subject really thinks he sees?

Verbal skill is still necessary to describe the interpretation of the material.

IQ testing is surrounded by uncertainty, and the design engineer certainly does not need more uncertainties to consider!

2.3.3 Personality Traits

Creative individuals will probably possess an ability for divergent thinking, a belief in their ideas, a desire for constructive change and a positive view of life. They are probably also adventurous, sensitive, perceptive and have a capacity for intuitive thought (if such an attribute exists!).

2.3.4 Preventing and Promoting Creativity

Barriers to inventiveness include habit, fixed methods, repetitive behaviour, background, training, fear of criticism, tension, hostile reactions and inadequate knowledge.

Factors that help foster creativity include adequate background preparation, review of previous work, broad-based knowledge and interest, breaking of fixed mental processes and work habits, recording and assessing all possibilities and use of analogous situations.

2.3.5. Developing Creativity

Your answer to Exercise 6 on page 23 may well show whether you are about to give up reading this book.

Your answer will also be an indication of whether you are, or might become, a creative design engineer.

2.3.6 Brainstorming or Forced Creativity

There is a technique known as brainstorming for helping groups of individuals create and recognise innovative ideas. It has been used with varying degrees of success (the successes having been more widely publicised than the failures), and much has been written about it. The basis of the method is that a group is given a problem to be solved and, without any prior preparation or time for evaluation of ideas, the individuals put forward their suggestions.

As with any technique there are rules which should be observed if the best results possible are to be obtained. The rules for brainstorming sessions are few and can be simply stated as follows. The problem to be considered should be presented in general terms rather than as a narrow specification, for instance 'to identify new products for a company to produce' rather than 'how to reduce fuel consumption by 10% by reducing the weight by 8% and the wind resistance by 11%'. The participants should be encouraged to suggest unusual ideas. Innovative and unusual solutions to the problem are required, as well as the more obvious alternatives. A conventional solution to a problem may ultimately be selected, but the aim is to consider all possible alternatives, especially unusual ideas which would often be rejected out of hand or not even considered.

Other requirements of a brainstorming session are that quantity is required and a record of all proceedings must be made, usually a tape recording. All that is needed is a basic idea or possible solution without any evaluation, discussion or analysis; that will come later. Finally, the most important rule, which must be strictly observed, is that no criticism or judgement of any kind is allowed, not even favourable comments. A leader should be elected for this purpose, as comment in any form will only tend to inhibit the responses of the people involved. The best results will probably be obtained if the group comprises people from different backgrounds who do not know each other well. The group should not be so large that ideas cannot be expressed immediately they are conceived. After the session has finished then *all* the ideas should be evaluated objectively and none should be rejected out of hand.

This method of obtaining ideas for possible solutions to a problem will be most effective if it is well planned and if these suggested rules are observed, especially regarding no judgement or comment. Be aware that brainstorming may not be warmly embraced by other people! However, the only thing to lose is time, and the possible rewards are high.

2.3.7 Creativity Exercises

The following are some suggestions of problems that are not rigidly defined. Add some more to this list:

(a) a manufacturer of medium-priced toiletries needs to produce some new items;

(b) the weight of a family car is to be reduced;

(c) private transport is to be excluded from a city and alternative modes of consumer travel are required;

(d) unauthorised private video recording must be prevented;

(e) personal service in banks, post offices or department stores needs to be more efficient;

(f) identify alternative energy sources;

(g) possible food and dietary changes, without adverse effects on nutritional requirements;

(h) improving job satisfaction of employees;

(i) reducing unemployment or inflation;

(j) alternatives to nuclear weapons;

(k) preventing shark attacks on bathers.

Select a group to discuss a particular problem in a brainstorming session. When this is complete, discuss the session itself and the ideas that were obtained. Make recommendations. Compare the findings with the experiences of other groups. Assess the usefulness of the brainstorming method.

Divide the class into groups of three or four persons, each person having a pen and a large piece of paper. Everyone is to consider the same problem (to be specified) and must write down a suggested solution in one sentence (10 words) in *1 minute* only. The papers are then rotated around the group and new solutions are written down. The process is repeated until each person receives their original paper for the second time, i.e. two complete revolutions.

Identify how many 'essentially different' suggestions are obtained from each group, and from the class as a whole.

Repeat this method, sometimes called *forced creativity*, where each group considers a different but related problem. For example, consider automobile design features:

Group 1— weight reduction;
Group 2— increased fuel economy;
Group 3— reduced exhaust emissions;
Group 4— improved safety features;
and so on.

Assess the effectiveness of this method compared with a brainstorming approach.

Devise other methods of obtaining creative or innovative ideas. (Try using brainstorming to produce forced creativity methods!)

- *Now go to Section 2.4.*

- *From page 16.*

Exercise 1—Suggestions

Some of the questions you should have asked:
(a) What are the sizes and masses of the power source and the disc?
(b) How are they connected?
(c) Is information available concerning the principle of operation?
(d) Is the power source rechargeable in the home or must it be returned to the company?
(e) Are there any dangers from using the product (short or long term)?
(f) What materials or conditions have an adverse or beneficial effect upon the operation of the device?
(g) How much does it cost to make?
(h) Can it be mass-produced?
(i) How are costs likely to change for mass production?
(j) What is the expected service life?
(k) Are competitors active in this field?

Answers to these questions, and any others, may be obtained from three sources:

(a) decide yourself;
(b) ask your tutor;
(c) turn to page 40.

If you do not use source (c) above, then continue the project but at the next *complete* impasse, or perhaps when you consider the exercise finished, study the comments on page 48.

● *From page 16.*

Exercise 2—Suggestions

Some of the questions you should have asked:

(a) What is the present market size, i.e. demand?
(b) How is this broken down, is there one (or more) large purchaser?
(c) Is this demand expected to change, and, if so, by how much?
(d) Are some uses of the product going to be replaced?
(e) If demand is estimated to increase is this due to growth or new uses?
(f) How is the gas required, i.e. piped direct (low or high pressure?), as gas in high pressure cylinders, as liquid, etc.?
(g) What is the sales/supply distribution network?
(h) Are any other companies expected to enter this market?
(i) Are there problems in the production, transportation or use of the gas, e.g. environmental, safety or carcinogenic effects?
(j) Is the gas required pure or as a gas mixture?
(k) What purity is required?
(l) What impurities are allowed?

The first question asked will often be:

What process will be used to produce the gas?

This question will certainly have to be answered, and if a suitable process is not available the exercise will not proceed. However, even if a highly suitable process exists, if some of the answers to the above questions (a to l) are unfavourable then the project will not be a success.

The selection of a production process should give rise to another long list of questions.

Answers to these questions, and any others, may be obtained from three sources:

(a) decide yourself;
(b) ask your tutor;
(c) turn to page 41.

If you do not use source (c) above, then continue with the project but at the next *complete* impasse, or perhaps when you consider the exercise finished, study the comments on page 48.

● *From page 17.*

Exercise 3—Suggestions

Some of the questions you should have asked:
(a) What tasks can it perform, e.g. writing, lifting or gripping?
(b) What is the market size and estimated sales?
(c) Can it be adapted for other limbs?
(d) Is the control panel removable, delicate, etc.?
(e) Are there one or two (i.e. up and down) panels?
(f) Are the controls simple, comfortable, safe (if swallowed?), etc.?
(g) What is the power source?
(h) What is the life of the power unit?
(i) Is the power source rechargeable?
(j) Is the work secret and covered by other contracts?
(k) Are sales going to be restricted owing to the secret nature of the operating mechanism?
(l) What is the expected competition?
(m) What standards relate to this product?
(n) What will be the maintenance requirements?
(o) What are the design and development costs?
(p) What is the estimated selling price?
(q) Are there guaranteed sales, i.e. to government hospitals, military personnel, etc.?
(r) Is the product to be custom-made or mass-produced?
(s) What is the timescale for development?
(t) What materials are to be used, and what will be their appearance, weight, strength?
(u) What conditions will have an adverse effect, e.g. heat, water, swimming?

Answers to these questions, and any others, may be obtained from three sources:
(a) decide yourself;
(b) ask your tutor;
(c) turn to page 41.

If you do not use source (c) above, then continue the project but at the next *complete* impasse, or perhaps when you consider the exercise finished, study the comments on page 48.

• *From Section 2.3.7.*

2.4 THE NATURE OF THE DESIGN PROCESS

There are two methods that are normally used to describe the nature of the
design process. The first approach is to examine the life cycle of the product
and to define each stage through which it passes. This is known as the
morphology or structure of the design process, and the appropriate stages are
detailed below.

2.4.1 The Morphology of Design

(a) Identifying the problem
The first step involves identifying a problem, or a requirement of society.
However good the final product there must be a need or a market for it. A full
specification of this need must be obtained in order to avoid extra work at a
later stage. Limitations and requirements of the design must be identified,
and stated in terms of constraints and criteria. Other relevant information
should be obtained and recorded such as the timescale, economics, estimated
life of the problem, whether the nature of the problem will change during its
life, ease of modifications, etc.

(b) Feasibility study
Having identified a need for the product, the next stage is to conduct a
basic feasibility study. It must be physically possible to produce the product,
it must be economically viable and it must be acceptable. If the alternative
solutions cannot be physically realised, because either appropriate materials
or fabrication methods are not available for example, then the project will not
proceed unless these problems are overcome by appropriate prior
development work. If the physical requirements appear possible then the
economics of the various alternatives can be investigated. The final stage of
the feasibility study will be to establish that any alternatives under
consideration will be acceptable, not only to the customer for whom the need
was identified, but also to society in general. If other members of the
community object to the product then it may not be successful.

(c) Preliminary design
At this stage, having defined the need or the problem and ascertained
that a solution may be feasible, it becomes necessary to select the best
approach to the problem. Where several alternatives exist they should all be
examined in an unbiased manner, with equal thoroughness. The personal
preferences of the designer should not be allowed to influence the evaluation
process. The preliminary design may be conducted entirely 'on paper', or
scale models may be used, or, if possible, experimental results may be

obtained. It may be found at this stage that a solution which appeared possible earlier can now not be achieved. At the conclusion of the preliminary design phase it must be decided which approach to the final solution is to be examined further and is most likely to be adopted. It should be remembered that this choice is not irreversible and results obtained later may require that this approach is abandoned and an alternative is developed. This may be a difficult decision to make, owing to the costs already incurred, but it will be cheaper than producing an unsatisfactory solution.

(d) Detailed design

An approach to the solution has been chosen and a full detailed design is now prepared. All components and systems are fully specified and all in-service requirements should be defined. A complete set of drawings is also prepared. At this stage it is usual to produce a scale model of the product, if this has not been done previously, and sometimes a prototype unit. The prototype may be a small-scale version of the final product or possibly a full-scale replica. The prototype is tested in order to evaluate its performance and effectiveness, and to decide upon any modifications, if the need for any has become apparent. Less alternatives are available to the designer at this stage and a significant economic commitment is usually required. Any major revisions, or the decision to abandon an approach entirely, should be made before progressing to the next stage. If not, considerable investment will be wasted. It may be that events outside the designer's control may cause the product to become unacceptable at a later stage. This possibility should be minimised by constant feedback and continual re-evaluation of the problem identification and feasibility studies.

(e) Production

At this stage, having finalised and approved the detailed design, the product is constructed. This will involve close co-operation between the designer and the production engineer. A commissioning engineer and/or a plant operations engineer may also contribute specialist advice at this stage. The capital investment for the project will have to be available to pay for materials, labour, fabrication, sub-contractors' fees, components, etc. If the product is a machine that will be used to manufacture mass-produced parts, then the raw materials to be used by the machine will also be required. The conditions concerning the repayment of the capital must also be agreed. Planning for the production phase should have been carried out in the detailed design, including specification of any specialised machines or equipment that will be required, quality control requirements, testing procedures and standards, tolerances, etc. The cost of the commissioning process should also have been considered. The detailed design should have included time schedules for each phase of the operation, and delivery dates. Assembly times should have been carefully calculated to avoid costly waiting time.

(f) Distribution, sales, usage

The distribution stage of a design project may be concerned with the supply of raw materials and services, or the distribution of the product to suitable sales outlets. This will include storage, packaging, transportation, etc. The distribution factors for a design process that does not produce a marketable quantity as the product, e.g. bridge construction or a waste disposal plant, will be concerned with ensuring that the inputs and outputs, e.g. vehicles or refuse in the examples quoted, have easy access and exit.

If the product is to be sold then the sales outlets must be conveniently situated, and if the market is not self-generating then suitable advertising and marketing techniques must be employed. The product must be priced to ensure that optimum sales and profits are obtained, so that the capital investment can be repaid.

The final product must be used, and it must satisfy the need it was intended to fulfil. If this is not achieved, then either the capital invested in the project will have been wasted or insufficient profits will be obtained. Feedback of public opinion can be useful for the designer to determine possible modifications or future developments which may increase customer satisfaction, and also company profits. The availability of suitable spare parts and maintenance facilities will also be important factors that will influence the final level of product use.

Although aspects of distribution, sales (if appropriate) and usage can be considered separately, it will make the overall process more efficient if close co-operation is maintained between these stages. This is due to the obvious interrelationship that exists between these activities.

(g) Obsolescence

The design of the product should ensure that it will not wear out or become prematurely obsolete, and also that at the end of its useful life it is not still capable of prolonged operation. The occurrence of either of these situations indicates that the design economics could have been improved.

There should be some consideration of whether at the end of the machine's life the product should continue to be produced on a new, improved machine, or if the need for the product has diminished or been satisfied. In the latter case, continued production may be uneconomic. The scrap or resale value of the components, their disposal and any special considerations must be taken into account.

At each stage in the design process there should be a constant recycling of new information and decisions back to previous stages, in order that previous decisions can be continually re-evaluated. In this way, unforeseen and unavoidable changes that affect the design will be detected at the earliest opportunity. For example, if a product becomes obsolete prematurely owing to unforeseen or rapid technical advances, then this will be unfortunate economically, but prolonged production will only harm profitability still further.

This description of the design process or of an engineering project as a series of time phases can be further classified as Design Stages (a to d) followed by Planning Stages (e to g). Many authors have identified different phases or activities which occur within the Design Stages.

Vidosic (see Bibliography) discusses the following eight phases:
      recognition
      definition
      preparation
      conceptualisation
      synthesis
      evaluation
      optimisation
      presentation

Jones identifies three phases:
      divergence
      transformation
      convergence

whereas Haefele chooses:
      (possibly incubation)
      preparation
      illumination
      verification

Woodson describes four main stages:
      feasibility study
      preliminary design
      detail design
      revision.

One of the clearest and simplest analyses of the stages in a design study is presented in an Open University Design Project Guide. This can be summarised as shown in Fig. 2.3.

Whichever description of the design stages is preferred, there must be adequate communication, discussion and agreement between *all* personnel involved in each phase, and in *every* previous stage (feedback).

* * *

Exercise

With reference to the design exercise you performed taken from the beginning of this chapter, identify the stages you carried out and compare these with the stages described as the morphology of design.
    What areas of investigation overlap?
    What are the disadvantages of breaking down the design process in this way?

* * *

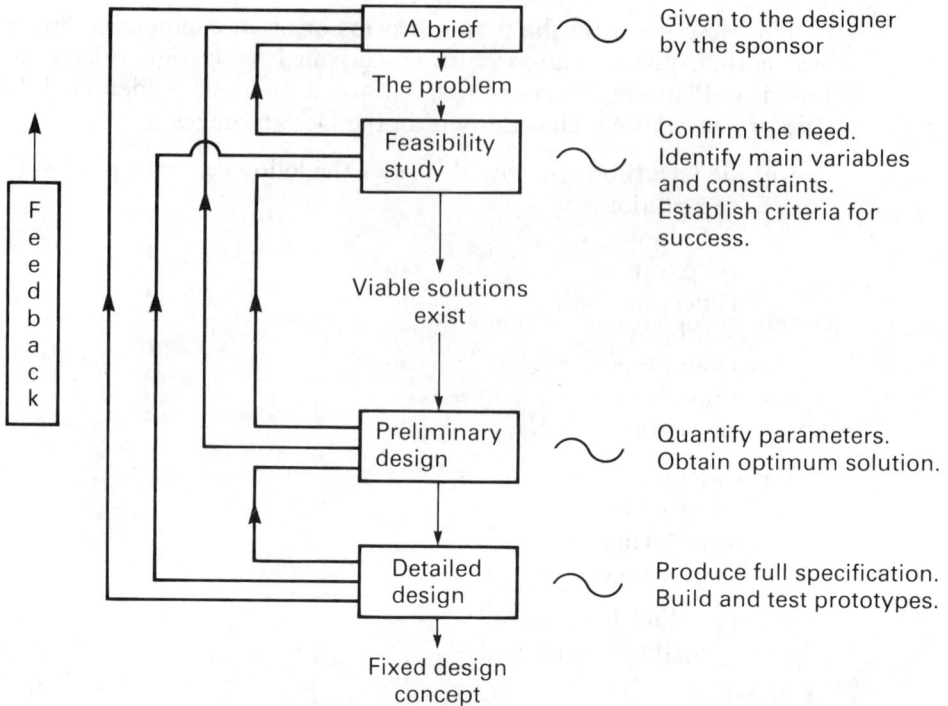

Figure 2.3   Flowchart showing Stages in a Design Process

(reproduced by permission of the Open University, *Design Project Guide Unit 12;*
T262 Man-Made Futures, Copyright © (1975) The Open University Press)

2.4.2 Example of the Morphology of Design

The Problem
A city congested by too much private transport, and the associated
environmental pollution problems.

Constraint
Any solution must be inexpensive (for whom?).

Possible alternatives
Promote and encourage alternative mode(s) of transport for the com-
muter. The following ideas are proposed: walk, roller skate, jet pack, bicycle.

Assessment/evaluation
Identify particular features and use a comparison/evaluation table as
shown below. Assign values from 1 (most desirable/advantageous) to 5 (most
unfavourable) to grade or rank the alternatives.

	Walk	Roller skate	Jet pack	Bicycle
Safety	1	4	5	2
Cost	1	2	5	3
Commuter acceptance	3	4	1	3
Maintenance	1	3	5	4
Convenience	2	4	4	2
Totals	8	17	20	14

Decision/selection

Decide to encourage walking—publicise the health advantages. In some cases walking may be too time-consuming, therefore also encourage the use of bicycles.

As an entrepreneur you decide to become a bicycle manufacturer and supplier.

Design

Production

Sales

Disaster!

The city authorities ban cars carrying less than three people from the city. This reduces congestion as commuters 'car share', or use the public bus service which can now travel around more quickly.

Obsolescence

Possible salvation

(a) Redesign the bicycle and sell as a static exercise machine.
(b) Export the bicycles to Third World countries. May be too expensive since they were designed for a developed country.
(c) Start a travel/holiday company specialising in cycling holidays— with a 'free' bicycle included in the price.

2.4.3 The Anatomy of the Design Process

The design process can also be examined by reference to the actions performed by the designer, from the initial evaluation to a final solution. This is known as the anatomy of the design process and generally contains several steps as discussed below. These steps will be carried out at each stage of the morphology approach. In this type of analysis the work of the designer is usually considered to be mainly concerned with the design, but only as far as the detailed specification. After this stage, specialist personnel become responsible for the project, although advice and feedback to the designer are still important.

(a) Identifying the problem and evaluating the need

At the outset of the design process the overall objective must be defined in terms of either the problem to be solved or the need to be fulfilled. The overall specification and any constraints that are imposed must be defined. This definition must be performed at each stage of the process as outlined in the morphology approach, in order for the project to progress as efficiently as possible. As the constraints and criteria are defined at each stage it may be that the original specification will require modification. Continuous and effective feedback of information is vital to the success of the project.

(b) Information retrieval and assessment

Information is collected and evaluated at all stages of the design process. This will include areas normally outside the scope of the engineer's work. Efficient retrieval of information, as well as identifying gaps in the knowledge, is necessary. The main sources of information are the customer, the user, experienced personnel, technical literature, patents, computer data-storage systems, observations, records, etc. Once obtained, any information should be analysed and assessed for its usefulness. The information should be categorised and referenced, and then filed for easy access. Information may be detailed and lengthy, and the establishment of abstracts from each source may be valuable. If the information is not collected, assessed and stored in an efficient manner, the design process will not proceed efficiently.

(c) Evaluating the alternatives

At each stage of the process new problems occur and there are often several alternatives available. Progression to the next stage can only be made when these alternatives have been analysed and the best solution selected. Certain constraints will affect the analysis and the selection, the main limitations being time and cost. It may be that a final design solution is required very quickly, but only a limited amount of information has been obtained. The immediate decision will be to select those areas that will yield the necessary information quickly, e.g. technical literature, patents and information from other companies. The time available for this stage will be specified, and also the consequences of proceeding with inadequate information. The alternatives will be assessed and decisions made after consideration of the physical limitations, and the constraints imposed by society. At each stage the specifications and constraints imposed by the overall problem must be fulfilled, or reassessed and modified.

In order to make a decision, the alternatives available will need to be tested and analysed. This may be carried out in the laboratory using scale models or prototype units, or it may be in more symbolic form using diagrams, mathematics or computer simulation. All modelling techniques require simplifying assumptions to be made, and these should be clearly stated and reviewed when results are analysed and decisions made. A whole field of study known as decision theory has been developed and if the alternatives are numerous or complex, specialist techniques and assistance may be valuable. A starting point may be to list the input, output and solution variables which will occur during the search for a solution to a problem. This type of analysis can also be useful at each stage of the project to help clarify complex situations. The approach can be illustrated by considering some of the variables that exist when evaluating the solution to a product distribution problem, as shown in Table 2.1.

Table 2.1 Variables in a product distribution problem

Input variables	Number of products, quantities of each product, arrival schedules of each product, shelf life of products, etc.
Solution variables	Storage times, location of storage site, method of transportation, number of storage sites, etc.
Output variables	Storage cost, transportation cost, cost of product shortages, fuel cost, etc.

Within the decision-making process compromises often have to be made between what could be required in an ideal situation and what can be achieved in the real world. This requires optimisation of the available alternatives. As an example, a company may require the maximum possible profit from a product. This will need production at the lowest cost and sales at the highest price the customer will pay. However, although smaller sales at higher prices may result in larger profits than increased sales at lower prices, this strategy may causes other companies, attracted by these high profits, to enter the market. The selling price of the product will then fall and the original company may be unable to attract sufficient sales. In this case a compromise, or an optimal solution, to the problem of profit, price and competition should have been made.

(d) Communication and implementation

At various stages in the design process, information, results and recommendations will have to be communicated to other personnel involved in the process. If communication is not performed effectively then the project will not proceed efficiently and wrong decisions may be made. Communication is the transfer of information between a transmitter and a receiver, and generally involves feedback. In order to transfer information correctly, considerable thought is required. This necessitates an assessment of the level of knowledge and understanding of the receiver. For example, a sales manager will require technical information in order to prepare a product specification, based on technical performance and consumer requirements. The quantity of information must be such that only essential details are presented for immediate evaluation, although the full product details should be available. Finally, the quality of the information and its presentation should be acceptable to the receiver, otherwise it may not be fully appreciated or accepted.

It is important that decisions are made and accepted by members of the project team at each stage, and in the following stage where recommendations previously made will have to be implemented. All personnel involved in the design process, from its original conception to final solution, should be kept adequately informed of the progress of the project and the decisions that are made at all stages, even though they may not be directly affected at that time. This will avoid decisions being made which create unacceptable conditions at a later stage. In this aspect, feed-forward of information is as important as feedback for the success of the project.

* * *

Exercise

Compare the design stages you performed and the morphology of the design (previous exercise), with the stages identified as the anatomy of design.

* * *

2.4.4 Example of the Anatomy of a Design

The problem
The need to reduce atmospheric pollution due to automobile exhaust gases. Data concerning the quantities of these gases present in the air (especially in large cities), the quantities of gases (especially carbon dioxide) higher in the atmosphere and medical studies of levels of lead and the effect on children are available.

The need for pollution control is obvious and can be tackled in one or more of the following ways:

(a) prevent (or reduce the amount of) these gases being produced;
(b) prevent exhaust gases escaping into the environment;
(c) counteract or neutralise the effects of these gases when released into the atmosphere.

Assessment of the problem
Obtain the following information:

(a) amounts of different gases produced by different types of vehicles;
(b) maps showing automobile density, average usage, level of emissions, etc.;
(c) possible future changes in the data in (b);
(d) existing legislation regarding automobile exhaust gases;
(e) available technology which could be used;
and so on.

Evaluating the alternatives

Counteracting or neutralising released gases is probably expensive and difficult to implement on a large scale, and it also does not directly encourage those responsible for this pollution to take action to reduce it. It may be difficult to raise the required revenue from appropriate sources, e.g. petrol tax, oil company profits tax, and this type of system may not operate equitably.

Preventing or reducing the gases being produced will probably need a complete redesign of the automobile engine which will take a long time. This alternative may only be possible by using fuels other than petrol, and these may not be readily or economically available. There is the problem of existing vehicles which would require conversion, otherwise restricting successful implementation of this approach.

Preventing or reducing the quantities of gases released suggests designing a small replaceable absorber unit to be fixed to the automobile exhaust system. This appears to be the most acceptable solution.

Communication and implementation

Use the media to increase public awareness of the problem, its effects and the need for a solution. This will increase the level of consumer acceptance of the solution.

If the technology is available to design such a unit, then the unit should have most of the following features:

(a) easy to attach and replace;
(b) cheap;
(c) appropriate lifetime, typically corresponding to automobile servicing intervals;
(d) easily rechargeable or disposable;
(e) no effect on automobile performance;
(f) one unit suitable for most if not all vehicles;
(g) include an indicator showing when replacement is required, e.g. smoking or reduced engine speed.

Legal action should be taken to ensure that automobile manufacturers, garages and the owners of older vehicles carry out the requirements.

The effectiveness in reducing levels of exhaust gases due to implementing this solution to the problem should be monitored, and the technology required to implement the other two alternatives developed.

Evaluate new ideas and alternatives as they become available/known/recognised.

* * *

Exercises

1 Identify two design problems from your local environment, either actual or imagined. One problem should be for a consumer-type product and the other for an improvement in society. Consider both the morphology and anatomy approaches to the solutions of these problems.

2 The morphology and anatomy stages of a design project are shown for comparison in Table 2.2. Which approach provides the better understanding of the stages involved in a design problem, and the work of the design engineer?

* * *

Table 2.2 Comparison of the Morphology of Design with the Anatomy of Design

Morphology of design	Anatomy of design
Identifying the problem	Identifying the problem and evaluating the need
Feasibility study	Information retrieval and assessment
Preliminary design	Evaluating the alternatives
Detailed design	Communication and implementation
Production	
Distribution, sales, use	
Obsolescence	

- *Now go to Section 2.5.*

- *From page 28.*

Exercise 1—Suggestions

These are not answers, only suggestions; accept or reject them as you wish!

Power source: length 5 cm; cross section either 2 cm diameter or 2 cm square.

Light disc: 2 cm diameter or 2 cm square; 1 cm thickness.

These are minimum dimensions.

Density: of both parts approximately 750 kg/m^3.

Materials: power source enclosed in thermosetting polymer or light aluminium alloy casing.

Light disc is a silicon-based, opaque, glass-type sintered material.

All information about the principle of operation is highly confidential, not to be disclosed even to the design team. This policy is to be maintained for as long as possible.

Operation is adversely affected by high humidity (how high?), and carbon dioxide gas in concentrations greater than 10% by volume.

The torch can be mass-produced.

Estimated service life (before recharging) is approximately 1000 hours.

This will be reduced by frequent on/off operation.

Competitors are thought to be at least 12 to 18 months behind in their development of this type of product. However, if technical details become available they could compete directly in 14 to 16 weeks.

Using these suggestions and other ideas, continue with the project as outlined in the instructions on page 17.

At the next *complete* impasse, or perhaps when you consider the exercise finished, study the comments on page 48.

● *From page 28.*

Exercise 2—Suggestions

These are not answers, only suggestions; accept or reject them as you wish!

Answer the questions so that the exercise can proceed. Make positive estimates, but do not assume there are no problems. For example, you may specify adverse pollution or safety aspects but either show or assume that solutions are available.

Select your own process, then ask:
(a) What equipment/chemical plant is required?
(b) Does the large parent company already produce certain items, e.g. valves, regulators, cylinders, pumps?
(c) What raw materials are required?
(d) What about their price, availability, toxicity, etc.?
(e) What by-products are produced?
(f) What about their market potential, toxicity, etc.?

Specify one form of supply, e.g. high pressure gas cylinders, distribution radius 320 km. Cylinders are rented to customers and returned for re-use.
(g) Is the country's economy or the world economy in a recession or growth phase?
(h) Does it matter for this product?

Using these suggestions and other ideas, continue with the project as outlined in the instructions on page 17.

At the next *complete* impasse, or perhaps when you consider the exercise finished, study the comments on page 48.

● *From page 29.*

Exercise 3—Suggestions

These are not answers, only suggestions; accept or reject them as you wish!

Consider the movements required for simple writing and then decide if this is possible.

Can individual fingers be manipulated?

Determine maximum load that can be lifted, and whether this is limited by the hand, or by the arm connection.

Decide how many push buttons each control panel needs.

Does the hand need five digits?

What operation for each digit, e.g. up, down, forward, backward, sideways, etc.?

Assume a power source superior to batteries is currently available, e.g. smaller, more powerful and with a longer life.

Make decisions which, if not over-optimistic, do not hinder the project. For example, assume it will be available in 12 months and any operating secrets can be preserved by use of a sealed unit with tamper-proof self-destruct features!

Costs will be reduced if the hand and controls can be mass-produced as far as possible. However, consider the design features necessary to fit one model on different arms, and the length required.

Your project should aim to produce as versatile a product as possible. Advantages should be the co-ordination of human and artificial hands, or even two artificial hands, in order to perform complex tasks. Also the ability to perform a wider range of tasks than currently available models.

How important are the aesthetic features?

Decide what development costs should be recovered for this product. What company costs related to research work for similar projects should be allocated to this product?

Consider your reaction and personal position if:

(a) Product sales are to be restricted owing to an overlap of technology with current secret research.

or

(b) The work and finance for the project is taken over by the government and is to be the responsibility of the Defence Department. Sales are then to be restricted to military personnel and details limited by the official secrets declaration.

Ethical and moral conflicts will be discussed in Chapter 11.

Using these suggestions and other ideas, continue with the project as outlined in the instructions on page 17.

At the next *complete* impasse, or perhaps when you consider the exercise finished, study the comments on page 48.

● *From Section 2.4.4.*

2.5 FACTORS INFLUENCING A DESIGN STUDY

The design process and the role of the design ʼengineer have so far been analysed by considering the actions of the designer and the methods used to solve the problem. It now becomes necessary to consider some of the more specific aspects of design such as planning, materials' selection and manufacturing processes, and their integration into the detailed design phase of the project.

The senior design engineer will probably spend the majority of his working time on two aspects of design work. First, the creative work that is required to obtain a solution to a problem (this is sometimes assumed to be the actual design process). Second, the organisational responsibilities associated with the overall design process and the functioning of the design department. The successful completion of a project will require logical sequencing and planning of the design operations, and suitable allocation of individual design tasks. These organisational aspects will ensure that personnel are assigned tasks commensurate with their abilities and experience. Parts having a more intricate design and longer manufacturing times are prepared sufficiently early so that the progress of the project is not impeded.

A junior design engineer will spend the majority of his working time engaged in the creative and developmental aspects of the design process. He will probably begin his career as a detail designer specialising in a narrow design field. With time and suitable training his experience will broaden and he will become competent in other design specialisations. As a detail designer he will be a member of a team, and concerned only with certain aspects of the design method. Often the problem specification, information collection and evaluation of the alternatives will have been completed, and the detail designer will be asked to prepare a detailed design for a particular specification.

Some of the factors influencing a design, and their interrelationships, are listed here in order to illustrate the diverse nature of design problems. This is not an exhaustive list and not all of these factors will be applicable or of equal importance for all problems. The diversity of the factors which the designer may have to consider shows why only exceptional persons are successful in the field of creative design.

The factors influencing a design problem can be placed in various categories. Factors due to the product criteria and constraints are shown diagrammatically in Fig. 2.4, and factors caused by manufacturing methods are shown in Fig. 2.5. There is a certain amount of overlap between Figs. 2.4 and 2.5. Although different design factors are identified and shown separately, during the design process no single factor can be considered in isolation. Each design factor or stage in the process will be dependent upon other considerations, and these need to be considered together before a decision can be made. Certain design factors are directly related, such as the availability of materials and the design specification. It would be pointless to produce a design specification using materials that are not available. Some factors are only related indirectly, such as delivery date and design where sufficient time must be allowed to complete the design process. However, the delivery date will become more important during the manufacturing stage and will have a significant influence on the selection of the manufacturing process to be used.

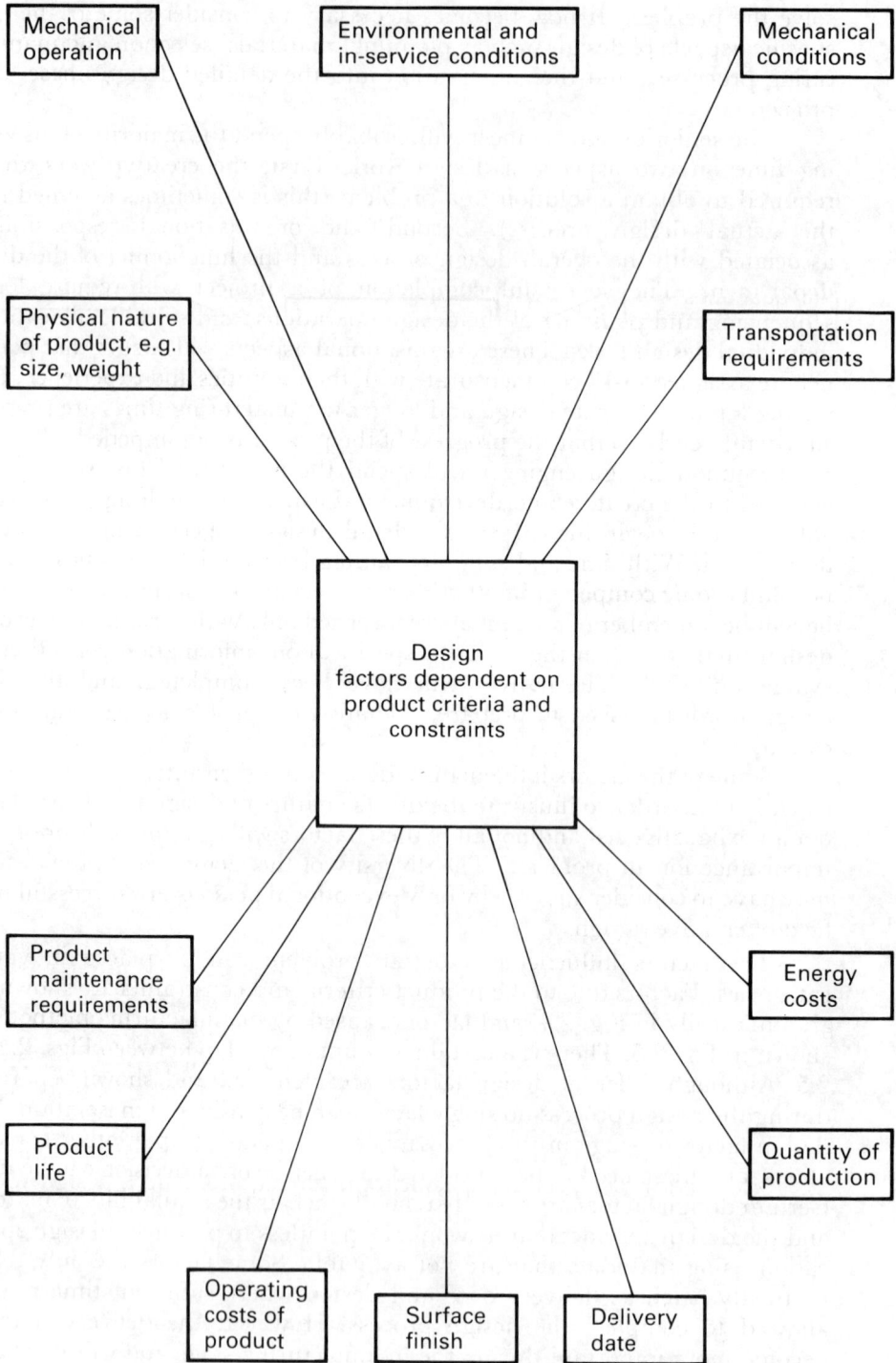

Figure 2.4   Design Factors Dependent on Product Limitations

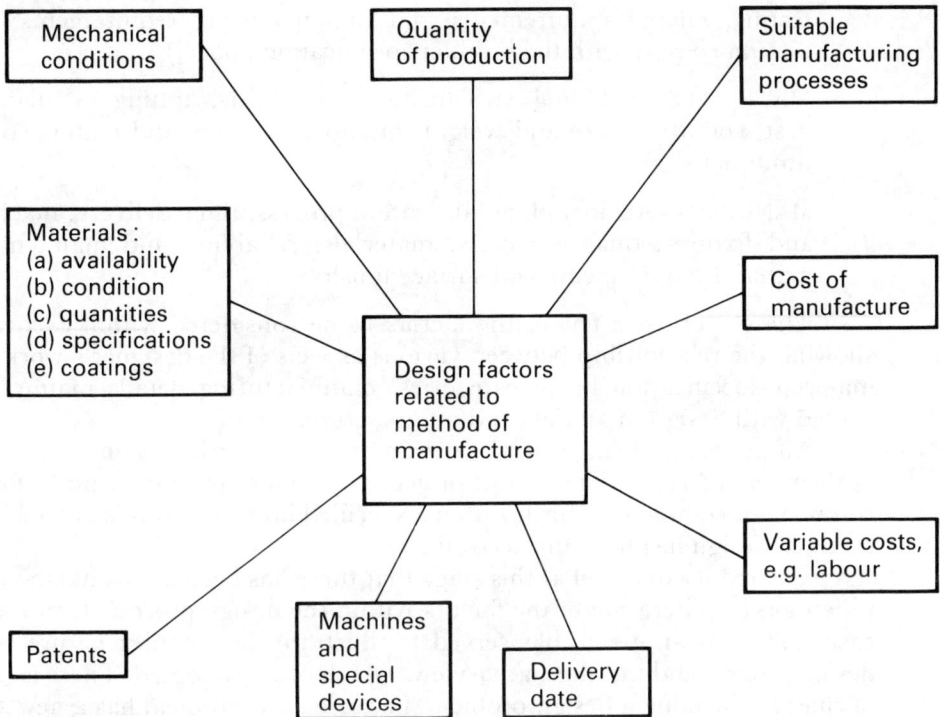

Figure 2.5   Design Factors Dependent on Manufacturing Method

Each stage within the life cycle (morphology) of a product (Section 2.4.1) can be considered as a part of the design process. The junior designer will be concerned mainly with the early stages of the process, usually as far as producing the detailed design. As his experience increases he may become more involved in the latter part of the cycle. Considerable interaction will take place between the designer and the production engineer, requiring the designer to be familiar with details of manufacturing processes. The problems or tasks confronting the designer early in his career or the undergraduate embarking on a design study, are often divided for convenience into four main categories. These are product specification (or requirements), design specification, materials and manufacturing. These areas are interrelated as can be clearly seen by consideration of their individual factors, detailed as follows.

PRODUCT SPECIFICATION: including operational requirements in-service conditions, size and weight, maintenance, life, reliability, quantity, delivery and operating costs.

DESIGN SPECIFICATION: including operational details, in-service conditions, size and weight, sub-components, existing designs, standards, testing, energy requirements and manufacturing requirements. (NB Should comply with the product specification above.)

MATERIALS: including specification, standards, testing, availability, cost, condition, size and weight, quantity, delivery and manufacturing limitations.

MANUFACTURE: including fabrication process, time, delivery, machines and fixtures, quantity, costs, materials, reliability and maintenance, material requirements and surface finish.

These are just a few of the factors to be considered within each area, showing the relationship between various aspects of the designer's work, e.g. material specification in all categories, manufacturing details mainly concerned with design, materials and manufacture.

All factors and stages in the design process must be evaluated in terms of their related costs. The overall project must either produce a profit for the company or complete the task within a specified budget. If this is not achieved then the design has been unsuccessful.

The reader may feel at this stage that there has been a certain amount of repetition or reiteration of the factors within the design process. If this is the case, then this approach has served to illustrate the complex nature of the design process and the alternative views which can be adopted. There is no set of rules for tackling a design problem, because each problem has a new set of requirements and limitations. Suggestions for solving a problem can be provided but each situation will have different priorities. The reader should not assume that this chapter represents a specification for a design engineer, or a 'blueprint' for solving design problems. It will be relevant to many problems, but few problems would be solved satisfactorily merely by following these suggestions and observations. In most jobs the essence of success is experience and practice, and this is especially true of design engineering.

2.6 EMOTIONS, STRATEGIES AND PROBLEMS

At the beginning of this chapter, the instructions accompanying the exercises included recording your reactions and personal feelings as the investigation proceeded. If you did this, now read through your chronicle of events (or try to recall what you think your feelings were). I suspect they started with feelings of helplessness and hopelessness which recurred as unfamiliar conditions or questions occurred. However, as problems were solved, questions answered and help became available these feelings probably ranged to the other extreme of enthusiasm and self-satisfaction.

Part of this see-sawing of emotions is caused by lack of familiarity with this type of project work, but it is also due to a lack of confidence and a realisation of the lack of knowledge in these specialised areas.

Table 2.3 Problems and Strategies in Design

(reproduced by permission of the Open University, *Design Project Guide Unit 12; T262 Man-Made Futures.* Copyright © (1975) The Open University Press)

Problem or pitfall	*Possible way out*
Being over-rigid about problems and solution ideas.	Abandon your pet idea for a while.
Being too ambitious for available time and resources.	Ask what can I do practically tomorrow, the next day, etc.
Having no strategy.	Make one!
Having too thorough a strategy.	Cut down at least one action from each stage.
Not following methods through to completion.	Completing methods takes time but it usually pays off.
Not knowing what to do next.	Review your strategy. Do anything relevant you can immediately.
Developing a fixed solution concept prematurely.	Abandon solution idea and explore problem further.
Solving a non-problem or a trivial aspect of a wider problem.	Apply the criterion of worthwhileness. Talk the problem over with someone else.
Not being able to transform the mass of data collected during analysis into solution ideas.	Allow time for the problem to become defined in your mind. Write summaries of the data.
Collecting data without knowing what to use them for.	Do not collect data at random unless you have a 'hunch' they may be useful somehow.
Lacking confidence to test ideas in real situations.	Courage needed. Do it!
Getting disheartened when things are going badly.	Plod on, this is to be expected (and can be a sign that you are discovering something new).
Being afraid to talk to others about the project.	Drag in everyone you can and excite their interest, or bore them to death with your enthusiasm.
Following your strategy too rigidly.	Review your strategy regularly against your timescale and spontaneous thoughts.
Designing for remote, inaccessible clients (e.g. Third World peoples).	Orient your project to yourself or to people whom you can actually contact.

Another major reason is inadequate identification of the various phases in the design stage of the project. The work to be carried out and its sequence in each phase of the project must be clearly identified.

Compare the scheme of work you prepared with the generalised description of the design process in Section 2.4.1. Perhaps you can see why some of your problems occurred, e.g. trying to do parts of the detailed design work without sufficient preliminary studies, or performing the design work without proper feedback provision and subsequent action and revision. Each stage of the design process (Table 2.2) should involve problem exploration (analysis), solution generation (synthesis) and selection (evaluation). The selection of a particular course of action should always be thoroughly checked against the design criteria before proceeding to the next task. For various stages of the project, identify these three activities in the work which you performed.

Next time you tackle a design problem or project consider the following suggestions first. Outline your strategies and tactics for 'solving' the problem. Decide the general direction of the work and the preferred sequence of actions. Constantly review the work already performed, record information, compare alternative sources of information and assess the actual work performed with the original plan.

Refer to the generalised stages in a design project (Table 2.2) and try to follow these stages, or keep them in mind during the investigation. Again record your personal reactions as the project proceeds and review these feelings at the end of the project. Compare this with your reactions for the first exercise performed and determine whether the teaching of design has made the task easier. Remember that the work should still be intellectually stimulating and challenging because your expectations and abilities are now higher. Hopefully, a structured approach eliminates many organisational errors and allows more time for creativity.

Finally, some ideas on what to do when you reach a point from which you find it impossible to proceed. Table 2.3 contains a list of problems and strategies which may help. (The original list was presented in an Open University Design Project Guide.) It will be useful to review this list periodically.

- *Now go to Section 2.7.*

- *From pages 28, 29, 41 and 42.*

Exercises—Comments

These are textbook exercises but are also typical of the type of project work assigned to graduate engineers in their first years in industry. The new graduate will not immediately assume the mantle of responsibility belonging to the project engineer, but company policy on training and experience will normally make this an early priority.

The project engineer is not necessarily a senior engineer or manager, but he must be experienced and capable of carrying out his particular tasks. In the design and development of a new car there will be several project teams, each considering different aspects of the overall problem, e.g. the engine, braking system or safety features. The work of all groups will be co-ordinated and supervised by a senior engineer.

The exercises given here are examples of small projects requiring a small project team, at least in the early (pre-production) stages of the work. Initially this team may comprise two or three engineers, a draughtsman and a technician. Numbers will expand as the project develops to include a production engineer, marketing manager, safety advisor, etc., as appropriate. Some of these may not be full-time members of the team.

The project engineer will be responsible for planning and co-ordinating the design of equipment, liaison with other departments, assembly and testing of prototypes, purchasing, etc., and also the transition from this stage to production and sales. The scope of each of these exercises is sufficiently narrow that a design team, or the student group, can keep in view the central theme or main objective and appreciate all aspects of the work. For large projects, such as commercial exploitation of a new oil field or a spaceship flight, there will be several large design teams and, within these, many smaller ones. It is important that each team appreciates its interrelationship with and dependence on others, and also (as far as possible) its place in the overall project.

From these exercises and associated discussion and background reading, the student should begin to appreciate several aspects of design work, and his future role in industry. First, the use of different titles for departments performing similar types of work (depending upon the policy and structure of the company), such as Research and Development, Process Design and Development, Engineering and Technical Services, Project Engineering, etc. Second, although the engineer is trained and experienced in certain aspects of his particular discipline, he is expected to have some knowledge of other areas and an appreciation of the relevance of many factors to the success of the project.

Consider the general nature or philosophy of these exercises:

EXERCISE 1: the exploitation of a new technology where the most important aspect is maintaining secrecy from competitors. Producing the 'best' design is in some ways secondary to this requirement. Here, the necessary research has been performed, now the appropriate engineering and design work has to be implemented.

EXERCISE 2: this project begins at the formulation or feasibility stage. Many of the early decisions are concerned with economics, and predictions or estimates of the future market size and sales. Many of these decisions will be based on engineering principles, e.g. the possibility of new developments that will use carbon monoxide and pollution and safety requirements. The engineering decisions and the design will only proceed if the economic factors are positive,

EXERCISE 3: this involves adaptation of an established technology (biomechanical) with advances in remote control techniques. The way the exercise is formulated it is uncertain how much previous research will be of use immediately, and how much extra development and testing still needs to be done. The unknowns in this case are directly related to the engineering aspects. Secrecy is again a prime consideration, but here it may be easier to achieve, depending upon the level of guaranteed government sales.

In each exercise, there are certain common factors or items to be identified. In every project the costs associated with the work, and hence the selling price and profit, have to be determined. Exercise 2 is fairly straightforward in that all development and testing costs would normally be recovered directly from sales revenue, determined over a period of several years. However, in Exercises 1 and 3, the apportionment of costs becomes more difficult. For the torch it is hardly practical to attempt to recover all research and development costs if this new technology is to be used as the basis for other projects. Similarly with the artificial hand: it would be difficult to determine what proportion of previous research is being utilised, and also what aspects of the further research required will be utilised in other products. Attempts to recover too many research costs too quickly can kill off an otherwise viable project. However, if the money spent in research departments is not recovered, then the company's profitability will suffer.

For each product it will be necessary to establish the selling price. This will be based upon the costs incurred and the required profit. In some cases, the price may initially be set very low, even below cost price, in order to establish a share of the market, certainly for repeating sales such as the carbon monoxide gas.

Alternatively, for a product requiring a high initial investment, either in research (Exercises 1 and 3) or production plant (Exercise 2), the initial price (and profits) may be set very high to maximise the return while there is no competition. However, this strategy may encourage competitors to enter this field (owing to the high profit margins) and then prices will inevitably be reduced. This type of strategy—high initial price followed by drastic price cutting, e.g. calculators, tape recorders, video equipment—can create an adverse company image and reduce the demand for its other products over a longer period. It may be more advantageous to have a lower selling price in order to discourage new competitors for a longer time.

The marketing and sales strategy for each product will have to be carefully planned. The approach should not be too aggressive otherwise sales may be *too successful*. This is an odd statement but if initial demand grows too quickly and production cannot keep pace, customers will be lost and competitors will benefit directly from your efforts.

Throughout the investigation, the students should have remained aware of all available alternatives, although they may not have been considered originally or may have been rejected. With Exercise 2 it is always possible (prior to plant construction!) to abandon the decision to build a production plant, and either buy the existing plant (or company!) or negotiate for a long-term, low-price imported supply. The old plant is probably becoming uneconomic, in both maintenance and energy costs, and the existing supplier may agree to produce a smaller amount of gas in exchange for a fixed market share. However, this type of arrangement may be difficult to enforce and may be contrary to government monopoly laws.

By now the student may:

(a) be exhausted;

(b) consider the problem exhausted.

It is appropriate to consider which aspects of design and the design process have been revealed. The remainder of this chapter, which describes the stages in design problem-solving, can then be studied in detail.

The student should have identified the following stages or activities which occur during the design process from inception to sales.

IDENTIFICATION: establish the need, the product, the problem, etc.

SPECIFICATION: formalise the requirements, the brief, the product specification, etc.

LIMITATIONS: list the constraints, actual, probable, possible, those that are probably solvable, those that appear insurmountable.

DESIGN: the technical stage, data acquisition, calculations, decisions, materials, etc.

PRODUCTION: custom-built or mass-production.

MARKETING AND SALES.

These are hopefully your first ideas. You can compare these ideas with the stages in the design process described in Section 2.4.

There is much, *much* more required for successful design and project engineering.

● *Now go to Section 2.1.*

● *From Section 2.6.*

2.7 DESIGN OF A STRUCTURAL COMPONENT

Engineers of all disciplines are frequently required to design structural components or machine parts which must support a load. The considerations

Designer's
actions

↓

Specify: | Functional requirements | | Operational requirements | | Constraints |

Establish: | Design conditions |

Consult: | Design codes |

Define: | Operational limitations |

Select: | Materials |

Determine: | Preliminary layout |

Perform: | Design calculations |

Specify: | Final design |

Ask: | Fulfilment of all requirements and limitations |

Accept
and
implement: | Manufacture |

Stages

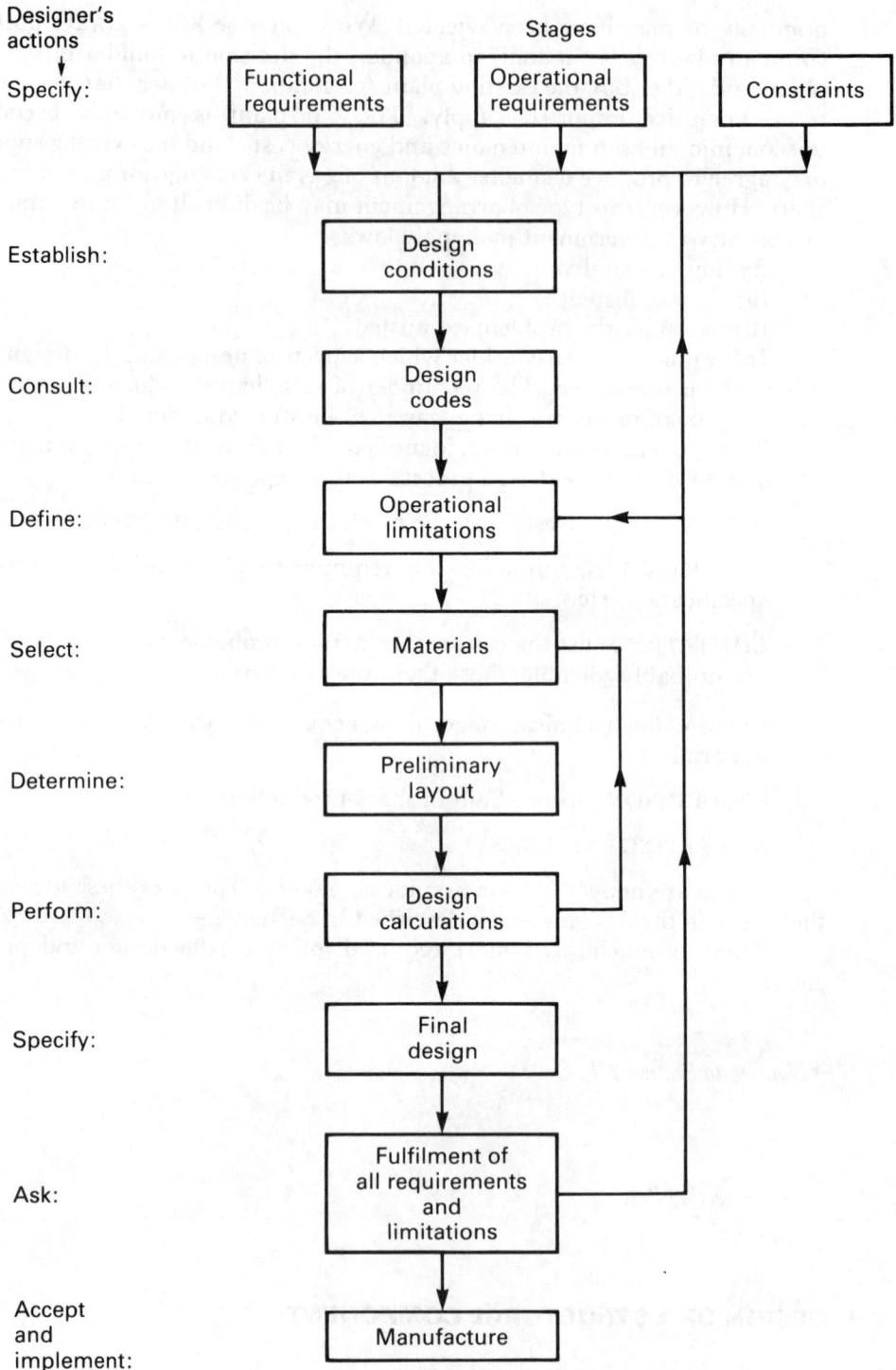

Figure 2.6   Design in Terms of Actions and Stages

made in obtaining a solution to this type of problem can be used to illustrate the points discussed in this chapter. The approach will be outlined in general terms but the reader could use these suggestions while preparing the design for a simple component. This would illustrate the simplifying assumptions necessary when attempts are made to define the design process, or to explain it in written terms.

The steps to be taken by the designer (yet again!) and also the actions that the designer makes are shown in Fig. 2.6. For a structural component, the main technical problems which need to be resolved are:

  (a) determination of the general shape or form of the component;
  (b) evaluation of the loads that the component must carry during its expected life;
  (c) selection of a suitable material and determination of the maximum allowable stresses;
  (d) expression of the maximum stress, or other criteria, in terms of the loads and dimensions;
  (e) modification of dimensions to comply with space, strength and fabrication requirements.

Definite procedures have been developed and are taught which enable solutions to these individual problems to be obtained. However, the problems are interrelated and a complete solution will require the designer to exercise judgement, experience, flexibility and innovation. The use of the component and its final appearance must be considered during the design stages, as must the structural properties required and the cost and availability of specified materials. The component must possess sufficient hardness and rigidity to perform satisfactorily, but excessive strength may mean unnecessary weight and excessive material cost.

The relationships between the specific factors that need to be considered for the design of a structural component are shown in Fig. 2.7. Usually the shape will be determined by the required function, and the loads to which it will be subjected are then estimated. The next problem is to determine the necessary size. This can only be achieved by consideration of the total stresses and displacements, and hence the unit stresses and strains. The maximum stress which will occur during operation and the normal working stress must be determined. Allowances must be made for uncertainties of loading, approximations in calculations and any uncertainties in the quality of the material to be used.

The effects of a component failure must also be taken into account. Whether the component can be easily and cheaply replaced or whether large costs will be incurred, will significantly influence the design. Failure usually means that the component will no longer perform its intended use satisfactorily. In certain cases failure may result in impaired performance but may not necessarily require immediate action. The component must be designed so that failure will not occur as a result of excessive distortion, cracking or rupture of the material. Therefore calculation of the strains developed within the material becomes necessary. The most common types of failure are due to excessive elastic deformation, slip, fracture and creep.

The failure of a component depends not only on the magnitude of the applied load but also on the manner in which the load is applied. Loading can be steady or static, impact or dynamic, repeated or fluctuating. The action of the loading may be uniform or variable with respect to time.

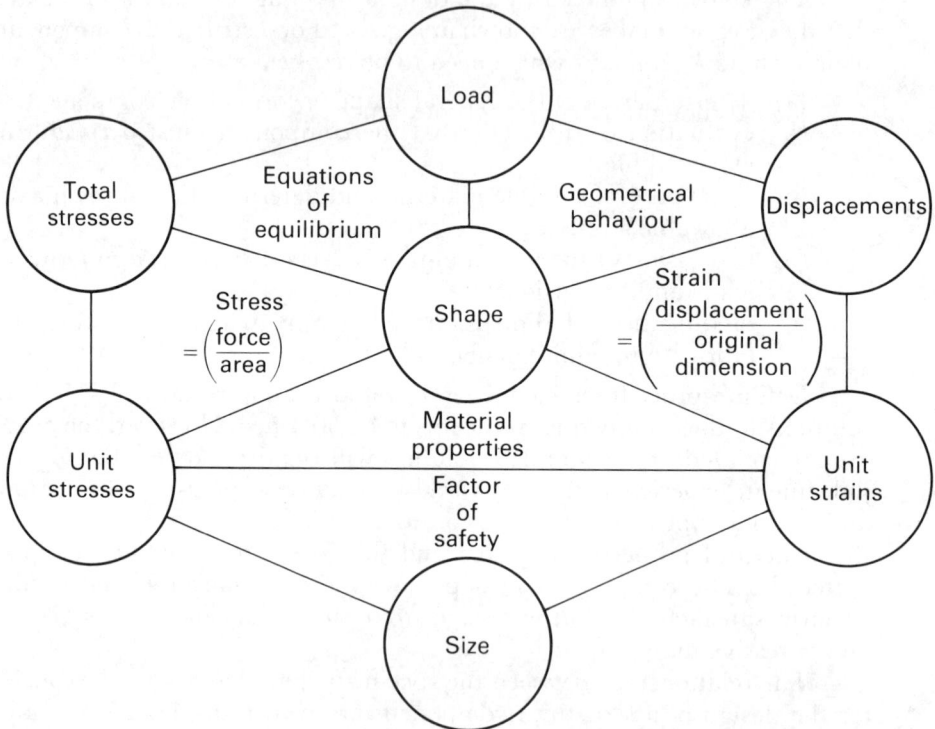

Figure 2.7   Design Factors for a Structural Component

* * *

Exercise

Select a simple structural component, preferably one that is in everyday use. Perform the calculations required in order to prepare a basic specification for this item *without* regard to its shape, size or appearance. Then consider the modifications that become necessary in your design when these factors are considered as constraints. Either decide your own values or determine those that the particular product is required to possess in a practical situation.

* * *

KEYWORDS

These keywords should be used as a revision aid. They are not in alphabetical order but in the order they first appear in the chapter (for easy reference).

development design   detailed design
adaptive design   prototype
creative design   production
conceptual ability   schedules
logical thought   obsolescence
invention   anatomy of design
creativity   problem evaluation
IQ   information retrieval
inventiveness   input variables
innovation   output variables
brainstorming   solution variables
morphology of design   product criteria and constraints
product life cycle   project engineering
feasibility study   design codes
preliminary design

BIBLIOGRAPHY

Refer to the general comments in the Bibliography for Chapter 1. All books and journals listed in Chapter 1 are applicable here.

Astley, P., *Engineering Drawing and Design II*, Macmillan Press Ltd, London (1978).
Just one good book from a multitude covering graphics, engineering drawing, design, etc.
Beakley, G.C., and Chilton, E.G., *Introduction to Engineering Design and Graphics* (1973) and *Design: Serving the Needs of Man* (1974), Macmillan Pub. Co. Inc., New York.
The second book contains extended material from the first!
Both are excellent, wide ranging and visual.
Dixon, J.R., *Design Engineering: Inventiveness, Analysis and Decision-Making*, McGraw-Hill Book Co., New York (1966).
French, M.J., *Engineering Design: The Conceptual Stage*, Heinemann Educational Books, London (1971).
A basic textbook.
Gibson, J.E., *Introduction to Engineering Design*, Holt, Rinehart and Winston Inc., New York (1968).
As title—basic textbook.

Gregory, S.A., *Creativity and Innovation in Engineering*, Butterworths, London (1972).

A good book showing where creativity is, and should be, in the engineering profession.

Haefele, J.W., *Creativity and Innovation*, Van Nostrand Reinhold Pub. Co., New York (1962).

As title.

Jones, J.C., *Design Methods—Seeds of Human Futures*, John Wiley and Sons Inc., New York (1980).

Highly recommended—read it and find out why.

Krick, E.V., *Introduction to Engineering and Engineering Design*, 2nd Ed. (1969) and *An Introduction to Engineering: Methods, Concepts and Issues* (1976), John Wiley and Sons Inc., New York.

Excellent books covering basic introductory courses; interesting and visual material.

The Open University, *Design Project Guide, Unit 12; T262 Man-Made Futures*, The Open University Press, Milton Keynes (1975).

Very short, paperback. Full of good ideas for starting projects. A manual of what to do.

Simon, H.A., *A Student's Introduction to Engineering Design*, Pergamon Press Ltd, Oxford (1975).

Two sections; first, the design textbook and second, technology and society. Recommended.

Vidosic, J.P., *Elements of Design Engineering*, Ronald Press Co., New York (1969).

A good textbook with useful case studies and examples.

Woodson, T.T., *Introduction to Engineering Design*, McGraw-Hill Book Co., New York (1966).

Comprehensive textbook (427 pages).

AUTHOR'S NOTE: CHAPTERS 1 AND 2

In the first two chapters of this book, I have attempted to do several things.

First, to study the process of design and identify common activities, features and tasks that have to be performed, whatever the nature of the problem. The intention is that the student will acquire and develop the skills necessary to obtain satisfactory solutions to design problems. I believe that design course tutors should help the students recognise the abilities they already possess, so that they can use their skills in situations where it is confidence and experience that are needed, rather than more knowledge and information.

I fully agree with the following observations reported by J.R. Dixon in the preface to his book *Design Engineering: Inventiveness, Analysis and Decision Making* (McGraw-Hill, New York, 1966).

'Experience with these design courses led me more and more to the conclusion that my students knew more than they understood or could use. I took it as my task to help them learn how to use purposefully what they had been taught ... the *way* they had been taught was interfering with their ability to use what they had "learned".'

The way to achieve these goals is not by more theory and lectures but by actively involving the students in design-type problems. Only then will they appreciate the interaction of the different factors involved in design work, and the usefulness of analysing the design process.

Second, to make the student aware that problems in the real world are not presented as 'pre-packaged' units. All the necessary information and data are not readily available, if at all! Often there is no obvious correct route to success. At the end of the solution search, the answer is not *the* solution, but just one of many alternative solutions obtained by making particular choices throughout the problem-solving process.

Third, to make the student think about the role of engineering and design in society, to view engineering in the broadest possible way and not to be constrained by particular disciplines and specialisations. Hopefully, the student who develops these thoughts will obtain a better understanding of his function as an engineer and the contribution he can make towards improving society.

AUTHOR'S NOTE: CHAPTERS 3 TO 11

The remaining chapters of this book are concerned with particular aspects of the design process. These include areas of background knowledge and awareness which are essential if the engineer is to achieve the best possible results. The chapters are deliberately brief and require little specialist knowledge. Most of these topics will, or should, also be taught separately within a degree course. This book encourages the student to use knowledge from all parts of the syllabus and to question the relevance of this material to the work of the engineer.

The students should be sufficiently self-motivated, and also encouraged, to identify, investigate and use particular knowledge relevant to the problems. There is no shortage of suitable textbooks and journals to provide the technical information. There is no shortage of ability in the students. What is often needed is interest and enthusiasm.

Decision-making

Chapter Objectives

To help the student begin to appreciate that:

1 design involves alternatives and therefore decisions must be made;

2 a single correct solution rarely exists;

3 the 'best' or 'optimal' solution requires compromise;

4 mathematical techniques are useful but do not provide absolute answers;

5 decision-making should be based upon reliable information and experience.

* * *

Exercises

For each (or some) of the following examples *list* the factors to be considered, the possible alternatives available and the decisions to be made before a final course of action can be recommended.

1 A car is to travel between two towns connected by a six-lane highway. The total distance to be covered is over 1000 km. Recommend the speed at which the car should be driven.

2 A department store is having a clearance sale of a discontinued washing machine model. The old machine can be purchased at 30% less than the selling price of the new, improved, replacement model. Which machine would you buy?

3 The Cost Control Department of a medium-sized engineering company observes large differences between claims for refund of travel costs incurred by its employees while on official business. Make proposals for a company policy which stipulates the mode of transport to be used by senior executives, servicing and trouble-shooting engineers and other personnel.

4 The lighting for a factory is provided by fluorescent strip lights. Determine a replacement policy for when these lights fail, e.g. individual replacement or all together.

5 The engineers and production managers employed by a large manufacturing company complain that telephone contact and communication is difficult because they are frequently needed outside their own offices. Investigate the possible use of a Tannoy system to overcome this problem.

6 A factory uses machines of type P and type Q, and has three machines of each type in operation. Work is fed continuously from P to Q. Machine type P has four component parts, each of which may fail and need replacement; type Q has only two such parts. A machine cannot function if a component fails. How would you determine the number of spare parts the company should keep in stock?

7 What factors will influence the choice of material(s) to be used for the production of a common, simple item?

8 Make proposals for the introduction of word processors in an engineering company.

9 Choose between the construction of a coal-fired power station and a river dam for the production of hydro-electric power in a region having a relatively low population density, except for one major town. There are local sources of coal, and the area is renowned for its outstanding natural beauty and abundant wildlife. Local unemployment is high and there are few opportunities for developing local work.

10 A chemical company produces chemical A for which there is a declining market. It also produces chemical B as a by-product which has little commercial value. Recent development work has identified three commercial products that could be made from these chemicals plus a chemical C. These products are identified as:

New product	Chemical composition (weight %)		
	A	B	C
I	20	50	30
II	60	20	20
III	85	15	—

Chemical C is a common chemical with many uses, but it is expensive.
    How would you determine which products to produce, and the quantities of each?

• *Ideas and suggestions are available on page 69.*

*   *   *

3.1 ALTERNATIVES AND DECISIONS

Some days seem to involve nothing but decisions, decisions and then more decisions. Our personal life is full of decisions, many of which we make almost automatically. For example, whether to get up or stay in bed 5 more minutes, to go to work by car or train, and so forth. Having made a decision we can then accept it and the consequences, or try to change it, if that becomes necessary and is still possible. There are many people who refuse to make decisions (unless they have to) or who always procrastinate as long as possible. Perhaps they are afraid of the possible consequences if they decide upon a course of action, but there may also be adverse effects by not making a decision.

The engineer does not usually have this choice of postponing decision-making, at least not for very long. If decisions are not made as early as possible (or practicable) then a project will not proceed efficiently and the engineer may have (eventually) decided himself out of a job. In order to make a decision there must be an alternative available. This may be either to select another course of action from a range of possibilities, e.g. perform task A or B or ..., make item P or Q or ..., etc., or to perform a task or not, e.g. to replace a component or not.

The task of the design engineer is to identify *all* possible alternatives and then select the one that is most satisfactory. The decision should be made after considering all the relevant available information, the constraints that exist, and the possible consequences of each alternative decision. All possible courses of action should be written down for consideration, even the most ludicrous and impractical, including the negative option. Ideas that would normally be rejected 'out of hand' should be recorded and considered. They may still yield useful ideas. After the alternatives are analysed in detail, some 'far out' ideas may appear more feasible. Also, today's impractical ideas may provide working solutions in the future, in which case a record will be useful.

A decision should only be made after the consequences of *each* alternative have been determined. These may be definite or possible consequences. A decision to replace an expensive worn component will include the cost of the component and also the cost of lost production (accurately estimated). The consequences of deciding not to replace the component may be higher costs if it fails (uncertain estimates). Some consequences are known exactly, some can be estimated and others are completely unknown. Unfortunately, the engineer must make many decisions where the outcome is not certain. However, techniques have been developed, and are being constantly refined and improved, which help the engineer select an alternative. These will be discussed later in the chapter.

The engineer has three functions to perform, not only in design work but in every task he performs. These are:

(a) Identify *all* possible alternatives. This will be based upon experience and will require creativity and inventiveness (as discussed in Chapter 2), and a wide-ranging view of the problem.

(b) Determine (quantitatively if possible) the consequences arising from selecting, and also *not* selecting, each possible alternative. The remainder of this chapter will discuss methods that are concerned with decision-making in this context.

(c) Make a decision. Is the decision irreversible? What are the costs of changing your mind? Continue to bear in mind the original alternatives even after the decision is made, and a course of action begun.

Before considering decision-making methods, a brief note about the role of the tutor.

My experience is that engineering students do not like making decisions. They prefer to be told what to do. This is due to a lack of experience in the subjects they are now studying and, hence, a lack of self-confidence. It is also due to a fear of the consequences resulting from incorrect or poor decisions, including low esteem from their peers and the tutor and also low marks. It is often thought that lower grades will be obtained by making poor decisions rather than by getting the tutor to disclose the preferred course of action, and then merely following this suggested approach.

The ramifications for the teacher should be obvious. Engineers constantly make decisions, and students should therefore get as much practice as possible at selecting alternatives. This means that problems should be more open to interpretation and less structured towards one particular solution. However, it should be explained to the student why his solution or suggestion does not represent the most efficient strategy, if this is the case. Such an explanation needs to be presented in a sympathetic and encouraging manner. Hopefully, the student will then feel inspired to learn from his mistakes and succeed next time, rather than instilling more fear of failure and decision-making. Do not denigrate the student's work, emphasise the positive aspects and encourage further study and improvement.

3.2 DECISION-MAKING

The design process involves various stages and activities which have been described and discussed in Chapter 2. At the start of a project, the engineer must be creative and inventive. He must identify many alternative strategies and he must be prepared to consider new ways of solving the problem. From these alternatives must emerge a design method.

Skills of analysis are then required which eventually lead to the final product. Between the beginning and the end of the project, decisions must be made. The decision-making process will involve information acquisition, value assessment and optimisation. Each decision is a step closer to the final single 'best' answer.

There are many alternatives to be considered in an engineering design. There are the technical factors such as materials' selection, fabrication method, energy sources and component specification. There are also more general considerations such as economics, market forces, government controls, safety and reliability. As the number of factors involved increases, so the range of alternatives and the possible decisions multiply rapidly.

The problem of designing a writing implement requires that certain decisions are made, and choices selected from a range of options. Typical alternatives for this particular problem are given in Fig. 3.1, and the way in which the information can be presented on a 'decision tree' is also shown. Figure 3.1 illustrates the way in which the number of possible solutions to the problem increases rapidly, even when only a few decisions need to be made.

Decision-making involves compromise. It will usually require a 'trade-off' in one factor in order to gain improved performance in another variable. For example, increasing the number of parts in a machine to give greater versatility will increase the cost. The use of cheap imported materials in a

Decision	Select	Number of possible products (cumulative)
D1 General type:	Ink(i) or biro(b)	2
D2 Material:	Metal(m), plastic(p) or combination(c)	6(i.e. 2 × 3)
D3 Operation:	Refill(r), disposable(d) or 'everlasting'(e)	18(i.e. 6 x 3)
Size:	Various—specific or multitude	
Mass:	Lightweight or substantial	Increasing rapidly — some choices may be mutually exclusive
Appearance:	Cheap or expensive	
Particular features:	Cap or retractable point	
Specialist features:	Removable nibs, etc.	

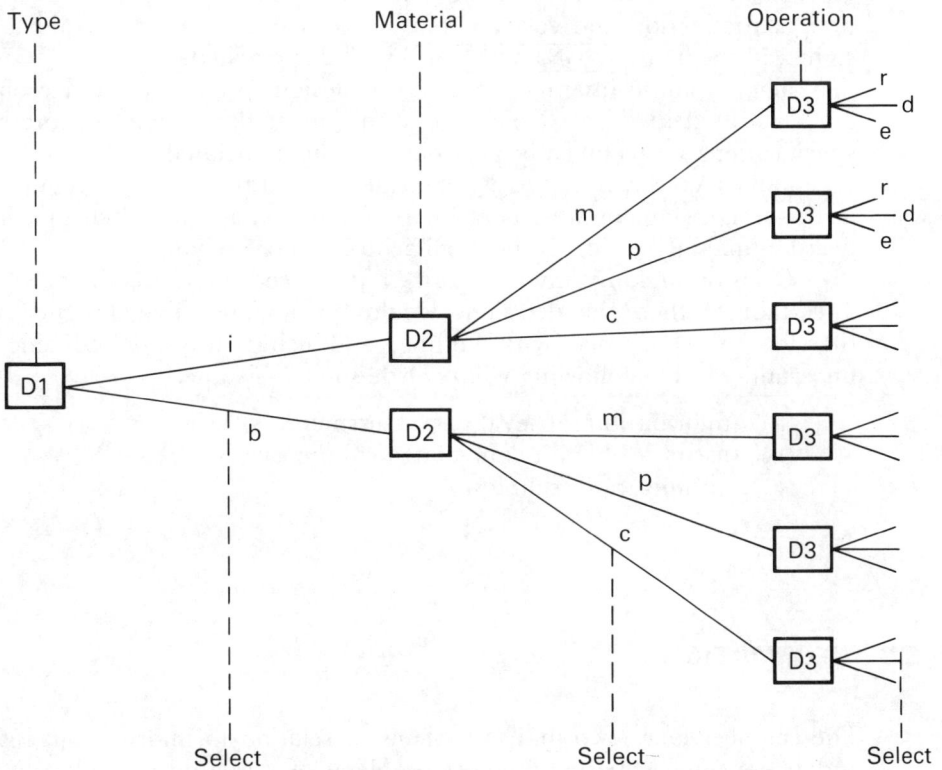

Figure 3.1   Possible Choices, Decisions, and Products for the Design of a Writing Implement

production process will have to be considered against the uncertainties of future supplies, government restrictions and economic forces beyond the control of the purchaser.

The search is to establish the decisions that will produce the most satisfactory solution. In this context, the terms 'best' or 'correct' really refer to an optimum solution which produces the most satisfactory balance between all the relevant factors. However, there is a difference between the *optimum* and the *optimal* solution. The optimum solution is the technical best that could be obtained within the constraints imposed by the relevant variables. The optimal solution is the one that is actually attained after considering the total constraints that exist. For a component design it will be necessary to consider the quantities of materials required and their cost, the cost of fabrication and many other factors. The optimum design specification will then be obtained. However, the final decision will also depend upon market demand, government policy, union action, etc. The final product is therefore the optimal solution — the best in the circumstances, but not the technical best.

The relationship between the design and production stages and the economic factors to be considered is shown in Fig. 3.2. The sequence of activities to be performed, and the questions to be asked upon completion of each activity, are illustrated using a flowchart. Only when all three aspects of design, production and cost are satisfactory can a solution be said to be in sight and the point   CONTINUE   in Fig. 3.2 is reached.

The dynamic interactions between design, production and economics are illustrated in Fig. 3.3. Here, not only must the design and production specifications be technically possible, but the associated costs must also be acceptable. As shown in Fig. 3.3, consideration of the economic requirements at the design stage as well as at the production stage should ensure that the decision-making process is more efficient and more reliable.

Decision-making involves risk and uncertainty. If there were no uncertainties then the decisions would be obvious. The designer should therefore be concerned with finding, and using, ways of reducing these uncertainties. The following will provide some assistance:

  (a)   information retrieval and assessment;
  (b)   use of probability and statistical theory;
  (c)   optimisation techniques.

3.3 INFORMATION

The engineer faces two main problems in relation to information: retrieval and assessment. Retrieval involves problems of availability, the location and nature of information sources, and also problems of access, cost, delay, etc. Table 3.1 lists the more readily available sources of information and the particular types of information that each source can provide.

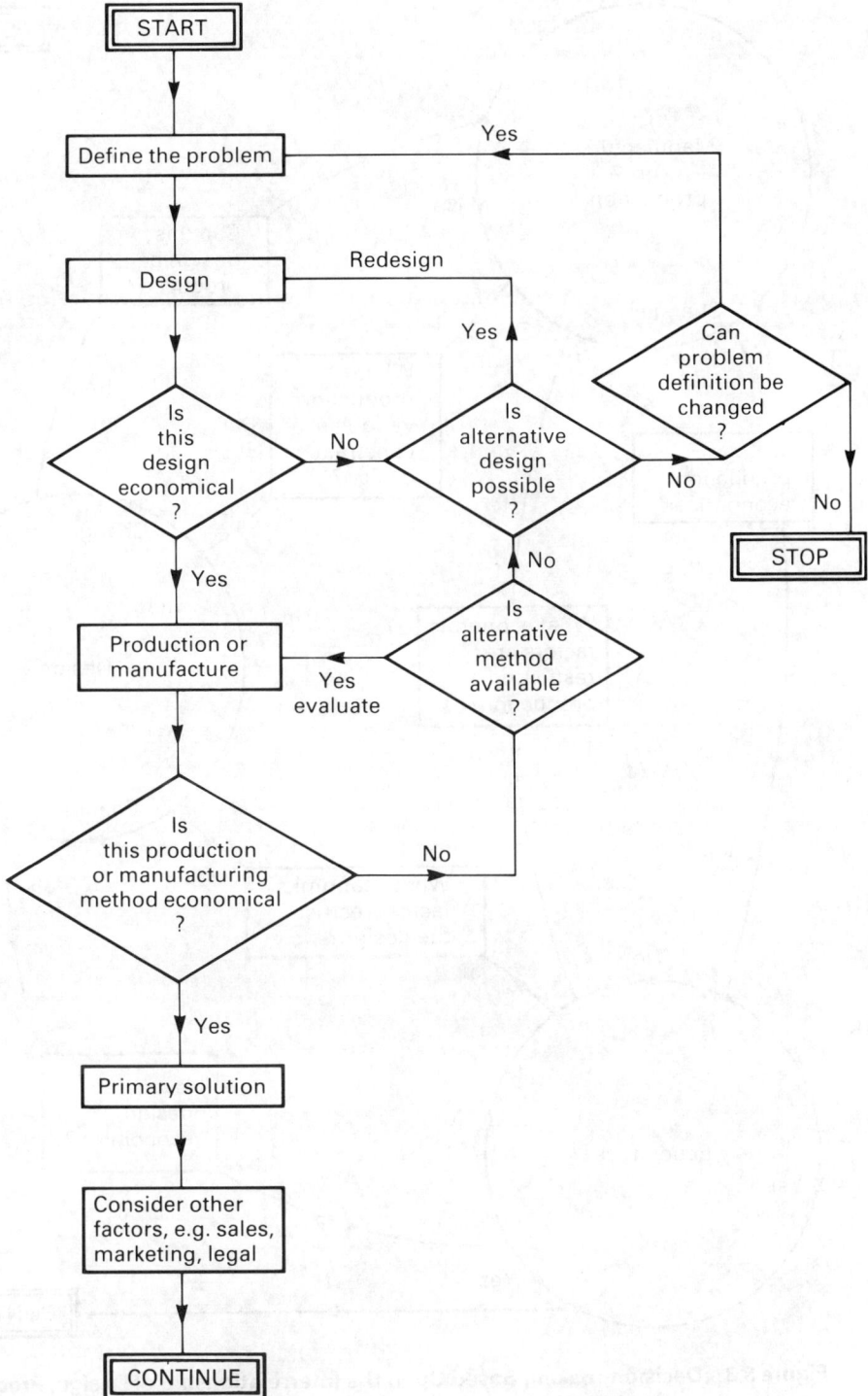

Figure 3.2   Decisions Concerning Design, Production and Cost

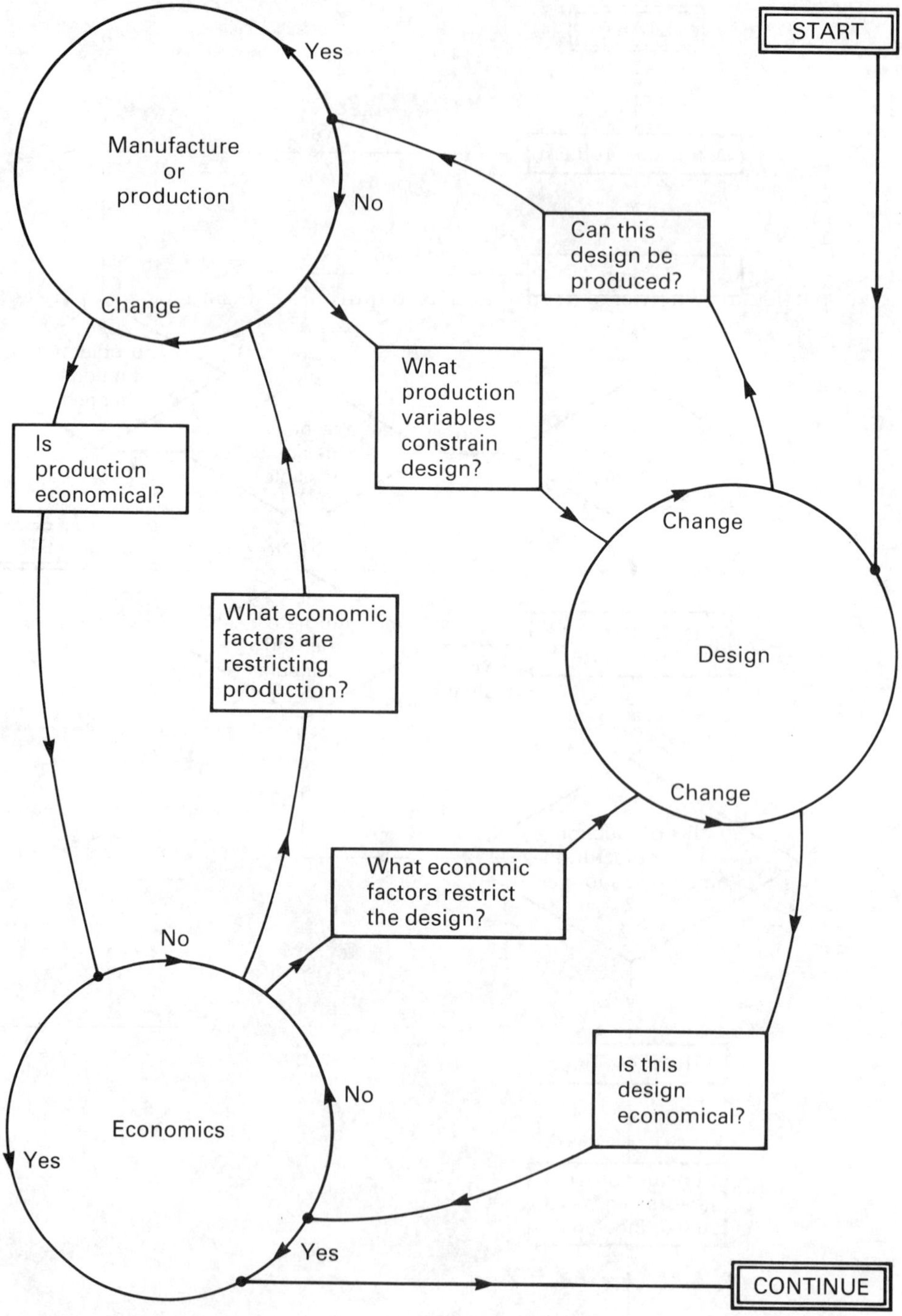

Figure 3.3   Decision-making Based Upon the Interrelationship of Design, Production and Economics

(N.B. There is a continuous negative loop!)

Once the information and the appropriate source are identified, the problem then becomes one of access, time and cost. Suggested procedures to obtain information are given by Woodson (pp. 71–3; see Bibliography, Chapter 2). The most effective and efficient methods are by personal contact or telephone. The time required to obtain information will depend upon the source and its nature, i.e. whether it is confidential or a translation, the date of publication, etc. In general, the more time available to obtain information, the greater the quantity obtained, although quantity is not necessarily proportional to usefulness. The cost will also depend upon the source and in particular whether a translation is required. Translations are particularly

Table 3.1   Sources and Types of Information

Information Source	Type of information
1 Libraries: public, company, university, government, private	Books, pamphlets, journals, newspapers, magazines, standards (BS, ASTM, etc.); reference sources e.g. Engineering Index, Citation Index, Engineering Abstracts.
2 The Patent Office, 25 Southampton Building, London, UK, or US Patent and Trade Mark Office, Box 9, Washington, DC 20231, USA	Source of patents and related information and may have a technical library.
3 British Lending Library, Boston Spa, Yorkshire, UK or Library of Congress, 10 First Street, SE, Washington DC 20402, USA	*All* information in 1.
4 Government	Publications and reports of government departments, e.g. Trade, Industry, Agriculture. Government agencies, e.g. (in the UK) the Statistical Office, National Physical Laboratory, The Design Council. In the USA, the National Science Foundation, National Research Council, National Academy of Engineering, NASA (all in Washington, DC). Government research centres. Foreign government's embassies or consulates.
5 Universities and various professional institutions	Various. Can approach departments directly.
6 Private consultants	Private/confidential information.
7 Private companies	Advertising and product literature, technical data sheets, technical advice and quotations.
8 International organisations	Publications (reports) of UNESCO, WHO, IAEA, etc.
9 Personal experimental study	Specific.

costly and time-consuming, so try to ascertain whether a translation will produce relevant information. If you think it will, try to get a 'rough' translation of the abstract and determine the nature of the original material, e.g. experimental results, theoretical analysis.

The student and the engineer will probably find that when they begin a project some information is already available. However, it will often be insufficient and the full extent of the information required will not be obvious at this stage. Time is the main constraint — to identify the information necessary and then to locate it.

Once information becomes available the problem then is assessment. The engineer must interpret the data and decide upon its usefulness for the project requirements. The information must also be assessed in terms of accuracy, experimental method, credibility, and so on. Some data look too good (or bad) to be true! Other information may be so inconsequential as to be of no use at all! To make these decisions it will be necessary to use probability and statistical methods. These are discussed in Section 3.4.

* * *

Exercises

1 Select one item from each of the following categories:

(a) a common household item, e.g. pressure cooker, toothbrush, light bulb;
(b) a standard laboratory machine, e.g. hardness tester, refractometer, gas chromatograph;
(c) a consumer item, e.g. aerosol deodorant, freeze-dried coffee;
(d) a chemical, e.g. toilet cleaner, rat poison, aspirin;
(e) any item you wish.

Obtain a reference list (not actual references) relating to your item. Decide upon a particular aspect of the product in which you are interested and make a short list related to this feature. Try to obtain suitable information related to this particular feature, such as:

(a) a patent;
(b) two papers from journals or conferences;
(c) a paper in a foreign language;
(d) company literature/technical data;
(e) a government publication (just the abstract if it appears costly and lengthy).

Assess this information in terms of usefulness and validity. Obtain an estimate of the cost and the time required to translate a technical paper from a foreign language (German, Japanese, or any other) of approximately 20 pages. See how much preliminary information *you* can elicit from the foreign paper you obtained.

2 Woodson (pp. 44, 45; see Bibliography, Chapter 2) classifies information as 'hard' or 'soft', and also 'necessary' or 'dispensable'. Define these terms yourself and categorise the information you have obtained accordingly. If possible, compare your classifications with those given by Woodson.

* * *

● *Now go to Section 3.4.*

● *From page 59.*

Exercises — Ideas and Suggestions

Some questions to be asked, and answered, in order to make a decision.

Exercise 1

Main decision — time available. If it is this simple, travel at maximum legal speed subject to constraints of driver fatigue, etc. However, cost may also be important especially if the driver is paying the petrol bill. Fuel consumption is dependent upon speed and the type of vehicle. Decide the maximum time available for the journey and calculate this minimum speed. Compare this with the most efficient speed based upon fuel consumption. What about rest periods and the increased accident risk with continuous travel? The route is a highway but is the region mountainous? Will refuelling be necessary? Where are petrol stations situated, and will this affect the chosen speed?

Exercise 2

Decide what functions you want the machine to perform. Can the old machine perform all of these?

Are you prepared to forego some needs?

Similarly, can the new 'improved' machine perform all the required functions?

How is it improved?

Why is the old machine being withdrawn? Is it unreliable or just not 'modern' enough?

Will spare parts and repair/servicing facilities still be available for the old machine?

Does it carry a guarantee?

Is the store or the manufacturer responsible for dealing with complaints and/or servicing?

Will the new machine be reliable or is it likely to suffer from 'teething' problems?

Try to obtain independent opinions from people in the trade, not only for these items, but for the full range of the manufacturer's products.

Exercise 3

Obtain data concerning recent claims. Include type of transport, distance, cost, any special considerations. Make a list of 'classes' of company personnel, e.g. executive, service engineers. Estimate their number of trips per year and the distances. List factors that are important in the mode of travel, e.g. comfort, ability to sleep, ability to take equipment, versatility of location visited, chance of delay or accident.

Can costs be reduced by making contracts with particular travel specialists, e.g. travel agencies, car rental companies?

What about the use of personal, company or rented vehicles?

Make recommendations for type of travel for each 'class' of personnel but do not make this a rigid company rule. Allow flexibility and consider examples of where this may be necessary.

Exercise 4

This decision will be mainly based upon costs. If all the strip lights are replaced at the same time, there will be the cost of the unused life. However, if replacement is on an 'as fail' basis, then maintenance labour costs and also the effect of disturbance on the workforce will be higher. This is particularly true if work cannot continue satisfactorily with a failed light. Data are needed for the frequency of failure against number of operating hours before a decision can be made. Consider whether it will be less costly to install dual lights at each point, with one set on standby in case of failure. Failed lights can then be replaced when production stops, e.g. weekends or holidays.

Exercise 5

The decision will be based upon the distraction the system will cause to other workers, and its cost and effectiveness compared with alternative systems. The disturbance effect can be estimated by trials and subsequent feedback, or observing an actual operating system. Determine the efficiency of a Tannoy system.

The alternatives may include use of personal radio telephones or employing an office junior to take messages, such as the caller's name and number, and to locate the engineer if it is urgent. Staff could inform the switchboard of their location, or use answering and message recording machines.

Exercise 6

$$\left(\begin{array}{c} \text{P} \\ 4 \text{ component} \\ \text{parts} \end{array} \xrightarrow{\text{work}} \begin{array}{c} \text{Q} \\ 2 \text{ component} \\ \text{parts} \end{array} \right)$$

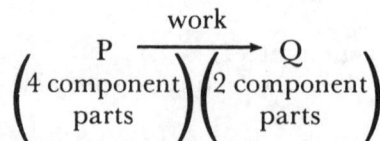

Stock level problems are included in standard textbooks. However, a few points are worth making. Determine whether a machine can handle extra work if another machine fails. Determine the probabilities of components and machines failing. P and Q failing together is less critical than P and P (and P?) together! Look up reliability theory.

If component A fails in machine Q1 followed by component B in machine Q2, then only one machine is completely inactive! If the problem seems to be getting out of control, try rationalising and putting in some numbers. Consult texts dealing with optimisation of stock levels.

The final decision will depend upon the cost of a breakdown, i.e. lost production, against the cost of holding spare parts. Determine which parts are needed, how many, the probability of failure, and also the time required to obtain spare parts.

Exercise 7

Before a material can be selected, all the functional requirements must be determined, e.g. shape, size, strength and environment. Then prepare a short (or long!) list of materials which could be used. Gradually eliminate the less favourable materials by consideration of:

(a)   material cost (plus wastage) — £ or $ per unit volume or unit weight;
(b)   availability — it is no good using/specifying a material which is not readily available (security of future supplies);
(c)   ease and cost of fabrication methods involved;
(d)   ease and cost of machining;
(e)   suitability and cost of joining;
(f)   surface finish;
(g)   service life;
(h)   other — toxicity, flammability, etc.

If more than one material appear suitable then select the material that produces the technical 'best' product at the lowest cost. This may not be possible and the best 'all rounder' with a satisfactory mix of the above factors will be selected. This may require a 'trade-off' in one value for improvement in another area, e.g. a heavier product but easier to weld.

Exercise 8

Compare the advantages and disadvantages of a word processor and a secretary. Consider cost, reliability, versatility, the functions to be performed, number required, cost when not in use, loyalty, ease of replacement, etc. The main considerations are cost and efficiency, although many other factors will influence the decision.

Exercise 9

This type of decision will never be clear-cut because it involves people, not just producers and consumers.

It also has many unquantifiable features such as the aesthetics and the price of protecting the environment. The main factors, in random order, will be economic, social (e.g. unemployment), political, public opinion, government and media involvement.

If this problem seems familiar, it occurred in Tasmania in 1983. Consult international magazines and newspapers for more details from January and February 1983 when public opinion was first aroused, e.g. *The Economist*, 29 January 1983 (p.45).

Exercise 10

This is an optimisation problem requiring the use of linear programming (see page 95) to obtain a solution. First choice would be product III which does not require purchase of the additional chemical C. The final decision will preferably use quantities of A and B that minimise any remaining chemical. This may not be possible with product III if A and B are not produced in approximately the ratio of 85:15.

As usual, costs and profits will also be prime factors in the final decision. It is most likely that the final choice will be to produce all three, or possibly two, of the products in certain quantities. The following factors or constraints must be taken into account:

(a) minimum cost of chemicals used — mainly a minimum of C, although B may have recoverable or allocated costs now it can be sold (previously it was waste material);
(b) maximise profits generated;
(c) use as much of A and B as possible.

The following data will be needed before a numerical solution can be attempted:

(a) quantities of A and B available, i.e. produced (kg per day);
(b) quantity of chemical C available (kg per day);
(c) cost of A, B and C per kg;
(d) maximum quantities of new products I, II and III which can be sold (kg per day);
(e) selling price of products I, II and III per kg;
(f) other costs of producing products I, II and III per kg;
(g) costs associated with sales of each product, e.g. transportation per kg;
(h) any other constraints that can be quantified.

Numerical values will be needed in order to proceed with the problem. Either obtain these from the tutor or make up your own. Decide which values will not lead to a solution, e.g. costs exceeding profits.

Either use the skills taught in an optimisation course or select a suitable text and try self-tutoring.

Alternatively, attempt a trial and error or iterative method of solution. 'Adjust' the numerical values accordingly and determine which ones are the most sensitive. (The use of a microcomputer should significantly reduce the time required.)

● *Now go to Section 3.1.*

- *From Section 3.3*

3.4 PROBABILITY AND STATISTICS

Probability and statistics are closely related sciences. Probability provides a way of quantifying uncertainties, and statistics is concerned with the evaluation of data. Statistics also includes the collection and assessment of data, and ways of using a limited amount of data to draw reliable conclusions. Probability theory is used in statistical analysis where an actual statistical distribution is approximated by a particular probability model. Numerical probabilities cannot be produced out of thin air: data or observations are needed in order to assign an accurate value to the chance of an event occurring. Hence the usefulness of statistical analysis to probability theory.

All students should have some basic knowledge of probability and statistics when they arrive at university. They should be familiar with simple statistical averages such as the mean, median and mode, and also, hopefully, the standard deviation. They should possess a basic appreciation of the meaning of probabilities, and be able to calculate the probability of a simple discrete event occurring, perhaps even appreciating the 'and' and 'or' laws of probability!

The student's knowledge will be extended by studying particular course units in statistics and probability. The use of these topics, and others in the latter part of this book, in design projects will depend upon the overall course structure. If design is taught at the earliest opportunity in the syllabus, then certain topics will not have been covered at that time. The student can then either apply self-teaching (with the guidance of the project tutor) within certain areas, or those aspects of a project can be ignored. The latter course of action would obviously detract from the well-balanced or 'total' view of a project, but may be necessary within obvious time constraints. However, it is hoped that students will consider at least rudimentary aspects of associated and, as yet, untaught topics (with some assistance). The material generated can then provide useful discussion topics when these subjects are taught formally.

It is not my intention to provide textbook theory or detailed examples of statistics and probability in this chapter. That could occupy a separate textbook. Instead, I have included a summary of common statistical terms and formulae in Section 3.4.1, and the following assortment of situations and questions which will illustrate areas where statistics and probability theory should, or may, be applied in a design project. The rest is up to you!

* * *

Exercises

There are no suggestions or ideas associated with these problems (except for Exercise 3), or to be more precise I have not included any! One reason is the space required to do this adequately. However, if you finish a problem in a short time or with little written working, then you have not been fully utilising your powers of creativity! Each problem should produce pages of notes and ideas if developed fully.

The reader should have realised that the approach advocated is one of self-learning by guided discovery. You should decide the direction and scope of the problems. The longer you live on a diet of analysis, the harder it will be to apply diverse thinking to problems and to the many alternatives that exist. Do not expect the problem data to be absolute, or to lead directly to a particular and correct solution!

1 What do probability values of 0, 1 and 0.5 mean?

2 A box contains four 2 amp fuses, seven 5 amp fuses and nine 13 amp fuses, all unmarked. What is the chance of picking a 5 amp fuse at the first attempt?
   What is the chance of selecting a 5 amp fuse with two selections?
   What is the chance of taking out two 13 amp fuses?
   What is the chance of taking out a 5 amp and a 2 amp fuse?
   What probability rules emerge from these problems?

3 You are attending a business meeting in a foreign country, and at the end of the meeting you intend to travel by taxi to the airport and then fly home. The schedule is such that you will miss your plane if either the taxi is delayed or there is a delay in completing formalities at the airport. You estimate that the probability of the taxi being delayed is 0.6 and the probability of a delay at the airport is 0.2.
   Calculate the probability of catching the plane.
   Consider the alternatives and the decisions to be made.
   Select a particular course of action.
   After attempting this problem, refer to page 83 for some suggestions.

4 A company manufactures separate radio and tape recorder units. It is decided to produce a combined radio-tape model, where each unit can operate independently. Production and testing, by continuous operation for 12 months, of 100 prototypes is carried out. At the end of this time, the radio has failed in 4 machines, the radio and tape in 13 machines and the tape in 9 machines. As the development engineer, you need to calculate and make decisions based upon the following probabilities:
   (a) that both units will fail;
   (b) that one part will fail but the machine will still have some use.
   This problem should produce much discussion and comment. If not, then your ideas are still too rigid!

5 You are considering investment in an intensive fish farming project. The aim is to produce high quality (and high price!) salmon for exclusive restaurants. Initial testing entails rearing a group of 100 baby fish each approximately 5cm long, for 60 days. This has been determined as the 'turn-around time' to satisfy demand without over-production.

   After this time 88% of the initial batch survived and they were all measured. It was found that the mean length of the fish was 70cm and the standard deviation was 20cm. Information relating to consumer preference, fish quality (i.e. taste) and ease of packaging has shown that only fish between 40cm and 100cm long can be considered a useful, saleable product.

Reliable data from other fish farms show that after 60 days no surviving fish are less than 30 cm, or more than 120 cm long. Also, there are very few fish present of these extreme dimensions.

Analyse the data and decide upon a future policy for this venture.

6 You are developing a machine that requires ball bearings to be seated in a circular groove with a hemispherical cross section. The groove is accurately machined to 10 mm ± 0.0001 mm diameter. Each groove is to contain 50 ball bearings. A manufacturer produces 'standard' ball bearings in a suitable metal and his specification is 1 cm ± 0.1 mm. This machine can be adjusted so that the mean ball diameter changes in steps of 0.01 mm, with the same tolerances. It is vital that balls less than 1 cm in diameter should not be packed in the groove as they cannot be easily identified or removed. What recommendations would you make based upon this information?

7 The planning stage has commenced for the construction of a steel box girder bridge with a span of 350 m. It is to provide a road crossing over a river and upon completion the only other easy access, a car ferry, is to be withdrawn. However, cross-winds occur and for a conventional bridge design, speeds in excess of 80 km/h (50 m.p.h.), represent a real hazard to cars travelling faster than 32 km/h (20 m.p.h.). Statistics are available for the number of days the cross-wind speed exceeded 80 km/h in the last 4 years (year 4 is the most recent), and the data are presented in Table 3.2.

What recommendations would you make concerning this problem?

Table 3.2 Number of Days Wind Speed Exceeded 80 km/h

Month	Year 1	Year 2	Year 3	Year 4
January	6	3	8	11
February	5	0	5	4
March	8	8	19	6
April	2	1	3	2
May	1	0	0	2
June	0	1	2	1
July	3	5	6	4
August	1	3	0	3
September	1	0	5	1
October	10	0	0	1
November	6	11	9	18
December	12	9	4	15

8 You are the personnel manager for an engineering company and executive management has identified positive advantages in the use of flexible working hours. You have been assigned the task of finding and collating the opinions of the workforce in a particular section; this is not a production or manufacturing section. Prepare a questionnaire to be issued to the workers and discuss the factors that are important in obtaining and evaluating the data.

9 What is a normal distribution?

In what engineering situations will it occur or be useful?

What are the differences between the binomial, Poisson and normal distributions?

In what applications/situations is each distribution most useful?

10 A new drug has been developed and extensively tested on animals with no adverse reactions. It has also been tested on a limited number of human volunteers. Out of a group of 100 'healthy human guinea pigs', 28 reported some form of reaction. This ranged from mild nausea to vomiting and headaches. The drug was also tested by a group of very ill patients for whom it was originally developed. Out of this group of 100, after short-term (2 weeks) treatment one person had died and after prolonged treatment (3 months) four patients had died.

How would you decide whether to make this drug available or withhold it pending further testing?

* * *

3.4.1 Summary of Statistical Terms and Formulae

This section provides only a summary of terms and formulae commonly used in statistics. It is intended to provide a reference source for the student attempting engineering design problems. Hopefully those students who are not familiar with statistics will be tempted to investigate for themselves such topics as probability, sampling and correlation. At the same time the students should be asking what applications these techniques have for solving design problems, and discussing some of their own ideas of typical situations.

Frequency distributions
Data are often presented in frequency distributions and some examples are shown in Fig. 3.4.

Measures of average and dispersion
The **mode** (**modal value, modal class interval**) is the most popular measured value, i.e. the value occurring most often or with highest frequency. The modal class interval and modal value are indicated on the histogram and the frequency polygon (Fig. 3.4.).

The **median** is the $((n+1)/2)$ observation or measurement when n measured values are arranged in order of magnitude. The median value corresponds to the *50th percentile value*.

The range of measurements below the *25th percentile* is known as the **lower quartile**, that above the *75th percentile* is known as the **upper quartile**. Between the 25th and 75th percentiles is the **inter-quartile range**.

For a set of n values, $x_1, x_2, \ldots , x_n$, the **arithmetic mean** $(\bar{x})$ is given by:

$$\bar{x} = \frac{\Sigma x}{n}$$

The **variance** (σ^2) is:

$$\sigma^2 = \frac{\Sigma x^2}{n} - \left(\frac{\Sigma x}{n}\right)^2 = \frac{\Sigma x^2}{n} - (\bar{x})^2$$

The **standard deviation** (σ) is:

$$\sigma = \sqrt{\left(\frac{\Sigma x^2}{n} - (\bar{x})^2\right)}$$

The **geometric mean** is:

$$\sqrt[n]{(x_1 \times x_2 \times ... \times x_n)}$$

The values of the arithmetic mean, the median and the mode for a set of measurements are usually close together. The value of the arithmetic mean is greater than that of the geometric mean.

Permutations and combinations

The number of **permutations** of n items selected in groups of r is:

$$^nP_r = n(n-1)(n-2) ... (n-r+1)$$
$$= \frac{n!}{(n-r)!}$$

Note: $n! = {}^nP_n$

For permutations, the group ABC is considered a separate entity from BAC, etc.

The number of **combinations** of n items selected in groups of r, where the order of the items is immaterial (e.g. ACB is equivalent to BCA, etc.), is:

$$^nC_r = \frac{n!}{r!(n-r)!}$$

The binomial probability distribution

The probability of an event occurring at a single trial is p, and the probability of the same event not occurring is q (where $p + q = 1$). Then the probabilities $P(0), P(1), ... , P(n)$ of $0, 1, ... , n$ occurrences of the event in n independent trials are given by the terms of the **binomial expansion**:

$$(q + p)^n = q^n + nq^{n-1}p + \frac{n(n-1)}{2!}q^{n-2}p^2 + ... + p^n$$
$$= P(0) + P(1) + P(2) + ... + P(n)$$

This is known as the **binomial probability distribution**, with mean value $= np$ and variance $= npq$.

The Poisson distribution

In a particular situation, if the average of a number of events is μ, then the probabilities of $0, 1, 2, ... , n$ events occurring are given by the consecutive terms of the **Poisson distribution**:

$$e^{-\mu} + \mu e^{-\mu} + \frac{\mu^2 e^{-\mu}}{2!} + \frac{\mu^3 e^{-\mu}}{3!} + ... + \frac{\mu^n e^{-\mu}}{n!} = 1$$

i.e. P (zero events) $= e^{-\mu}$

The mean (μ) $= np$, as for the binomial distribution.

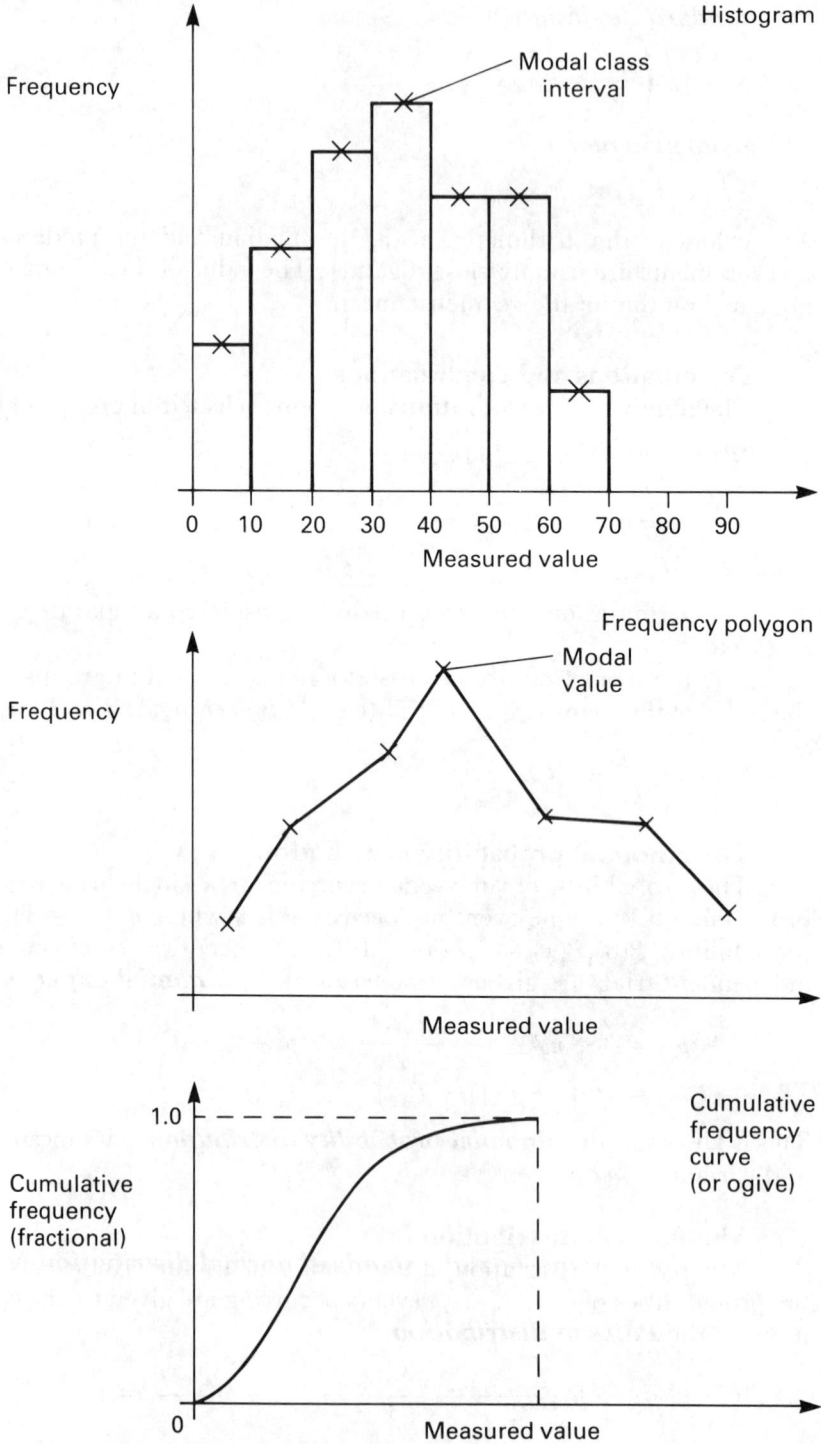

Figure 3.4   **Presentation of Data — Frequency Distributions**

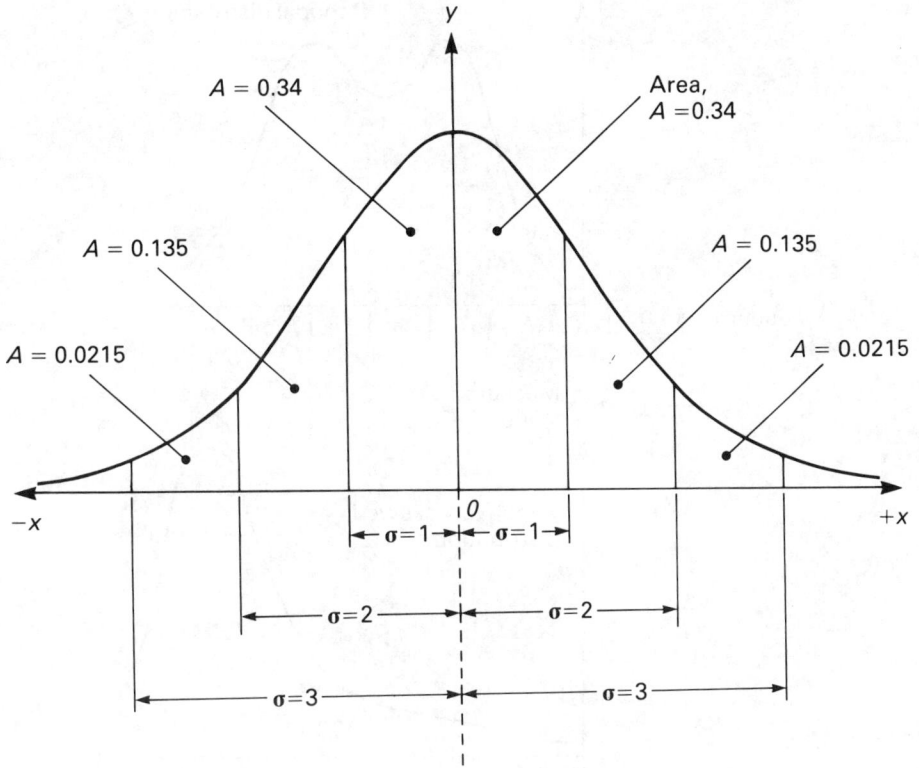

Figure 3.5   The Normal Distribution Curve

(Mean = 0; Variance = 1)

The normal distribution

When n is large, the **normal (probability) distribution**, or error curve, or Gaussian distribution, is indistinguishable from the binomial distribution. The characteristic shape of the normal distribution curve is shown in Fig. 3.5.

This curve represents a probability distribution with a mean = 0 and standard deviation = 1. The total area under the curve = 1. The areas shown under the curve are probabilities, and the probability that a random value falls within ±1 standard deviation of the mean is 0.68. Similarly, within ±2 or ±3 standard deviations the probabilities are 0.95 and 0.993, respectively.

This curve is known as the **standard normal distribution**. With actual data with a mean value $\bar{x}$ and standard deviation σ, then a **normalised variable** Z is defined as:

$$Z = \frac{x - \bar{x}}{\sigma}$$

Actual values are then *normalised* and tables used to determine probabilities.

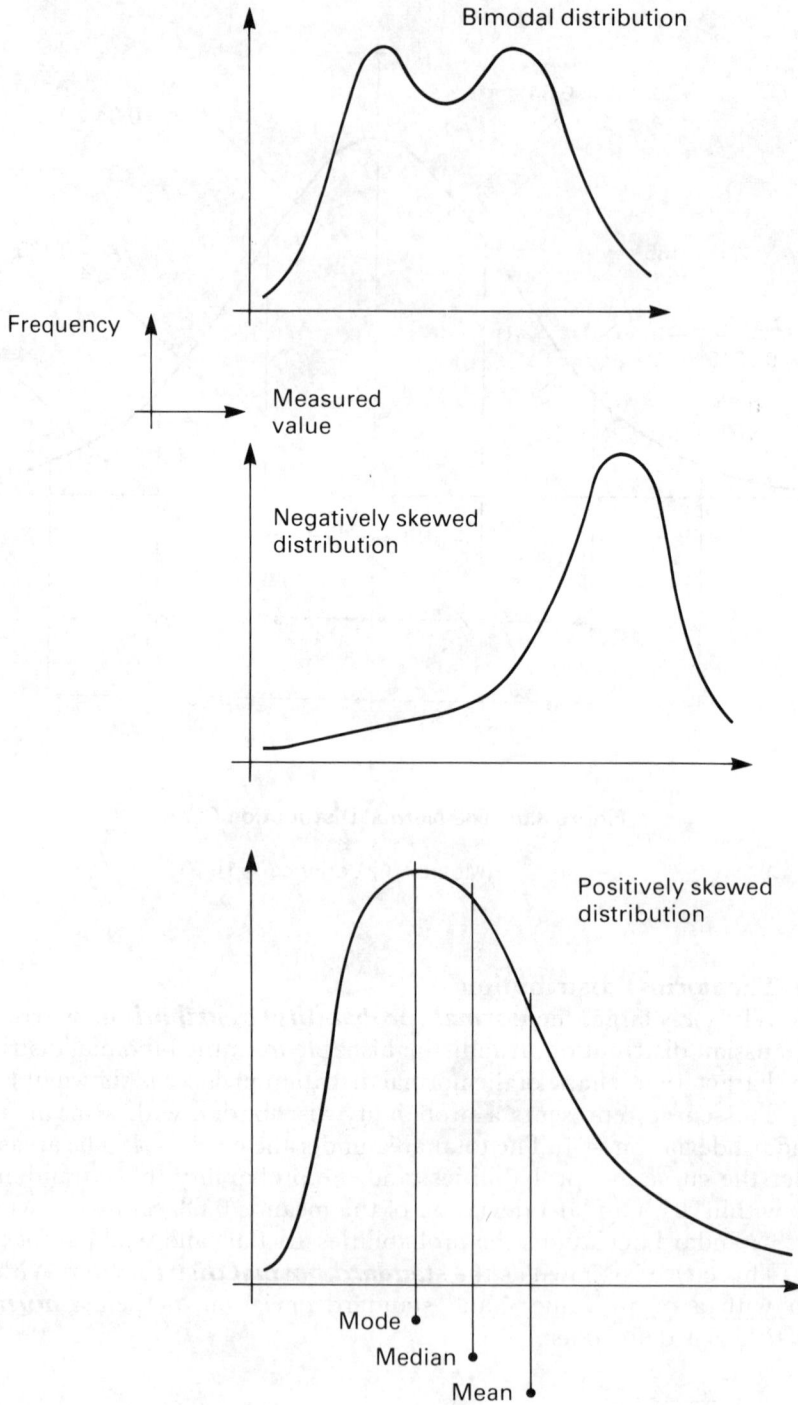

Figure 3.6   **Effects on the Normal Distribution**

The general equation describing any symmetrical normal distribution is:

$$y = \frac{1}{\sigma\sqrt{2\pi}} \exp\left\{-(x - \bar{x})^2/2\sigma^2\right\}$$

With actual data, effects may occur that change the shape of the normal distribution as shown in Fig. 3.6.

Samples and estimation

The population mean (μ) is estimated by the sample mean ($\bar{x}$):

$$\frac{\Sigma x}{n} = \bar{x} = \mu \text{ (estimated)}$$

The population standard deviation ($\hat{\sigma}$) is estimated *not* by the sample standard deviation but by:

$$\hat{\sigma} = \sqrt{\left\{(\Sigma x^2 - (\Sigma x)^2/n)/(n-1)\right\}}$$

The standard error of the sample mean is:

$$\sigma_m = \frac{\hat{\sigma}}{\sqrt{n}}$$

Significance tests

In general, t is calculated from the ratio:

$$t = \frac{\text{difference}}{\text{standard error of the difference}}$$

and the probability of obtaining this value of t by chance is found from tables of the t distribution.

ONE SAMPLE: When comparing the sample mean with a standard value (mean of a known population, μ):

$$t = \frac{|\mu - \bar{x}|}{\sigma_m}$$

and the degrees of freedom (d.f.) = $n - 1$.

TWO SAMPLES: When comparing the sample means to see if they are from the same population the symbols are as given in Table 3.3.

Table 3.3 Two Samples-symbols

Sample	Number	Mean	Estimated standard deviation	Standard error of difference	d.f
1	n_1	$\bar{x}_1$	$\hat{\sigma}_1$	$\sigma_{\bar{x}_1 - \bar{x}_2}$	$n_1 + n_2 - 2$
2	n_2	$\bar{x}_2$	$\hat{\sigma}_2$		

Then for large samples (n_1 and n_2 over 30):

$$(\sigma_{\bar{x}_1 - \bar{x}_2})^2 = \frac{\hat{\sigma}_1}{n_1} + \frac{\hat{\sigma}_2}{n_2}$$

For small samples, replace both $\hat{\sigma}_1$ and $\hat{\sigma}_2$ above by:

$$\hat{\sigma}^2 = \left(\Sigma x_1{}^2 + \Sigma x_2{}^2 - \frac{(\Sigma x_1)^2}{n_1} - \frac{(\Sigma x_2)^2}{n_2} \right) \Big/ (n_1 + n_2 - 2)$$

In both cases t is calculated from:

$$t = \frac{|\bar{x}_1 - \bar{x}_2|}{\sigma_{\bar{x}_1 - \bar{x}_2}}$$

CHI-SQUARED: The value of χ^2 is given by:

$$\chi^2 = \Sigma \frac{(f_o - f_e)^2}{f_e}$$

where f_o is the obtained and f_e the expected frequency. The degrees of freedom (d.f.) $= (r-1)(c-1)$ where r and c are the numbers of rows and columns in the contingency table. The probability of obtaining this value of χ^2 by chance is found from tables of the χ^2 distribution.

Correlation and regression

For n pairs of measurements of two variables, x and y, the corrected sum of squares are given by the formulae:

$$\Sigma' x^2 = \Sigma x^2 - \frac{(\Sigma x)^2}{n}$$

$$\Sigma' y^2 = \Sigma y^2 - \frac{(\Sigma y)^2}{n}$$

The corrected sum of products is calculated from:

$$\Sigma' xy = \Sigma xy - \frac{(\Sigma x)(\Sigma y)}{n}$$

The ***Pearson product moment correlation coefficient*** is:

$$r = \frac{\Sigma' xy}{\sqrt{\{(\Sigma' x^2)\ (\Sigma' y^2)\}}}$$

and its significance when the population value is assumed zero is tested by calculating:

$$t = r \sqrt{\left(\frac{n-2}{1-r^2} \right)}$$

with d.f. $= n - 2$, and using tables of the t distribution. If $r = 0$, there is no correlation between the variables; if $r = \pm 1$, there is perfect correlation.

For the correlation between rankings, the ***Spearman rank correlation coefficient*** (R) is calculated using:

$$R = 1 - \frac{6\Sigma d^2}{.n(n-1)(n+1)}$$

where n is the number of pairs of measures making up the rankings and d is the difference between the numbers that make up each pair of positions. When n is greater than 30, its significance is tested by calculating the standard error:

$$\sigma_R = \frac{1}{\sqrt{(n-1)}}$$

and using tables of the normal distribution or tables of percentage points for significance of R.

The ***regression equation*** of y on x is:

$$(y - \bar{y}) = \left(\frac{\Sigma' xy}{\Sigma' x^2}\right)(x - \bar{x})$$

$(\Sigma' xy)/(\Sigma' x^2)$ is the slope of the line when the equation is plotted on a graph and $\bar{x}$, $\bar{y}$ are the arithmetic means of the x and y distributions respectively.

Similarly, the regression equation of x on y is:

$$(x - \bar{x}) = \left(\frac{\Sigma' xy}{\Sigma' y^2}\right)(y - \bar{y})$$

- *Now go to Section 3.4.2.*

- *From page 74.*

Exercise 3 — Suggestions

The probability of an event occurring is denoted by $P(\text{event})$.

$P(\text{taxi on time}) \quad = 1 - 0.6 = 0.4$
$P(\text{no airport delay}) = 1 - 0.2 = 0.8$

Therefore,

$P(\text{catching the plane}) \quad = P(\text{taxi on time } and \text{ no airport delay})$
$= 0.4 \times 0.8$
$= 0.32$

THOUGHT: the probability of catching the plane depends upon *both* events occurring and should be less than either individual probability. This is the case (i.e. multiply two fractions).

CHECK: by calculating the probability of *not* catching the plane, which should be:

$$1 - 0.32 = 0.68$$

$$P(\text{missing the plane}) = P(\text{taxi delay}) \ or \ P(\text{airport delay}) \ or \ P(\text{both delays})$$
$$= 0.6 + 0.2 + (0.6 \times 0.2)$$
$$= 0.92?$$

THOUGHT: the combined probability of alternative (*or*) events should be higher than that of individual events because more options are available.

However, not only do the answers not agree, but if $P(\text{taxi delay}) = 0.7$, then:

$$P(\text{missing the plane}) = 0.7 + 0.2 + (0.7 \times 0.2)$$
$$= 1.04!$$

and $P(\text{event})$ must be $\leq 1!$

CONSIDER ALTERNATIVE CHECK:

$$P(\text{missing the plane}) = P(\text{taxi delay } and \text{ no airport delay})$$
$$or \ P(\text{no taxi delay } and \text{ airport delay})$$
$$or \ P(\text{taxi delay } and \text{ airport delay})$$
$$= (0.6 \times 0.8) + (0.4 \times 0.2) + (0.6 \times 0.2)$$
$$= 0.48 + 0.08 + 0.12 = 0.68$$

This is the *expected* answer.

WHAT WENT WRONG WITH THE FIRST CHECK? Examining the numbers, if $P(\text{both events})$ is subtracted, the expected answer is obtained:

$$P(\text{missing the plane}) = 0.6 + 0.2 - (0.6 \times 0.2)$$
$$= 0.68$$

IS THIS PECULIAR TO THESE VALUES OR IS IT A GENERAL FEATURE? Assume $P(\text{taxi delay}) = 0.7$, then:

$$P(\text{catching the plane}) = 0.24$$
$$\text{and } P(\text{missing the plane}) = 0.7 + 0.2 - (0.7 \times 0.2)$$
$$= 0.76, \text{ as expected.}$$

HINT: is $P(\text{both events})$ subtracted because these probabilities are already included/accounted for in the first two probabilities? Consider the alternative check again, using the knowledge that:

$$P(\text{taxi delay}) = 1 - P(\text{no taxi delay})$$

Can the subtraction of $P(\text{both events})$ be achieved in the alternative check by substitutions and manipulation of the data?

COMMENT: this may seem to be an over-long approach but it is meant to emphasise the importance of checking, and the possibility of using erroneous logic in probability problems.

Evaluation of the data

The probability of catching the plane (0.32) appears quite low—only a one in three chance! It is of course totally dependent upon the individual probability values used. Decide whether these are reliable values or estimates based on limited data.

Care should be taken not to amend these values in hindsight in order to obtain a preferred higher probability.

The decision regarding a course of action will depend upon such factors as:

(a) availability of later flights;
(b) alternatives to the taxi — helicopter, motorbike, etc.;
(c) essential home commitments;
(d) cost of missing the plane, e.g. if another plane ticket can be issued at no extra cost, overnight accommodation;
(e) possibility of holding the business meeting at or near the airport;
(f) accurate timetabling of the meeting agenda;
(g) leaving early.

In this case the final decision will probably mainly depend upon the consequences of missing the plane and, in particular, the availability of later flights, accommodation and the costs incurred.

3.4.2 Using Probability Data

The design and building of a system to mix two chemicals, liquid A and liquid B, is required. There is a significant amount of heat evolved when the chemicals are mixed in the proportions needed to produce the exit stream. This heat is to be used to raise the temperature of the product stream.

The chemicals can become unstable if the liquids are not thoroughly mixed or if there is an insufficient quantity of liquid B, in which case the temperature rises rapidly.

For these reasons it is decided that a cylindrical tank with a conical base and incorporating a stirrer will be used. This design will reduce or eliminate regions of slow-moving, unmixed liquids. A typical schematic illustration of the equipment is shown in Fig. 3.7.

Automatic control action must be taken if the flow of liquid B into the tank is restricted. A temperature-sensing and measuring device is to be included inside the tank and its signal will then be used to close a valve (V_1), situated in liquid A pipeline, if the temperature exceeds the safe limit.

As an additional safety precaution a large diameter pipeline, containing a valve (V_2) which is normally closed, is connected to the base of the vessel.

Valve V_2 opens upon receiving a high temperature signal from the controller. The liquids in the tank are then dumped.

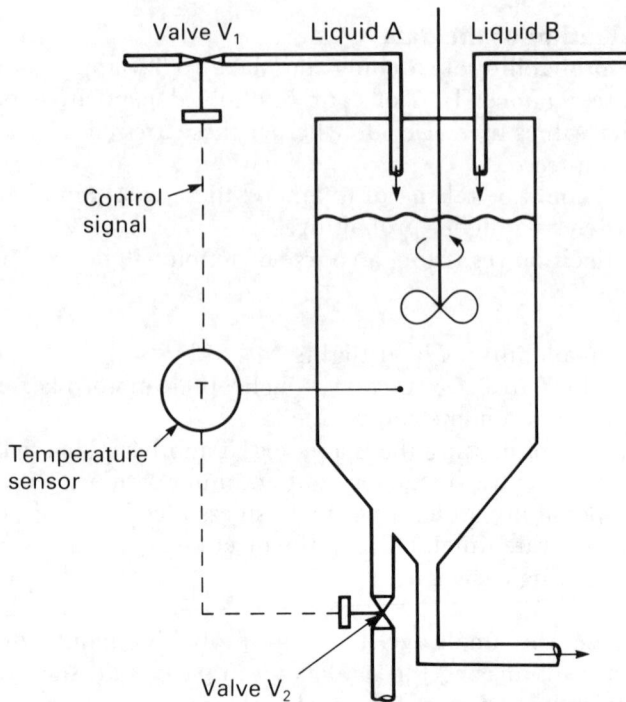

Figure 3.7   Liquid Mixing with Temperature Control

Consider other automatic control features which could be included in the design to improve the safety of operation.

There are two main considerations when deciding which safety features, and how many, to include in the design of equipment. These are the increase in cost of the system and the consequences of failure.

Failure of one piece of equipment may have serious effects on other items, or on the environment. If the chemicals in this example were highly toxic it would not be advisable merely to dump them. In this case they could be transferred into a holding tank containing liquid B. Alternatively, a cooling coil could be incorporated in the mixing tank and a flow of cooling fluid could be controlled by the temperature sensor.

Before a final decision can be made regarding implementation of a safety control system, it will be necessary to assess the probability that the system will work, i.e. the probability that components will fail. If the probability of failure is high then duplicate equipment may need to be installed, or the entire approach to the problem reconsidered. In this example it would be possible to install extra valves in liquid A pipeline in parallel with valve V_1, to increase the probability that the line is closed.

The valves (V_1 and V_2) to be used in Fig. 3.7 are two-position (open or closed) valves; they are actuated by an electrical solenoid which receives a signal from the temperature sensor (T). In normal operation, V_1 is open and V_2 is closed. For an emergency, V_1 must close and if this does not happen then

V_2 must open shortly after. (Not both valves together! Why?) It will be necessary to incorporate a controller with two temperature settings or a time delay, or both.

The safety (control) system will not function correctly if the valves fail to operate because of either electrical failure or blockage by particles in the liquid.

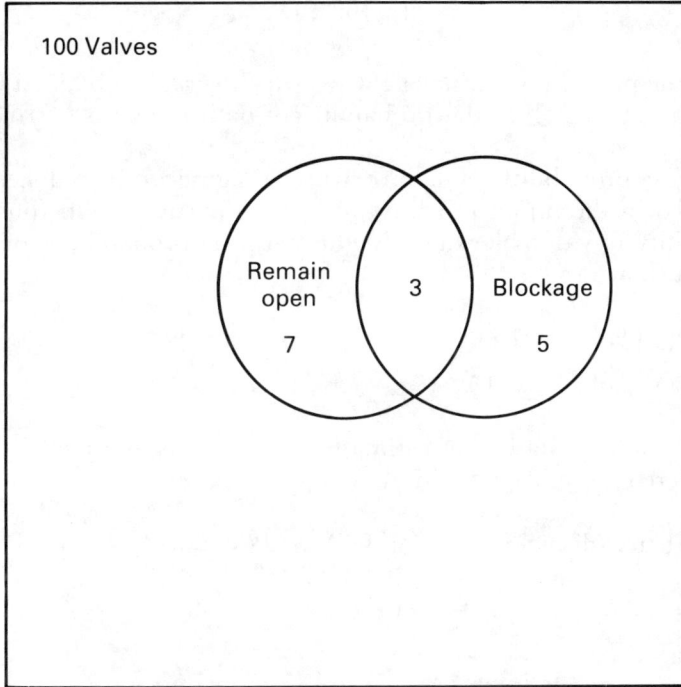

Figure 3.8 Results of Testing Valve Operation

Extensive testing of 100 valves of type V_1 has been carried out under actual operating conditions over a long period of time. The number of failures and the reasons were recorded. This information is presented in Fig. 3.8 and is to be used to estimate the probability that the safety system will fail.

For a valve the probabilities of failure, using Fig. 3.8, are:

P(electrical failure) $= 10/100$
P(blockage) $\quad = 8/100$

It might be concluded that:

P(disaster) $\quad\quad = 10/100 + 8/100 = 18/100$

However, out of 100 valves only 15 failed and therefore:

P(disaster) $\quad\quad = 15/100!$

The probabilities of electrical failure and blockage are not mutually exclusive (cf. throwing a die and obtaining a 6: this excludes the possibility of a 3).

Then, P(disaster)　　$= P$(electrical failure *and not* blockage)
$+ P($*not* electrical failure *and* blockage)
$+ P$(both events)
$=$　7/100 + 5/100 + 3/100
$=$ 15/100

The probability value of 18/100 which *might* be obtained (above) should be reduced by 3/100, the probability of both events occurring having been included twice.

The probability of failure will be significantly reduced if filters are installed in the liquid pipelines to remove particles, and thus eliminate the possibility of valve blockage. In this case the probabilities of failure for the system shown in Fig. 3.7 are:

$P(V_1 \text{ fail})$　$= 1/10$

$P(V_2 \text{ fail})$　$= 2/10$

$P(V_2 \text{ fail})$ is a higher estimate because V_2 is a much larger valve and is considered more likely to fail than V_1.

Then P(disaster)　$= P(V_1 \text{ fail}) \text{ and } P(V_2 \text{ fail})$
$= 1/10 \times 2/10$
$= 2/100$

CHECK:

P(successful operation)　$= P(V_1 \text{ fail } and \text{ } V_2 \text{ operates})$
$+ P(V_1 \text{ operates } and \text{ } V_2 \text{ fail})$
$+ P$(neither fail)
$= (\frac{1}{10} \times \frac{8}{10}) + (\frac{9}{10} \times \frac{2}{10}) + (\frac{9}{10} \times \frac{8}{10})$
$= \frac{98}{100}$ — OK!

Yet again, if P(success) is assumed to be $P(V_1$ operates) *or* $P(V_2$ operates), then an erroneous answer is obtained, in this case

9/10 + 8/10 = 17/10!

The designer must decide whether the probability of success (0.98) is acceptable or whether additional features can be justified on the basis of cost and the possible consequences.

The probability of failure of the temperature sensor has not been considered. It would be better to install two temperature sensors, one for each valve, depending upon the cost and reliability of these items. It may also be possible to incorporate a temperature controller which, if it fails, will automatically operate a valve. (V_1, but not V_2?)

3.5 RELIABILITY

When engineering and technology were developing more slowly, the reliability of a product was improved mainly by redesign based upon experience. The introduction of more stringent safety requirements, the rising cost of new equipment and rapid technological advances, all made the 'learning by failure' approach uneconomical and unacceptable. Reliability engineering is now an accepted separate field of study, and quantitative techniques have been developed.

    Reliability has been defined as the *probability* that a device will perform its intended function for a specified time under specified operating conditions.

*   *   *

Exercise

Define quality and reliability so as to identify the difference(s) between these two attributes.

*   *   *

Quantitatively, the reliability (R) of a component after a certain time can be defined as:

$$R = \frac{\text{number of failures at that time}}{\text{original number of components tested}}$$

The **failure rate** (F) of a component is:

$$F = \frac{\text{number of failures during a particular time interval}}{\substack{\text{average number of working components during} \\ \text{that particular time interval}}}$$

    The **mean time between failures** (M) is the reciprocal of the failure rate.

    A typical curve showing the reliability of a component, or entire system, as a function of time is shown in Fig. 3.9(a). The corresponding failure rate curve is given by Fig. 3.9(b), where regions I, II and III represent preliminary testing, actual operation and wear-out periods, respectively. Random failures will occur during all three operating stages.

*   *   *

Exercise

Consider the general shape of a failure rate versus time graph for a new component and a well-established product. Explain the significance of different regions of the graphs.

*   *   *

The above definitions apply to single components; however, most practical systems rely on more than one part for satisfactory operation. It will often be uneconomic to test complete systems for failure, and the overall

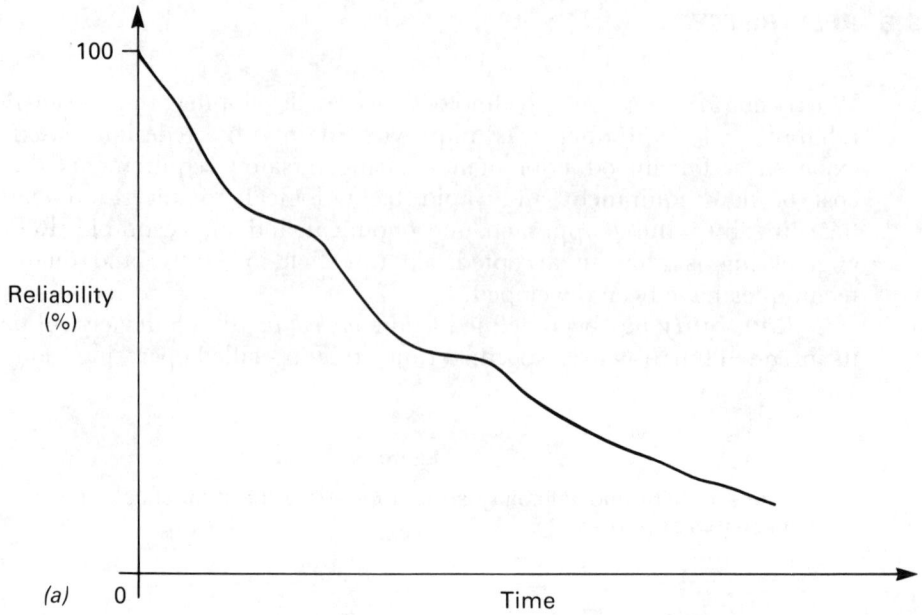

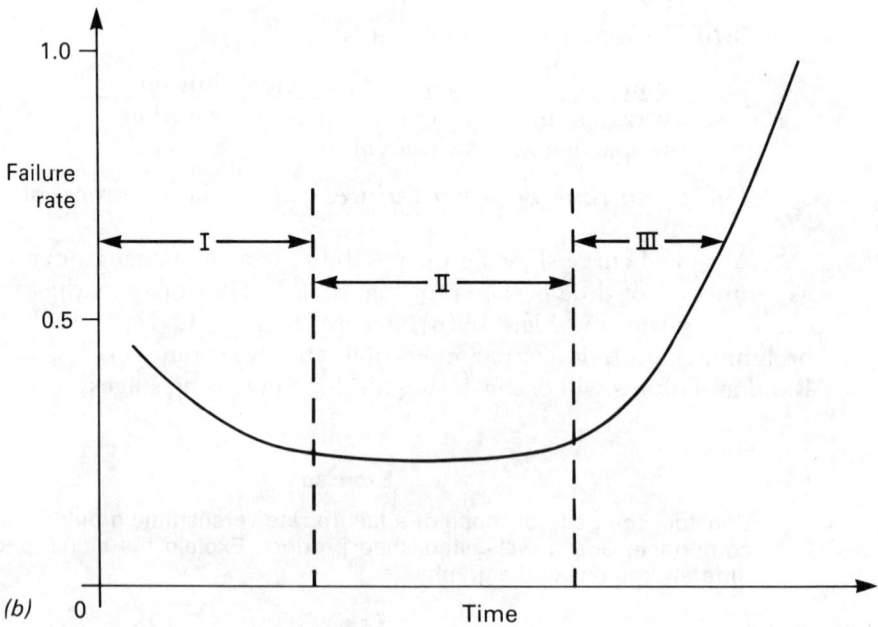

Figure 3.9   Presentation of Data for Component Reliability
(a) reliability, (b) failure rate; I = preliminary testing,
II = operation, III = wear-out

reliability is usually calculated from individual probabilities. When components are connected in series, the reliability of the entire system is the product of the individual reliabilities.

For two components (A and B) in parallel, the reliability of the system is the addition of the probabilities that:

   (a)   A fails but not B;
   (b)   B fails but not A;
   (c)   neither A nor B fails.

With reliabilities of R_A and R_B for components A and B respectively, the system reliability is given by:

$$R_A + R_B - R_A R_B$$

There are two design features that can be used to improve the reliability of multicomponent systems. First, a **standby** or **back-up** unit that only operates when a component fails can be incorporated into the system. Second, a duplicate component can be incorporated in the system. This is known as **parallel redundancy** because the additional component is not *necessary* for system operation.

In Fig. 3.10(a), the system reliability is given for the situation where *both* valves must close, for example the isolation of the vessel between the valves. For Fig. 3.10(b) the system reliability is determined assuming that either valve V_3, or valve V_4, or both valves operate correctly.

*   *   *

Exercises

1 Define a simple system and choose the reliability of each component. Calculate the overall reliability using parallel redundancy protection.

2 Compare this with the system reliability if an alternative standby component is used.

3 How do these reliabilities change if more than one redundant or standby component is used?

4 Determine (using numerical values) the reliabilities of more complex systems, e.g. several components in series and parallel, such as those shown in Figs. 3.10(c) and (d).

*   *   *

Experimental data for the failure of components can be compared with a theoretical distribution known as the **exponential reliability curve**. The failure rate (F) is calculated from experimental data, and the probabilities of 0, 1, 2, ... , n failures in the same time period are then calculated from the terms of the Poisson distribution (see Section 3.4.1). The reliability can be written as:

$$R = \exp(-F/t)$$

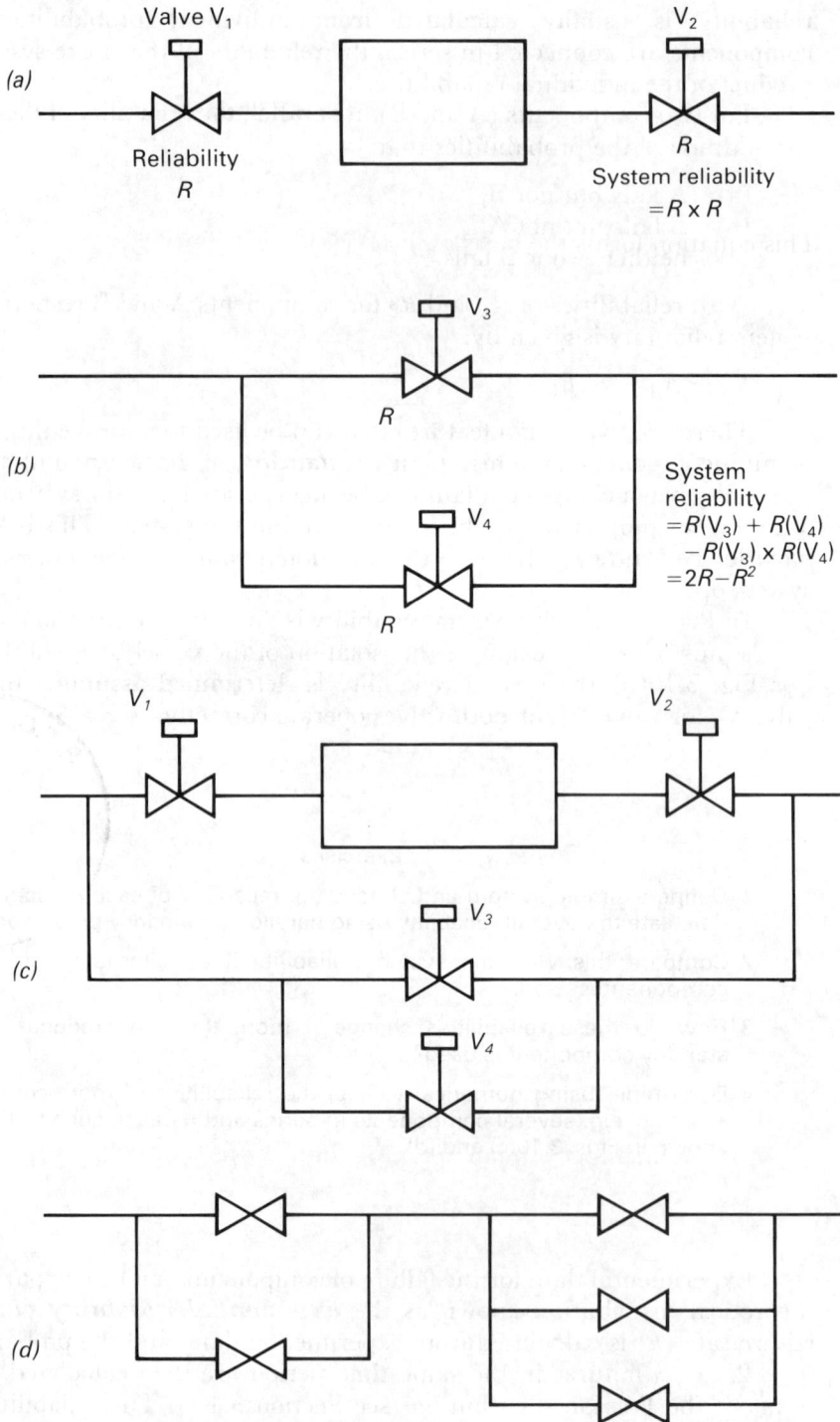

(a)

Valve V₁

Reliability
R

V₂

R

System reliability
= R x R

(b)

V₃

R

V₄

R

System
reliability
$= R(V_3) + R(V_4)$
$\quad - R(V_3) \times R(V_4)$
$= 2R - R^2$

(c)

V₁

V₂

V₃

V₄

(d)

Figure 3.10   Systems involving Automatic Valve Operation

This exponential law is a special case of a more general theoretical reliability curve known as the **Weibull distribution**. If experimental data are available then two parameters can be determined, the mean time between failures (M) and the **shape factor**(m). The reliability of an item can then be predicted from the equation:

$$R = \exp(-t^{m}/M)$$

This equation forms the basis of the Weibull distribution.

* * *

Exercises

1 Obtain data for the failure of components with time, either from a reference source, by experiment or using 'invented' figures.

2 Determine the exponential reliability curve and compare it with the curve shown in Fig. 3.9(a).

3 Use the same data to predict the Weibull distribution.

4 What effects do changes in the values of m and M have upon the shape of the distribution?

5 Under what conditions is the Weibull distribution the same as the exponential curve?

* * *

The actual reliability of a system may be higher than the predicted value if, despite one or more component failures, it continues to function, although not perfectly.

Reliability represents another factor the engineer must consider during the design process. There are many ways in which the reliability of a system can be improved: the following are some suggestions which can be incorporated into the design process.

(a) Less components means greater reliability (also lower cost!). However, fewer parts also means less versatility.

(b) Specify the use of standard components of known and proven reliability.

(c) Operate below full-loading, known as derating.

(d) Remember human operating limitations.

(e) Protect against vibration and shock.

(f) Specify running-in requirements.

(g) Specify debugging procedures.

(h) Consider both environmental and functional requirements.

(i) Maintain suitable temperatures.

The designer must produce a component that satisfies the requirements of performance and reliability. Reliability is a probability, and the data to be used are obtained by experiment and evaluated by statistical methods. Although components may be deemed to have a reliability of 0.99, there is always a chance that one (or even two) components will fail. Even horses with odds of 100 to 1 win races sometimes! However, if correctly applied, the use of reliability theory can significantly improve the performance and service life of a system.

3.6 OPTIMISATION

Optimisation is a term which is now in common use. It has resulted from the realisation that activities are not, need not and sometimes cannot be defined or promoted simply as win/lose, maxima/minima, profit/loss. The world's natural resources need to be conserved, the environment should be protected and preserved, big is not necessarily best or beautiful. The oil producing and exporting countries realised that maximum profit was not necessarily the ultimate goal. Who would buy the oil if many countries became bankrupt? Also, if the price was raised too much or too quickly, then perhaps the developed nations would pay serious attention to conservation and alternative energy sources. A company with a few products making large profits will not necessarily be as 'recession-proof' as a company with a wider market-base and lower profits.

The designer/engineer will be involved in a search for an optimum design, i.e. the technical best that could be achieved. However, this is often not attainable in reality. There may be conflicting objectives or unquantifiable but real restrictions, e.g. political or social pressures. What the designer usually accepts is the optimal design.

It will be useful if the student is at least familiar with some common terms before consulting a suitable optimisation text.

The *system* defines the boundaries of the investigation.

Subsystems (often interdependent) can often be defined within the total system.

The *parameters* are the factors or variables within a problem. The three main groups are functional, material and geometrical. Parameters that can be considered positive or desirable, e.g. safety and profits, are maximised, while negative (undesirable) aspects, e.g. costs and wastage, are minimised.

The *criterion function* is the primary design equation and represents the object of the optimisation process. It is expressed in terms of the parameters that are related by natural laws or empirical relationships.

Functional constraints are expressed by equations (and often equal zero);

Regional constraints are inequalities. These constraints define the parameter limits.

There are various optimisation methods in common use.

SUBJECTIVE JUDGEMENT: based upon design experience, intuition and a constant awareness of the continual need to optimise.

SCALE MODELS, GRAPHICAL SOLUTIONS OR SIMULATION PROCEDURES.

ANALYTICAL, I.E. MATHEMATICAL, OPTIMISATION: this requires that all or a sufficient number of parameters, variables and constraints can be expressed in explicit mathematical terms. Differentiation can be used if there are no functional constraints, or if all functional constraints can be combined in the criterion function. Sometimes a solution can be obtained by using a method of dual variables, or Lagrangian multipliers (if separate constraint equations exist).

LINEAR PROGRAMMING: this technique can be used where the regional constraints are simple linear inequalities, and the criterion function is also linear.

Linear programming is often applied to problems involving materials mixing or blending (see Exercise 10 on page 60), capacity allocation, production schedules, transport routes and schedules.

DYNAMIC PROGRAMMING: if the constraints and the criterion function are quadratic relationships and the problem cannot be split into a series of linear programmes, then a multistage optimisation process is needed. This is dynamic programming.

MONTE CARLO METHOD: this is a statistical technique applied when dynamic programming is not possible, using the steepest ascent method of solution.

Linear and dynamic programming are known as **_operational research_** (OR) methods of optimisation. All the above methods should be studied in detail in an engineering course.

* * *

Exercises

1 Identify some of the optimisation decisions you make in your day-to-day life. What are the trade-offs that make the results optimal rather than optimum?

2 Consider the optimisation that has, or should have, been used in the preparation of the syllabus and in the teaching of your engineering course.

* * *

3.6.1 Examples of Optimisation Problems

Satisfactory operation of machines, equipment, chemical plant, etc., where there are several operating variables, often requires optimisation of the operating conditions in order to obtain acceptable performance. The effect of changing one of the variables often influences other conditions in the equipment.

Consider the situation where a dilute feed must be concentrated by heating the solution to evaporate a proportion of the solvent (typically water). Equipment which could be used for this purpose is shown in Fig. 3.11(a), and

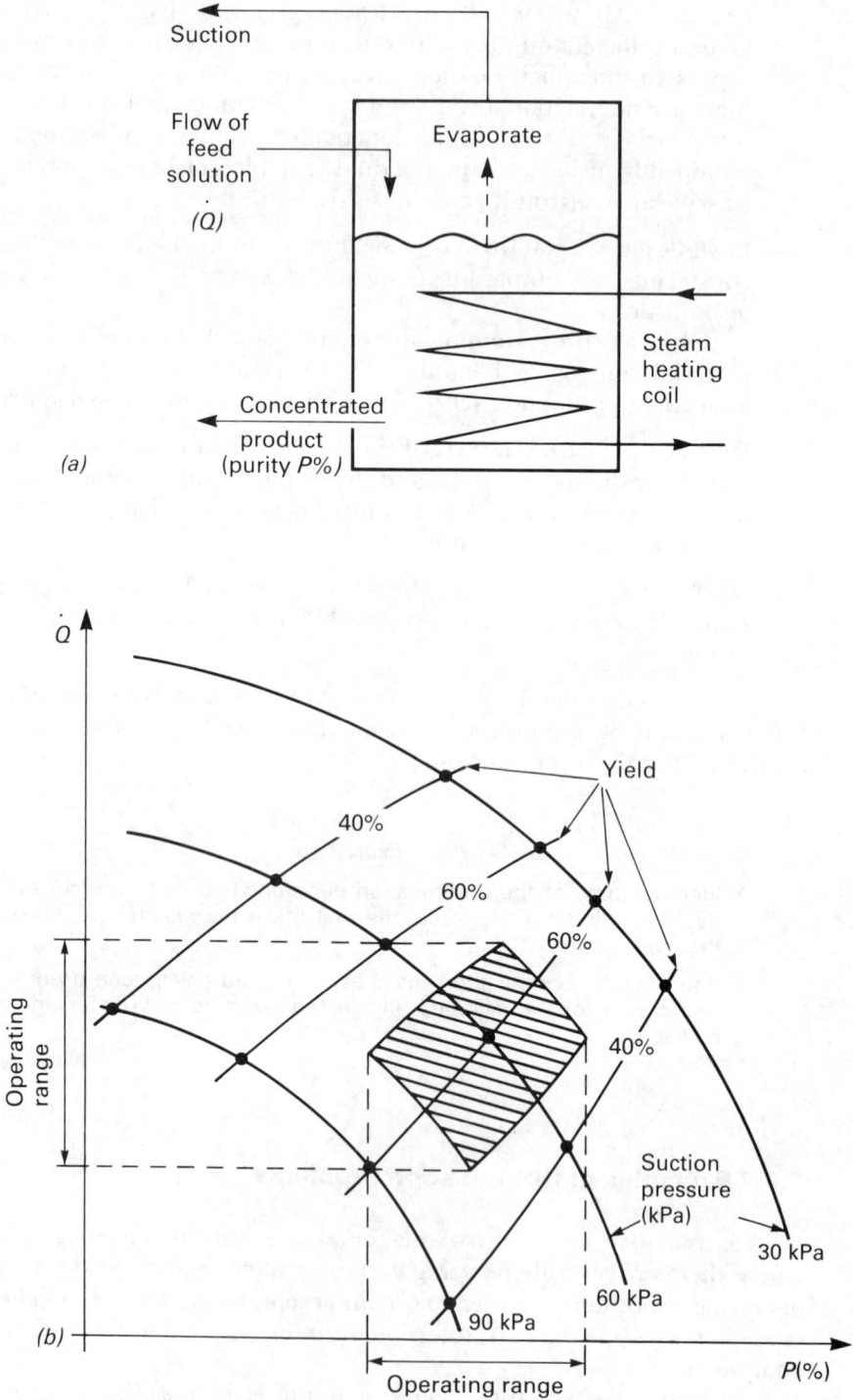

(a)

Suction

Flow of
feed
solution
$(\dot{Q})$

Evaporate

Steam
heating
coil

Concentrated
product
(purity *P%*)

(b)

$\dot{Q}$

Yield

40%

60%

60%

60%

40%

Operating
range

Suction
pressure
(kPa)

30 kPa

60 kPa

90 kPa

Operating range

P(%)

Figure 3.11   Identifying Optimum Operating Conditions

includes a steam-heating coil and a suction line to remove the vapour phase. If the equipment is operating at steady conditions then the effects on the process of changing the following variables, if all other conditions remain unchanged, are:

(a) increasing the flowrate of solution reduces the purity of concentrated product;

(b) increasing the energy content of the steam, i.e. using superheated steam, or a higher steam flowrate, increases the product purity;

(c) increasing the suction pressure, i.e. level of vacuum, increases the product purity.

Equipment is often designed to include some operating flexibility, and the possible range of conditions that the process must be able to satisfy should be determined. In this example fluctuations may occur in the feed flowrate, or the requirements for product purity may change. Results which may be obtained for the operation of such a plant (these are *not* actual results) are shown in Fig. 3.11(b). The curves show that as the flowrate decreases or the vacuum level increases, the product purity improves. Another parameter may be necessary in order to define the preferred operating conditions for such a process. This could be the yield (defined as mass of chemical (solute) in feed: mass of solute in product) which is shown in Fig. 3.11(b).

Alternatively, the operating costs for such a plant could be determined by evaluating the steam cost, vacuum cost and feed solution pumping cost under different operating conditions. The yield lines on Fig. 3.11(b) could then be omitted and operating cost lines plotted instead.

The evaluation of operating data such as that shown in Fig. 3.11(b) should enable a region of optimum operation to be defined, where costs and operating conditions are acceptable (shown shaded in Fig. 3.11(b)).

Another common problem which can be used to illustrate the search for optimum conditions is the determination of the appropriate insulation thickness for a pipeline. If a pipe carrying a hot fluid is not sufficiently insulated, heat will be lost, as shown in Fig. 3.12. However, if too much insulation is used then the cost of this material (including fabrication and installation costs) may exceed the savings in heat costs. The economic insulation thickness can be determined by plotting a graph such as Fig. 3.12, where the minimum region of the total cost curve (cost of heat saved plus insulation cost) is the required optimum range.

The curve in Fig. 3.12 typically exhibits a flat minimum region rather than a single, well-defined, minimum value. The upper limit of this region occurs where the insulation cost exceeds the cost of lost heat. However, if energy costs are likely to increase then it may be advantageous to install extra insulation to enable future energy savings to be made.

The cost of the heat saved will be a straight line in Fig. 3.12, but the cost of insulation will depend upon the quantity of material used, the cost of fabrication (if it is required in moulded form) and labour costs for installation.

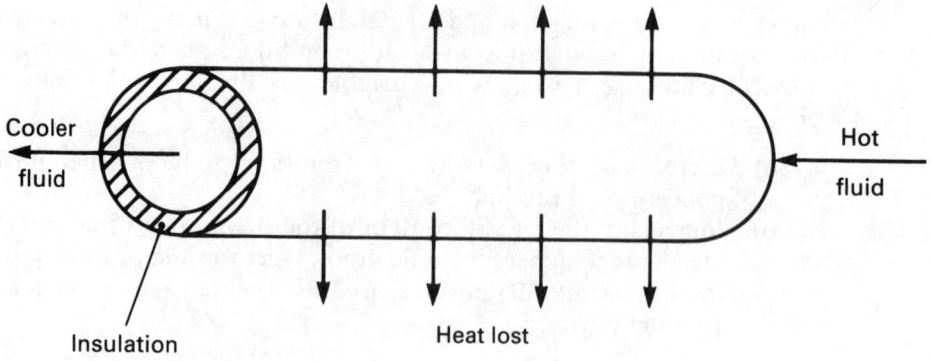

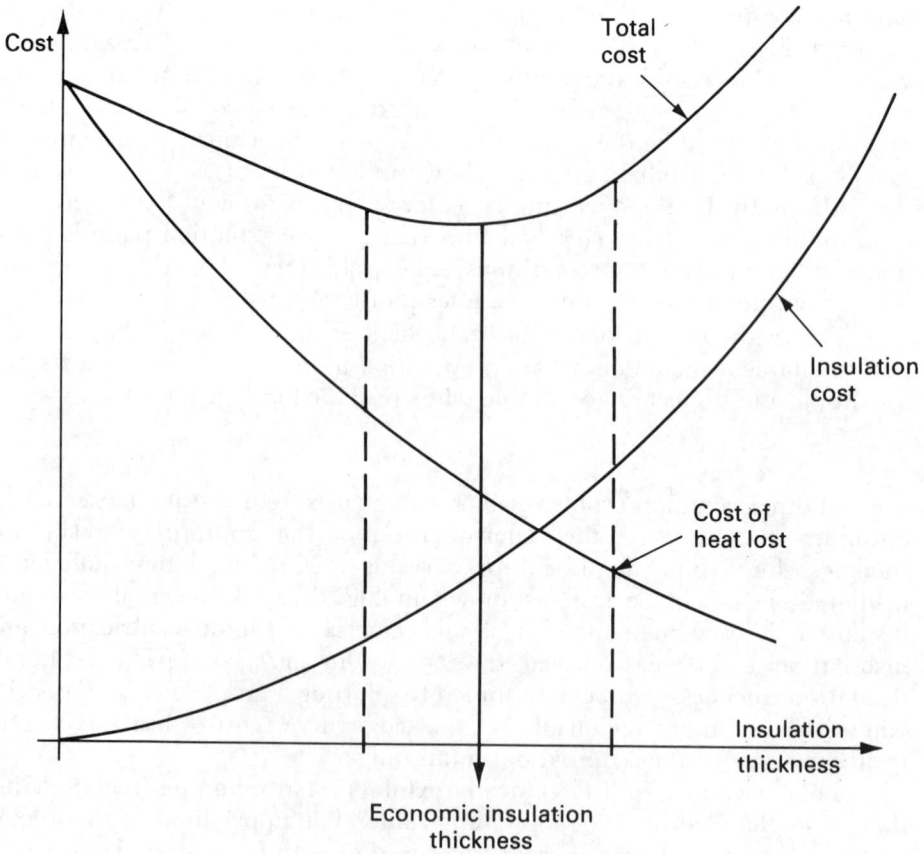

Figure 3.12   Optimisation of Insulation Thickness

Fabrication and labour costs will not necessarily increase linearly above a certain insulation thickness.

KEYWORDS

decision-making
alternatives
negative option
irreversibility
consequences
optimisation
compromise
'trade-off'
optimum solution
optimal solution
information retrieval
information assessment
probability
'hard' or 'soft' information
'necessary' or 'dispensable' information
statistics
statistical data
frequency distribution
histogram
frequency polygon
cumulative frequency curve (ogive)
mode
modal value
modal class interval
median
percentile value
lower quartile
upper quartile
inter-quartile range
arithmetic mean
variance
standard deviation
geometric mean
permutation
combination
binomial probability distribution
binomial expansion
Poisson distribution
normal distribution
error curve

Gaussian distribution
bimodal distribution
negatively skewed distribution
positively skewed distribution
standard normal distribution
normalised variable
sample
estimation
population mean
sample mean
population standard deviation
standard error of the sample mean
significance test
t distribution
standard error of difference
chi-squared
correlation
Pearson product moment correlation coefficient
Spearman rank correlation coefficient
regression
laws of probability ('and' and 'or')
reliability
failure rate
mean time between failures
standby unit
parallel redundancy
exponential reliability curve
Weibull distribution
derating
running-in
debugging
system
subsystem
functional parameters
material parameters
geometrical parameters
criterion function
functional constraints
regional constraints

subjective judgement
scale models
simulation
analytical methods

linear programming
dynamic programming
Monte Carlo method
operational research (OR)

BIBLIOGRAPHY

Bell, D.E., Keeney, R.L. and Raiffa, H., *Conflicting Objectives in Decisions*, John Wiley and Sons Ltd, London (1977).
A mathematical approach.

Bompas-Smith, J.H., *Mechanical Survival: The Use of Reliability Data*, McGraw-Hill Book Co. (UK) Ltd, London (1973).
The use of data and statistics, including some good examples.

Burrington, G.A., *How to Find Out about Statistics*, Pergamon Press, Oxford (1972).
A good general source book, useful for locating statistical data.

Crowley, M. (Ed.), *Energy: Sources of Print and Non-Print Materials*, Neal-Schuman Pub. Inc., New York (1980).
One particular example of source books available for a wide range of subjects.

Dano, S., *Nonlinear and Dynamic Programming, An Introduction*, Springer-Verlag, New York (1975).
As title.

Fischer, F.E., *Fundamental Statistical Concepts*, Canfield Press, San Francisco, California (1973).
Good, straightforward introduction, including probability and statistical inference.

Hall, A.H., *An Introduction to Statistics*, Macmillan Press Ltd, London (1978).
A very basic text, a good place for the complete novice to begin.

Howard, K., *Quantitative Analysis for Planning Decisions*, Macdonald and Evans Ltd, Plymouth (1975).
Would be improved by more practical examples.

Johnson, L.A. and Montgomery, D.C., *Operations Research in Production Planning, Scheduling and Inventory Control*, John Wiley and Sons Inc., New York (1977).
For the mathematically inclined!

Kapur, K.C. and Lamberson, L.R., *Reliability in Engineering Design*, John Wiley and Sons Inc., New York (1977).
A mathematical treatment.

Kim, C., *Introduction to Linear Programming*, Holt, Rinehart and Winston Inc., New York (1971).
As title.

La Valle, I.H., *Fundamentals of Decision Analysis*, Holt, Rinehart and Winston Inc., New York (1978).
Strictly for mathematical analysts.

McMillan, C. Jnr, *Mathematical Programming*, 2nd Ed. John Wiley and Sons Inc., New York (1975).
For linear and non-linear programming examples.

Murdoch, J., *Queueing Theory. Worked Examples and Problems*, Macmillan Press Ltd, London (1978).
As title.

O'Connor, P.D.T., *Practical Reliability Engineering*, Heyden and Son Ltd, London (1981).
A good textbook, containing practical examples.

Parsons, S.A.J., *How to Find Out about Engineering*, Pergamon Press, Oxford (1972).
An information/reference source book.

Robertson, A.G., *Quality Control and Reliability*, Thomas Nelson and Sons, London (1971).
Good presentation of statistical data and examples.

* Romano, A., *Applied Statistics for Science and Industry*, Allyn and Bacon Inc., Boston, Massachusetts (1977).
A good basic textbook.

Schuring, D.J., *Scale Models in Engineering. Fundamentals and Applications*, Pergamon Press, Oxford (1977).
Interesting case studies.

* Smith, C.O., *Introduction to Reliability in Design*, McGraw-Hill Inc., New York (1976).
A good place to start.

Tanur, J.M., *Statistics: A Guide to the Unknown*, Holden-Day Inc., San Francisco, California (1972).
Quote from the preface: 'describing important applications of statistics and probability in many fields'. True.

Whitehouse, G.E., and Wechsler, B.L., *Applied Operations Research: A Survey*, John Wiley and Sons Inc., New York (1976).
Good examples and text covering linear programming, simulation, networks and decision analysis.

Computer-aided Design

Chapter Objectives

1 To become familiar with the common terms used to describe computer applications in engineering.

2 To appreciate the meaning and scope of CAD.

3 To become conversant with the computer equipment and techniques which are used by engineers.

4 To recognise the importance of computers for solving engineering problems, and their uses in all stages of the design process.

5 To become aware of activities such as CAM, NC and robotics and their importance in engineering design, manufacture and production.

The student who has some experience, however elementary, of writing and implementing computer programs will gain most from this chapter.

Computer facilities should be easily accessible for the students to evaluate the problems they encounter.

4.1 SELECTED GLOSSARY

CAD	computer-aided design
CAD/CAM	computer-aided design and manufacture, also CADCAM, CADAM, CADM
CADD	computer-aided design and draughting
CAE	computer-aided engineering
CAM	computer-aided manufacture
CNC	computer numerical control
DNC	direct numerical control
NC	numerical control
APT	automatically programmed tools
BASIC	a programming language, an acronym for Beginner's All-purpose Symbolic Instruction Code
FORTRAN	a programming language, an acronym for FORmula TRANslation

ALU	arithmetic and logic unit
CPU	central processing unit or central processor
CRT	cathode ray tube
CRTDU	cathode ray tube display unit
IC	integrated circuit
VDU	visual display unit
DMA	direct memory access
RAM	random access memory
RBT	remote batch terminal
RJE	remote job entry
ROM	read only memory
Access time	the time taken to reference an item in storage.
Algorithm	a set of rules for performing a task or solving a problem, e.g. a formula.
Alphanumeric	characters composed of letters and numbers and sometimes arithmetic signs, punctuation marks, etc.
Binary	arithmetic system using base 2.
Bits	BInary digiTS (a 1 or 0).
Bug	computer jargon for an error in a program or system.
Bytes	a number or word comprising several (usually 8) bits.
Chip	a small piece of material (typically 6 mm square) containing electronic circuits.
Databank	a collection of stored data.
Data-base	an organised pool of shared data, typically a series of regularly updated applications files.
Disk	circular metal plate coated with magnetic material and used to store data on concentric tracks (a floppy disk is flexible).
File	an organised collection of records.
Graphics	visual display of data in graphical form, or as two- or three-dimensional representations.
Integrated circuit	a circuit in which all the components are chemically formed upon a single piece of semiconductor material.
Interactive	system where the operator obtains the processing results immediately so that further action can be taken or when the program prompts the operator for data by asking questions.
Interface	a boundary between two pieces of equipment across which all the signals that pass are carefully defined.
Light pen	a photo-electric device used to detect or modify images on the surface of a CRTDU.
List	a printing operation detailing a series of records on a file or in a store.
Mainframe	a large computer with a wide range of facilities.

Memory	the data storage unit (e.g. RAM, ROM).
Microcomputer	a computer utilising a small number of chips.
Microprocessor	a CPU on a chip, part of a microcomputer.
Minicomputer	medium-sized computer.
Peripheral	device used with a computer to display, store or convert data, e.g. printer, keyboard, VDU.
Real time	system where the results of processing are produced with sufficient speed to be used for control purposes.
Software	computer programs.
Tab(ulation) character	a character used to control the format of printed output, may be required as input with data.
Terminal	device allowing communication between a computer and an operator.
Timesharing	sharing CPU time and system resources between several tasks or operators.
Wafer (or slice)	thin disk of semiconductor material.

* * *

Exercises

1 List situations in everyday life where computers are now used. Identify the advantages and disadvantages of these applications. Suggest improvements and extensions in the functions performed by computers.

2 Carry out a survey of currently available microcomputers and their associated equipment.
Present a performance–cost–benefit analysis using the information obtained and recommend a computer:
(a) if the 'best' value computer is to be purchased, and cost is not a constraint;
(b) for purchase with your own limited funds.
What other factors influence your final decision?

3 Repeat Exercise 2 for hand-held, programmable calculators.

4 Identify particular engineering areas or applications which could utilise computers in the next 10 years. List the main reasons for their acceptance or rejection.

5 Assess the main social and economic implications of the widespread use of computers, both at work and in the home.

6 Consider how computer technology may be used to overcome some of the social problems which its widespread acceptance may have created.

7 Compare the life cycle of the microcomputer (thus far) with that of a previous high-technology innovative product, e.g. cassette tape-recorder, video equipment.
What predictions and recommendations could be made for the microcomputer industry?

8 Design an automobile that is to contain a completely integrated microcomputer.

* * *

4.2 COMPUTERS AND ENGINEERING

The appearance of computers in schools, being used even by five-year-olds, and in homes is no longer unusual. The widespread use of computers by companies and individuals is further evidenced by the media, and in particular by advertisements directed at various levels of society. For these reasons terminology and exercises have been presented at the beginning of this chapter. The student should have some computing experience before commencing this course and should be familiar with such terms as '64K memory', 'available software and peripherals', etc. It will be worth spending a short time reading through the terminology to help clarify the meaning of some of the jargon. The exercises are intended to make the student think about the possible applications of computers in engineering situations and the advantages for the engineer–user. This chapter will be mainly concerned with the use of computers for purposes of improved engineering design.

The computer industry was only born after the Second World War and its growth has been dramatic, if not phenomenal. Technological advances resulting in physical size reductions, improved reliability, more complex operations and shorter response times have helped to create a wider market for computers since the 1960s. The development of time-sharing computer systems and storage-type, cathode ray tube (CRT) displays lead to even wider applications. The integration of computers into engineering design was mainly due to the success of applying an interactive mode of operation and significant improvements in on-line computer graphics.

The student should be aware of two important aspects of computer technology. First, the successful miniaturisation of component parts and the associated improvements in computer operational ability. Second, the rapid decrease in cost and price of computers, especially the personal micro-computer.

A computer is an assembly of electronic and electromechanical devices which obeys instructions provided by an operator. It is a system comprising a **central processing unit** (CPU) and other equipment known as **peripherals.** These physical components are known as **hardware.** The computer receives its instructions on computer programs and these are known as **software.** The elements that make up a computer system are shown in Fig. 4.1.

The CPU contains three sections:

(a)   the controller;
(b)   the arithmetic and logic unit (ALU);
(c)   the core store.

It is sometimes more economical to store data in peripheral devices using magnetic tape, cassettes or hard or floppy disks. The peripheral device is known as backing store.

Before an item of data or a program instruction can be processed it must first be converted into a binary or two-digit machine code. The machine operates with the binary (base 2) arithmetic system using only the digits 0 and 1.

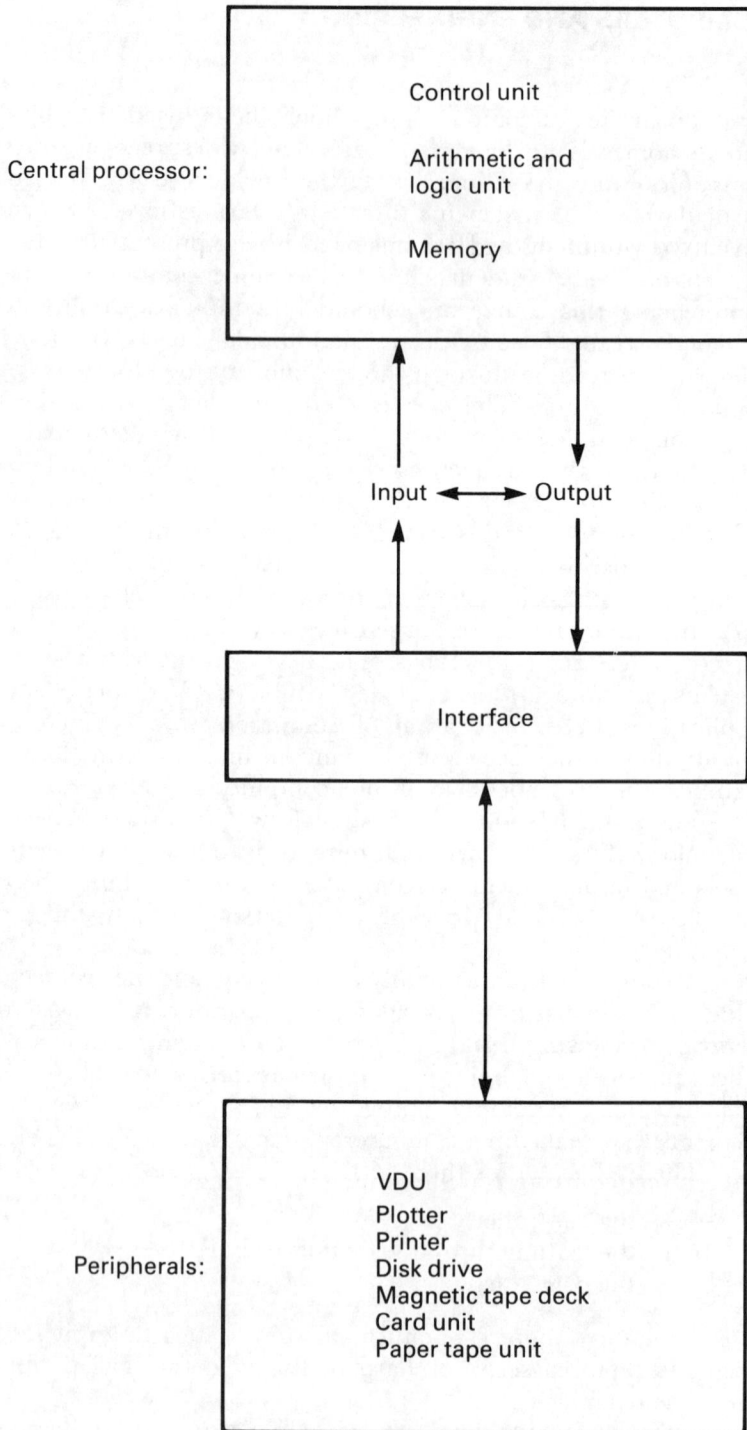

Central processor:

Control unit

Arithmetic and
logic unit

Memory

Input ⟷ Output

Interface

Peripherals:

VDU
Plotter
Printer
Disk drive
Magnetic tape deck
Card unit
Paper tape unit

Figure 4.1   Elements of a Computer System

Binary digits or **bits** are the smallest units that the computer can recognise, and computer 'words' are made up of groups of bits. The **byte** is defined as:

1 byte = 8 bits = 1 alphanumeric character

A computer word may represent a character e.g. A to Z or 0 to 9, an instruction, e.g. 'print', and a floating point number up to a value of 10^{128}. Computers may use variable word lengths in multiples of 8 bits (1 byte) or have fixed word lengths of 16 bits or 32 bits.

Communication between the user and the computer is now frequently carried out using a teletype or alphanumeric visual display unit (VDU). Instructions are typed on a keyboard and displayed on a screen, and the answer is obtained either as printed copy from a line printer or as a VDU display.

A time-sharing or multi-access system incorporates several teletypes or VDUs so that more than one operator can use the computer at the same time. The computer may also be used by **remote access** from a telephone connection.

A computer can only provide a solution to a problem if it receives instructions in the correct form. Operations to be performed must be presented in the correct sequence, and the planning of such a sequence is known as **programming.** The actual plan is known as a **program.** The instructions are prepared and given to the computer using a specified code which is known as a **programming language. FORTRAN** and **BASIC** are two programming languages. In order to 'run' a program the computer must be able to accept (understand) the particular language used, and a **program language compiler** is fed to the CPU before it is operated. A system program relates all the peripherals to the CPU.

The information given in this section is presented in general terms so that it will not be quickly dated by the rapid changes and developments within the computer industry. It is intended to provide an introduction to the subject. It is perhaps worth mentioning the difference between a mainframe computer, a minicomputer and a microcomputer.

A **mainframe computer** is a large machine housed in a purpose-built room or building, capable of performing many and varied tasks and supporting a wide range of equipment.

A **minicomputer** is a smaller machine, but the differences between it and a **microcomputer** are becoming increasingly blurred. The functions that can be performed by a microcomputer are continually being increased and improved. The power (and cost) of a computer system is mainly determined by its operating speed and the size of the internal working store. Machines are often designated by their main storage capacity, e.g. 64K (eight bit bytes) where K is nominally 1000 (actual value 1024). In general, the larger the internal storage capacity of a machine, the greater the number of functions that can be performed and the terminals and peripheral devices which can be supported.

However, increasing the storage size of a computer is extremely costly and it may be cheaper to use backing storage, such as magnetic tape or disks.

The computers described are known as 'general-purpose, stored program, digital, electronic computers'. ***Analog computers*** are also used by engineers to simulate physical systems or for automatic control of a system. They are usually electrical, mechanical or electronic devices and their output is an electrical voltage, in graphical form or as a CRT display. Analog computers cannot store large amounts of information and their logical capabilities are smaller than digital computers. In certain situations, microprocessors are now replacing analog computers.

4.3 WHAT IS CAD?

CAD is the acronym for computer-aided design. Other common terms used in computer/engineering activities are defined in Section 4.1.

It is important that the work of the engineer and the tasks performed by the computer should be integrated to provide the most efficient combination for design problem-solving. The computer and the engineer should each perform the work for which they are best suited, but the engineer will be responsible for his own work and acceptance of that produced by the computer. CAD can simply mean that a computer is being used in design studies. The computer may merely be used to perform the numerical calculations quicker than they could be carried out by hand, or it may be carrying out draughting work. Computer-aided design and draughting (CADD) describes these combined functions.

By studying Chapter 2 in detail the student will be aware of, and may have some ideas and suggestions about, particular aspects of design that can be improved by the appropriate use of a computer.

Table 4.1 Potential for Computer Utilisation in a Design Study

Design stage	*Computer potential*
Recognition of a need and problem definition	Desirable, but few applications
Feasibility study	Limited application
Preliminary design	Certain specific applications and use of specially prepared programs or packages
Testing, evaluating and improving the design	Yes, using numerical methods and optimisation
Final design and documentation	Yes, preparation of standard specification and draughting
Communicating the design	Computer graphics
Manufacture	CAM and NC
Feedback	Storage on files of subsequent results and reports

The stages in the design process and the possible uses of a computer are illustrated in Table 4.1. The early stages of a design are not particularly suitable for computer applications. These stages represent the more creative aspects of problem-solving and are best performed by the designer using a combination of past experience, judgement and intuition, and having the ability to locate various sources of data.

The use of computers in these preliminary stages is desirable, but the range of possibilities that are available is so wide that vast amounts of data would need to be stored in the computer memory. The ability of the computer to extract significant or appropriate information would require a systematic search of all data, and would be less cost-effective than the intuitive human approach.

The latter stages of a design which require a larger amount of repetitive and iterative working as well as significant computational time, are ideally suited to computer applications. The design tasks that can be performed using a computer are:

(a) computations to be repeated many times, similar to hand calculations;

(b) computations that are complex and time-consuming, extensions of hand calculations requiring greater accuracy; and

(c) the manipulation of data, including storage and retrieval of old designs and recent modifications.

The computer can be used as a design aid not only by the designer working on one particular aspect of a project, but by all members of the design team. Data can be made available to other engineers either by multiple access or remote access to the computer, or by storing data on devices such as disks which are portable and may be compatible with other machines.

* * *

Exercises

Consider the problem of designing a pipeline for the transportation of sewage. A specific solution is required, and then a general design method must be developed.

1 Identify the stages in the design process.

2 List all the variables.

3 Select (realistic) values or specifications for each unknown item except the pipe thickness.

4 Calculate the pipe thickness required.

5 Change one or more of the values specified and repeat the calculation of pipe thickness.

6 Compare the two solutions obtained.

7 What is the (relative) cost of each solution?

8 What are the effects on other aspects of the design, such as the estimated life, reliability, capacity for overload?

9 Assess the possibility of using a computer to solve this type of problem.

10 Now prepare a general design approach to this problem and assess the use of computer-aided design at each stage.

11 What factors restrict the use of a computer in solving this problem?

12 Is it possible to include such factors as customer preference or bias (e.g. specifying common materials) into a computer-aided design method?

*   *   *

Computer programs or software packages may be written in two ways, either as a straightforward input–output operation or as an interactive or conversational mode of performance. All problems require data in order to obtain a solution, but whether these are numerical values to use in calculations or a choice between alternatives is immaterial. The data are provided by the operator. There are three ways in which data can be provided for use in the operation of a computer program, and these are now described.

(a) Data are presented and incorporated as part of the computer program. They are available in the program when first required in the sequence of operations. The disadvantage of this method is that different applications of the program will require changes to be made at different places in the program. This may be time-consuming and liable to mistakes, especially the possibility of overlooking a value to be changed.

(b) All data are supplied as a complete 'block' or 'package' at one section of the computer program, usually the start or finish, and are provided before the program is executed (i.e. run). The data must then be in the correct order, as read (required) in operation of the program. This requirement may cause problems but can be partially alleviated by situating all 'data read' or 'assign' statements at the beginning of the program. Also all data must be in the correct form, e.g. integer or real, and in the required units. Data provided for density may be kg/m^3 or g/cm^3 and it is important always to use a specified system. Checks can be implemented by telling the computer to print an appropriate error message if the data value is very different from an expected value, e.g. if the density for liquids is less than 700 (kg/m^3) or greater than 10 (g/cm^3). An alternative is to obtain a 'hard copy' (print-out) of all data values after a program run.

(c) Data can be provided by the operator during the execution of a conversational or interactive computer program. The program requests data (and may specify the form) as required. The advantage of this programming method is that it can make the user 'feel' more integrated with the computer, rather than considering it merely as a plug-in machine. The user also feels in control of the operation and the machine. Requests for data can be used as a sequential check on the actual program operation, rather than supplying data in one block and then attempting to identify the source of any error messages.

* * *

Exercises

1   Write a computer program to investigate the effects of changes in the process variables for the ethanol synthesis reaction using the information provided. The program should be written in three different ways:
(a) data included in the program where first required;
(b) all data to be provided in one 'block'; and
(c) conversational or interactive mode.
Include checks in the program, i.e. error messages or printed statements, relating to the use of the data.

2   Write the program to obtain a particular set of process conditions: temperature, pressure, percentage conversion, etc.

3   Assess the possibility of using or modifying your program as a general software package to provide information for a wide range of process conditions.

4   Is your program 'user friendly'?

The following information will provide the basis of the computer program:
Synthesis reaction: $C_2H_4 + H_2O \rightarrow C_2H_5OH$.
The change in free energy for the reaction (ΔG_T J/mol) as a function of absolute temperature (T K) is given by:

$$\Delta G_T = -42920 - a\,T\ln T - (bT^2/2) - (cT^3/6) - (dT^4/12) + 31.1T \quad (4.1)$$

where $a = -16.3$
$\quad\quad b = +5.1 \times 10^{-2}$
$\quad\quad c = -3.1 \times 10^{-5}$
$\quad\quad d = +6.0 \times 10^{-9}$

The equilibrium constant (K_a) for the reaction is given by:

$$\Delta G_T = RT\ln K_a \quad\quad\quad\quad (4.2)$$

where R = universal gas constant.

Considering particular process conditions:
Initial moles: C_2H_4—12; H_2O—6; C_2H_5OH—0
Final moles: $(12-x)$; $(6-x)$; x
Total moles at equilibrium: $(18-x)$.
Then the equilibrium constant based upon partial pressures (K_p) is given by:

$$K_p = \frac{x\,(18-x)}{P\,(12-x)\,(6-x)} \quad\quad\quad (4.3)$$

where P = total pressure in the system.

At low operating pressures:

$$K_a = K_p \quad\quad\quad\quad (4.4)$$

At high pressures, activity coefficients must be determined to account for non-ideal behaviour of the gases.

The percentage conversion of ethylene (C_2H_4) in the process, using the above conditions, is given by:

$$\left(\frac{x}{12} \times 100\right) \%$$
(4.5)

It is not necessary for the student to have any knowledge of chemical processes or thermodynamics in order to write a computer program to solve this problem. It will obviously increase the student's understanding of the nature of the problem and promote more interest if some courses in chemistry or thermodynamics have been studied, as will usually be the case for engineering students.

An evaluation of the ethanol synthesis reaction from a thermodynamic viewpoint involves specifying certain process variables, in this case temperature (T) and pressure (P), and solving equations (4.1) to (4.5) to obtain the percentage conversion of ethylene.

Industrial process conditions are, typically, 300°C and 6.5 MPa. However, this pressure is high and fugacity corrections should be applied. It may be easier to investigate the process for operation at 1 MPa pressure (say).

The value of the universal gas constant (R) can be obtained from an engineering handbook, in appropriate units. The calculated conversion of ethylene is typically 6 to 20%.

5 Decide how to modify your specific program to investigate the effects of changes in a range of process variables, including temperature, pressure, C_2H_4 : H_2O mole ratio (12:6 in this example).

6 Decide how your general program could be adapted for use with other reacting species. The values of a, b, c and d used in equation (4.1) are based upon data for the variation of specific heat capacity of each gas with temperature. These values will depend upon the gases used, and any changes will require some knowledge of thermodynamics.

A suitable introductory text is *Concise Chemical Thermodynamics* by J.R.W. Warn, which also describes the thermodynamics of the ethanol synthesis reaction in detail (pp. 112–115). The corrections to be made for high-pressure operation and values of a, b, c and d to be used in equation (4.1) are also given in this textbook.

* * *

4.4 COMPUTER GRAPHICS

The training and expectations of designers and engineers, as well as the nature of the work they perform, make it essential that design data and solutions can be represented graphically. This is also necessary to ensure the effective communication of ideas. The designer's output will be further improved if the graphics display can be easily and quickly modified by the designer.

In the 1950s all output from a computer was on paper, either printed in alphanumeric form or plotted slowly on an 'xy' recorder. In the 1960s, facilities were developed that enabled results to be displayed in graphical form, although this was mainly used for data plotting. Complete systems were expensive and therefore not widely used, and also the time required to obtain a display represented a significant constraint on the design process. A further development was the use of graphics to assist the input of data which meant that the operator required some means of indicating a position on the screen. Various devices were produced, including keyboard, lightpen (a sensitive photo-electric device), 'mouse', crosshairs, joystick and digitising tablet.

The **digitiser** is a particularly useful device for the conversion into digital form of the large amount of data which a computer may require in order to produce a drawing from a two- or three-dimensional display. Magnetic tape, floppy disks, etc., are then used to store the digital data. A pen or cursor can be used to trace details of an engineering drawing, using a flat tilting table with a programmable controller and a keyboard.

A **VDU** is capable of displaying alphanumeric data only, whereas graphics display units use **cathode ray tube (CRT) devices** for data presentation. The **direct view storage tube display** uses a CRT which stores a drawing in its phosphor coating. The drawing cannot be modified unless it is first entirely erased. The **refresh graphics display** is more expensive but allows repeated modification of the image without complete erasure. A **raster scan display** uses a television monitor, although this requires scanning of the entire screen rather than a discrete line, and it produces a coarse image.

Hard-copy output from a graphics system is generally produced by a plotter, available in various forms. The **digital incremental plotter** (drum plotter) is the least expensive and is widely used. It consists of a variable speed, paper-carrying drum with a pen that moves along a line parallel to the drum's axis. It is available in different sizes and specifications, and may be operated from a peripheral unit (off-line) or directly from a computer (on-line).

An alternative machine is the **flat bed plotter** which is available in sizes from A4 to that of a draughting table (125 by 200 cm). It may also be operated in an off-line or on-line mode, and may be used by remote access from a telephone connection.

Finally, an **electrostatic plotter** can be used. This employs an electronic matrix-scanning technique to print high density dots on specially prepared paper in a raster-type format. The disadvantages are that special paper and chemicals are required, the data must be pre-processed, the quality and accuracy are inferior, and multi-colour or over-drawn plots cannot be produced. However, high plotting speeds are possible and the plotter can also be used as a printer.

Until the mid-1970s the development of computer graphics techniques and software was relatively slow owing to high capital costs and time-consuming data preparation. The breakthrough occurred with the availability of cheap microprocessors and cheaper computer memory. This was reinforced by a shortage of draughting skills (and hence higher salaries), more competitive markets, falling costs and prices, and the development of more flexible graphics software.

Software is available that allows the computer to draw two-dimensional objects, or three-dimensional objects in a 'wire-line' drawing form. The planes can then be smoothed out, the surface coloured and hidden areas removed from view. The shape displayed can be rotated, enlarged or panned, and areas subtracted or added as they leave or come into view. This is carried out using a light pen or digitising tablet.

The development and availability of interactive computer graphics were also major influences on the range of computer applications. The increased speed of computer response and introduction of direct input/output devices meant that the designer could see almost immediately the graphical plots of computed results, and could then amend the output as necessary. In this way the designer–computer combination became fully integrated as an interactive design team, each performing its particular expertise with support from the other.

* * *

Exercises

1 Prepare a sequence of instructions and, if possible, write a computer graphics program to perform the following tasks:

   (a) draw an octagon;

   (b) draw a regular pentagon such that all points (corners) of the shape are joined to each other;

   (c) draw a representation of a rectangular box;
   (d) draw a representation of a tetrahedron that has been cut into two parts;
   (e) show two cross-sectional views of a simple two-position flow valve.

2 Recommend appropriate computer graphics systems to be used with the following design problems:

   (a) thermal design of a heat exchanger;
   (b) mechanical design of a heat exchanger;
   (c) structural load support;
   (d) construction of a low-cost, dome-shaped house;
   (e) specification of an automobile body cross section, including floor panel design.

3 Assess the current and future uses of computer graphics within your particular engineering discipline.

* * *

4.5 CAD SYSTEMS

Computers are now widely used in the planning, design and construction stages of engineering projects. The expected benefits are greater efficiency, better standards of design and improved management planning.

CAD is not a specific term, it includes all aspects of the design process including draughting work. The present trend in CAD is away from systems performing only numerical calculations, and requiring a large computer memory, towards the analysis of complex problems requiring many hours of program developments. CAD is being increasingly adopted by architects, shoemakers, surgeons, industrial designers, etc.

Many requirements are for **turnkey systems** which are purchased as a complete unit: the purchaser turns the key and starts learning. The cost of such systems is declining as computer hardware costs fall and competition increases.

Before embarking on a CAD problem-solving approach it should first be ascertained that the problem is suitable for solution by a computer. Any problem which can be written as a sequence of logical steps and mathematical representations can then be expressed as a computer program. However, CAD is most economical with situations that are repetitive, and with complex problems. CAD is also useful for assessing alternative methods and possible situations since the time required to prepare a design specification will then be reduced and quotations will become more accurate. Other advantages are less tangible, such as the job satisfaction associated with using modern, sophisticated techniques or the extra time available for creative work.

A long-term benefit will be the creation of a data-base of information covering many design situations and products. Designers and engineers then use the same set of data, thus producing consistent specifications and standards, avoiding copying errors and making improvements and modifications easier. A common data-base also provides the opportunity to introduce computer-aided manufacture (CAM) into a company.

The decision to proceed with a CAD approach to a problem means that the appropriate hardware must be available. This can be a large, time-sharing computer system or a microcomputer with access to a larger data-base stored by a mainframe computer. However, avoid trying to combine computer equipment from different manufacturers—the reasons should be obvious! Certain conditions must be achieved, such as convenient and easy computer access, well-defined program input, ease of data revision and error correction, results quickly available in an appropriate form and expert assistance available when required.

It may be possible to use an existing CAD system or program to solve a problem. If this option is not available then it will be necessary to develop or obtain suitable software. There are four main types of software packages for use with CAD systems.

First, **interactive computer graphics** as described in Section 4.4. This should be an interactive program using CRT screen presentation and providing a complete draughting capability including labelling, dimensioning and geometrical analysis.

Second, programs that provide the engineering analysis of a design problem. This often comprises a set or 'suite' of programs that are integrated to provide a complete solution. These programs may perform a geometrical analysis, plant layout or piping diagram, stress analysis, temperature distribution, etc. **Finite element methods** may be used which divide the design problem into small components or elements, and then consider the solution in terms of the interrelationship of these elements.

Third, CAD programs that are written specifically to handle, store, retrieve and present the large amounts of data which may be involved in a design project.

Finally, software packages are available which produce final design drawings and instructions for **numerically controlled (NC) machines.** The communication of design information via a network of terminals is becoming increasingly popular.

The use of CAD systems in particular industries will now be considered.

The aircraft industry is probably the largest user of CAD systems. The Boeing Co. is one of the leaders, having used NC equipment for production of machined parts since the 1950s, and APT technology and surface analysis programs since the late 1960s. From the mid-1970s, Boeing has utilised interactive computer graphics and now produces about 30% of the total design drawings for an aircraft using CAD systems. These drawings are for the design of parts representing about 90% of an aircraft's structural weight. All the design information is stored in a common data-base and the computer checks the entire assembly for clearance and fit. Individual assemblies and units may also be checked.

The automobile industry also embraced CAD at an early stage, mainly because of the wide range of interrelated variables that influence an automobile design. However, in the USA the rapid introduction of stringent requirements relating to safety, the environment and fuel economy has now made CAD a necessity. Interactive graphics can be used to compare different overall shapes for a vehicle and then make alterations and improvements. This is known as **concept surfacing.** CAD can be used for the design and analysis of component parts and to show the interaction of moving parts using animation techniques. Computer graphics have been used to simulate the shape and requirements of a human model to improve the ergonomic features of a design.

Many of the techniques developed by the aircraft and automobile industries have also been used for the design of construction equipment and agricultural machinery, such as earthmovers, road rollers, combine harvesters, etc. Companies that manufacture machine tools have extended their involvement in CAM and NC to the design of the tools themselves. Computer simulation can be used to investigate tool paths, machine vibration and other operating variables, and then to modify the specification until a satisfactory tool design is achieved.

* * *

Exercises

Make recommendations for the use of CAD systems in the following cases:

1 The design of a milk bottle (Fig. 4.2).

2 The design of two intermeshing toothed gear wheels (Fig. 4.3).

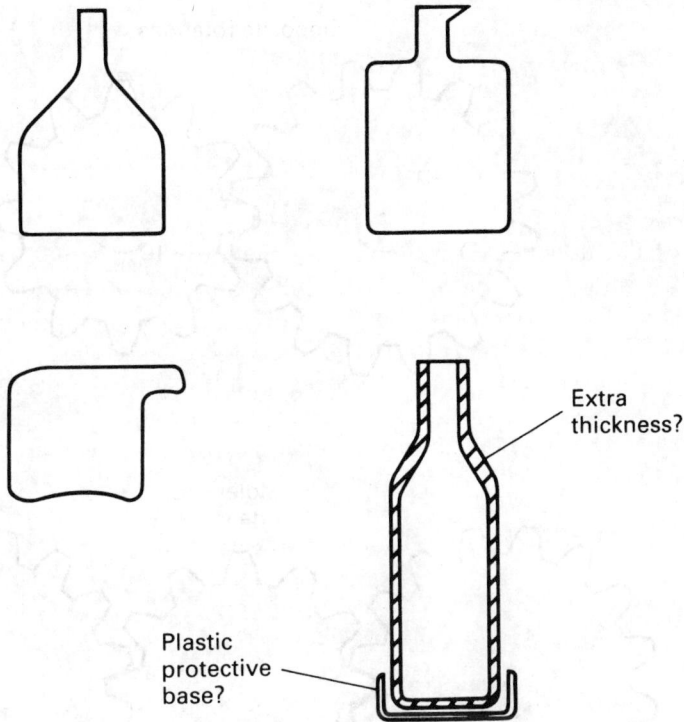

Extra
thickness?

Plastic
protective
base?

Figure 4.2   Alternative Designs for a Milk Bottle

3 The arrangement or layout of several units of a chemical plant, machining tools or a production process, where the units are connected in a sequential operation (Fig. 4.4. What about vertical space?).

4 The design of an excavator digging arm (Fig. 4.5).

5 The connection of two moving parts, e.g. two rotating shafts of different diameters.

6 Joining together individual components to form an assembly that is to be joined to, or fitted inside, another unit.

7 Selecting a body shape for a new automobile (Fig. 4.6).

8 Comparing the possible sheet car body skins for the alternatives in Exercise 7 (Fig. 4.7).

9 Deciding whether to use a pressed shell or welded panels for an automobile body.

Figures 4.2 to 4.7 provide illustrations relating to these exercises and may be used for clarification or as starting points.

*   *   *

Opposite rotations

Idler
gear

Would
they work?

Worm
gears

Figure 4.3   Design of Gears

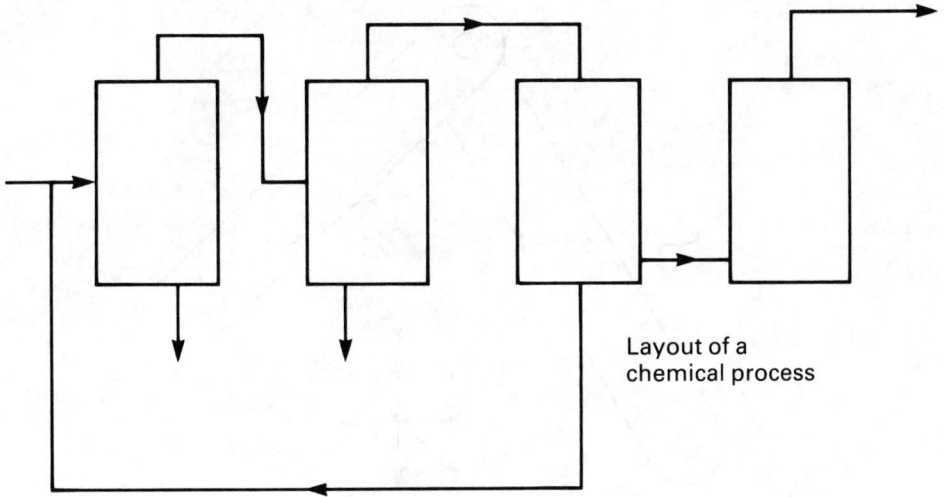

Layout of a
chemical process

Machining process — 4 machines
(MT—machine tool operation)

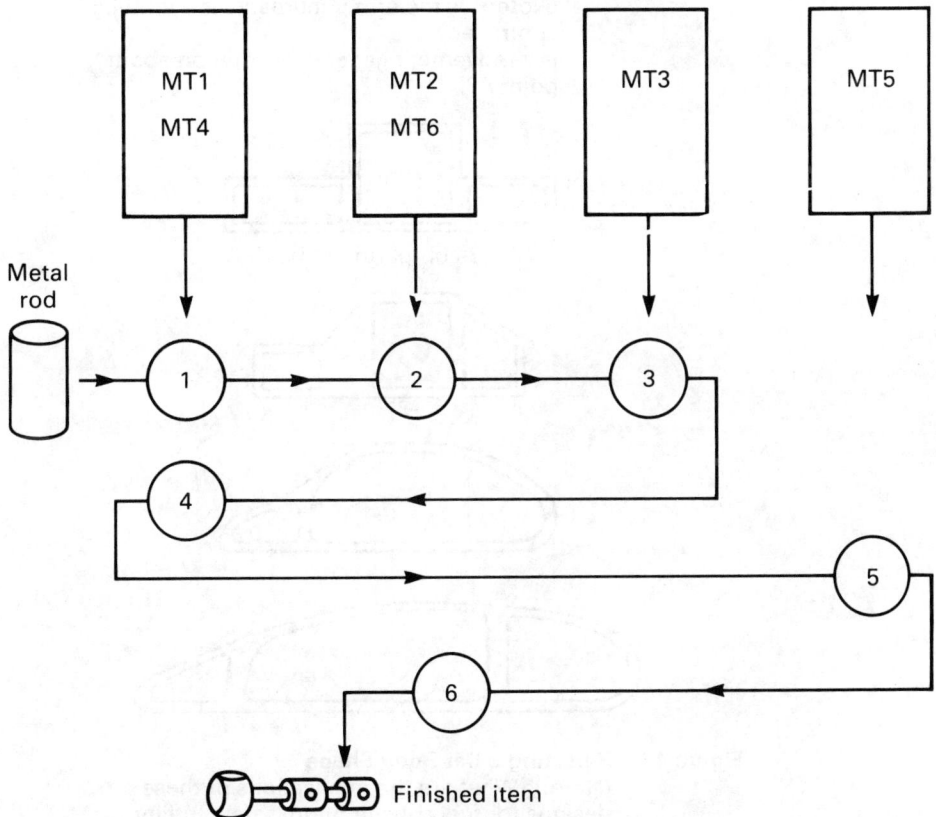

Figure 4.4　Plant Layout and Operating Sequence

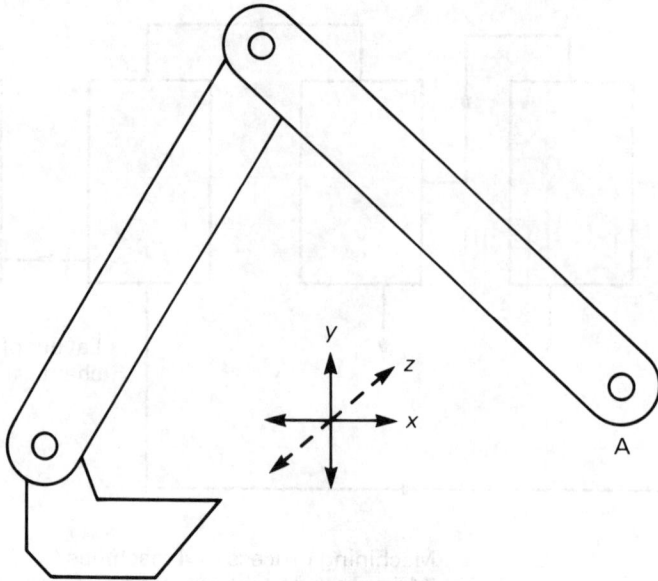

Figure 4.5   Excavator Digging Arm
(Note: Is the arm pneumatic or hydraulic controlled?
Is z-movement general, or rotation about point A?)

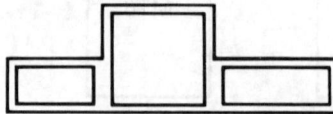

A   3-compartment box!

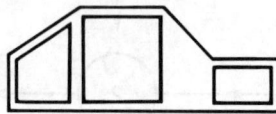

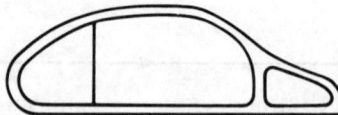

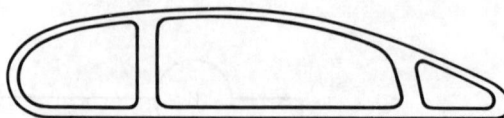

Figure 4.6   Selecting a Car Body Shape
(Note: What are the implications of these designs for fuel consumption, safety, utility, consumer preference, corrosion, repair, mass production, etc., etc.?

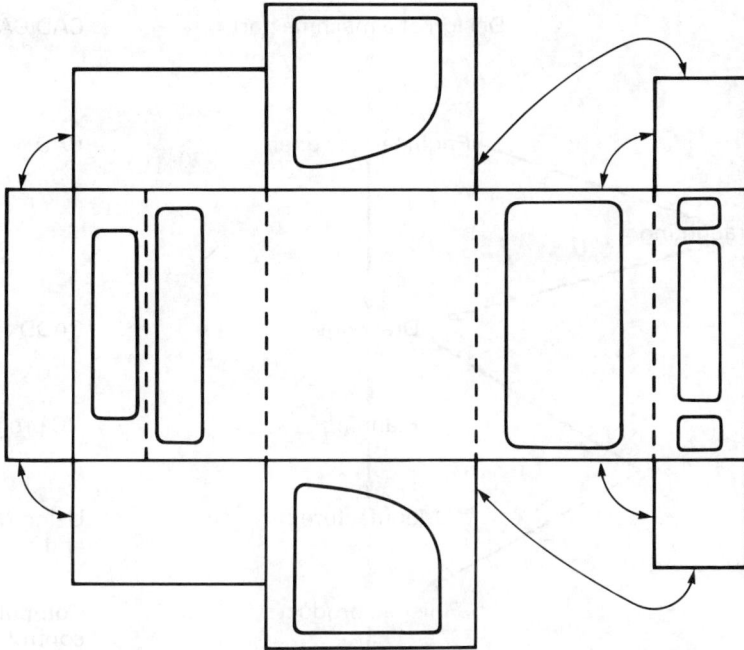

Figure 4.7   Possible Automobile Body Shell

4.6 COMPUTER-AIDED MANUFACTURE

The influence of rapid developments in computer-aided manufacture (CAM) has been most significant in the areas of manufacturing and assembly. The earliest applications of CAM involved numerical control (NC) of machine tools. Pre-recorded, coded information and instructions were usually stored on punched paper tape and used to control the operations of a tooling machine, welding machine, etc. Significant improvements in speed, efficiency and reliability were achieved, especially when many functions were to be carried out in a programmed sequence using multiple tools.

   The importance of CAM now extends beyond NC and it is being utilised in other areas such as process planning, robotics and factory management. The division between CAD and CAM has gradually narrowed as new developments have emerged and as attempts have been made to integrate both functions. It is now possible to obtain a fully integrated CAD/CAM system where all the functions are interfaced with a common data-base. Such a system is sometimes referred to as computer-aided engineering (CAE). Figure 4.8 illustrates the stages in the design of a machine part from inception to production; also shown are the CAD/CAM features that can be included in an integrated system.

Design of a machine part   CAD/CAM activities

Engineer/designer   CAD

Draughting

Drawings   CADD

Planning   NC programming

Manufacture   Using NC/robotics
                                            and microprocessors

Finished product   Computer-aided quality
                                            control, measurement
                                            and testing

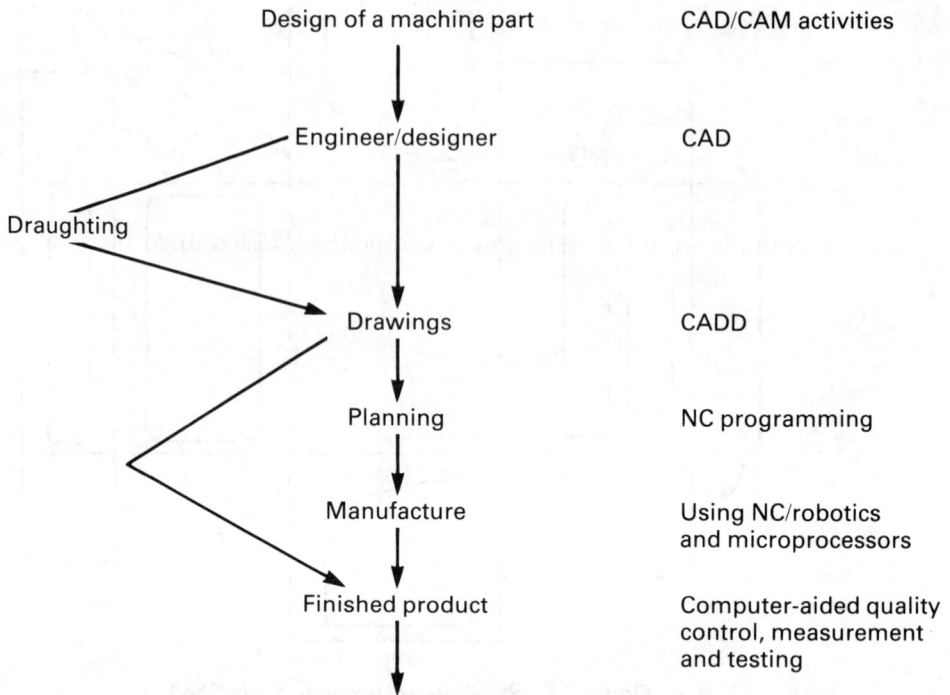

Figure 4.8   CAD/CAM Activities for the Design of a Machine Part

Instructions for NC are usually written in either the APT or COMPACT II language. The use of paper punch tape can be eliminated by storing the operating instructions in a mini- or microcomputer, and then connecting this computer directly to the machine tool. This is known as **computer numerical control** (CNC). By using time-sharing, several machine tools can be controlled directly from a mainframe computer. This is known as **direct numerical control** (DNC). Advances in CAM and NC have been achieved by using computer simulation to check the operation of a machine tool program, rather than actual operation and subsequent adjustment. A computer can also be used to produce NC instructions directly from the geometric modelling data-base, although this is usually restricted to flat parts, symmetric shapes or special parts.

Hierarchical CAM systems have been developed where a mainframe computer controls the operation of a set of microcomputers, each of which controls the action of a machine tool.

An important extension of CAM is the development of industrial robots. A robot consists of a programmed manipulator arm with grippers that move and position materials, parts, tools or other objects. Robots may be simple pick-and-place machines with limited movement, or may be controlled by a servomechanism providing more versatility by use of a programmable controller (the 'teach' mode). There are even computer-controlled 'smart' robots. Robots are used mainly for tedious or dangerous tasks such as heavy loading, handling toxic chemicals or checking for terrorist bombs. However, advantages and applications continue to be found, e.g. welding, parts assembly, fabric cutting, material handling, etc. The automobile industry in particular has adopted robotics for spot welding and assembly functions.

Research studies are now attempting to produce robots with an increasing amount of artificial intelligence. This is not in the form of consciousness or creativity, but a problem-solving ability using sensory input. An example would be selection of components based upon their shape. The development of a robot with television camera 'eyes', so that the image received can be compared with that stored in a computer memory, is considered the most promising application.

The introduction of CAM systems into a production process should improve the productivity of that operation. There are still improvements to be made in the integration between CAD and CAM systems, and in the sophistication of CAM compared with CAD.

* * *

Exercises

1 Comment on the following statement: 'The use of robots in manufacturing processes will create more jobs than it will eliminate'. (NB First considerations might be increased productivity and higher profits.)

2 Prepare sketches for the design of an industrial robot to be used for specific tasks.

3 List the possible control features to be incorporated in a machine required to sort and assemble several components.

4 Sketch a component with several features which is to be machined from a single cylinder of metal. Outline the machining processes, the tools and the sequence of operations to be used. Consider the use of a CAM system in this situation.

* * *

4.7 SUMMARY

The use of computers and the development of computer-aided design techniques have resulted in significant reductions in design project costs, working hours, inconsistencies and mistakes. Major technological advances in computer technology, reductions in computer costs and improvements in computer 'power' indicate that the uses of CAD will continue to increase.

The computer is a tool to be used by the designer and it should only be used for tasks where it will result in higher efficiency. It is important that CAD becomes integrated within the activities of a design process and is accepted by all persons in the design team.

All computer applications, including CAD, CAM, NC and any other method, should be compatible with each other and with all other aspects of the design, manufacture and production process. An appropriate CAD/CAM system should provide the opportunity to reduce costs and improve profits. However, an inappropriate system will have the opposite effect. It is therefore important that the designer is familiar with CAD/CAM applications and is capable of making recommendations relating to their purchase and subsequent use.

KEYWORDS

(Also refer to Section 4.1.)
64K memory
software
hardware
program
core store
multi-access system
remote access
programming
programming language
program language compiler
analog computer
conversational program
program data
hard copy
interactive computer graphics
keyboard

light pen
digitiser
refresh graphics display
raster scan display
digital incremental plotter
on-line/off-line modes
flat bed plotter
electrostatic plotter
finite element methods
turnkey system
concept surfacing
animation
process planning
robotics
computer simulation
geometric modelling
hierarchical CAM systems

BIBLIOGRAPHY

Adey, R.A., and Brebbia, C.A., *Basic Computational Techniques for Engineers*, Pentech Press Ltd, Plymouth (1983).
As per title.
Besant, C.B., *Computer-aided Design and Manufacture*, 2nd Ed., Ellis Horwood Ltd, Chichester (1983).
An excellent book providing an introduction to the topic, later material and examples in depth. Biased towards mechanical engineering.
Brebbia, C.A., and Ferrante, A.J., *Computational Methods for the Solution of Engineering Problems*, Pentech Press Ltd, Plymouth (1978).
As per title.
Goetsch, D.L., *Introduction to Computer-aided Drafting*, Prentice-Hall Inc., New Jersey (1983).
Excellent introduction, many and varied examples, visual presentation.
Hodge, B., *Computers for Engineers*, McGraw-Hill Book Co., New York (1969).
A useful introduction despite its age.

Krouse, J.K., *What Every Engineer Should Know About Computer-aided Design and Computer-aided Manufacturing: The CAD/CAM Revolution*, Marcel Dekker Inc., New York (1982).
This is volume 10 in the 'what every engineer should know' series which I highly recommend to engineering students. This volume is excellent, a good introduction and many industrial applications, also visual material courtesy of many companies and institutions.

Owen, D.R.J., and Hinton, E., *A Simple Guide to Finite Elements*, Pineridge Press Ltd, Swansea (1980).
As title—simple; includes applications and a computer program for a particular problem.

Roberts, S.K., *Industrial Design with Microcomputers*, Prentice-Hall Inc., New Jersey (1982).
A comprehensive coverage, more suitable for systems engineers.

Ryan, D.L., *Computer-aided Graphics and Design*, Marcel Dekker Inc., New York (1979).
Mainly for graphics and draughting applications.

Ryder, R.A., *The Engineer's Computer Handbook*, Thomas Telford Ltd, London (1980).
A good introduction, concerned with computers and computing, rather than design or application to particular engineering problems.

Smith, W.A. (Ed.), *A Guide to CADCAM*, The Institution of Production Engineers in association with the Numerical Engineering Society, London (1983).
An excellent introductory text emphasising mechanical and production engineering applications.

Voisinet, D.D., *Introduction to Computer-aided Drafting*, McGraw-Hill Inc., New York (1983).
An excellent introduction, but *not* CAD.

Warn, J.R.W., *Concise Chemical Thermodynamics* Van Nostrand Reinhold Company Ltd, London (1969).
Reference source for the exercises at the end of Section 4.3.

Reports, Papers and Conference Proceedings

Microprocessors in Industry—Selected Papers, The National Computing Centre Ltd, Manchester (1981).
A good introduction to a variety of topics.

CAD in Medium Sized and Small Industries, North-Holland Publishing Co., Amsterdam (1981). Proceedings of the First European Conference 'CAD in Medium Sized and Small Industries', MICAD, Paris, 23–26 September 1980.
As per title, many papers, mainly applications and installation of computer systems. Good background (selected) reading.

CAD 82, Butterworth and Co. (Publishers) Ltd, Guildford (1982). 5th International Conference and Exhibition on 'Computers in Design Engineering' at Brighton Metropole, Sussex, UK, 30 March–1 April 1982.
  With such a wide range of topics there must be a few papers of interest!

Managing Computer Aided Design, Papers from the Conference at the Institution of Mechanical Engineers, London, 19 November 1980.
  Emphasising industrial applications of CAD.

Making a Start in CAD, IPC Business Press Ltd, Sutton (1982). A collection of papers originally published in *Engineering Materials and Design* magazine.
  A good place for the novice to begin.

Information Technology and *Computer Aided Design and Manufacture,* Her Majesty's Stationery Office, London (1980).
  Reports from the Advisory Council for Applied Research and Development.

Journals and Magazines

The growth rate of publications concerned with computers is probably higher than in any other field. The following are just an example of journals related to the use of computers in engineering in general and CAD in particular.

CADCAM International, EMAP Business and Computer Publications Ltd, London.

CAD Computer-Aided Design, Butterworth Scientific Ltd, Guildford.

Engineering Computers—Hardware, Software and Commonsense, Innopress Ltd, Horton Kirby, Kent.

Many other journals now regularly publish articles and papers concerning computers in engineering and CAD. The journals listed in the bibliography for Chapter 1 will be useful sources of information for recent applications of computers in industrial situations, especially those journals dealing with engineering and design in general. Journals with a mechanical engineering bias are also a useful reference source because of the early involvement of that field in CAM and NC and later in CAD.

Company Literature

Many private companies now exist that can provide complete CAD, CAM and NC packages, including hardware and software for standard applications, available 'off-the-shelf' as turnkey systems. These companies also provide design and consultancy services (at a fee) for other non-standard situations. The literature produced by such organisations is often extensive, well-presented and a useful source of background reading; it is also often available free upon request. Such companies usually advertise in the computer and CAD journals and in engineering and design journals, and they are listed in buyers' guides and the business section of the telephone directory.

5

Materials' Selection

Chapter Objectives

To appreciate that:

1 materials' selection requires consideration of several factors including technical suitability;

2 ease of fabrication, use of an appropriate joining method, economics, etc., must all be carefully assessed before a final material choice is made;

3 the technically 'best' material will not necessarily be chosen if other factors are unsatisfactory;

4 'trade-off' or compromise is often required to obtain a suitable product that satisfies all the requirements.

* * *

Exercises

1 Select a common household item, e.g. can opener, spoon, chair, door handle, ball point pen. Identify the component materials. What other materials could be used? Which materials would you choose and why?

2 In the specification of an engineering component, which requirements will influence the choice of materials?

3 Which factors do you think are important when selecting particular materials?
   Compile a list of factors when you have finished studying this chapter, and compare it with your answer to this question now.

4 List different categories of materials, with examples of each, to assist in the selection process.

5 Which material properties are desirable for particular common working conditions such as tensile loading, atmospheric exposure, vibration, etc.?

6 For some everyday items such as a washing machine, a bicycle or a child's toy, decide whether the materials used represent the technical 'best' that is available. If not, why not?

* * *

5.1 ENGINEERING MATERIALS

Every aspect of our lives and our society brings us into intimate contact with a wide range of different materials. These may have different forms and different uses, but they will influence our comfort, safety, progress and our survival. In contrast to the global view of the importance of materials and materials' selection, the task facing the engineer or designer appears quite straightforward. However, as emphasised in Chapter 3, engineering requires decision-making based on compromise. In this situation, a poor or incorrect choice can result in an unsatisfactory design, and in some cases may lead to disaster.

Although there are approximately 100 known, naturally occurring elements, only 10 of these (including oxygen) are present in either the Earth or its crust in any significant quantity, i.e. more than 1% by mass. The relative abundance of these elements is illustrated in Table 5.1. The figures in Table 5.1 show that the Earth possesses a light crust which is depleted in iron and enriched in oxygen, silicon and aluminium. The availability, cost and future supplies of a material will depend upon the geographical location of its sources, their distance from the surface and problems of extraction and transportation. If one country is the main supplier of a material then political factors may also be important. The cost of a material will depend not only upon the cost of mining but also upon the cost of extracting the metal from the ore, refining and subsequent fabrication or welding. For these processes the temperatures which are required significantly affect the costs incurred; typical temperatures for particular operations are shown in Fig. 5.1.

Table 5.1 Relative Abundance of Elements by Mass (%)

Element	Whole Earth	Earth's crust
Iron	35.0	6.0
Oxygen	30.0	46.0
Silicon	15.0	28.0
Magnesium	13.0	4.0
Nickel	2.5	—
Sulphur	2.0	—
Calcium	1.0	2.8
Aluminium	1.0	8.0
Potassium	—	2.4
Sodium	—	2.2
Others	<1.0	<1.0

One problem for the design engineer when selecting a material is the wide range of available materials, increased by the continual development of new materials. Although we appear to have reached our full complement of

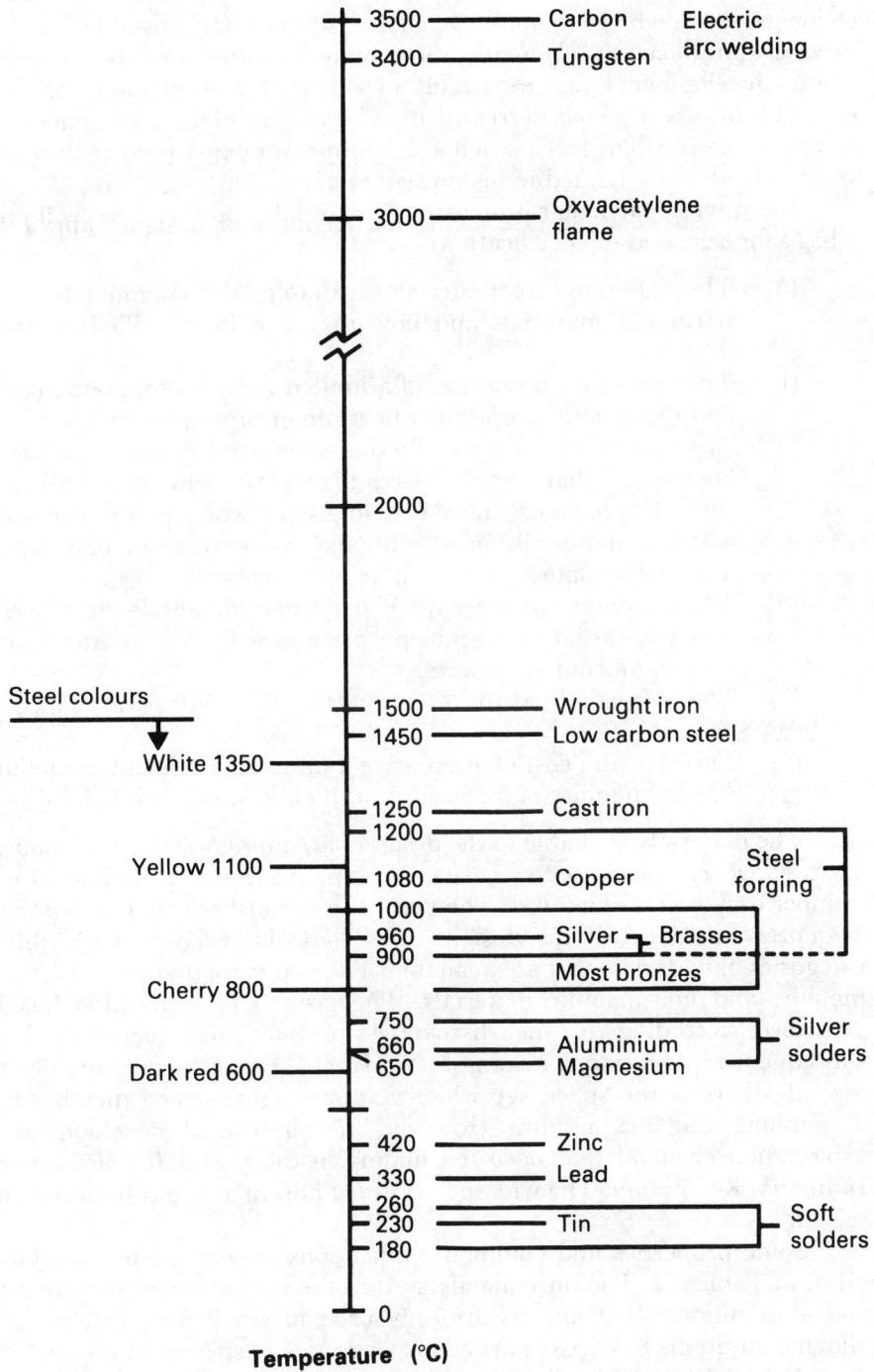

Figure 5.1 Melting Points and Process Temperatures (°C)

elements from which to make a choice (except for unstable, artificially produced, radioactive elements), new materials are developed every day. These developments are the result of different combinations of existing materials or new methods of treatment. The successful design engineer will be aware of many such developments and improvements, even if they do not seem to be directly related to his present work.

Identifying particular properties of materials and their uses will provide a basis for selection or specification.

(a) The full range of materials available, the common features of particular materials and how they can be classified or grouped together.

(b) The important properties of a limited range of common materials, and the general properties of certain groups of materials.

(c) The 'nature' of materials. By this we mean the underlying scientific knowledge that can be used to explain why materials exhibit certain behaviour. This is also useful when predicting ways in which a material's behaviour and properties can be altered, or 'tailored' to satisfy particular requirements.

(d) The common applications of particular materials either because they possess advantageous properties, or their adaptation and use offer significant economies.

(e) The effects of corrosion upon the life and performance of a material.

(f) The ease and cost of alternative joining and fabrication methods.

(g) The availability of materials and their cost per unit volume.

The materials available to the designer are numerous: several thousands exist. However, those used extensively for engineering applications probably number only several hundred. The task of materials' selection will be easier, or more systematic, if the designer has some knowledge of the different categories of materials that are available. The two main divisions are between metallic and non-metallic materials. The growth of civilisation has been described according to the discoveries of particular metals and their subsequent exploitation, for example, the Iron Age and the Bronze Age. More recently there is the space age which requires light, strong metals such as aluminium and magnesium. However, the industrial development and exploitation of metals has occurred mainly in the past 100 years, since the Industrial Revolution. The widespread use of non-metals has been even more recent.

Some properties and common applications of selected pure metals are given in Table 5.2. The pure metals used in the largest quantities are copper and aluminium. Many metals are only used in small amounts, mainly as alloying additions to impart particular properties to special alloys.

Although pure metals may possess particular advantages, e.g. good corrosion resistance (aluminium) or good electrical conductivity (copper), they generally lack the strength required for engineering applications, and alloys are often preferred.

Table 5.2 Properties of Some Metals

Metal	Melting point (°C)	Specific gravity	Specific heat (0–100°C) (J/kg°C)	Coefficient of linear expansion ($\times 10^{+6}$)	Poisson's ratio	Relative electrical conductivity	Relative thermal conductivity at 20°C	Young's Modulus 'E' (N/mm²) ($\times 10^{-3}$)	Tensile strength (N/mm²)	Modulus of rigidity (N/mm²) ($\times 10^{-3}$)	Crystal* structure	Properties and uses
Aluminium	660	2.7	980	24.0	0.35	64	61	70	60	26	FCC	The most widely used of the light metals. Common in the Earth's crust.
Copper	1083	8.9	390	16.6	0.34	100	100	125	160	48	FCC	Used mainly in the electrical industries, because of its high conductivity. Also in bronzes, brasses and cupro-nickel.
Iron (pure)	1535	7.9	450	11.9	0.29	17	15	206	270	82	αBCC < 908°C γFCC 908–1388°C δBCC > 1388°C	A fairly soft metal when pure, but hard and strong when alloyed to form steel.
Lead	327	11.4	130	29.1	0.44	8	9	16	15	5	FCC	A soft, heavy metal.
Magnesium	651	1.7	1030	26.1	0.30	39	39	45	100	17	CPH	Used with aluminium in the light alloys.
Nickel	1458	8.9	450	12.8	0.28	25	17	200	370	78	FCC	Used to toughen steels and many non-ferrous alloys; also for plating.
Tin	232	7.3	220	21.4	0.33	15	16	40	13	19	BC tetragonal	A rather expensive metal. 'Tin cans' carry only a very thin coating of tin on mild steel. Very resistant to corrosion.
Titanium	1660	4.5	450	9.0	0.32	3	4	114	460	43	αCPH < 880°C βBCC > 880°C	Structural uses due to excellent strength – weight ratio, good corrosion resistance, and can be used up to 500°C.
Tungsten (wire)	3410	19.3	140	4.5	0.34	34	39	400	4500	140	BCC	The very high melting point makes it useful for electric lamp filaments. Also used in high-speed and heat-resisting steels.
Zinc	419	7.1	380	33.0	0.25	29	27	90	155	35	CPH	Used widely for galvanising mild steel and as a basis for a group of die-casting alloys. Brasses are copper-zinc alloys.

Some values are approximate because many properties depend upon past treatment

*BCC—body centred cubic; CPH—close packed hexagonal; FCC—face centred cubic.

Table 5.3 Classification of Metals

Ferrous metals	Non-ferrous metals				
	Used in large quantities*	*Used in small quantities*	*Refractory*	*Precious*	*'New' or 'space age'*
Irons: cast, wrought, grey, ductile, malleable	Copper Aluminium Magnesium Nickel	Chromium Molybdenum Vanadium *mainly as alloying additions*	Tungsten Tantalum Molybdenum Columbium	Platinum Gold Silver	Hafnium Zirconium Beryllium Titanium
Plain carbon steel: cast, wrought	Zinc Tin *may be used as pure metal, but are generally the base metal*				
Low alloy carbon steel: cast, wrought	Lead Titanium Beryllium *(solvent) for alloys*				
Special-purpose steels: corrosion resistant (stainless) steels, heat resistant steels, high strength steels, low temperature steels, tool steels					

* Copper and aluminium are the most widely used. Common copper alloys are brasses, bronzes and cupro-nickels. Aluminium alloys may be wrought or cast.

Table 5.4 Classification of Non-metallic Materials

Polymers	Ceramics	Others
Elastomers: natural and synthetic rubbers	Glass	Wood
Thermoplastic polymers	Refractories	Fibres: natural and synthetic
Thermosetting polymers	Cermets	Leather
Others, e.g. silicon-based polymers	Concrete, cement, clay	Rope
	Crystalline materials	Cork
	Abrasives	Asbestos
		Diamond
		Oil and grease

Alloys are combinations of metals present in a solid solution. The major and minor constituents are called the solvent and the solute(s), respectively. The aim is to modify the properties of the base metal (solvent) by the addition of suitable alloying elements in appropriate quantities. An example is the strengthening of copper wire by the addition of aluminium, itself a good conductor of electricity. This means that although the conductivity of the wire may be reduced, thinner cables can be used because of the improved strength.

The categories shown in Tables 5.3 and 5.4 can be used to distinguish between groups of materials with common features.

Alloying is only one way in which the properties of a material can be altered. Other methods such as heat treatment, surface treatment, cold working, hot rolling, etc. can also produce significant changes in the properties of a material. These changes may affect the suitability of a material for its intended use. Careful consideration should be given to processing and manufacturing methods such as welding, shaping, machining, etc. which can cause changes in the material properties during fabrication and may produce effects that were not intended.

Material properties have been frequently mentioned and it will be useful to identify particular categories of properties. Four main categories have been arbitrarily chosen and are listed, with examples, in Table 5.5.

Values of some of these properties can be obtained from data books, handbooks, journals, standards, etc. Others may have to be obtained by experiment.

Two important points need to be made:

(a) A given material has a range of property values, rather than the single value often obtained from a data book. The appropriate value, or range of values, will depend upon the purity of the material and the conditions, e.g. temperature and environment, under which it was determined.

(b) In-service conditions are sometimes subject to change or fluctuation, perhaps in a random or unpredictable manner.

It is not always possible to predict accurately the effects of changes in temperature or the sudden application of a load upon material properties.

Table 5.5 Categories of Materials' Properties

Physical properties	Chemical properties	Mechanical properties	Fabrication properties
Density	Composition	Strength	Casting defects
Melting point	Structure	Hardness	Cold working hardening
Thermal expansion	Chemical reactions	Brittleness	Hot working hardening
Specific heat	Corrosion	Elasticity	Porosity due to powder metallurgy
Electrical conductivity	Electromachining	Plasticity	Surface finish
	Chemical coating	Fatigue	Weldability
		Creep	Machining heat effects
		Impact strength	

However, these effects will have to be considered and appropriate safety factors or safety features incorporated into the design.

A wide range of mechanical tests has been developed and standards are available which, if used correctly, make it possible to obtain reproducible and comparative values. The most common mechanical test is the tensile, or static tension, test. In this test a standard, accurately machined specimen is subjected to an increasing tensile force, as shown diagrammatically in Fig. 5.2. The force and the extension produced are measured at regular intervals. The force and elongation are usually converted to values of stress and strain, and a diagram is obtained which is similar to the examples shown in Fig. 5.3 for particular metal alloys. However, the shape of the curve depends upon the type of material under test. The ideal stress–strain curve is shown in Fig. 5.4, but very few engineering materials actually exhibit (clearly) all the features depicted on this diagram. Various material properties can be calculated from this particular test, hence its common use.

The results obtained from a static test cannot usually be applied to situations where the in-service conditions are very different. When a load fluctuates in magnitude or direction, then an *S–N* curve (stress vs. number of reversals) is required in order to determine the fatigue properties. A typical *S–N* curve is shown in Fig. 5.5; the nature of the metal surface when fracture has occurred is also shown. Results relating to sudden loading or impact can be obtained from an Izod or Charpy test, shown diagrammatically in Fig. 5.6, but accurate predictions require careful interpretation of the data.

There are several tests and machines available for determination of the hardness of a material. Several comparative hardness scales exist, often related to particular testing procedures, e.g. Rockwell and Vickers. Most tests determine the hardness of a material based upon the size of a small indentation made using a known load. Two such tests are illustrated in Fig.

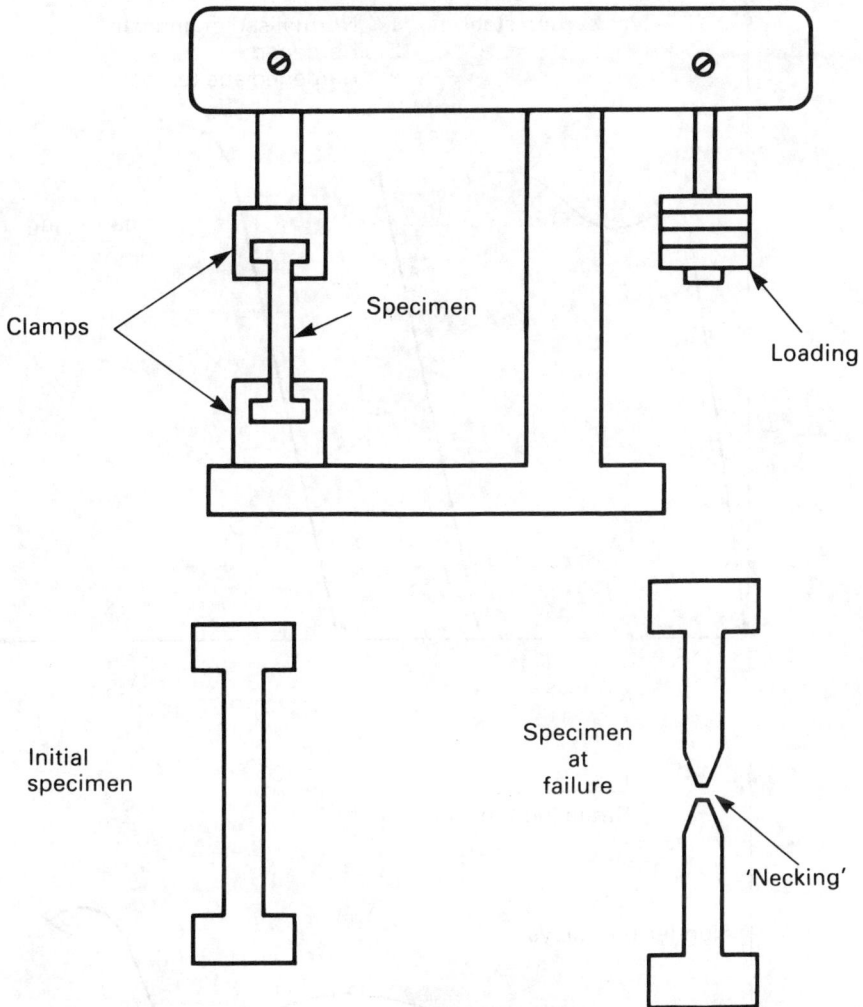

Figure 5.2 Tensile Testing Equipment

5.7. Other tests, that entail measuring the height of rebound of an object, are available.

The influence of temperature must be determined at an early stage in the design. Low temperatures usually increase the strength but reduce the ductility of a material, and extreme brittleness may occur at temperatures only slightly lower than normal. At elevated temperatures, the possibility of creep failure must be investigated. Many other factors affect the material properties such as the environment, vibration, residual stress, thermal stress, corrosion fatigue, hot working, etc. Unfortunately, some of these factors require many hours of testing to establish the likely consequences of changes in the operating conditions.

Carbon steel
1   Normalised or annealed
2   Hardened
3   Hardened and tempered

Stress

Strain (nominal origin)

Non-ferrous alloys

Annealed

Cold worked

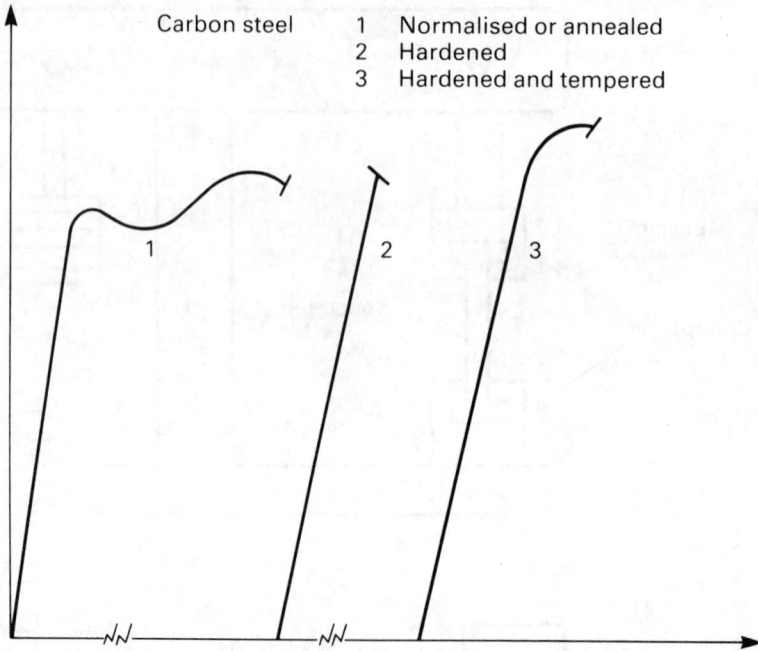

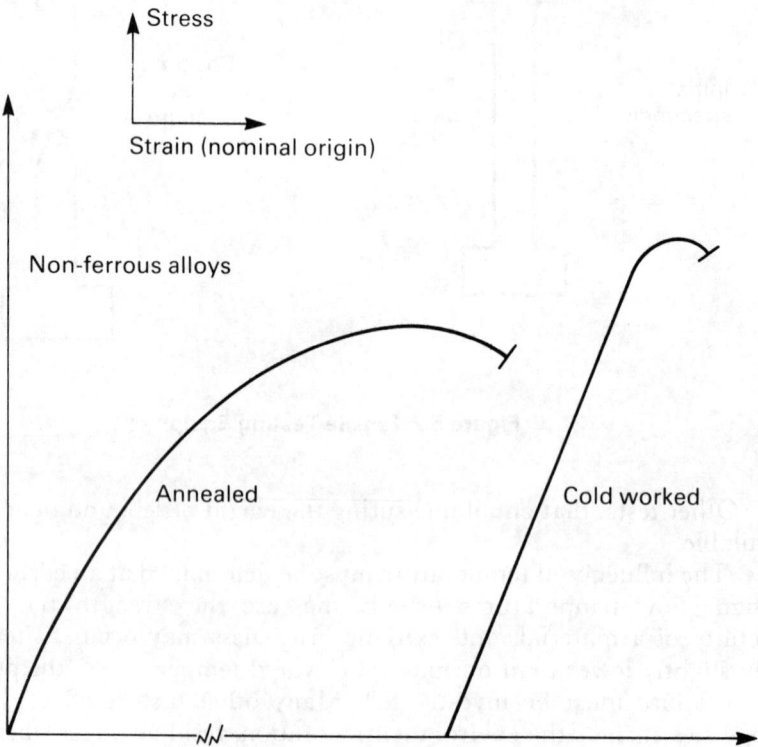

Figure 5.3 Typical Tensile Test Results

Stress

$\left(\dfrac{\text{Force}}{\text{Area}}\right)$

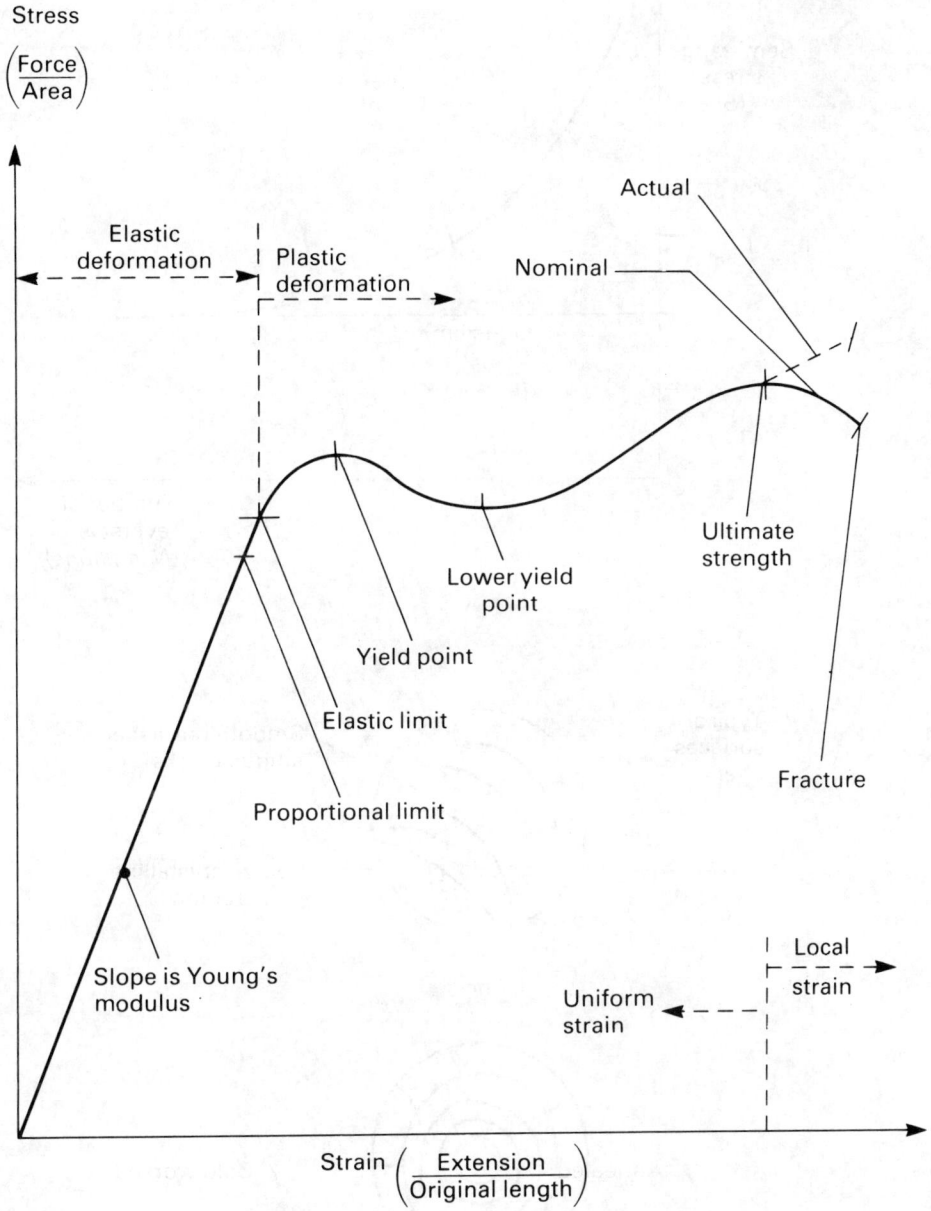

Figure 5.4 Ideal Stress–Strain Curve

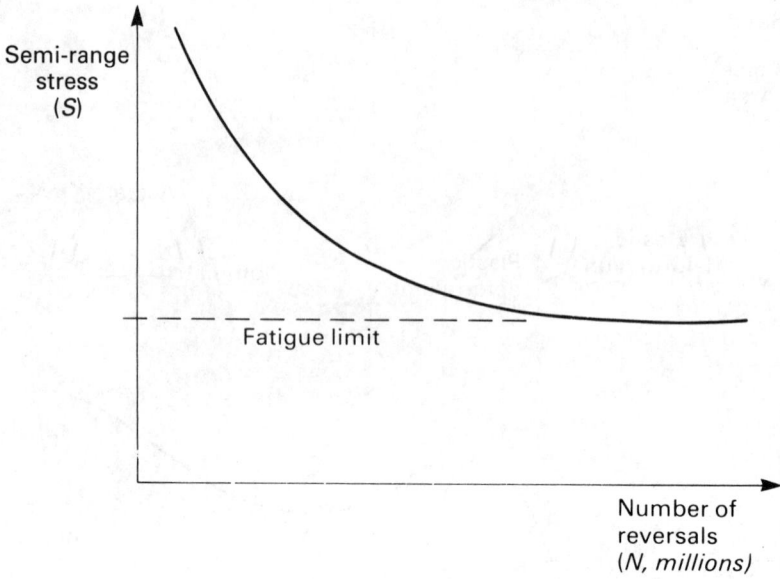

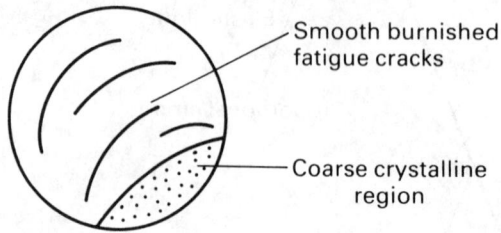

or

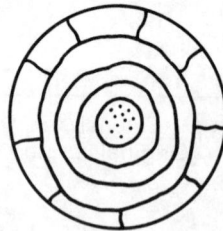

Figure 5.5 Results from Fatigue Testing

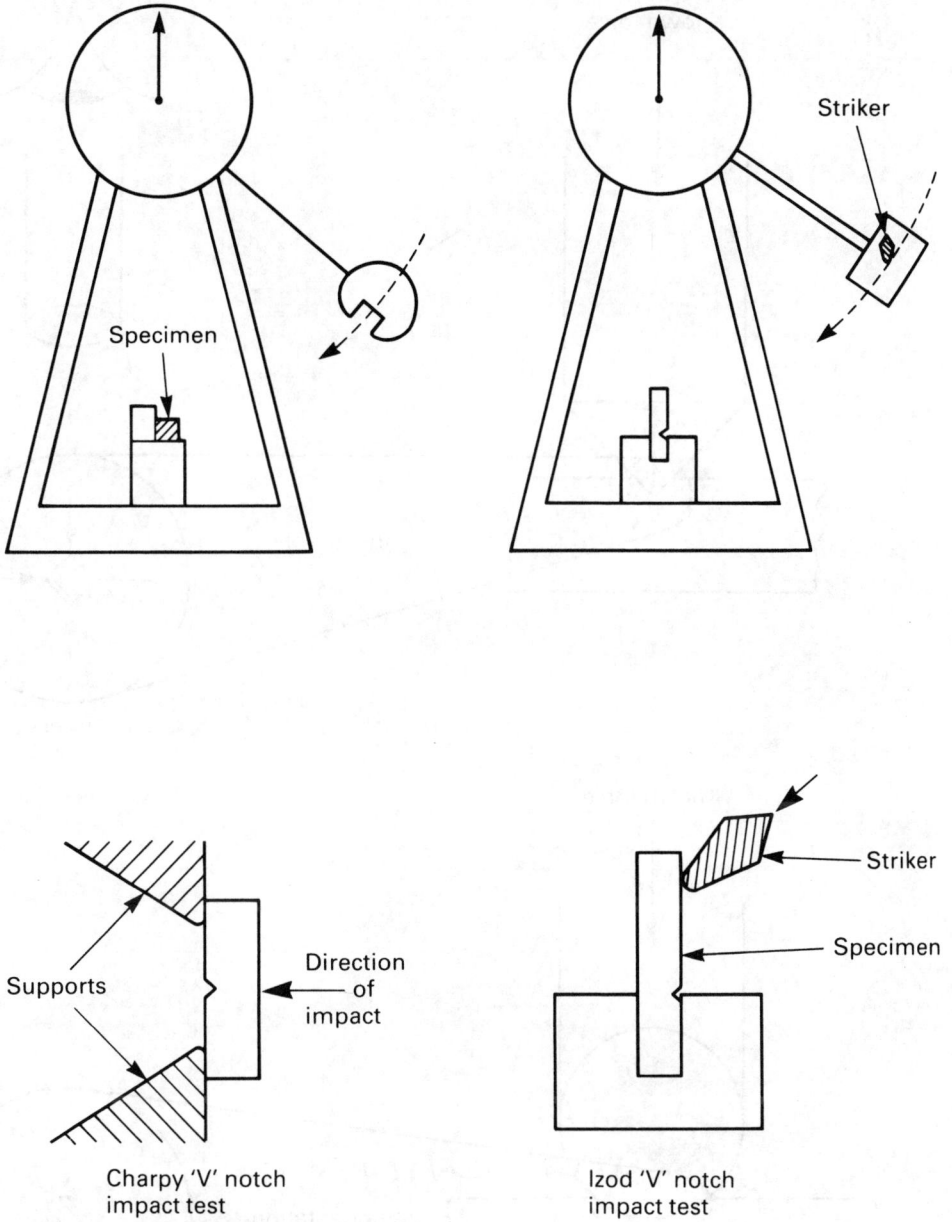

Striker

Specimen

Supports

Direction
of
impact

Striker

Specimen

Charpy 'V' notch
impact test

Izod 'V' notch
impact test

Figure 5.6 Impact Testing of Materials

Known force

Indentation

Vickers

Known force

Indentation

Brinell

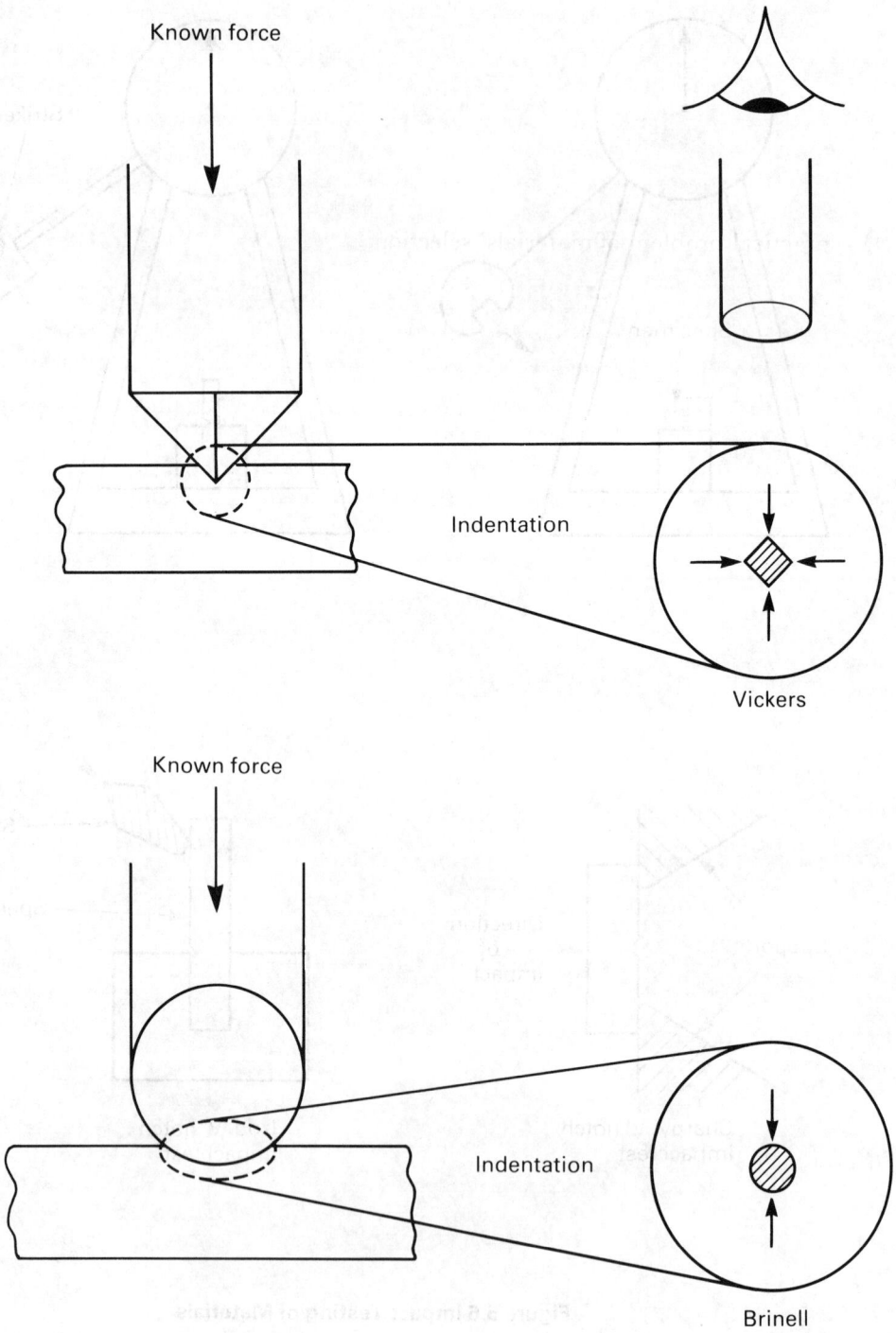

Figure 5.7 Hardness Testing Methods

Comment

An engineering course will contain a unit related to materials. This may be 'Science of Materials', 'Applied Metallurgy' or 'Materials for Engineering'. Often the title reflects the syllabus and, more important, the emphasis that is placed upon the topics that are included. Hopefully such a course will not only teach the 'science' of materials but will also consider the important practical problem of materials' selection.

When students are asked to specify a material for a particular problem the answers are often fairly predictable. The choice of material is usually based upon consideration of only one property, rather than an overall view. The following are typical answers, depending upon the property required:

(a)   For ease of fabrication, cheapness and lightness—a polymer material.
(b)   For lightness and improved strength—aluminium.
(c)   For good electrical or thermal conductivity—copper.
(d)   For low temperatures—a polymer.
(e)   For strength and *any* corrosive environment—stainless steel.

If only materials' selection was really this easy! But students too often have this incorrect, simplistic and dangerous view of the materials available and used by the engineer. Polymers are often seen as the immediate choice if cheapness is an important consideration. Stainless steel is considered the panacea for all problems!

Fortunately, many courses present a more realistic view of engineering materials to the students. This is only possible where the emphasis of the course teaching is placed firmly on the technology and applications of engineering materials. Each separate topic, whether it is corrosion or welding or polymers, should allow time for the students themselves to consider practical problems. Only by consulting data books, standards, etc., will the student appreciate that, like many aspects of engineering and project work, materials' selection is mainly an exercise in compromise and property 'trade-off'. Information and knowledge should not be considered attributes in themselves, but rather as useful tools to be used in design problems.

* * *

Exercises

Metals and plastics are important engineering materials but large quantities of other common materials such as concrete, wood and glass are also used by engineers.

1 Determine the annual production or usage of these materials, and compare these figures with comparable data for some common metals and polymers.
   (a) How have these figures changed in recent years?
   (b) What conclusions can be drawn from the data?
   (c) What industries, and products, are the main consumers or users of these materials?

2 Phrases such as 'the plastic age', 'the plastic revolution', 'the plastic society', etc. are frequently employed by the media and the layman.
   (a) What are the main industries, products and applications which use polymer materials?
   (b) Which are the more important polymers?
   (c) What are their uses? Why?

3 Obtain statistical data for the production and/or usage of polymers compared with other materials since 1945.

4 Identify the major uses and users of irons and steels. Obtain annual figures for iron and steel production.

5 For which products could polymers replace irons and steels and other metals, if necessary?
   (a) Would this be technically possible?
   (b) Would this be possible when social effects and political pressures are taken into acount?

A useful starting point for ideas and data are the books by Cameron and Darwent listed in the bibliography.

*   *   *

5.2 MATERIALS' SELECTION

The process of selecting a material must ensure that the product specification for strength, hardness, etc. is satisfied. However, there are many other factors that must be taken into account. Some of these will now be discussed.

5.2.1 Cost

If the material selected is too expensive, then the product will not be a commercial success and the project will founder at some stage in its development. The cost is directly related to the quantity of material used, which itself depends upon the specifications of strength, size, etc. Different quantities of alternative materials will be required to satisfy the same specification. Direct comparisons must be made and the most suitable (economical) materials identified. Cost is a prime consideration but it must be determined in conjunction with other factors.

5.2.2 Availability

The cost of a material will depend upon its availability. A material that is in short supply, whether because of technical factors, e.g. zirconium, or the influence of a cartel, e.g. oil, will command a higher market price. However, there are many, many examples of materials that have been considered as stable commodities because of abundant supplies, only to have this situation reversed overnight, often because of political influences.

There are few materials that have consistently commanded a high price; the 'precious metals', gold, silver and platinum, are the exceptions. Materials with limited availability usually have their price advantages reduced (eventually) by one or more of the following factors:

(a) substitution by cheaper, acceptable materials;
(b) discovery of new sources;
(c) exploitation of sources previously considered uneconomic;
(d) redundancy of the products or applications that are the main uses for the material;
(e) recycling of scrap materials;
(f) political influences.

* * *

Exercise

Copper possesses the advantageous property of excellent electrical conductivity, and is traditionally considered to be an expensive and scarce material. Obtain statistical data for the last 10 or 20 years, showing the fluctuations in the price, production and usage of 'pure' copper.

Discuss why these variations have occurred and what trends can be predicted for the next 5 years.

* * *

The design engineer must investigate the availability of the materials intended for use in a design, *before* the final specification is prepared. It will be no use specifying a material that is not available either in sufficient quantity or at an economic price. A mass-produced item must be assured of stable, long-term supplies of the raw materials required. If at all possible, materials should be obtained from local sources and suppliers since transportation costs are often significant and also difficult to control. Materials to be obtained from other countries should, if possible, be purchased in bulk for protection against sudden changes in price and availability. This will require a larger initial, or periodic, cash outlay and will also mean that any benefits of subsequent price reductions will be lost.

Finally, consider any constraints that may not be present during the design stage but that may become significant during manufacture or production. These are often legal or political constraints resulting from environmental considerations or government-imposed trade embargos. These constraints are generally expensive, time-consuming and outside the control of the company itself. They may occur after a production process is fully operational or a component is assembled and working, hence making a solution more difficult to achieve. However, engineers who value their position and salary would do well to consider possible ways of countering such problems, should they occur, rather than assuming that future problems and solutions are not their responsibility.

5.2.3 Technical Factors

These factors are related mainly to the mechanical properties of the material mentioned earlier in the chapter. A primary consideration is the strength based upon the loads that the component must withstand, the appropriate margin of safety and any particular or unusual conditions, e.g. high temperatures or fluctuating loading. The second major criterion in a design is usually the weight (or mass!). For handling and transportation, lightness combined with strength is of considerable importance. For aircraft, the strength-to-weight ratio becomes all-important. Reduction in weight can be achieved by designing all components to be uniformly stressed up to the permissible maximum working limit, or using materials possessing a higher strength-to-weight ratio. However, weight is sometimes a required property of a design, either to absorb vibrations or to counteract impact loads.

The in-service conditions detailed in the design specification such as corrosive environments and changes in temperature, are all technical factors. They define the overall design requirements that the design engineer must satisfy. Unusual design features or technical innovations should all be based upon a detailed evaluation of the inherent technical factors to be satisfied.

* * *

Exercise

Consider the simple, basic design of a common item such as a laboratory stool, a car jack or a bicycle, ignoring any requirements for the weight of the product. How is this design modified when practical weight limitations are imposed?

* * *

5.2.4 Joining Methods

The design, shape or size of a component may mean that it is impractical, or impossible, to produce the part in one operation from the desired material. It may be necessary to produce several sub-components, which must subsequently be assembled and secured together. Even if a component can be produced as a single unit, it may still have to be joined to items that already exist. In some cases, the component will have to be fixed in a particular location and position, or secured in order to minimise vibrational effects.

The three main methods of joining materials are:

    (a)   metallurgical processes;
    (b)   mechanical methods;
    (c)   adhesive joining.

Metallurgical joining represents the largest group, both in terms of industrial usage and the number of alternative processes which are available.

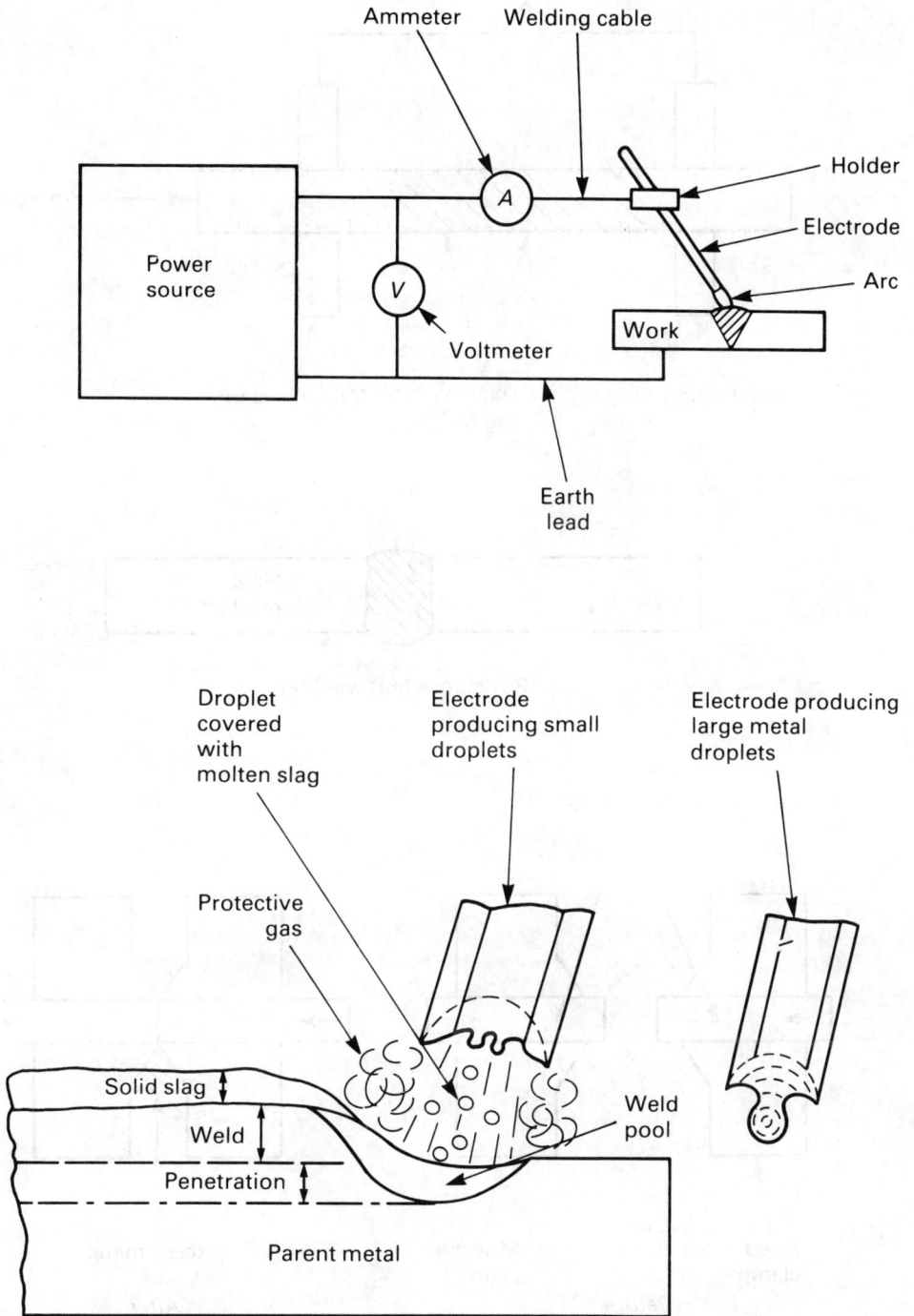

Figure 5.8 Principles of Manual Metal Arc (MMA) Welding

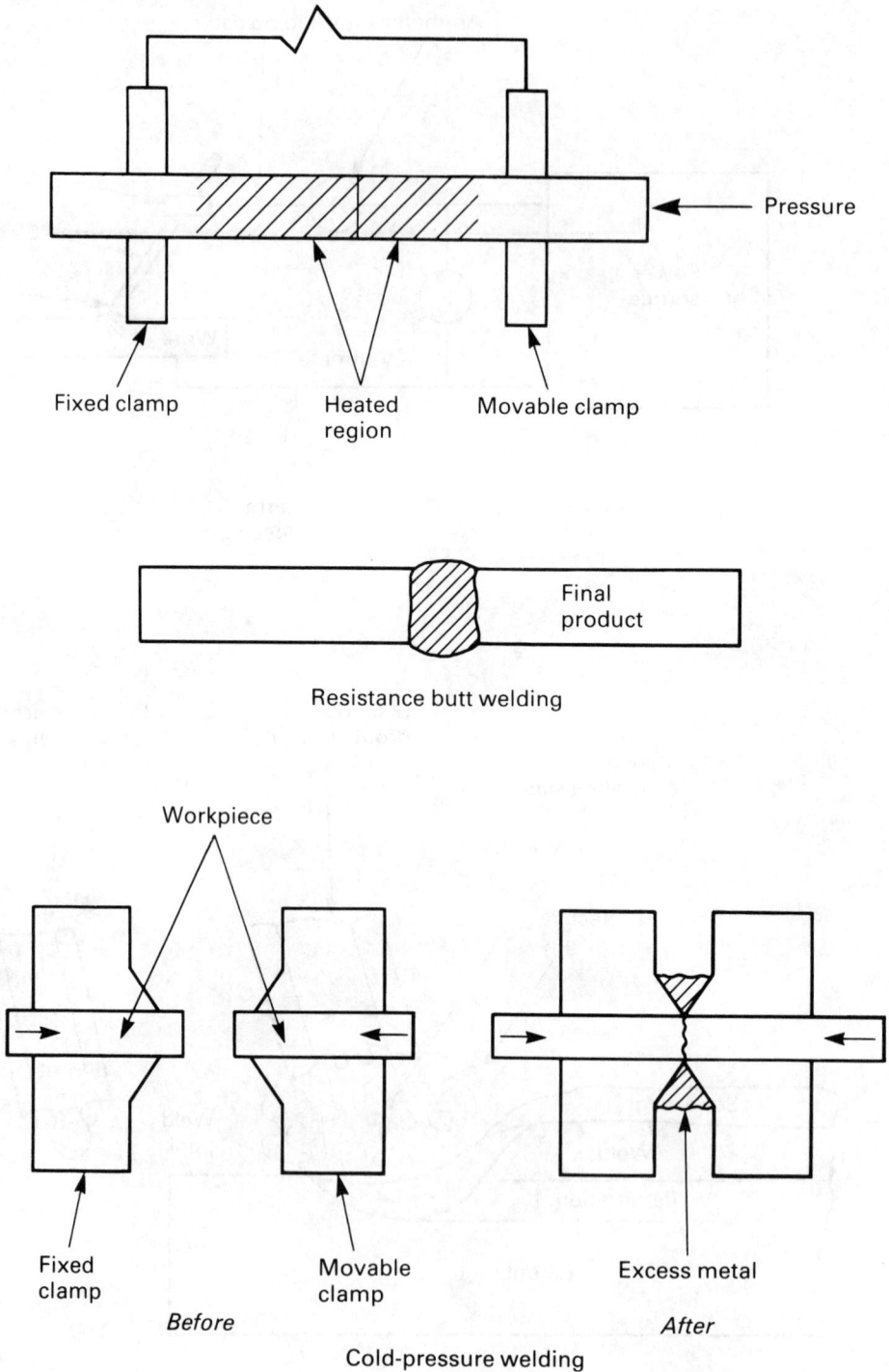

Pressure

Fixed clamp   Heated
                     region   Movable clamp

Final
product

Resistance butt welding

Workpiece

Fixed
clamp   Movable
                     clamp   Excess metal

Before   *After*

Cold-pressure welding

Figure 5.9 Examples of Pressure (Resistance) Welding

The three main metallurgical joining processes are:

(a)   soldering;
(b)   brazing;
(c)   welding.

These processes differ mainly in the temperature used to join the metal(s). Soldering and brazing are lower temperature processes, used for joining metals with low melting points. Soldering is generally performed below 500°C. For brazing, joining is achieved by capillary attraction, and for soldering, by adhesion between the parent metal and a fusible filler-metal alloy (the solder). Welding is a high temperature process usually near or above the melting temperature of the parent metal.

Welding techniques can generally be categorised as either heating, non-pressure (fusion) processes, or as resistance (pressure) welding processes. For heating processes the heat is usually provided either by gas combustion or an electric arc between the work and an electrode (or two electrodes). The filler metal and part of the parent metal are molten. Fluxes or gas shielding, or both, are used to produce and maintain the surfaces that are as clean as possible. The principle of one particular welding method, namely manual metal arc (MMA) welding, is illustrated in Fig. 5.8.

With resistance welding, the fusion temperature is achieved by the resistance to passage of an electric current, and the pieces are then forced together under pressure to produce the join. Examples of resistance butt welding, involving some local heating below the melting point of the metal, and cold-pressure welding are shown in Fig. 5.9.

Spot, stud and seam resistance welding are common industrial techniques. Other specialised methods such as thermit, laser, plasma arc, friction and metallising have been developed in recent years.

Mechanical joining methods have been used for centuries in their basic forms. Riveting is a common industrial operation, where a headed cylindrical piece of metal is heated and passed through a hole in each of the metal pieces to be joined, as shown in Fig. 5.10. The unheaded end is then shaped to form the opposite head, thus providing a seal and a join as the rivet cools and contracts. Many types of mechanical fasteners are available, such as nuts and bolts, self-tapping screws, wing nuts, etc. A selection of such fasteners is shown in Fig. 5.10.

Adhesive bonding is a relatively new process using polymer resins to produce a strong, reliable join. A recent development known as 'weld bond' combines the advantages of both adhesive joining and spot welding at selected positions. Adhesive bonding is often considered a less reliable and less durable joining method. However, this is not the case if the following precautions are observed:

(a)   correct selection of an adhesive for the particular materials to be bonded;
(b)   follow recommended cleaning and preparation instructions for the surfaces;

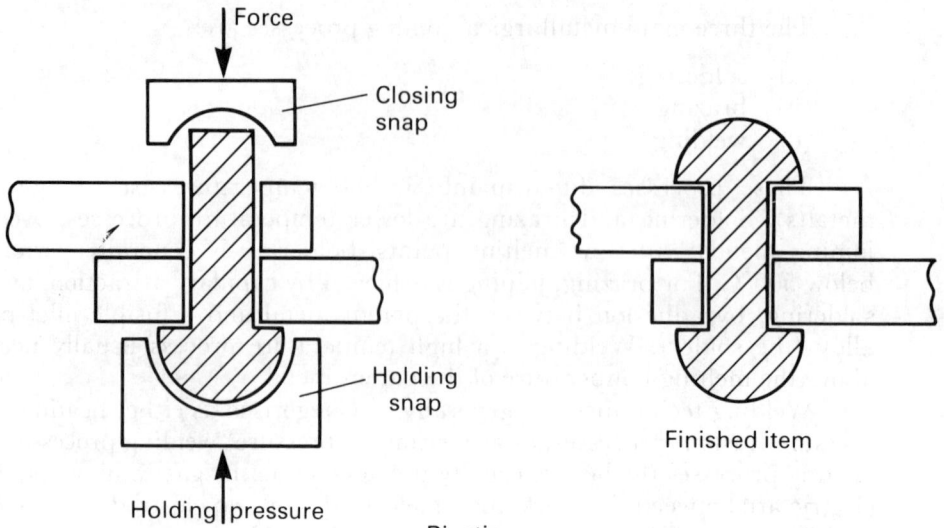

Force

Closing snap

Holding snap

Holding pressure

Finished item

Riveting

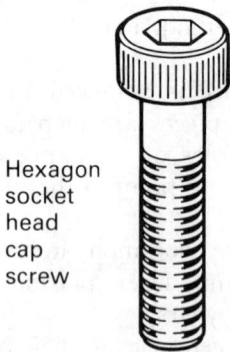

Hexagon
socket
head
cap
screw

Thumb screw

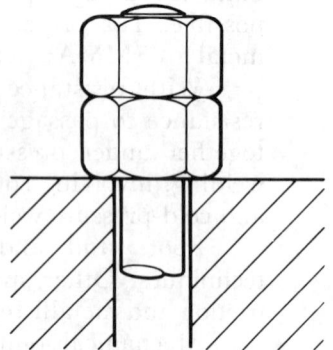

Lock nuts

Tap stud and nut

Die-cast wing nut

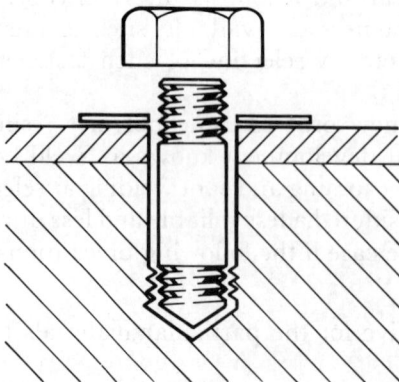

Figure 5.10 Non-welded Joining Methods

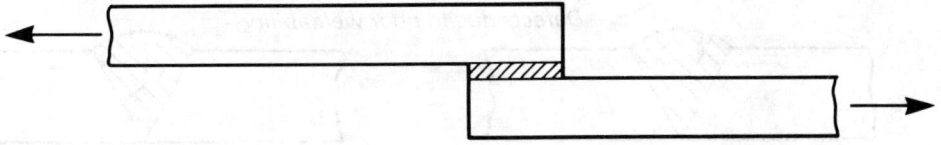

Shear stress

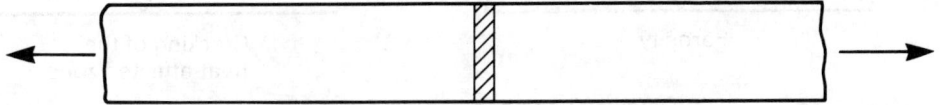

Tensile stress

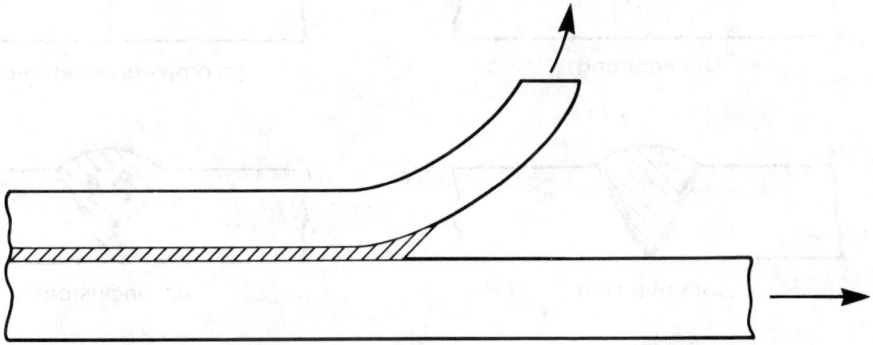

Peel stress

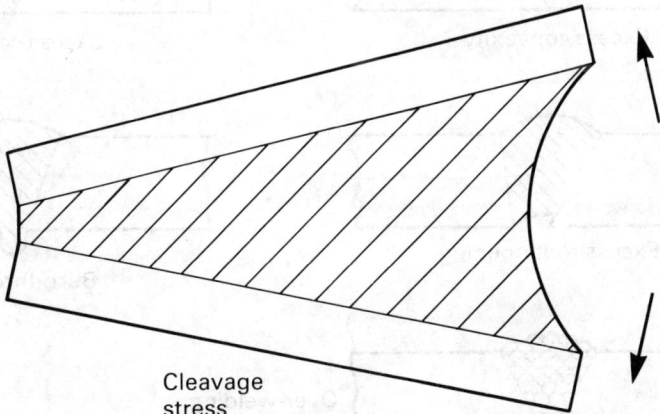

Cleavage
stress

Figure 5.11 Basic Stress Situations with Adhesive Bonding

Defects due to poor weldability

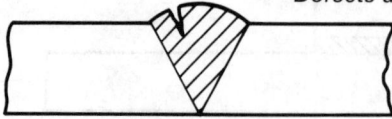

Weld cracking

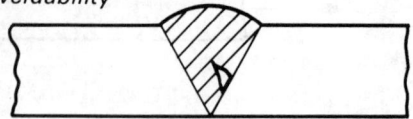

Underbead cracking

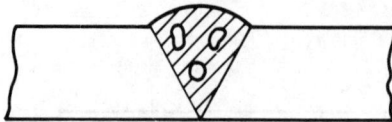

Porosity

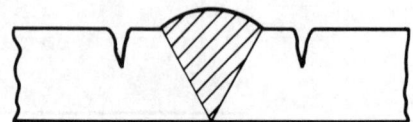

Cracking of the
heat-affected zone

Defects due to poor welding technique or design

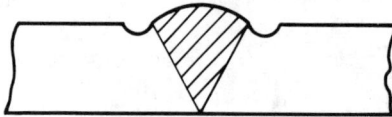

Undercutting

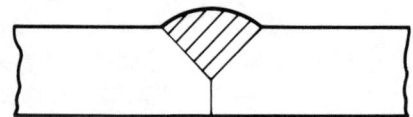

Incomplete penetration

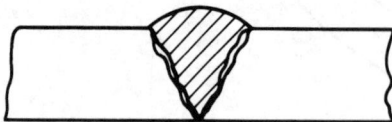

Lack of fusion

Slag inclusions

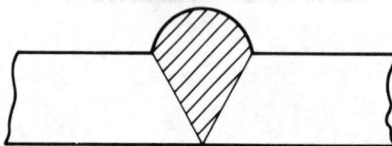

Excess convexity

Excess concavity

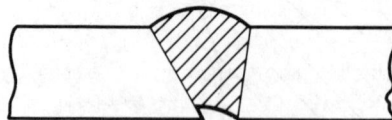

Excess root opening

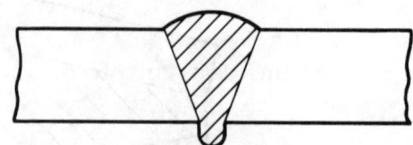

Burn-through

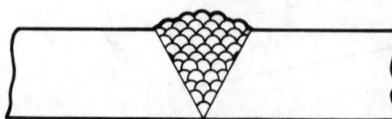

Over-welding

Figure 5.12 Examples of Welding Defects

    (c)   do not use adhesives when the bond is to be subjected to peel or cleavage forces shown in Fig. 5.11;

    (d)   investigate in detail the effects of in-service conditions and the environment on the adhesive to be used.

The choice of a particular joining process will depend upon many factors. Welding should produce a joint that is pressure-tight and as strong as the parent metal, if the process is carried out correctly. Many defects can occur, owing to either poor welding technique, poor design or poor weldability of the parent metal(s). Some welding defects are shown in Fig. 5.12. Techniques are available which can be used to detect welding defects, but these are usually expensive and slow, at least if reliable results are required!

The main disadvantage of metallurgical joining is that the joint is permanent, or should be! If a structure may need to be dismantled then mechanical joining will be the preferred process. However, access may be required from both sides of the joined structure and it may not be possible to guarantee a 100% effective seal. If failure of the assembled unit may occur due to excess pressure, then repair may be possible if mechanical fastening is used. Obviously this situation should be avoided by correct design. When selecting a joining process an important consideration will be the costs incurred in the alternative methods. Costs will include labour (if automation is not possible), materials, equipment, electricity, etc. The designer will be seeking the most efficient process at the lowest cost, conflicting aims requiring an optimal solution.

There are several excellent books and very many acceptable books available, which describe all aspects of metallurgical joining, and welding in particular. Books giving details of mechanical and adhesive joining are less numerous and some of these books, although good, are more suitable for students training to be technicians and workshop or machine personnel. There are some references and comments in the bibliography for this chapter.

5.2.5 Fabrication

The production of a component or a product will necessarily involve some aspects of fabrication. The fabrication method(s) selected will depend upon the materials used, the shape and size of the component and its intended use, the costs involved, the time available, the number of parts to be produced, the original product specification, etc.

The designer must decide whether to select a material and then specify suitable fabrication methods, or to decide upon the fabrication process (based upon such factors as mass production and the design intricacies) and then identify appropriate materials. A problem reminiscent of the chicken and the egg!

The answer is by now familiar—compromise.

Some materials cannot be used because they cannot be easily fabricated, and some fabrication techniques cannot be used with certain materials owing to inappropriate physical properties. The probability of success of a design product will often be significantly improved if fabrication methods, and the associated production stages, are considered in some detail during the preliminary and detailed design stages.

An innovative and creative design that cannot be economically fabricated and produced, does not deserve the accolades 'innovative or creative'—at least in the industrial and commercial sense.

The techniques and operations commonly called fabrication methods can be divided into two categories. These are methods that produce a semi-finished product requiring extra working, and methods producing a finished (final) item. Semi-finished components are generally produced by such techniques as:

(a)   casting;
(b)   hot working or cold working, e.g. rolling, drawing, coining, spinning, extrusion;
(c)   powder metallurgy.

Casting is an old technique and has been used by man since he discovered reasonably pure sources of metals, and began to appreciate the usefulness of fire! It is a slow process and is also labour-intensive. Sand casting is still used, although other casting techniques have been developed that are quicker, easier, require less labour and can produce more intricate and accurate shapes with less likelihood of serious defects. Apart from the points mentioned above, the main disadvantage of casting is that it requires molten metal and, hence, higher heating costs. It is not an economical process for producing large numbers of parts, and the dimensional accuracy of the product is not consistent.

Many hot and cold working methods, e.g. rolling, are very similar in operation, except for the temperature of the metal being shaped. The distinction between hot and cold working is arbitrary, cold usually meaning approximately ambient temperature and hot meaning below the melting temperature of the metal. The main disadvantages of cold working are the adverse effects upon the material's mechanical properties, such as increased brittleness and hardness. Only a limited amount of cold working can be carried out before these effects must be reversed by heat treatment. These problems can often be avoided by choosing a suitable hot working temperature, although care must be taken to ensure that undesirable effects such as ageing shrinkage, precipitation hardening and grain growth do not occur. The hotter a material, the easier it will be to shape, and less working, i.e. mechanical energy, will be required. However, initial heating (thermal energy) costs will have been incurred and cost–benefit calculations will have to be performed.

An example of material shaping by drop forging is shown in Fig. 5.13. The shape to be produced is accurately machined in two sections of a die. The material is placed in, or above, the bottom die and the top die falls onto it, thus forging the required shape. Excess material is forced into a surrounding flash gutter and must be removed (machined) from the component, as shown in Fig. 5.13. If the original material is thin or only small changes in its shape are necessary, then the process may be performed with cold materials. Often the material to be shaped is heated before being subjected to drop forging.

   Two other examples of hot and cold working processes are shown in Fig. 5.14. All commercial wire is produced by cold drawing, even though heating the feed material would reduce the pressures exerted on the die and hence extend its life. Extrusion processes (see Fig. 5.14) usually use heated materials or incorporate a heater element into the equipment itself. Extrusion is a particularly useful fabrication method for polymer materials which become fluid at temperatures lower than for most metals, and are generally weaker materials. The selection of a fabrication process to be used with a polymer material often depends upon the shape to be produced; two common examples are illustrated in Fig. 5.15.

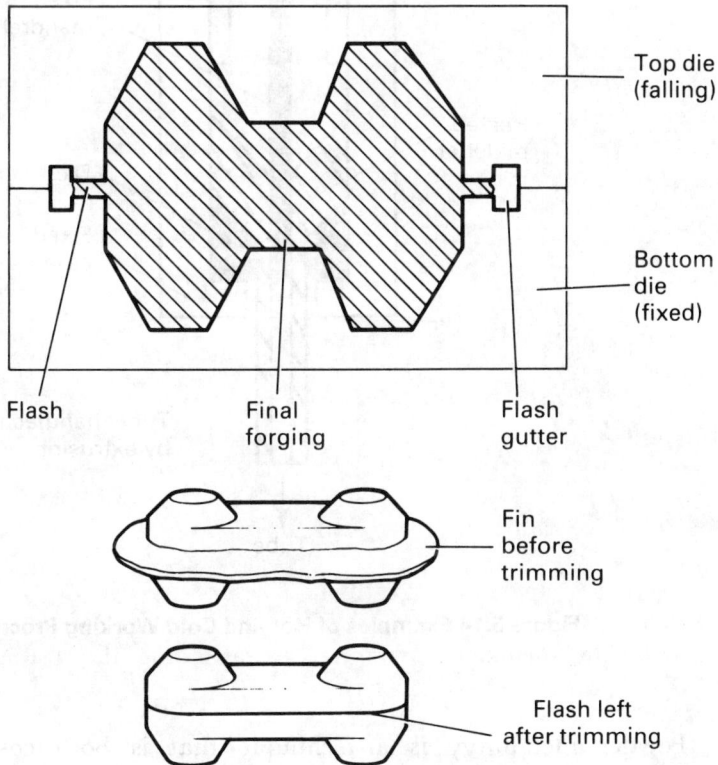

Figure 5.13 Drop Forging Process

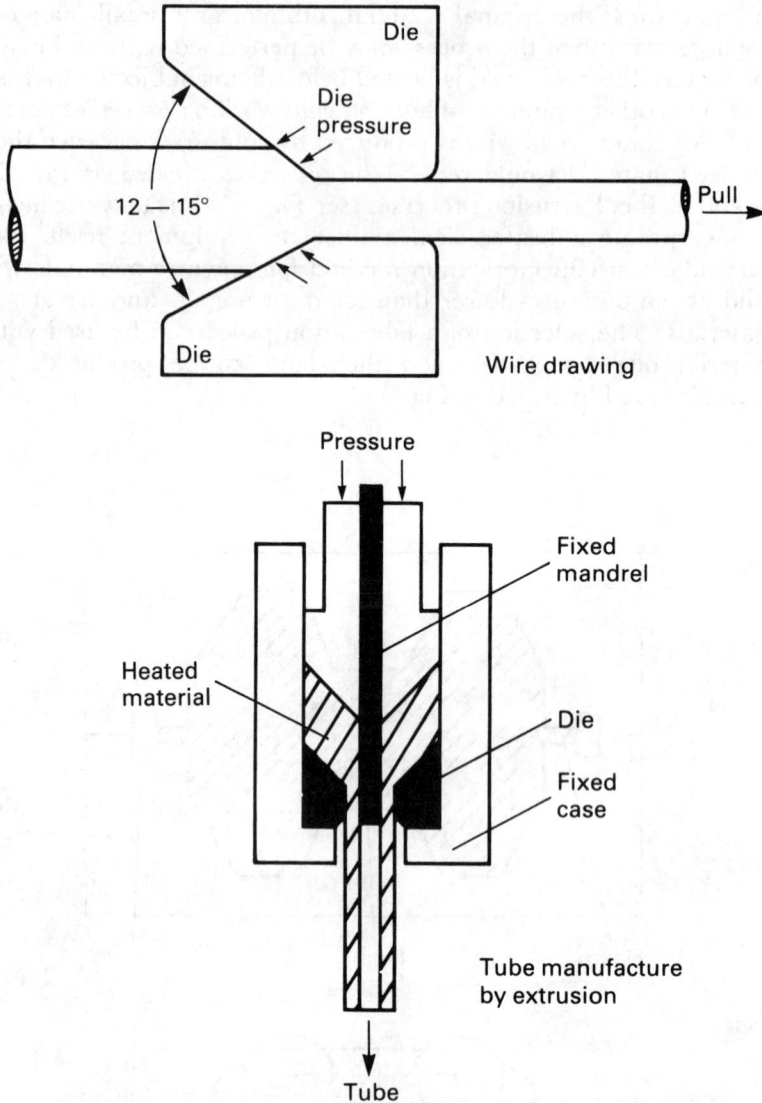

Figure 5.14 Examples of Hot and Cold Working Processes

Powder metallurgy is a technique that is both costly and time-consuming. It is normally only used for materials that cannot be shaped by other methods, because of either their extreme brittleness or hardness. However, it can produce a component with extremely accurate dimensions—this is important, since further working is often impossible. The process is shown in Fig. 5.16 and involves grinding the material to a powder, compacting it in an accurately shaped mould, and then sintering (heating) to

Compression (transfer) moulding

Injection moulding

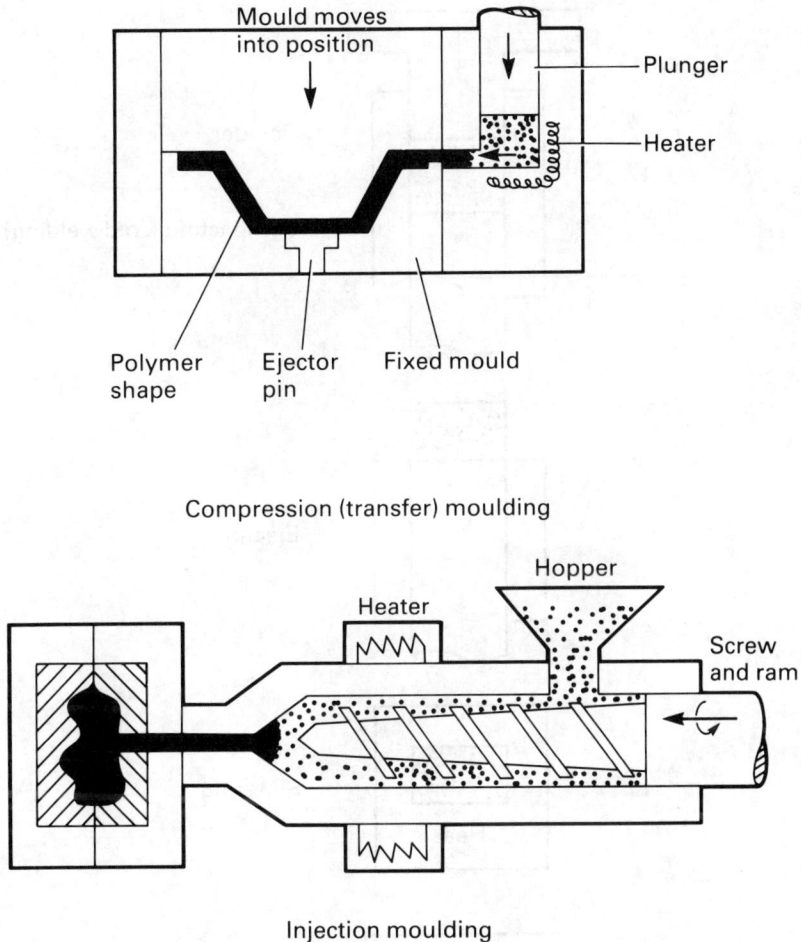

Figure 5.15 Examples of Polymer Fabrication Processes

produce a component of fused material. The product may possess the disadvantage of porosity depending upon whether other materials (fillers) are added, but it is a very useful method for producing hard, brittle, ceramic components.

These fabrication methods, except possibly powder metallurgy, usually produce components that are considered semi-finished. Further work is required to obtain the desired final dimensions (hopefully by removing excess material from an oversized part, rather than the opposite task!), or to achieve the specified surface conditions. Finished products usually require one or more of the following operations to be performed:

(a) machining;
(b) surface preparation;
(c) surface finishing.

Powder

Compacting (cold welding)

Ejection

Sintering

Heat

Final sizing

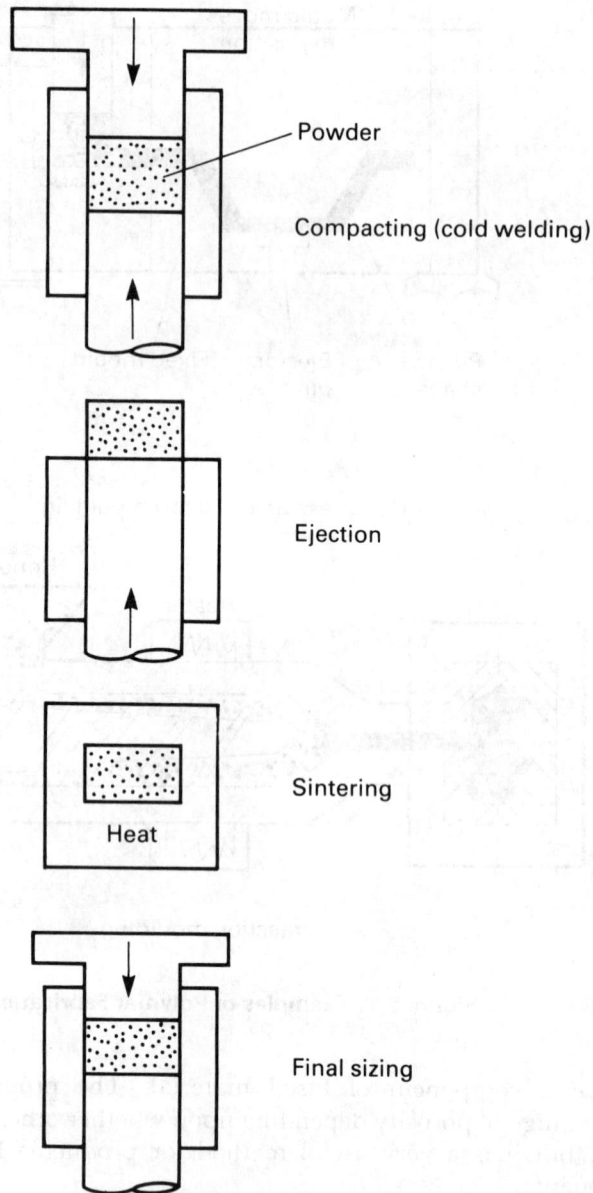

Figure 5.16 Stages in Powder Metallurgy

There is some overlap between these operations; the finishing operations performed on a component will largely be determined by its ultimate use and the product specification. A hot working process may produce a satisfactory final product if tolerances and appearance are not critical.

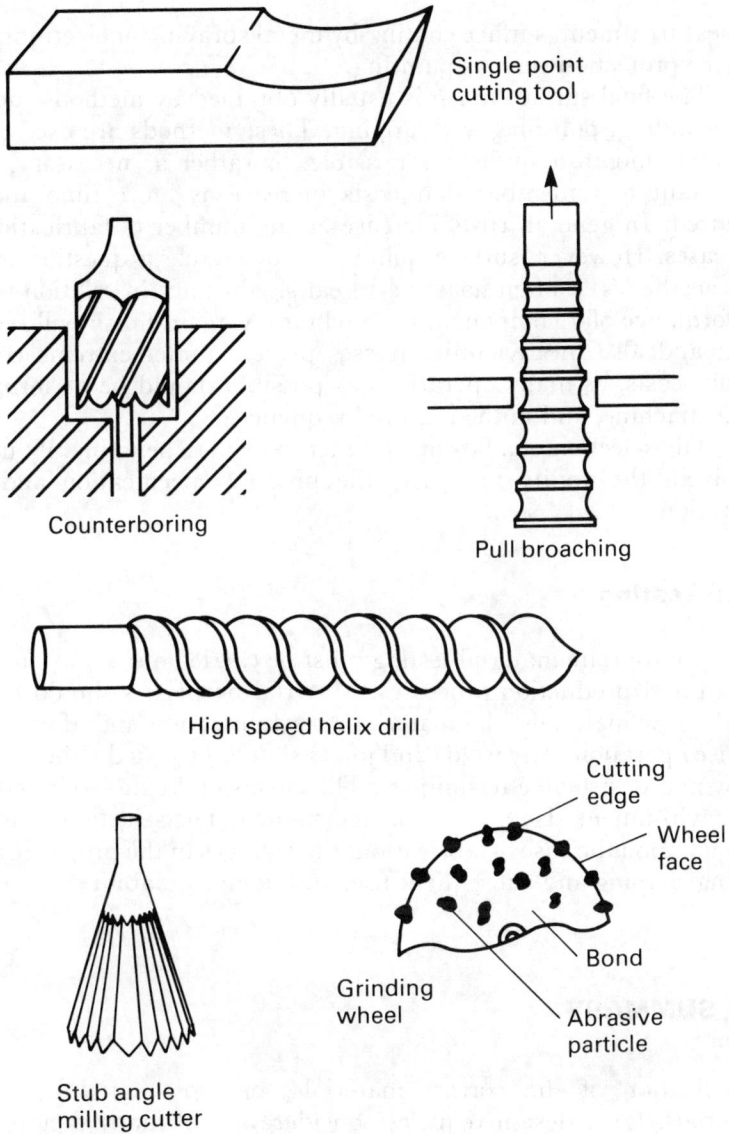

Single point cutting tool

Counterboring

Pull broaching

High speed helix drill

Stub angle milling cutter

Grinding wheel

Cutting edge

Wheel face

Bond

Abrasive particle

Figure 5.17 Machining Tools

Machining also produces a semi-finished product. Some common machine tools are shown in Fig. 5.17. Often a shape produced by sawing, turning, milling, etc., will require additional surface treatment. However, as more accurate and versatile machines and tools become available it becomes easier to obtain an acceptable finished product.

Surface preparation encompasses all operations that change the nature or composition of the surface of the component. This includes surface hardening, case hardening, flame hardening, changes in the microstructure by heat treatment, surface coating by metal spraying or electrodeposition, and surface protection such as painting.

The final surface finish is usually obtained by methods such as honing, fine grinding, polishing and lapping. These methods are used to produce as flat and smooth a surface as possible, or rather as necessary, because it is important to remember that costs increase as more time and labour are required. In general, costs increase as the number of fabrication operations increases. However, surface quality is not merely a question of appearance and aesthetics, it often has a very real significance in relation to the life and performance of a component or machine. A micro-finish will reduce friction, wear and also the dynamic stresses present under extreme conditions. To reduce costs, as many operations as possible should be performed using the same machine, and in the required sequence.

The selection of a fabrication method will depend mainly upon the costs involved, the ability to satisfy the product specification and the ease of operation.

5.2.6 Testing

Appropriate material testing must be carried out at various stages in the design and production processes. Material properties should be determined for the raw materials, the material after fabrication and after a period of in-service operation. Any welds and joints should be tested either under pressure or by non-destructive techniques. The effects of the in-service conditions and the environment also need to be determined. The selection of a material will depend upon the ease of fabrication, the changes in the properties which occur during shaping and subsequent use, and its machinability.

5.3 SUMMARY

Specification of the correct materials, or most suitable or appropriate materials, for a design requires consideration of many factors. Among the most prominent will be scientific knowledge, technical skill, economic assessment, political influences and general adaptability. The final decision will attempt to combine the technical 'best' at the lowest cost, with the minimum of problems due to outside influences. An impossible task, which generally results in a technically acceptable product at an acceptable cost. However, despite the compromises the product specification must still be achieved.

KEYWORDS

element
metals
alloy
solvent
solute
base metal
alloying addition
major constituent
minor constituent
ferrous metals
irons
plain carbon steels
low alloy carbon steels
special-purpose steels
non-ferrous metals
refractory metals
precious metals
'new' or 'space age' metals
non-metals
polymers
elastomers
thermoplastic polymers
thermosetting polymers
ceramics
cermets
physical properties
chemical properties
mechanical properties
fabrication properties
range of property values
in-service conditions
safety factor
safety feature
tensile (static tension) test
tensile force
extension or elongation
stress
strain
load fluctuation

S–N curve
impact
Izod test
Charpy test
corrosion
property 'trade-off'
product specification
cost
availability
technical factors
strength-to-weight ratio
joining methods
metallurgical processes
mechanical methods
adhesive joining
soldering
brazing
welding
fusion welding
pressure welding
electric arc
electrode
parent metal
filler metal
flux
riveting
'weld bond'
welding defects
fabrication
semi-finished products
casting
hot working
cold working
powder metallurgy
finished products
machining
surface preparation
surface finishing
non-destructive testing

BIBLIOGRAPHY

Andrews, D.R., *Soldering, Brazing, Welding and Adhesives,* The Institution of Production Engineers, London (1978).
A good, practical textbook.

Begeman, M.L., and Amstead, B.H., *Manufacturing Processes,* 6th Ed., John Wiley and Sons Inc., New York (1969).
Comprehensive.

Cameron, E.N. (Ed.), *The Mineral Position of the United States. 1975–2000,* Proceedings of a symposium sponsored by the Society of Economic Geologists Foundation Inc., at Minneapolis, Minnesota, November 1972. Published by the University of Wisconsin Press (1973).
Fascinating data and ideas.

Darwent, T. (Ed.), *World Resources, Engineering Solutions,* Proceedings of the third joint conference of the American Society of Civil Engineers and the Institution of Civil Engineers (UK) at Harrogate, Yorkshire, 30 September – 3 October 1975. Published by the Institution of Civil Engineers, London (1976).
A collection of fascinating papers on set themes.

Doyle, L.E., Keyser, C.A., Leach, J.L., Schrader, G.E., and Singer, M.B., *Manufacturing Processes and Materials for Engineers,* 2nd Ed., Prentice-Hall Inc., New Jersey (1969).
Very wide depth and coverage of topics.

Gourd, L.M., *Principles of Welding Technology,* Edward Arnold Ltd, London (1980).
As title, well presented.

Houldcroft, P.T., *Welding Process Technology,* Cambridge University Press, Cambridge (1977).
As title.

Lancaster, J.F., *Metallurgy of Welding,* 3rd Ed., George Allen and Unwin Ltd, London (1980).
As title.

Li. Jberg, R.A., *Processes and Materials of Manufacture,* 2nd Ed., Allyn and Bacon Inc., Boston, Massachusetts (1977).
Comprehensive and in-depth coverage of joining and fabrication.

Lindberg, R.A., and Braton, N.R., *Welding and Other Joining Processes,* Allyn and Bacon Inc., Boston, Massachusetts (1978).
As title, good reference source.

Lipsett, C.H., *Metals Reference Encyclopedia,* Atlas Publishing Co. Ltd, New York (1968).

Little, R.L., *Metalworking Technology,* McGraw-Hill Inc., New York (1977).

Moore, H.D., and Kirby, D.R., *Manufacturing: Materials and Processes,* 3rd Ed., Grid Publishing Inc., Columbus, Ohio (1982).
Excellent comprehensive text.

Neely, J.E., *Practical Metallurgy and Materials for Industry,* John Wiley and Sons Inc., New York (1979).
Excellent visual presentation of material.

Neely, J.E., *Practical Machine Shop*, John Wiley and Sons Inc., New York (1982).
> Excellent visual introduction to workshop tools and practices. Student workbook and instructor's manual also available.

Niebel, B.W., and Draper, A.B., *Product Design and Process Engineering*, McGraw-Hill Inc., New York (1974).
> A good, general textbook, covering a wide range of topics.

Parkin, N., and Flood, C.R., *Welding Craft Practice*, 2nd Ed., Volume 1, Oxy-acetylene gas welding and related studies. Volume 2, Electric arc welding and related studies. Pergamon Press Ltd, Oxford (1980).
> As titles, well-presented material.

Peck, H., *Allocating Tolerances and Limits*, Longmans, Green and Co. Ltd, London (1968).
> Practical workshop topics.

Roberts, A.D., and Lapidge, S.C., *Manufacturing Processes*, McGraw-Hill Inc., New York (1977).
> Wide range of topic coverage including fabrication, joining and machining.

Ross, R.B., *Handbook of Metal Treatments and Testing*, E. and F.N. Spon Ltd, London (1977).
> As title.

Ross, T.K., *Metal Corrosion, Engineering Design Guide Number 21*, Oxford University Press, Oxford (1977).
> Brief and to the point.

Woods P.F., *Fundamentals of Welding Skills*, Macmillan Press Ltd, London (1976).
> A practical approach, very visual material.

Corrosion. Attack and Defence, The British Steel Corporation (BSC), Market Promotion Department, Special Steels Division, Sheffield (1975).
> Brief and to the point, with practical examples.

Which Metals? The British Steel Corporation (BSC), Market Promotion Department, Special Steels Division, Sheffield (1976).
> Useful for materials' selection and examples of particular applications, easy reading.

Selected Papers

Graff, G.M., 'Engineering Plastics: Primed for Key C.P.I. Role', *Chemical Engineering*, pp. 42–5, 23 August 1982.

Kirby, G.N., 'Corrosion Performance of Carbon Steel', *Chemical Engineering*, pp. 72–84, 12 March 1979.

Kirby, G.N., 'How to Select Materials', 29th Biennial Report on Materials of Construction, *Chemical Engineering*, pp. 86–149, 3 November 1980.

Redmond, J.D., and Miska, K.H., 'The Basics of Stainless Steels', 30th Biennial Report on Materials of Construction, *Chemical Engineering*, pp. 78–118, 18 October 1982.

Journals and Newspapers

Economist, Economist Newspaper Ltd, London.
  Weekly. Wide topic coverage including economics, commerce, business, industry, technology and politics.
Financial Times, Financial Times Ltd, Cannon Street, London.
  Material prices, share prices, company news, etc.
Heat Treatment of Metals, Wolfson Heat Treatment Centre, University of Aston, Birmingham.
Journal of Materials Science, Chapman and Hall Ltd, London.
Journal of Mechanical Working Technology, Elsevier, Amsterdam.
  An academic journal but includes 'reviews' and 'recent trends' type papers.
Journal of Metals, Metallurgical Society of AIME, Warrendale, Pennsylvania.
Metalworking Production, Morgan Grampian Ltd, London.
Modern Plastics International, McGraw-Hill Publications, Overseas Corp., Lausanne, Switzerland.
Plastics Today, Imperial Chemical Industries PLC, Welwyn Garden City.
Powder Metallurgy, Metals Society, London.
Stainless, British Steel Corporation, Sheffield.
Welding Journal, American Welding Society, Miami, Florida.

6

Human Factors

Chapter Objectives

To develop an understanding and appreciation of:

1 the needs, desires and limitations of man and society;

2 the human factors that will influence the success of a design;

3 the relative importance of different aspects of the man–machine relationship;

4 the limitations that human factors impose, and the improvements to be obtained by their incorporation into a design study.

* * *

Exercises

1 Design a bicycle from considerations of anthropometric factors only.

2 Make recommendations for particular features to be included in a secretary's desk, and also the positioning of various important and common items.

   NB. Begin by sketching the envelope for the reach of your arms including cross-over regions. Discuss how the use of a swivel chair, rather than a fixed chair, will affect your decisions.

3 You are responsible for setting up a large warehouse containing automobile spare parts to be sold to the public. It has been decided that all parts will carry a four digit code using all the basic digits—zero to nine. Items will not be on display on shelves, but customers will obtain the appropriate code from a catalogue and then telephone this to the warehouse. Consider possible errors that may occur, especially those due to incorrect recording of the desired part code by the warehouse. Decide whether less errors are likely if 10 letters (rather than numbers) are used, or combinations of letters and numbers.

   What problems will occur if this type of system is to be used nation-wide, or in different countries?

4 Select the layout of an operating panel with combinations of (say) six levers, handles or buttons. What factors will affect the final choice of positions?

   Consider a similar problem for the layout of an instrument panel.

5 Study a car jack and suggest improvements in the design to make it easier for human operation.

6 Design a can opener to be used by a person with only one arm.

163

7 The night shift operator for a small, but vital, effluent treatment unit is required to monitor and record the performance of the unit. The operator must also make emergency repairs, or shut down the unit if necessary until the maintenance engineers arrive next morning (in this situation the effluent is diverted and stored in a large holding tank). Consider the psychological factors that become important if this work is to be performed efficiently.

8 How may political, religious and moral influences be related in a society?
 What effects might they have upon the manufacture and subsequent sale of a product?

* * *

6.1 HUMAN FACTORS

Man is both the designer and the user (or consumer) for any product. Man is also the only client for any engineering design. The purpose of a design is to satisfy the needs of individuals, and of society as a whole. The successful designer will appreciate these needs and will possess some basic understanding of the factors that must be considered. This appreciation and understanding will be reflected in the nature of the product and its success. The importance of human factors in a design project will be determined by the extent of the man–machine interactions. By this we mean the combined functioning of man and machine for the achievement of a common objective. The influences that human factors impose upon a design may be either human- or machine-oriented; an example of each would be safety and rapid operation, respectively.

However, there are certain aspects of human factor engineering of which the designer should be aware, and beware! The designer should aim to produce what man wants, needs and can use. A satisfactory and successful design usually incorporates all three of these attributes. Remember that need does not necessarily mean want, nor does require equate with desire. A good marriage contains compatibility and love, and an operator–machine relationship must contain physical suitability and desire. The designer should also avoid using himself as the model for the typical final operator.

The designer's relationship with the machine is too personal and specialised, and field trials with actual operators should be conducted. However, an element of unpredictability exists in human behaviour, and the old maxim that 'whatever can go wrong, will go wrong' is useful advice.

Problems involving human factors and their solutions, are interdisciplinary. They involve applications of principles and knowledge normally outside the scope of engineering and the related sciences. The areas which are most commonly investigated and applied by the designer are:

(a) anthropometry — the physical sizes and limitations of humans;
(b) physiology — the reactions, functions, limitations and capabilities of the human body;

(c)   ergonomics — dynamic interactions of operator and machine;
(d)   psychology — the influence of mental activities.

Other human factors may influence a design, but these are usually less significant. They are social, cultural and emotional influences (sometimes also classified as psychological), health and biological aspects (sometimes also classified as physiological), effects of the climate and the environment, etc. The designer should aim to acquire an appreciation of these human factors, rather than attempting to become an expert in any of these fields.

The human factors most often considered in design problems will now be discussed in more detail.

6.2 ANTHROPOMETRIC FACTORS

Anthropology is the science (or study) of man. In order to design a component that can satisfy the requirements imposed upon it, and also be safe and comfortable, it will be necessary to take into account anthropometric factors. These factors are concerned with the physical size of the human operator and how he can be accommodated by a machine. In this sense, the relationship is static. Many studies of the shapes, weights and dimensions of men and women have been carried out. These can provide the designer with a 'data-bank' of information related to human dimensions, or a product specification for all current human models!

Examples of some human measurements which may be required by the designer are illustrated in Fig. 6.1. The way in which these data are usually presented is shown in Table 6.1.

Table 6.1 Example of Anthropometric Data (Female)

Body measurements (female). Refer to Fig. 6.1	Percentile (dimensions in mm)		
	5%	50%	95%
1 Sitting height	863	905	962
2 Shoulder height	547	585	631
3 Thigh to foot	510	559	600
4 Arm reach	691	732	791
5 Elbow to finger	443	470	494
6 Buttock to knee	560	607	640

Such a table could also include the arithmetic mean, the standard deviation, and the range of the actual measurements.

The engineer should know, or be able to find, where such information is located. If you do not know, now is a good time to go and look! Many design projects start from a determination or estimate of man and machine sizes.

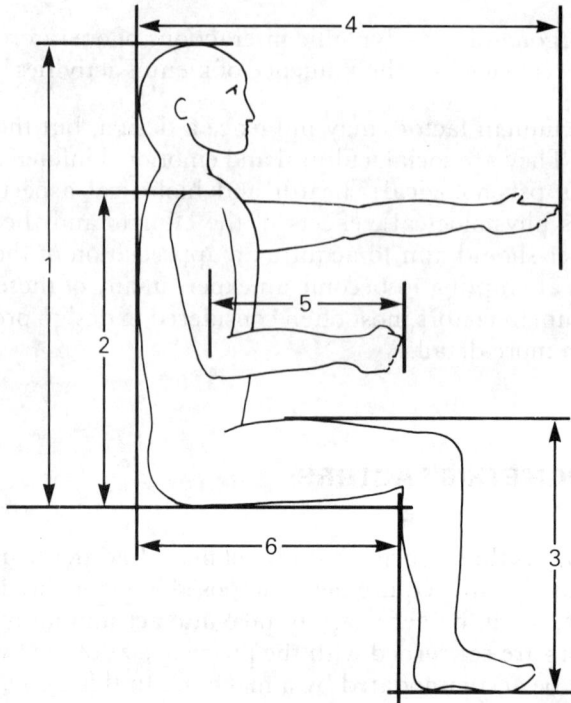

Figure 6.1 Basis of Anthropometric Data in Table 6.1

The main problem facing the designer is to decide for which size of human body he is designing! The published anthropometric data are for particular groups of people, i.e. samples from the whole population that could have been measured. Obviously, time and cost prohibit measuring the entire population, and even if this were done the designer would probably still not be able to design a product suitable for everyone. Therefore, when a particular sample size is chosen for investigation and measurement, statistical methods should be used to ascertain that biased results are not obtained. Check these methods yourself (see Section 3.4.1).

Even within a very limited or select group, there will usually be considerable variation (or 'scatter') in the values of the particular dimension to be measured. This is particularly true of any physical characteristic: identical measurements are only identical because of, or within, the accuracy of the measuring technique. Information can be conveniently presented on a statistical distribution diagram of frequency against measured value. This graph is often the familiar (or it should be!) bell-shaped normal distribution as shown in Fig. 6.2. The same information can be presented on a cumulative (frequency) distribution as illustrated in Fig. 6.3. When information is collected, various statistical values can be easily calculated. The first choice is probably the arithmetic mean value, i.e. the layman's 'average'. However, this value in isolation will have little relevance, and the standard deviation should also be known. Both these values should be given in anthropometric data tables.

Relative frequency

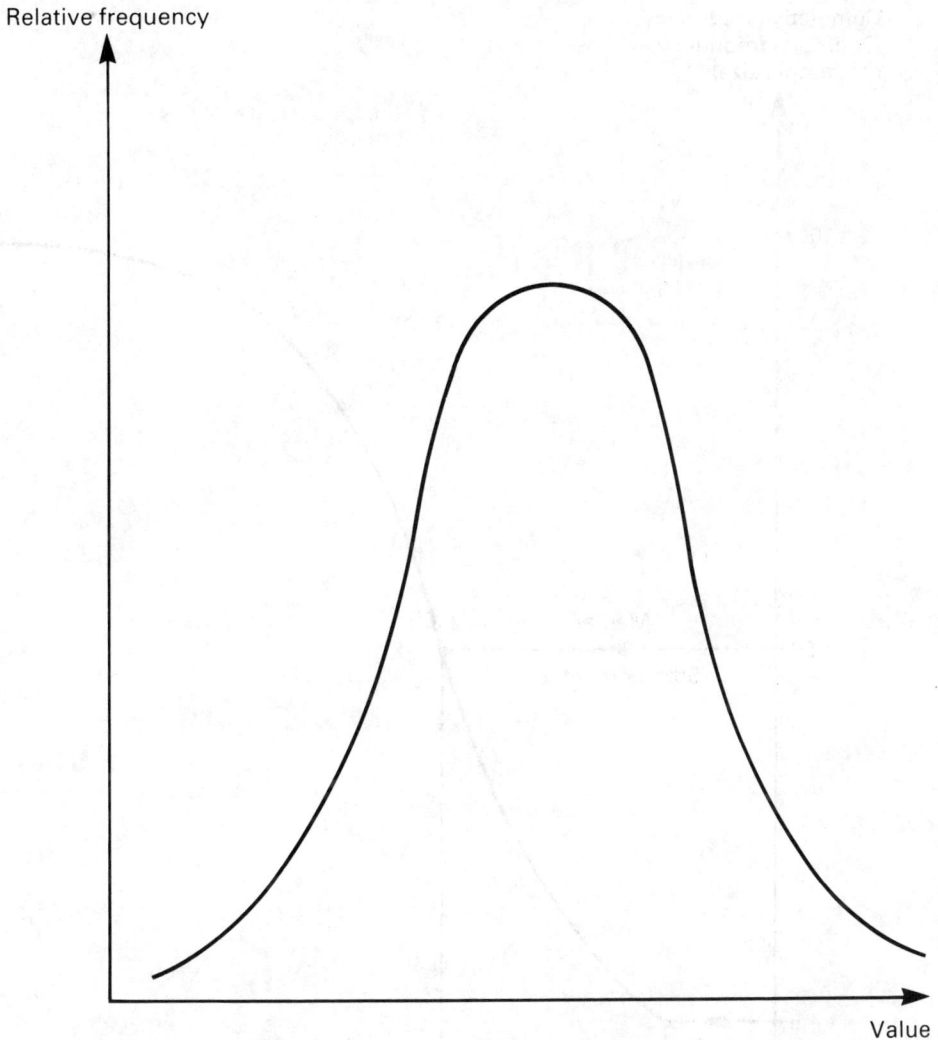

Value

Figure 6.2 Statistical Distribution Diagram (bell-shaped normal distribution)

* * *

Exercises

1 Determine how the arithmetic mean and the standard deviation of a sample can be calculated. What useful information does the standard deviation convey?

2 If time permits, decide upon a particular human dimension to be measured, define a suitable population and sample size, and obtain your own data. Calculate the arithmetic mean and the standard deviation. Draw the relative and cumulative frequency distribution diagrams.

* * *

Cumulative frequency
(% of total frequency,
i.e. sample size)

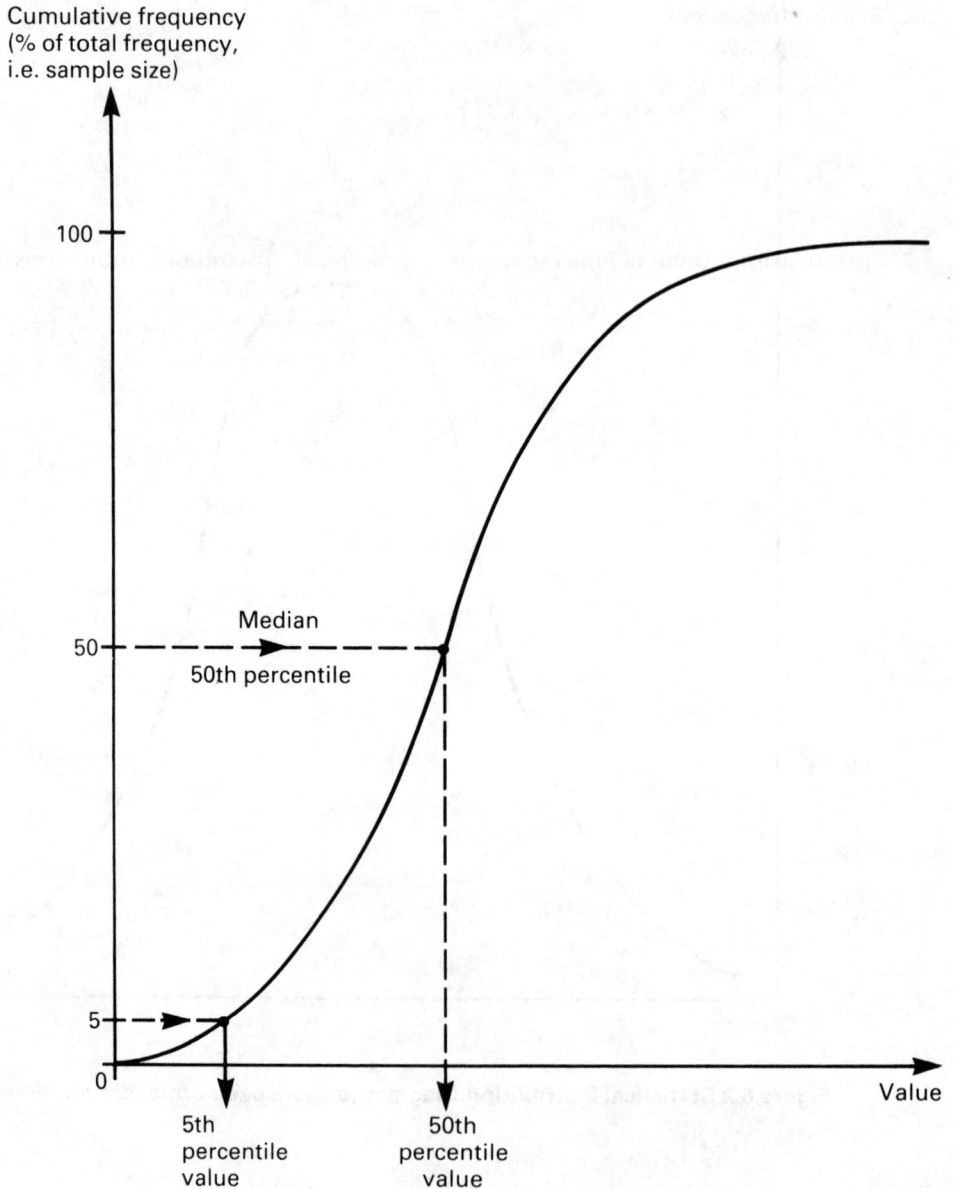

Figure 6.3 Cumulative Frequency Diagram

Using Fig. 6.3 another statistical average can be determined. This is the median, or the 50th percentile value. It is the middle value, i.e. the value above (or below) which 50% of the population sample is found.

* * *

Exercises

1 Obtain your own data and calculate the median value.

2 How does the median compare with the arithmetic mean?

* * *

If the designer chooses an 'average' value as the basis for his design specification, then he may exclude, or at least discomfort, many possible users. (The values of the mean and the median are usually close together.) It is clearly not practical to design in order to accommodate everyone, and usually the 5th percentile value is chosen as the minimum for design purposes, or the 95th depending upon the particular application. In *either* case, 5% of the population will be disadvantaged and must make their own arrangements!

A third 'average' value is sometimes useful, depending upon the particular feature being considered. This is the mode or modal value, and is the value (or value range) that has the highest frequency, i.e. the most popular measurement. For a large sample size, the mode will depend upon the class interval sizes which are used to group the data, e.g. 37 people weigh between 61 and 65 kg.

* * *

Exercises

1 Obtain your own data and determine the effect of changing the size of the class interval upon the value of the mode obtained.

2 In which examples/applications might the mode or modal value be the most useful average for design purposes?

* * *

Anthropometric data will be useful when designing such items as car interiors, domestic appliances, manual tools, clothes, furniture and human 'accessories' such as spectacles and building interiors.

The idea of 'Mr or Ms Average', or the 'average person' is a fallacy, and nowhere is this shown more clearly than in man–machine design studies. There are many examples of products that have not been satisfactorily designed for the needs and comfort of the operator or user. (Find and discuss some.) Not all of these have been failures, mainly because of human tolerance and adaptability, or necessity for the product.

6.3 PHYSIOLOGICAL FACTORS

Physiology is knowledge of the body in relation to particular stimuli and responses. The designer is concerned with body functions that influence the recording of an event, and any subsequent reactions. The body systems related to these functions are sensory, respiratory, vascular, muscular and neurological. For design purposes it is the causes and effects that are most important, rather than a complete understanding of the ways in which these systems function. The main physiological factors relevant to the designer are sensations of seeing, hearing and touching.

It has been estimated that humans gain 80% of their knowledge through sight, and human machine operators certainly receive much visual information. The eye can perceive form, colour, brightness and motion. This enables an operator to read instruments and written instructions, observe moving objects and react to colour and shape combinations. Certain factors influence the visual process, some of which can be controlled or altered. These are:

(a) acuity—the ability to perceive detail at various distances, i.e. the resolving power of the eye (or how well it can see);

(b) motion of an object;

(c) colours;

(d) brightness—the difference in light intensity between an object and the background (there is a definite relationship between brightness, contrast and acuity);

(e) illumination—light intensity (many human environments and particular tasks have recommended, or measured, levels);

(f) glare—light reflected directly from a solid, flat surface.

Problems such as glare and shadows can often be overcome by using suitable materials and colour selections, and also correct specification of lighting systems. The nature of the object and the intervening medium influences what the eye sees and an operator's interpretation and response.

Sound is an important source of information, and also a means of communication. The ear, like the eye, is a sensitive organ. The perception of sounds can be used by the designer as an emergency alarm system and as a warning device, e.g. equipment malfunction noises. Noise is unwanted sound and a large amount of time, effort and money is now spent trying to reduce or eliminate noise, as well as trying to correct the damage it has caused. For effective communication a sound must be sufficiently loud, clear and distinctive to be distinguished from the background noise. Noise levels are reduced because of their adverse effects on mental activities and the interpretation of speech, and the possibility of physical, mental and psychological damage. The effects of noise on an operator can sometimes be reduced by limiting the exposure, or by using personal protective devices or barriers for absorption or reflection of noise. Noise can sometimes be controlled or reduced by suitable design features or considerations, e.g. absorbing machine vibration, periodic lubrication of moving parts, situating items of equipment of similar noise intensity together.

Tactility is the sense of touch. The cutaneous senses are those relating to the skin, and are due to fine nerve fibres at the surface. They can detect temperature differences and contact pressures. This 'touching' sensation has also been utilised in the coding of knobs and handles according to size, shape and relative position, and to aid unconfused recognition. The kinaesthetic senses are those noting movement of the body, i.e. nerves in the muscles, joints, etc. They provide an awareness of body positions or arm and leg positions, without visual or tactile information. These considerations are used to help the designer select foot or hand controls, decide the position of hand controls for rapid operation, etc.

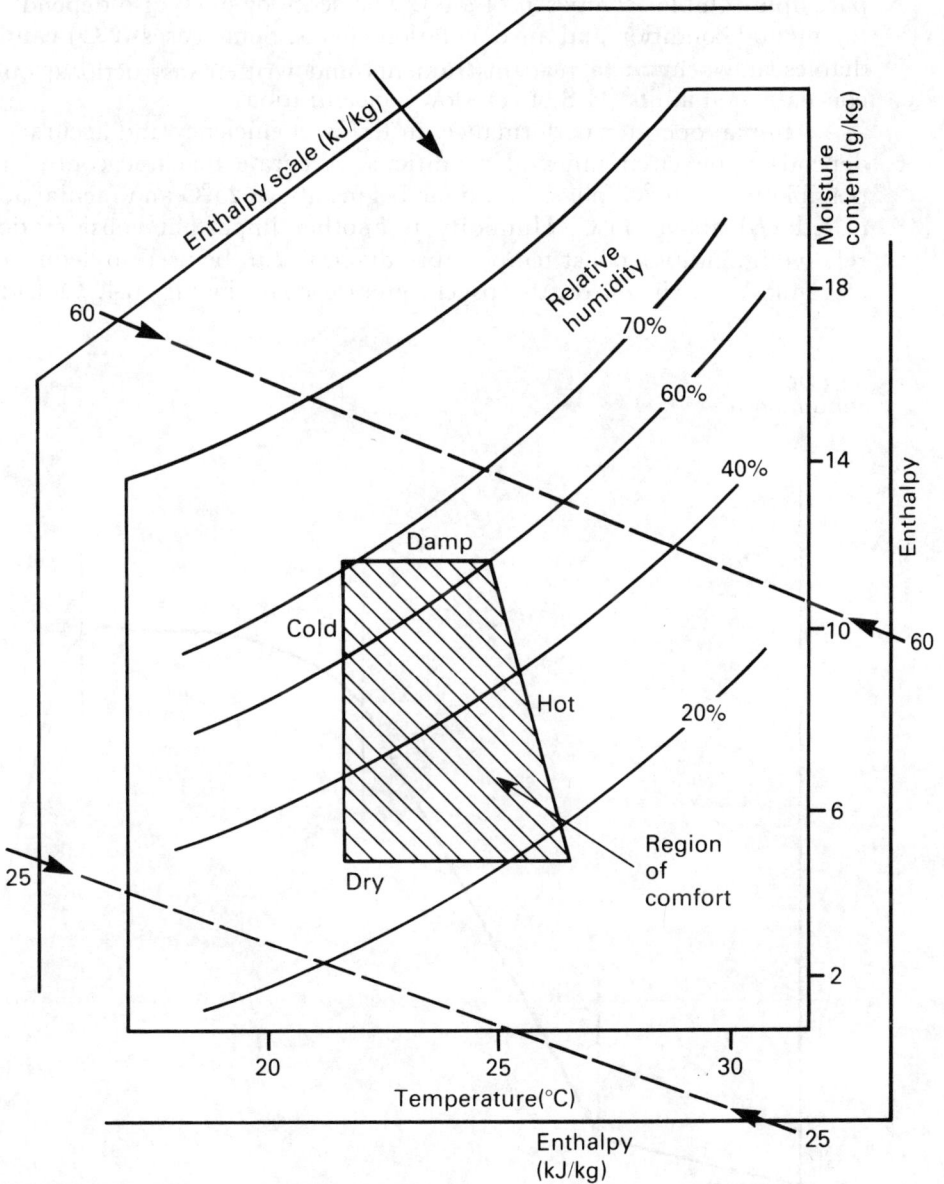

Figure 6.4 Region of Comfort using a Psychrometric Chart

Taste and smell are also very sensitive. Taste can be categorised as sweet, sour, salty and bitter. It is related to smell, hence the reduced sensitivity caused by holding the nose or because of a cold. The taste buds (nerve cells) are located primarily on the tongue. Smells are less easily categorised; one suggestion is spicy, flowery, fruity, resinous, putrid and burnt. For smells to be detected they must be volatile, part water-soluble and part lipid-soluble (composed of fats). The sense of smell also depends upon the mental condition and air circulation effects. Some gases (CO) cannot be detected by humans, and others are only detected within certain concentration limits (H_2S of very low concentration).

Human operator performance, in terms of efficiency and accuracy, also depends upon environmental conditions. Accurate machine control is not possible below $10\,°C$, physical fatigue begins above $25\,°C$ and mental activity is reduced above $30\,°C$. Humidity is another important consideration. A relative humidity against temperature diagram can be used to define a zone of comfort, as shown on the psychrometric chart in Fig. 6.4. Outside the

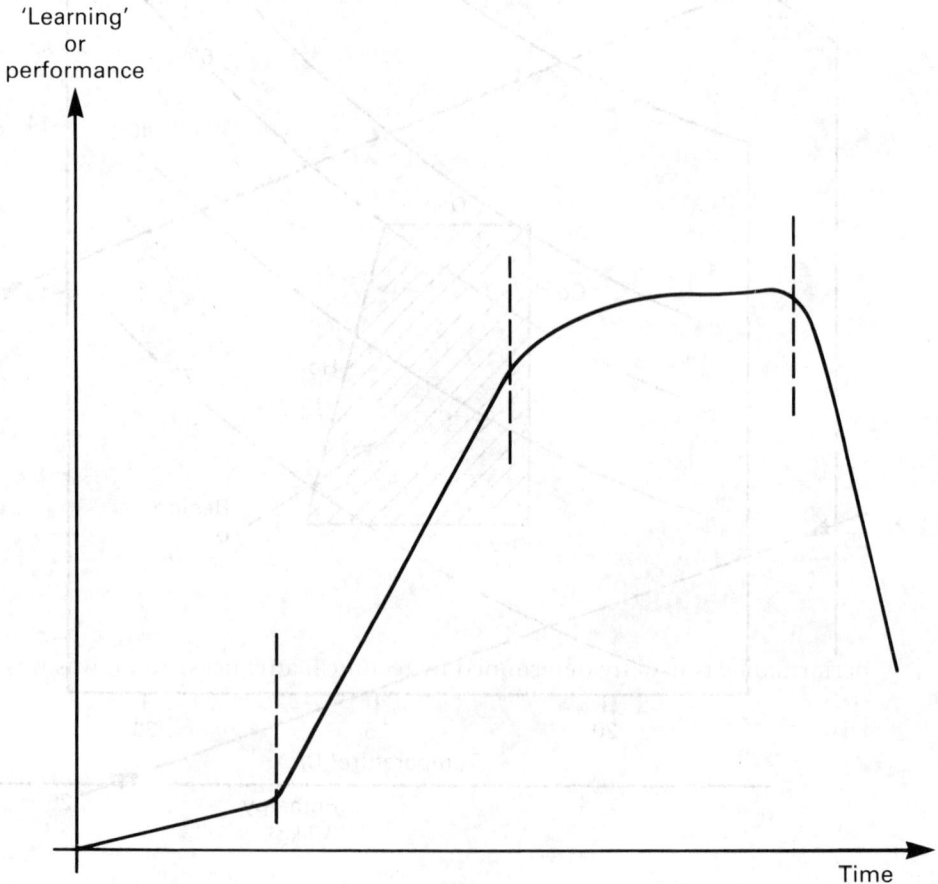

Figure 6.5 Typical Learning Curve

shaded region, feelings of hot, cold, dry or damp will be experienced. The body is continually generating and losing heat in order to maintain a constant body temperature of about 36.8°C. As body activity increases, more heat is generated. The rate at which this heat can be lost depends mainly upon the air temperature, the type of clothing worn and the ventilation. Humidity has little effect on this process unless the ambient temperature is high.

* * *

Exercises

1 Describe the ways in which the eye and the ear function as sensory receivers. If you were 'designing' a new human being, which existing features of these organs and their method of operation, would you consider to be good and bad designs? Which features would you retain? Which features would you change and in which ways? What considerations would influence your selection of either an auditory or visual method of conveying a message or a warning?

2 You are assigned to work in a noise abatement study group. Assess the importance of both the 'signal-to-noise ratio' and the 'speech interference level' in a variety of common working environments. Sketch a typical learning curve as shown in Fig. 6.5, and identify particular features or regions. What environmental or operational factors will affect this curve, and in which ways? What is the importance of this type of curve for the designer who is planning to introduce a new, human-operated machine into a factory?

* * *

6.4 ERGONOMIC FACTORS

Ergonomics is concerned with the dynamic interaction between operator and machine. It is the study of the factors to be incorporated into a design project, whenever a human operator is required for a machine. Even fully automated machines have to be assembled and maintained, tasks usually performed by humans. Ergonomics is therefore concerned with the factors involved in 'doing', whereas anthropometry is concerned with 'being', i.e. physical dimensions such as size, weight, volume and comfort.

The functions to be performed (satisfactorily) by the operator are usually a bigger problem than those required of a machine. The machine performance is usually determined by technical advances, and it was this area that received most attention and improvement after the Industrial Revolution. The importance of the operator was largely neglected until the 1930s, with improvements occurring particularly during and after the Second World War. Design had previously assumed that the operator would adapt to the machine, unless this was physically impossible! It was eventually realised that the operator was an essential element in the operator–machine relationship. Only by paying adequate attention to the operator could further improvements in operation and efficiency be achieved. The design emphasis was then shifted towards adapting machines to fit the needs of the operator.

The operator may be required to perform one, or more, of the following functions:

(a)   feed work (or materials) to a machine;
(b)   hold work in a machine;
(c)   remove work (or materials) from a machine;
(d)   perform certain physical actions, i.e. operate levers, handles, buttons;
(e)   observe instrument readings;
(f)   record instrument readings;
(g)   communicate with other personnel.

All of these tasks require movement by the operator, and sometimes exertion of physical force. The design system from an ergonomic viewpoint may be represented by the following system components:

THE OPERATOR: physical ability, mental ability, agility.

THE MACHINE: operational features, number and sequence of operations.

OPERATION: time required, speed of different operations, operating changes, the environment.

THE DESIGN: number and positioning of instruments, controls, operating handles, etc.

Particular features to be incorporated are comfort, safety and satisfaction. The operation of a machine should be easily taught and remembered!

The above features are not meant to be exhaustive, they represent just a few ideas. The reader should be able to extend these lists.

The consideration of ergonomic factors in the preparation of a design *should* be standard practice. Two (typical) examples of problems or situations where ergonomic factors need to be considered are shown in Fig. 6.6. The following situations have received particular attention and publication. The number of operations, the speed, the accuracy and the number of movements that an operator can efficiently perform have all been the subject of extensive study. The results of various studies are readily available as data for use by the designer; they contain such information as the maximum force that can be exerted by a forearm, the maximum reach for a seated operator, etc. This is the ergonomic equivalent of the anthropometric data discussed in Section 6.2. The same problem of defining or selecting the 'average' person exists, and again it is often taken as the 5th (or 95th) percentile.

Ergonomics is important for the design of factories and workplaces where the operator should spend the minimum time away from the work area. Components should arrive at appropriate times, in a particular sequence, in the correct manner and at the required position, especially if the task is subsequent assembly.

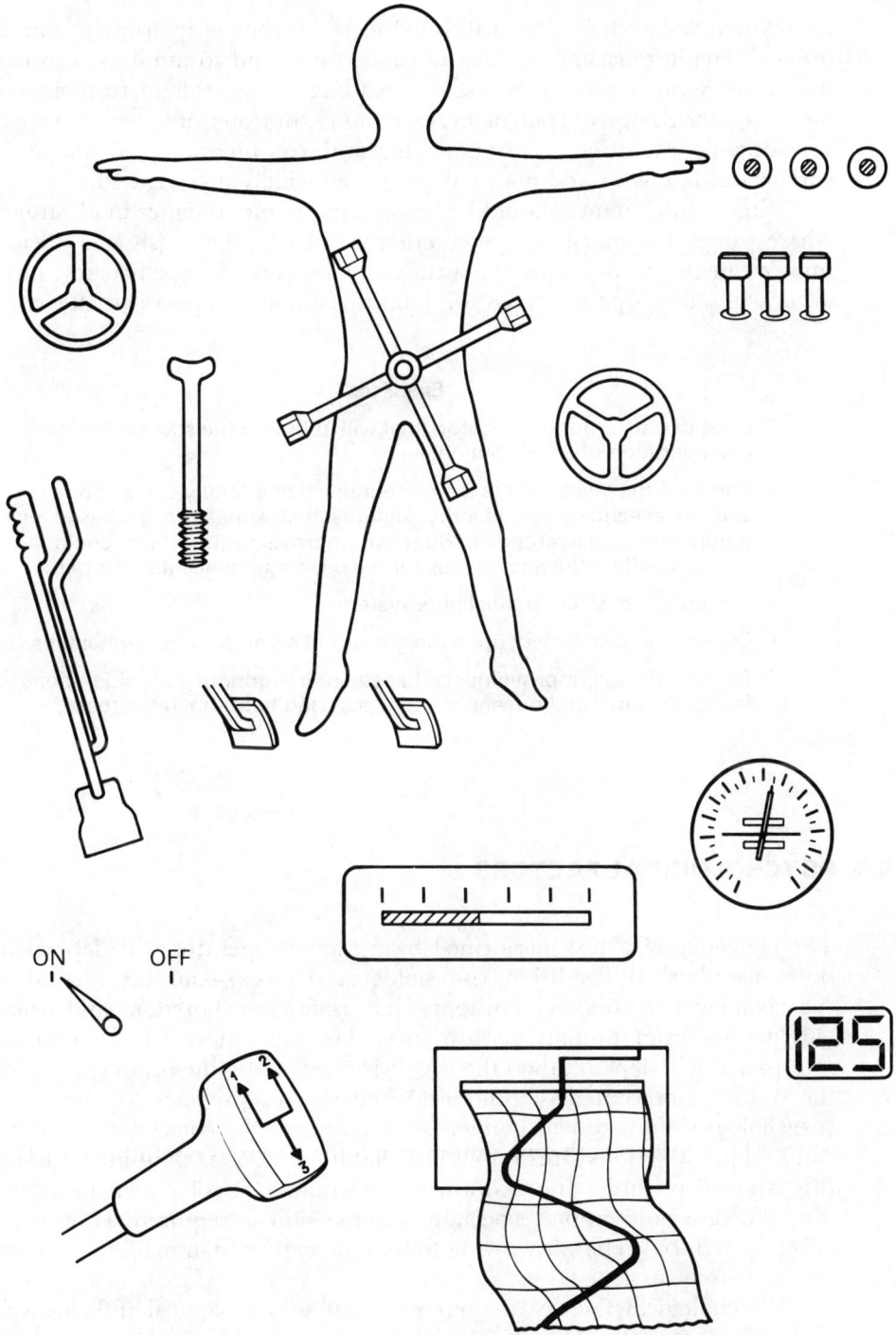

Figure 6.6 Situations Requiring Ergonomic Analysis

Much research has been devoted to the layout of instrument panels in order to facilitate easier reading and recording, and to minimise errors that may have serious consequences, e.g. reading an aircraft instrument panel. Similarly, the design of control levers, handles, buttons, etc., has also received considerable attention. Variations in, and combinations of, shape, size, surface finish, colour and material have been widely investigated.

Ergonomic factors should be considered by the designer in *all* situations where operator–machine interactions occur. This should lead to improvements in one, or more, of the following: operating efficiency, product accuracy, safety, operator comfort, job satisfaction and operator efficiency.

* * *

Exercises

1 Consider the ergonomic factors that will influence the successful design and operation of a wall can opener.

2 Compare the layout of the instrument panel in a family car, a sports car and an executive car. Identify features that appear to be based on ergonomic considerations. Suggest improvements which could be made, detailing the advantages for the user, and hence the company.

3 Design a pedestrian traffic lights system.

4 Compare and comment upon the designs of some common water taps.

5 Suggest design improvements, based upon ergonomic considerations, for some common household appliances and hand-operated tools.

* * *

6.5 PSYCHOLOGICAL FACTORS

The efficiency of a task performed by a human operator will depend upon both his physical condition (physiological factors) and his mental state (psychological factors). Conditions such as fatigue, boredom and isolation will detract from human performance. The way in which information is interpreted will depend upon the way it is received by the brain (physiology), the way it is presented (ergonomics) and the receptiveness of the operator (psychology). Fatigue can be mental or physical; the consequences of either can be equally serious. It is important that motivation is built into a design, if this is at all possible. Incorporating more human involvement in a process may increase motivation, especially if some skill is required. However, the effects of reduced reliability owing to human, rather than machine, operation will have to be assessed.

Psychological factors are often related to sociological influences, and their effects are often equally important. Certain limitations and patterns of behaviour are imposed upon the machine operator and the designer by the expectations of society. The designer will need to consider both the general cultural features and the class divisions within the society for which he is designing a product. This is particularly true if export markets are in very different countries. The human operator will also be influenced by group behaviour, both at work and outside. Finally, the designer should remember that each operator is an individual and will never be totally reliable or predictable, whatever testing is performed or data collected. He will, however, be able to make decisions based upon intuition, and apply judgement and reasoning to problems. He is also a flexible component in a design system.

* * *

Exercises

1 List and discuss those functions that are performed better by a human operator or those performed better by a machine.

2 What problems do you envisage when trying to sell consumer goods to a particular Third World country? (Consider human and social problems, rather than economic, political and geographical difficulties.)

* * *

6.6 SUMMARY

The human factors that influence a design are many and diverse. Certain factors, e.g. physiological and ergonomic, appear to be merely technical problems, until it is realised that humans are involved in machine operations. A human operator is never 100% predictable, and cannot be standardised. This chapter has outlined some important factors and hopefully stimulated the student designer to think more deeply and widely about the less technical factors influencing a design. The role and responsibility of the engineer in society is discussed in Chapter 11.

KEYWORDS

anthropometry
physiology
ergonomics
psychology
arithmetic mean
standard deviation
normal distribution
cumulative frequency distribution
the 'average' value
50th percentile
median
mode or modal value

stimuli and responses
sensory functions
respiratory functions
vascular functions
muscular functions
neurological functions
seeing, hearing, touching
brightness
acuity
illumination
glare
sound

noise   smell
tactility   fatigue
cutaneous senses   humidity
kinaesthetic senses   'signal-to-noise ratio'
taste   'speech interference level'

BIBLIOGRAPHY

Chapanis, A.R.E., *Man-Machine Engineering*, Tavistock Publishers Ltd, London (1965).
  Good examples and text, despite its age.
Edholm, O.G., *The Biology of Work*, World University Library, Weidenfeld and Nicolson, London (1967).
  Well written and easy to read.
Grandjean, E., *Fitting the Task to the Man–An Ergonomic Approach*, Taylor and Francis Ltd, London (1980).
  Well-presented material.
Hammer, W., *Product Safety Management and Engineering*, Prentice-Hall Inc., New Jersey (1980).
  Well-presented material, a 'handbook' approach.
Henley, E.J., and Kumamoto, H., *Reliability Engineering and Risk Assessment*, Prentice-Hall Inc., New Jersey (1981).
  A detailed coverage.
Murrell, H., *Men and Machines*, Methuen and Co. Ltd, London (1976).
  A good introductory text, easily read.
Murrell, K.F.H., *Ergonomics: Man in his Working Environment*, Chapman and Hall Ltd, London (1979).
  As title.
Roebuck, J.A., Kroemer, K.H.E., and Thompson, W.G., *Engineering Anthropometry Methods*, John Wiley and Sons Inc., New York (1975).
  As title.

Journals

Applied Ergonomics, IPC Science and Technology Press Ltd, Guildford, Surrey.
Ergonomics, the journal of the Ergonomics Society and the International Ergonomics Association, Taylor and Francis Ltd, London.
Health and Safety at Work, Maclaren Publishers Ltd, Croydon, Surrey.
Human Factors, the journal of the Human Factors Society, Santa Monica, California.
Noise Control Engineering Journal, Institute of Noise Control Engineering, Poughkeepsie, New York.

Economics

Chapter Objectives

To obtain an appreciation of:

1 some common economic terms, principles and ideas;

2 the importance of economic factors for the success of an engineering design project.

---

*   *   *

Exercises

1 Outline the main costs associated with constructing a bridge across a river. How do these costs compare with those incurred if a hovercraft is used as an alternative?

2 What is the value (in the economic sense) of a rail transportation system?

3 Detail the costs associated with producing a disposable ball point pen.

4 How would you determine the selling price of a laser disc record player?

5 Before reading this chapter, what do you understand by the Law of Supply and Demand?

6 What are the significant economic factors for the manufacture of a non-stick frying pan and an electric cooker? How do these factors differ between the USA, the UK, China and the USSR?

7 How effective are economic and/or trade sanctions imposed against particular countries?

8 How has the oil 'crisis' of the 1970s changed economic thinking in general, and the engineer's task in design?

9 Discuss the effects of a world recession upon the planning of an engineering design project.

10 You are concerned with the development of a new, consumer-oriented, technological innovation. Which has first priority: technical factors or economic considerations?

*   *   *

7.1 ECONOMIC TERMS

Ask a group of engineering students what is meant by economics and you will receive many different answers. Most will include using money, or producing more money; alternatively, the process of making a product and selling it at a price that exceeds the costs incurred during production; perhaps merely the idea of a business producing profits. For many students this is the extent of their knowledge, i.e. a connection between industrial activity, money and profit. Perhaps we should be reminded that economics, economies and being economical are all related. If we consider merely the use of money, then economics is the study of ways of using money profitably. Economies are reductions in costs to reduce financial losses and being economical means using money in the most efficient manner. There are, however, factors other than money that are involved in economics.

Economics and economists are often considered with some scepticism by engineers. This is partly because economists have not done a good enough job of 'selling' or explaining their expertise to the engineers. Engineers are required to make their designs work at the first attempt, with perhaps minor modifications and tests. However, it often seems that economists are allowed endless modifications to their ideas and theories, in order to explain past discrepancies. Part of this problem is the frequent use of economic phrases and references to economic theories by politicians. It has been said that 'if all the economists in the world lay down head to toe, they would still fail to reach a conclusion!' Perhaps a harsh, although amusing, view of economists.

There are several terms which are used when describing the commercial aspects of a project. All have a different meaning and a different emphasis in economics, although they are often confused by the layman.

Money is a means of exchanging or obtaining goods or services. Its *only* value is that it is easily used, as such it is referred to as a **liquid asset.**

Capital is money invested in a business; it is then used to make profits from the sale of products or services.

Interest is the compensation paid to a lender when he agrees to forego the use of his money and commit it to a business venture. The interest, or **return on investment,** must (or should) be paid out of profits.

Wealth is material items that are useful and scarce. These can be land, buildings, consumer products, chemicals, machines, agricultural produce, etc. Wealth will depend upon the market forces that determine the monetary valuation of these items.

Income is the payment received for services rendered or goods sold. Items such as rent, dividends and wages represent income.

Cost is the sum total of all the expenditure incurred during the production of a product. **Variable** or **running costs** are dependent upon the amounts of consumable quantities involved in the production process. Alternatively, **fixed costs** are not dependent upon the level of activity and must always be paid, even if no items are produced or sold.

Price may be either the **'asking' price** or the actual **selling price.** Both will depend upon such factors as market surveys, competition, advertising, profit margins, etc. It is not realistic merely to add up the costs involved in producing a product, and then include the required profit in order to arrive at the selling price. A profit will only be realised if the product can compete with the alternatives available, and if the prospective purchasers *want* the new product.

It may be necessary to achieve a certain profit in order to satisfy the investors and to stay in business, but this should be determined by calculation and market evaluation rather than as a predetermined rigid specification. If a required return on investment, or profit as a percentage of sales, cannot be achieved, then the product should not progress beyond the design stage.

Value is a difficult concept to define. It is related to the worth of a product as determined by both the producer/seller and the purchaser. There is no absolute value that can be assigned to a product. The most commonly used measure of value is the price that a willing buyer will pay to a willing seller, and this is known as the **market value**. The **fair value** is that which is determined by an independent arbiter. The **use value** is the value to the owner as an operating unit; this is often higher than the value ascribed by a prospective purchaser.

7.2 ECONOMIC CONSTRAINTS

The designer should estimate the costs that will be incurred in a design project, the profits that can be achieved and the profits that are required. These calculations should be performed at an early stage in the design study, and in more detail when the final specification is being formulated. It will be no use proceeding with a project if economic aspects are unsatisfactory. The designer should consider the following constraints when making decisions:

(a) the size of the total market for that particular type of product;
(b) the portion of the total market 'available' to this particular product;
(c) the portion of the market which the product is likely to achieve, known as the **market penetration**;
(d) the price at which competitive products are being sold;
(e) estimated profits which can be achieved (when costs are known) and the required return on the capital, also profit as a percentage of sales revenue;
(f) if profitability is too low, ways of reducing costs or increasing sales, and/or increasing the selling price must be examined.

7.3 THE PRODUCT LIFE CYCLE

The life of a product may vary from several months to several years. It may require very few changes during its life or many major modifications. The life

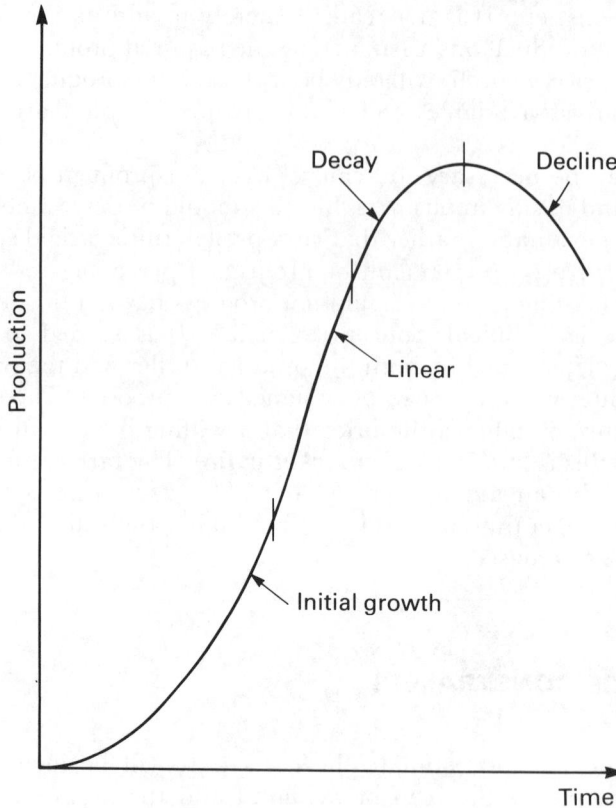

Figure 7.1 Typical Product Growth Curve

of many instant novelty foods is less than 1 year, mainly because of changes in consumer desires and a high level of competition.

An item such as a car jack or a battery torch will probably maintain constant sales, and changes in design occur mainly as a result of technological advances.

The life cycle of a successful product can often be depicted on a growth curve similar to that shown in Fig. 7.1. The initial stage may be either a slowly increasing rate, if a similar product is already available, or it may increase rapidly for a technologically innovative product. This first stage will be replaced (eventually) by a period of stable sales, followed by a reducing level of demand and finally a market contraction. The time taken by each phase will depend mainly upon the type of product, although other factors will also be influential, e.g. an economic recession, sudden increase in material prices, political decisions.

The promotion of new products must be timed so that company profits do not begin to decline. The use of growth curves will be helpful, as will early recognition of any changes in the factors that influence sales and production costs. The choice will be either to produce an improved version of an ailing product, or to abandon this product and concentrate on a totally different item. If an improved product is marketed too soon, then possible sales and profits will be lost from the original item. However, if delays occur, then the product may enter the undesirable falling sales (and falling profits) phase. Timing will be an important factor.

In order to prevent, or rather delay, a product entering the sales decay phase, the designer can concentrate on improvements and simplifications to the design, and also cost reduction features. It is important to demonstrate to the student that design is a continuous process. There are always new aspects and different problems to be considered. The successful designer is always looking to the future, to the unknown and to the unexpected. A company that does not improve and change its products will eventually run out of profits.

7.4 SUPPLY AND DEMAND

If economic equilibrium were ever to exist in a society, and competition followed idealistic trends, then the value and price of a product would become equal. The price would then be determined by the Law of Supply and Demand; the basis of this law is shown graphically in Fig. 7.2. This shows the market price as a function of both the supply of, and the demand for, a product. The selling price is obtained when supply just equals demand.

* * *

Exercise

Explain (in your own words) the reasons for the shape of the curves in Fig. 7.2, and the principle of the Law of Supply and Demand.

* * *

A more realistic, although still ideal, representation is presented in Fig. 7.3. The supply and (especially) the demand for a product are subject to many influences and the curves represent the ranges of the estimates that are made. It is, therefore, more realistic to think of these values as falling between certain likely, i.e. probability, limits as shown in Fig. 7.3.

However, society does not obey ideal laws and the sales and price of an item are influenced by many factors, including political, legal and social factors. The designer has little or no influence on the selling price of a product, only upon the costs involved. It will be his responsibility to ensure that costs are kept to a minimum, so that the asking price (cost plus required profit) is acceptable to both the company and the purchaser. Costs other than those associated with design and manufacture need to be considered. These will include transportation, advertising, marketing and storage.

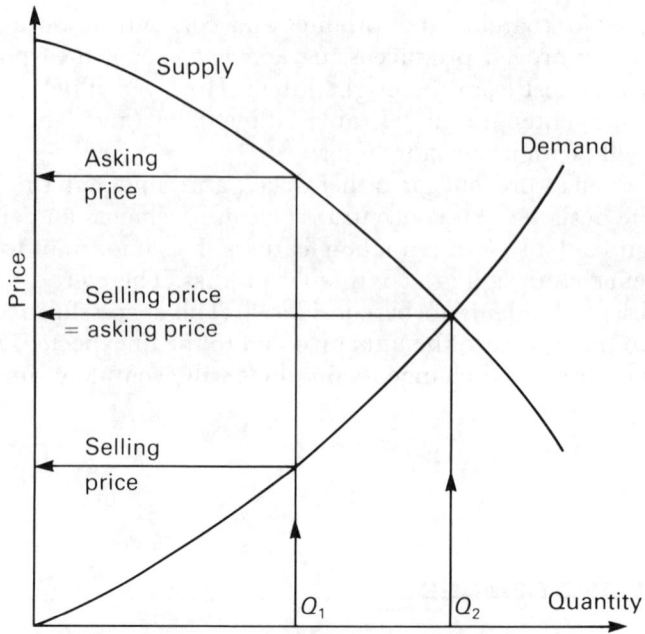

Figure 7.2 The Law of Supply and Demand

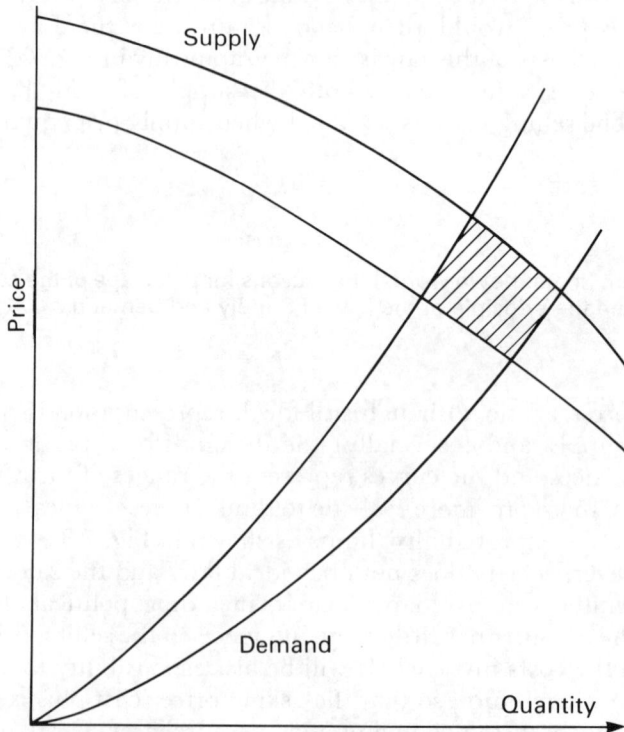

Figure 7.3 Estimated Limits of Supply and Demand

The final product must satisfy both the product specification and the wants (not necessarily the needs!) of society. In general, as the quality of an item improves, so the price and sometimes the demand will increase. However, costs will also increase. Beyond a certain limit, increased quality (and costs) will not yield sufficient increases in the market price (and hence profits) for this to be justified. This is called the **point of diminishing returns**. The designer should always remember that he is attempting to satisfy the needs and wants of society. Although product excellence is an admirable personal goal, what is required is a satisfactory or adequate product. Profitability is the important constraint.

*   *   *

Exercises

For particular products, consider how (and why) modifications to the Law of Supply and Demand have been necessary in the real world. NB Consider consumer items with widely different markets such as automobiles, engineering equipment, foods, oil, new technology items, etc.

*   *   *

7.5 PRODUCT AND PROJECT COSTS

In the design and production stages, there are often several alternatives that can be adopted. In the real world, these alternatives are not usually directly and easily comparable. For example, different types of pumps have different estimated operating lives; the strength, price and fabrication costs of materials are not directly related. Some projects require a large capital investment, e.g. a road transport bridge or tunnel, whereas others have larger operating costs, e.g. a car ferry. It will be necessary to make cost comparisons of the available alternatives that are suitable for a design solution.

If no capital expenditure is required and costs can be presented per hour, or (say) per 1000 items produced, then comparisons are usually straightforward and the **present economy method** is used. Common examples are materials' selection, component size, welding method, etc.

If different capital investments and operating costs are incurred by alternative projects, then a **present worth method** of annual cost comparison is required. The profit from each project should also be compared with the rate of return that could be obtained by investing the capital required in an interest-bearing deposit. For a more detailed explanation of these methods, reference should be made to the textbooks listed in the Bibliography for this chapter.

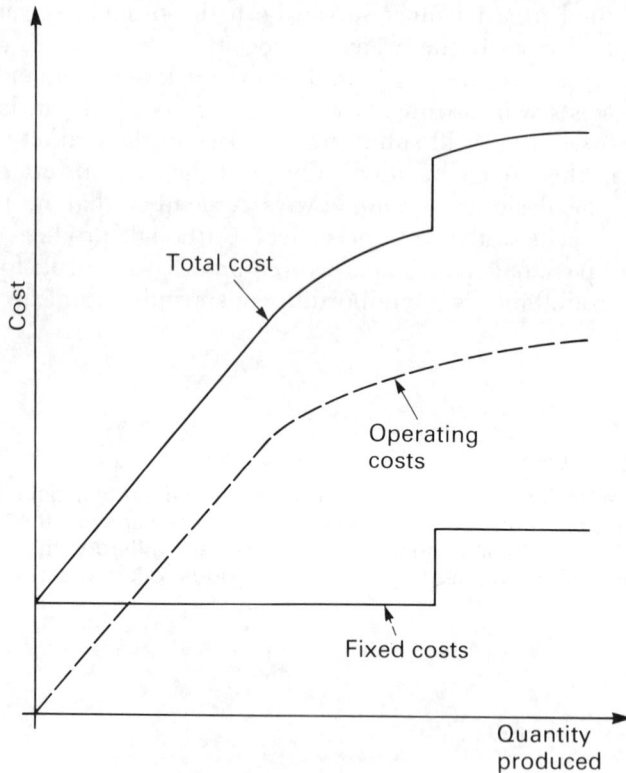

Figure 7.4 Influence of Level of Production on Project Costs

Machines wear out, new equipment is more efficient and economical to operate, the needs and wants of society change. All these factors mean that after a certain time, new investment must be made, or a project must be declared obsolete. Capital costs (less the salvage or resale value) must be recovered from profits during the lifetime of a project. Depreciation is the annual cost of this loss in value. There are various methods that can be used to calculate the depreciation cost. One method is a straight-line calculation of the total cost divided by the lifetime. Another method is a reducing balance method. However, it should be remembered that in times of inflation (are there any other times?), money spent today will purchase more goods than it will in 1 year's time. Therefore, depreciation charges should also take into account the loss in purchasing power owing to the (estimated) rate of inflation.

The total cost of a project should be broken down into different stages or activities in order to identify where cost reductions will be most effective. The main stages are design, manufacture, sales and distribution, and other. This last category (other) although not easily defined, may make a significant contribution to the total cost, and hence the possibility for major reductions!

Figure 7.4 illustrates ideal relationships between cost and quantity of production. Within each stage of a project there will exist both fixed and variable costs. The former do not change whatever the level of production, even if no items are produced! Most fixed costs can only normally be reduced if a project is abandoned or a company becomes bankrupt. Examples are rent, rates and interest charged on capital borrowed. Fixed costs may increase stepwise at a particular production level owing to the purchase of new machines, or even another factory. Variable costs depend upon the level of production, i.e. number of units produced, or the operating time. Therefore, they offer more possibilities for reducing expenditure, particularly if demand is reduced. Examples are hourly labour costs, material costs and electricity. The variable costs, running costs or operating costs usually increase steadily as production increases, although not necessarily in a linear relationship. The operating costs may level out at higher levels of production, as shown in Fig. 7.4, if discounts or off-peak tariffs then become available. The total cost curve for a product or a project is the sum of the fixed costs and the operating costs.

*   *   *

Exercises

1 For two widely different items, e.g. toothpaste, an automobile, a high speed passenger railway, a nuclear missile system, prepare detailed lists of the costs associated with their production. Identify fixed and variable costs.

2 What are the capital costs and running costs associated with these projects or production processes?

3 Which of these costs can be more easily reduced or controlled by the designer?

*   *   *

Research and development costs are often large and it will be unrealistic to attempt to recover all of these costs from the initial sales of a product. However, it will be equally unrealistic to make no attempt to recover such costs, and merely to assume that they will be provided out of the company's total profit.

The cost of maintenance must also be accurately assessed. It should be decided and specified early in the design phase whether maintenance is to be the responsibility of the producer/supplier or the user. This may be an all-inclusive guarantee by the producer to provide inspection, servicing and repair. Alternatively, it may be a clear statement of the extent and limitations of the responsibility accepted by the manufacturer. Maintenance costs will include the cost of components, their replacement, associated damage, and the cost of lost production. The designer must be aware of the probability of failure (see Section 3.5) and the costs associated with particular actions or events.

If more money is (correctly) spent on the design stage, then the quality of the product should improve, and the chance of failure be reduced. Manufacturing costs should also decrease and the total cost will exhibit a minimum, or optimum, value. However, the total cost is often fairly insensitive to design improvements.

The designer must analyse in detail costs and profits at an early stage in the design process. Ideally, this should be during the preliminary feasibility study, and these estimates should be revised and refined as the design proceeds.

7.6 SUMMARY

The importance of costs and profit has been discussed in this chapter. These factors have as much influence on the success of a product as the technical specification, or the engineering innovations.

When a product is to be used in industry, the price and the technical performance are the main considerations and the determinants of success. However, for sales to the general public, a wide range of less quantifiable factors becomes important. Often there is no specific relationship between the needs or the wants of the consumer and the inclination to purchase. The designer is aiming for consumer acceptance—often an elusive quality.

Apart from an awareness of the factors influencing the cost and profit of a product, the student (and the design engineer) should also be familiar with basic aspects of accounting. This will include profit and loss statements and balance sheets.

* * *

Exercises

1 Obtain a copy of the annual report, including the financial statement and balance sheet, of any large company for the last financial year. What conclusions and recommendations can be made from these data?

2 What are the main sources of finance for small businesses?

3 Determine how the rate of interest has varied over the last 12 months for different types of loans.

4 What advantages will large companies have when borrowing money, and why?

5 Students are often completely unaware of the taxation structure within their own country. This is a surprising neglect, especially when considering the amount of tax (direct and indirect) paid by an individual during his working life. Determine the average salary paid to new graduates in your particular discipline, and the amount in taxes you would be required to pay. How does this change if your salary increases by 30%, or 50%?

6 What taxation charges are made against companies in your country? What are the basic rules, regulations and laws governing company tax liability?

7 How do personal taxation and company taxation laws vary in different countries?

*   *   *

KEYWORDS

economics	fair value
money	use value
liquid assets	market penetration
capital	product life cycle
interest	product growth curve
return on investment	Law of Supply and Demand
wealth	point of diminishing returns
income	capital investment
variable or running costs	operating costs
fixed costs	present economy method
asking price	present worth method
selling price	rate of return
value	capital costs
worth	depreciation
market value	cost of maintenance

BIBLIOGRAPHY

Betts, R.J., *Business Economics for Engineers*, McGraw-Hill Book Co., New York (1980).

De Garmo, E.P., *Engineering Economy*, 6th Ed., Macmillan Press Ltd, New York (1979).

Dell'Isola, A.J. and Kirk, S.J., *Life Cycle Costing for Design Professionals*, McGraw-Hill Book Co., New York (1981).
Well presented and many engineering case studies.

Humphreys, K.K., and Katell, S., *Basic Cost Engineering*, Marcel Dekker Inc., New York (1981).
A practical approach with engineering examples.

Kaplan, S., *Energy Economics: Quantitative Methods for Energy and Environmental Decisions*, McGraw-Hill Book Co., New York (1983).
An energy perspective on engineering economics, particularly useful for chemical, petroleum, mining, agricultural and nuclear engineering students.

Leech, D.J., *Economics and Financial Studies for Engineers*, Ellis Horwood Ltd, Chichester, Sussex (1982).
   Note—'for engineers'. Describes cash flow techniques, including computer programs, and also cost estimation in manufacturing industries.
Miles, L.D., *Techniques of Value Analysis and Engineering*, McGraw-Hill Book Co., New York (1972).
Mitchell, R.L., *Engineering Economics*, John Wiley and Sons Inc., New York (1980).
   Useful for cost appraisal and cost–benefit analysis.
Resnick, W., *Process Analysis and Design for Chemical Engineers*, McGraw-Hill Book Co., New York (1981).
   Economic analysis, evaluation and forecasting in process engineering.
Riggs, J.L., *Engineering Economics*, 2nd Ed., McGraw-Hill Book Co., New York (1982).
Modern Cost Engineering: Methods and Data, papers reproduced from Chemical Engineering Magazine, McGraw-Hill Publications Co., New York (1979).
The Engineering Economist, joint publication of the Engineering Economy Divisions of the American Society for Engineering Education and the American Institute of Industrial Engineers.
   In-depth, specialist topics, useful to see the depth of analysis!

Legal Factors

Chapter Objectives

To develop an appreciation of:

1 the legal protection that may be possible for engineering products, processes and designs;

2 the legal factors and restraints to be considered when designing a product.

* * *

Exercises

1 Define an invention. What sets an invention apart from an application of previously available knowledge?

2 Obtain a copy of a utility patent (UK or USA). Examine it in detail and identify the main features.

   What aspects make this patented item 'original'?

   How would you defend this patent against possible legal action to declare it non-valid?

   What aspects of the patent do you think are most vulnerable to legal action?

3 For which items can a patent *not* be granted?

4 What is copyright? Find some common examples. What rules relate to copyrights?

5 What is a trade mark? Find some common examples. Identify some names that *cannot* be used as trade marks.

6 What is a design registration (UK) or design patent (USA)? Obtain a copy of a design registration (patent) and show how it differs from a utility patent.

7 Imagine you are Sir Walter Raleigh returning to Earth, having been the first person to visit Mars. You are bringing back potatoes and tobacco which are unknown on Earth. With present legislation and scientific knowledge, decide the best ways to 'protect' your discoveries and maximise their potential commercial exploitation.

8 What protection is provided to the author of this book against reproduction, use and sale of unauthorised copies?

9 Assume you have devised the principle of the internal combustion engine as an alternative to the horsedrawn, four-wheel carriage. Decide the best protection for your invention.

   How could you maximise the life of the protective cover, and any returns from licensing the use of your invention?

10 What legal protection extends to the Ford Motor Co. for use of the name Ford, and the car model name 'Mustang'?

Could you trade under the name Fort Motor Co. and produce a 'Muchzang' car?

* * *

8.1 PATENTS AND INVENTIONS

A letters patent for an invention arises as a result of a contract between the inventor and the Crown (in the UK) or the Congress (in the USA). The inventor is then granted a monopoly for the invention for a number of years. In return he must provide a full and complete specification of the patented article for the public. The inventor does not gain the right to make, use or sell the invention (he usually already has this right under common law), only the right to prevent others from so doing. In many cases, the contract is between the inventor's employer and the Crown (or Congress).

The advantages of this system are that ideas are revealed and society will benefit, and that the inventor has protection for the use of his invention.

Defining what constitutes an invention is a difficult task. It is often determined subjectively, or established negatively by deciding what does not constitute an invention. An invention is often defined as something that did not exist previously. To be patentable, an invention must be new, useful, fully disclosed and must not have been previously abandoned. The patent should also relate to a new and useful process, or machine, or method of manufacture, or composition of matter, or new improvement, or in certain cases a method of testing (if this leads to the improvement or control of manufacture). The requirement of novelty means that the invention should not have been used, sold, or details published, prior to filing an application for a patent at the Patent Office (London) or the United States Patent and Trademark Office (Washington, DC). If information does need to be disclosed, say to a supplier, then it should be clearly marked as 'confidential'.

Patentable articles should fall into one of the following categories:

(a) process—using the laws of nature to make useful changes in materials, environment or information;
(b) machine—an assembly of parts producing a useful result;
(c) manufacture—any useful article made by man;
(d) composition of matter—any useful mixture or chemical compound;
(e) design—ornamental appearance;
(f) hybrid plants—asexually reproducible, not tuber propagated.

Categories (a) to (d) form the basis of utility patents (the 'usual' meaning of the term 'patent'), items in category (e) are known as design registrations (UK) or design patents (USA).

Items that are non-patentable include products of nature, laws of nature or principles (these existed before they were 'discovered'), musical notation, accounting systems, computer programs, medical treatment ideas and agricultural and horticultural ideas. An aggregation of items is not patentable if the items were previously known and if they can work separately, e.g. an eraser attached to a pencil. Similarly, finding a new use for a material or machine, mere substitution of materials, and substitution of elements to perform the same task do not qualify as inventions. Patents may be refused for inventions whose publication or exploitation may encourage offensive, immoral, illegal or anti-social behaviour. There are various grounds on which a patent may not be granted. These include insufficient description, ambiguity, lacking invention or novelty, false suggestion or claim, existence of a prior patent, previous use or disclosure, cannot be used, or the applicant is not entitled to apply.

The granting of a patent does not guarantee that the invention will be a commercial success and produce profits for the inventor. That will depend upon its advantages compared with the alternatives, and its usefulness. Profits may be realised if the inventor uses the patented item himself, or if he sells it or licenses it to someone else. The granting of a patent does not mean that the government, the Patent Office, or anyone else will prevent others from illegally using or abusing a patent. It does provide the patentee with an established legal method of protecting his property through the courts.

The designer should ask: What are the advantages of the present patent system? The main answer is incentive. This is achieved by encouraging the inventor to make the invention, knowing that legal protection is available. The public are given the opportunity to use the invention and society benefits because information is openly available, thus time need not be wasted by working on items already discovered. Also, publishing details of inventions should help to promote further discoveries.

The designer who wants to obtain a patent for his invention would do well to remember the old saying 'Bad advice is cheap, good advice is expensive'. Nowhere is this truer than in the field of patent application. A patent specification that is incorrectly or poorly presented will not provide protection for the inventor, and will have wasted time and money. The services, in the UK, of a registered patent agent will be well worth the fee he will charge.

In the USA a patent attorney is a lawyer who is qualified and registered to practice patent law, while those who are not lawyers and are registered are called patent agents.

The following descriptions apply specifically to the UK and include certain details relating to the USA; similar procedures are followed in other countries. Before applying for the granting of a patent, it will be helpful and informative if the inventor contacts the patent (and trade mark) office in his country. He can then obtain advice, and copies of the relevant instructional booklets which are produced. Details of some of those available in the UK and the USA are given in the bibliography at the end of this chapter. Informative and easily read books are those by Capsey dealing mainly with the UK system, by Liebesny for most countries (before 1972) and by Muncheryan specifically for the USA.

The application procedure to be followed will now be described briefly.

The legislation concerning UK patent applications is the Patents Act 1977. A patent application is made using the appropriate forms, a fee is paid and a patent specification with necessary drawings is submitted. The specification and the drawings should contain as full a description as possible of the invention, and should conform to the formal requirements for their presentation.

If the inventor wishes to proceed with a patent application through to granting, then the specification must be fully detailed, so that a 'suitably competent and knowledgeable person' can carry out the invention. It should also be described in precise language—another good reason for obtaining the services of a patent agent. Typically, the specification begins by describing the idea of the invention in general terms, with an indication of how it differs from previous work. This is then followed by a more detailed description of one or more 'embodiments' or examples of the invention.

If the inventor is uncertain whether he wishes to pursue the granting of a patent he can file a less detailed specification, thus establishing the 'priority date' for the invention. A new application can then be filed within 12 months from the original date of filing. The original application is declared for priority of the new application, the old application is allowed to lapse and the new one proceeds.

An inventor cannot add new or extra material to an application on file: the specification can only be extended by filing a new application as described above.

If not filed with the original application, then within 12 months the inventor must file an 'Abstract of the Invention'. This provides a concise summary of the contents of the specification with particular emphasis on the inventive aspect. The inventor must also file a request for preliminary examination and search using the prescribed forms, and including payment of a fee. One or more claims of the invention must also be filed; this defines the exact scope of the monopoly that is being sought. If there is more than one main claim, then the claims should be linked so that the application relates only to one invention or inventive concept.

An examiner will then search through documents in the Patent Office to see if there exists any earlier published description of the invention described by the claim. The applicant must also file a statement of inventorship, naming all inventors and how he derived the right to be granted the patent.

If the applicant decides to proceed with the application after he has considered the preliminary search report, then he must file a request for a substantive examination. The examiner will then examine the specification in detail and write a report which is sent to the applicant. Any unsatisfactory features are notified to the applicant, and discussion and re-examination will result until the examiner is satisfied that there are no outstanding objections to the granting of a patent. This must be within 4½ years from the original filing date and a patent can then be granted.

The applicant is given an opportunity to submit amendments to his descriptions and claims, but they must not extend the matter disclosed in the specification originally filed. In the UK, the maximum life of a patent,

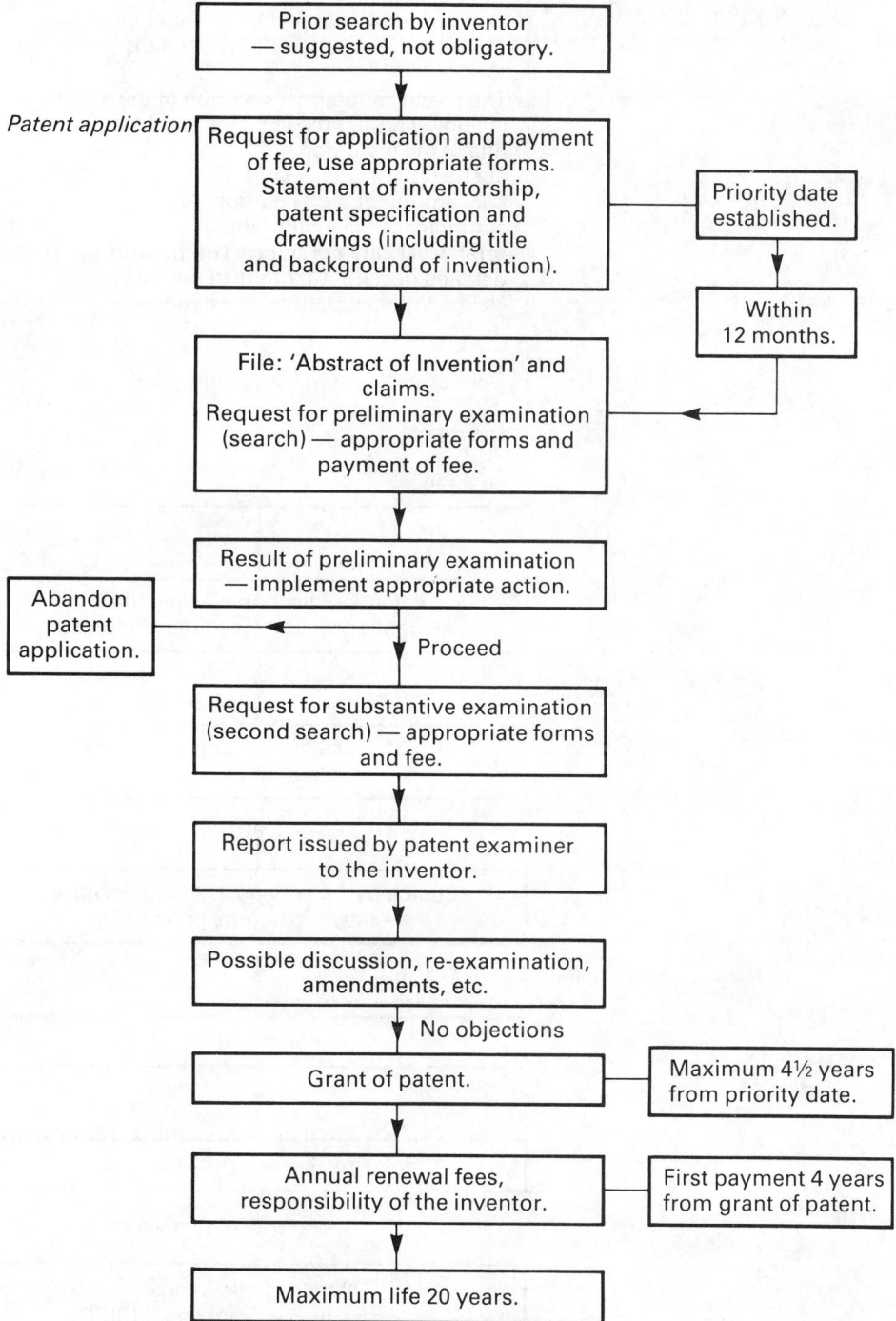

Patent application

Figure 8.1   Summary of the Stages and Requirements for Granting of a UK Patent

Application for filing:

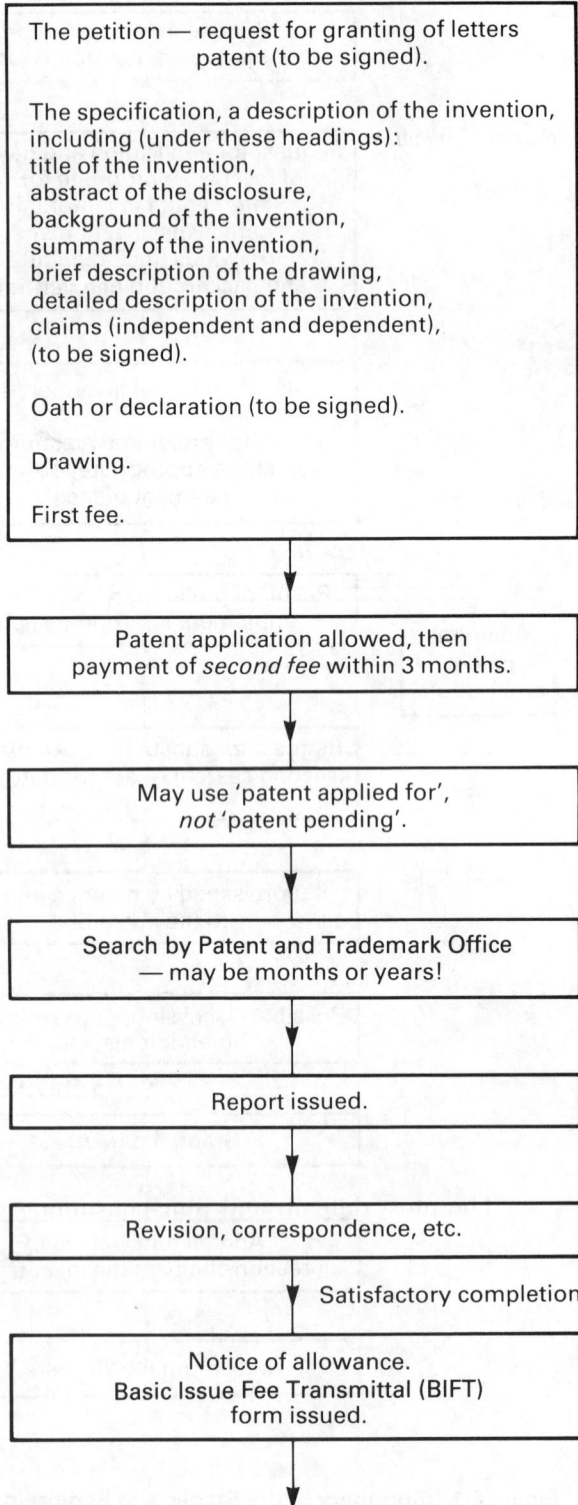

The petition — request for granting of letters patent (to be signed).

The specification, a description of the invention, including (under these headings):
title of the invention,
abstract of the disclosure,
background of the invention,
summary of the invention,
brief description of the drawing,
detailed description of the invention,
claims (independent and dependent),
(to be signed).

Oath or declaration (to be signed).

Drawing.

First fee.

Patent application allowed, then payment of *second fee* within 3 months.

May use 'patent applied for', *not* 'patent pending'.

Search by Patent and Trademark Office — may be months or years!

Report issued.

Revision, correspondence, etc.

Satisfactory completion

Notice of allowance.
Basic Issue Fee Transmittal (BIFT) form issued.

```
                    ┌──────────────────────────────────┐
                    │ Within 3 months return 'BIFT' form │
                    │ with payment of *final fee* (including │
                    │            cost of copies).          │
                    └──────────────────────────────────┘
                                    │
                                    ▼
                    ┌──────────────────────────────────┐
                    │  Fee received and patent issued.    │
                    │   The record of patent is now       │
                    │    available to the public.         │
                    └──────────────────────────────────┘
                                    │
                                    ▼
                    ┌──────────────────────────────────┐
                    │ First, second, third patent maintenance │
                    │  fees due 3½, 7, 14 years after     │
                    │   issue of patent, respectively.    │
                    └──────────────────────────────────┘
                                    │
                                    ▼
                    ┌──────────────────────────────────┐
                    │       Maximum life 17 years.        │
                    └──────────────────────────────────┘
```

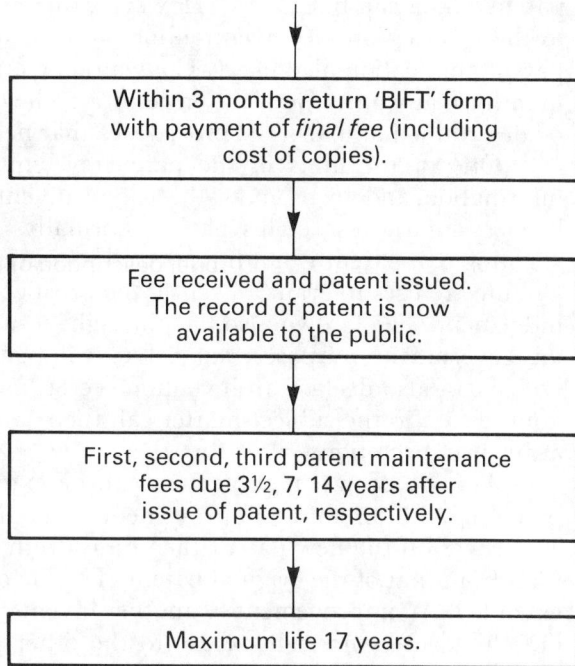

Figure 8.2   Summary of the Requirements and Stages for Granting of a US Utility Patent

provided renewal fees are paid each year, is 20 years from the date of the application. No renewal fees are due for the first 4 years of the life of a patent. The onus is on the applicant, not the Patent Office, to ensure that a patent is renewed each year.

The requirements and stages involved in obtaining a patent in the UK and the possible timescale are shown in the flowchart in Fig. 8.1. The corresponding process in the USA is shown in Fig. 8.2. Also included are details of the official requirements for the patent specification.

The most difficult and time-consuming stage in either country, will be that following the issue of the findings of the patent examiner, if he raises objections to the application. However, even if this occurs, there is still plenty of opportunity for reappraisal, evaluation, amendment, negotiation and explanation. All is not lost!

An example of a UK patent application is shown in Fig. 8.3 and an example of a US patent in Fig. 8.4.

There are a few important points which the intending patentee should bear in mind. The designer should decide at an early stage in which countries he requires patent protection for his invention. There is no 'international' patent which can be granted. However, there are many countries that belong to the International Convention for the Protection of Industrial Property. A patent application filed in any Convention country can also be filed within 12 months in any other Convention country. The cost of foreign patenting should be determined and also the procedures that need to be followed.

One of the aims of the patent system is to provide easy access to information and to promote the use of inventions. The patentee can grant licences for the use of his patent, normally in exchange for royalties. The granting of a patent is not intended to encourage restrictive practices and if a patentee refuses to grant a licence, the prospective user can appeal for a legal judgement. The Crown also has the right to use any UK patented invention for government purposes, subject to compensation payments. The Patent Office can also declare an invention secret, and the government department appropriate to the subject-matter can then cancel or renew the secrecy order at its own discretion.

The life of patent protection can be extended by patenting a separate patent of addition for an improvement or modification to an existing process. The cover on the new patent then dates from the date of application, rather than from that of the original patent. The life of the original patent cannot be extended. An improvement to another inventor's patent can also be patented, but this confers no right of use to the other patent for either party; it is a matter for negotiation.

The Patent Office resources can also be used to perform personal patent searches, and details of the most efficient method of performing these are given in the official booklet listed in the bibliography. It will be important during the detailed design stage, and at all future stages, to perform a thorough search of relevant prior patented ideas. This will prevent time and money being wasted in developing and implementing ideas already available. It will also prevent liability for compensation due to patent infringement.

8.2 DESIGN REGISTRATION

Design registration (called design patent in the USA) is also carried out by the Patent Office (and US Patent and Trademark Office), but it is very different from the grant of a patent. The important legislation in the UK is the Registered Designs Acts 1949–61 and the Design Copyright Act 1968. 'Industrial design' means the appearance of articles manufactured in quantities of more than 50, or goods produced in lengths or pieces. It includes the features of shape, configuration, pattern or ornament applied to an article by any industrial process. The features are judged by the eye, not including the method or principle of construction or the function it is required to perform. In the USA a design patent refers only to the artistic appearance of the item, not its utility. The item should be a new, original and ornamental design for a manufactured product, the utility feature of an item cannot be patented as a design patent. Three typical (but not actual) examples of items

(12) **UK Patent Application** (19) **GB** (11) **2 122 911 A**

(21) Application No **8317818**
(22) Date of filing **30 Jun 1983**
(30) Priority data
(31) **8218830**
(32) **30 Jun 1982**
(33) **United Kingdom (GB)**
(43) Application published
　　25 Jan 1984
(51) INT CL³
　　A61M 5/16
(52) Domestic classification
　　B1D 1107 1510 1806
　　1819 1821 1902 1906
　　2304 KC
　　U1S 1052 1296 B1D
(56) Documents cited
　　GB 1505589
　　GB 1501665
　　GB 1334555
　　GB 1297651
(58) Field of search
　　B1D
(71) Applicant
　　Ronald Bernard Baker, c/o
　　GB Biomedical Products,
　　19 Southgate Spinneys,
　　South Ranceby, Sleaford,
　　Lincolnshire NG34 8QF
(72) Inventor
　　Ronald Bernard Baker
(74) Agent and/or Address for
　　Service
　　Swindell and Pearson,
　　44 Friargate, Derby
　　DE1 1DA

(54) **Blood filter**

(57) A disposable blood filter
comprises a chamber having an inlet
(12) and an outlet (16) separated by
one or more mesh filter elements (18)
arranged in pleated preferably tubular

form with the folds of the pleats
running from top to bottom of the
chamber. The filter is generally made
from plastics, and the filter element of
polyester or nylon meshes
interspersed with supporting medium
(see Figs. 3 & 4).

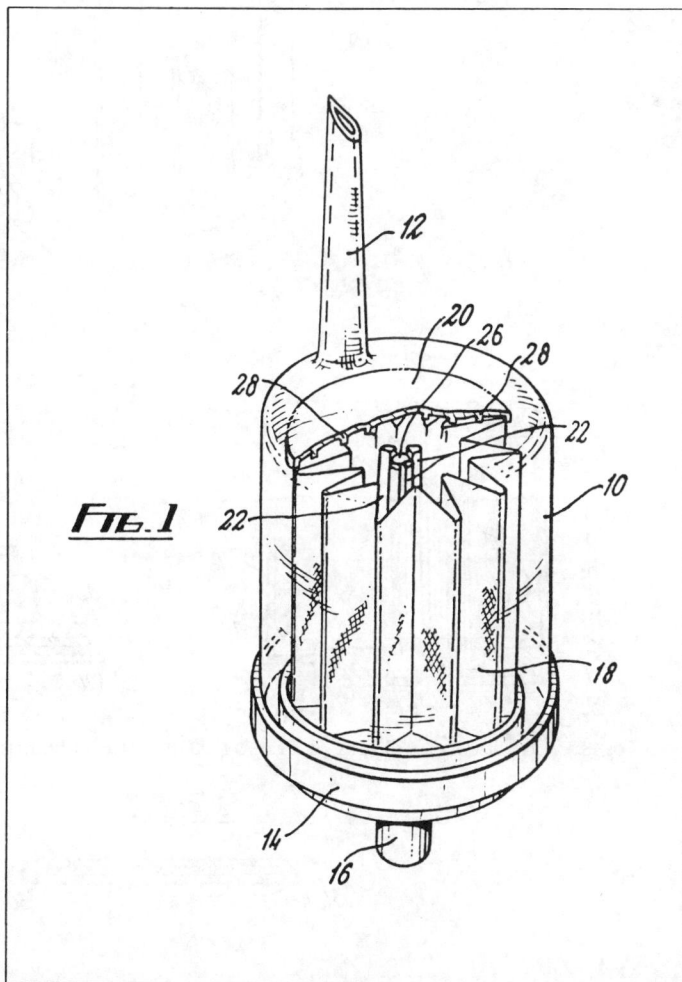

FIG.1

GB 2 122 911 A

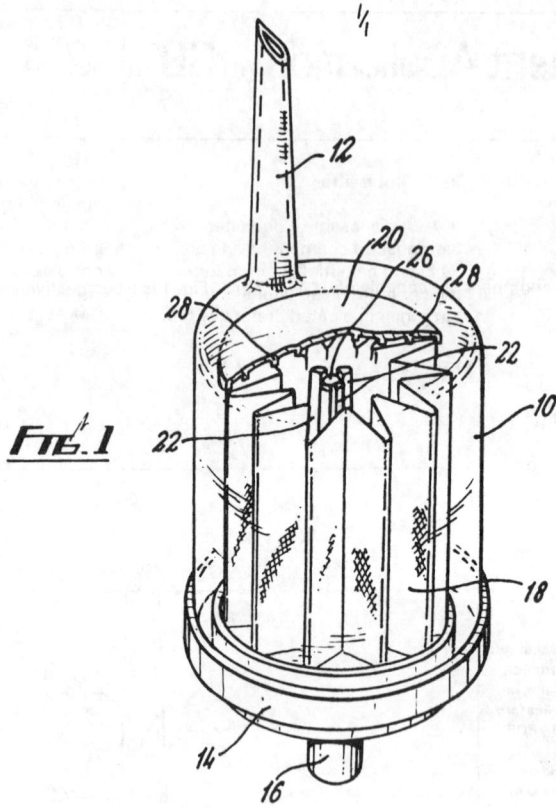

Fig.1

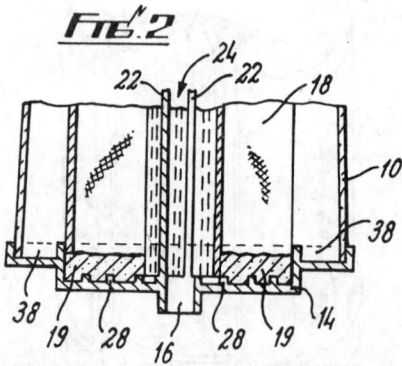

Fig.2

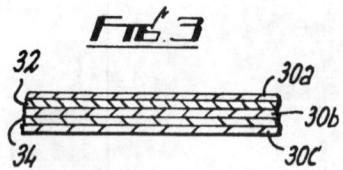

Fig.3

Fig.4

1　　　　　　　　　　　　　　　　　　　　　　　　　　GB 2 122 911 A　　　1

SPECIFICATION
Apparatus for use in blood transfusion

The invention relates to apparatus for use in
blood transfusion, particularly for use in filtering
5 micro-emboli from stored human blood before
administration to the patient.

Known apparatus of this type is either bulky in
order to maintain a suitable flow rate, a
disadvantage being that the volume of the device
10 is somewhat large, or compact but capable of
delivering blood at only small flow rates. Some
known apparatus utilises filter media lying in
horizontal planes and which thus quickly become
clogged with debris and micro-bubbles, whilst
15 others have filter media lying in vertical planes but
comprise filter media of the "depth" type
presenting a greater resistance to flow and
increased likelihood of damage to the blood
structure and being of small first surface area.

20 The present invention provides apparatus for
use in blood transfusion comprising a chamber, an
inlet for blood and an outlet for blood, the
apparatus being arranged in use with the inlet
higher than the outlet, the chamber being divided
25 into two parts separated from one another by at
least one mesh filter element arranged in pleated
form with folds of the pleats arranged to extend
substantially from a top region to a bottom region
of the chamber, the inlet communicating with one
30 part of the chamber and the outlet communicating
with the other part.

The term "mesh" as used herein is intended to
mean formed from monofilament, as distinct from
the "depth" type of element which comprises a
35 fibre mat of relatively large thickness (typically
40—200 μm) and volume.

Preferably the mesh is supported between two
layers of support material, and the mesh size lies
in the range 10—70 μm. Desirably two meshes
40 are provided, one of about 50—70 μm, desirably
60 μm mesh size, the other of mesh size less than
25 μm preferably 23 μm, the meshes being
supported between alternate layers of supporting
material. If a single mesh is provided, the mesh
45 size is preferably about 40 μm.

The or each filter element is preferably formed
into a pleated tube having a star-shape in cross
section, the star having a large number of limbs
but not so many that they have surfaces which
50 overlie one another or otherwise obstruct the
access of blood and blood clot material to the filter
surfaces. Preferably the area inside the star-
shaped element is substantially unobstructed and
forms part of said chamber, desirably the part
55 communicating with the blood outlet. The open
ends of the pleated tube are sealed to end plates
in fluid-tight manner, preferably by hot melt
adhesive, preferably one end plate providing a wall
of said chamber and having said outlet therein. If
60 the filter element is made from a flat piece of
material having two free ends which are joined
together to form the tube, the free ends are
preferably joined using a melt-sealing technique or
ultrasonic welding.

65 The supporting material may be polyester,
nylon or any other suitable material, and is
desirably net-like and of large mesh size relative to
the mesh of the filter element.

The filter element is preferably made from
70 polyester or nylon mesh.

Embodiments of the invention will now be
described by way of example only and with
reference to the accompanying drawings, in
which:—

75 Fig. 1 is a perspective view partly cut away of
apparatus according to the invention;

Fig. 2 is a schematic cross-sectional view of a
lower part of the apparatus shown in Fig. 1;

Fig. 3 is a schematic diagram showing the
80 construction of a filter element utilised in the
apparatus of Fig. 1; and

Fig. 4 is a schematic diagram showing the
construction of an alternative filter element to that
shown in Fig. 3.

85 The apparatus shown in Fig. 1 comprises a
transparent plastic dome 10 having a blood inlet
tube 12 therein, and a plastic base cap 14, which
together provide a chamber for blood. The base
cap 14 has a blood outlet tube 16 at its centre
90 which communicates with the interior of a tube-
shaped, pleated filter element 18 whose ends are
sealed by hot melt adhesive 19 to the base cap
and a top cap 20 respectively. The space exterior
of the filter element 18 communicates with the
95 inlet tube 12. Thus blood entering through the
inlet tube 12 can only leave the apparatus through
the outlet tube 16 after passing through the filter
element 18. The base cap 14 and dome 10 are
secured in fluid tight manner by acrylic sealer. On
100 an interior surface of the base cap 14 the inlet
tube terminates and three support rods 22 with
apertures 24, for passage of blood therebetween,
extend upwardly to the top cap 20, and are held in
spaced apart relation by a web member 26. The
105 top cap 20 has on its underside at the centre a
stud (not shown) of circular section which is
received in an interference fit between the ends of
rods 22 and is secured thereto by hot melt
adhesive, so connecting the base cap 14 and the
110 top cap 20. Lower and upper edges of the filter
element abut concentric circular ribs 28 at the
base cap 14 and top cap 20 respectively, these
ribs helping to form a key for the hot melt
adhesive.

115 As can be seen from Fig. 1, the filter element is
of tubular pleated shape and is constructed from
support mesh and filter mesh, the tube being
formed from a sheet whose ends are secured to
make the tube by ultrasonic welding in fluid-tight
120 manner. Fig. 3 shows a preferred filter element
construction comprising from the upper blood
inlet side, a first supporting mesh 30a of widely
spaced polyester monofilament, a first polyester
filter mesh 32 of 60 μm, a second polyester
125 supporting mesh 30b, a second filter polyester
mesh 34 of 23 μm, and finally a third polyester
supporting mesh 30c. Fig. 4 shows an alternative
filter element comprising first and second
polyester support meshes 30a and 30b

respectively between which is sandwiched a nylon filter element 36 of 40 μm. The supporting meshes are 800 μm mesh size.

In operation, blood flows through inlet tube 12
5 through filter element 18 where micro-emboli is separated out, tending to fall to the base of the apparatus in regions 38, through the apertures 24 between rods 22 and thence out through the outlet tube 16.
10   Various modifications may be made to the apparatus without departing from the scope of the invention, for example the filter element may comprise any suitable number of layers of mesh, and the various parts of the apparatus may be
15 secured together by any convenient means which does not adversely affect the blood being filtered.

The filter element should have surfaces which are generally upright when the apparatus is in use so that separated debris does not generally lie
20 over the mesh to impede blood flow. The volume of the apparatus is as small as possible consistent with a reasonable flow rate, and the filter is pleated to give a large filter surface area. The filter element is made with monofilament filter meshes
25 in one or more layers, and desirably the supporting meshes are made of polyester monofilament, the filter meshes being made of polyester or nylon. It is intended that the apparatus according to the invention will be disposable after use.
30   Whilst endeavouring in the foregoing specification to draw attention to those features of the invention believed to be of particular importance it should be understood that the Applicant claims protection in respect of any
35 patentable feature or combination of features hereinbefore referred to whether or not particular emphasis has been placed thereon.

When the apparatus is used with a second or subsequent batch of blood, the configuration of
40 the apparatus is such that any air micro-bubbles introduced into the apparatus will not impede the flow of blood to the filter.

The support mesh is desirably of polyester material. Where a single filter mesh is provided as
45 hereinbefore described the filter mesh is preferably of nylon material. Where two filter meshes are provided these are preferably of polyester material.

CLAIMS
50   1. Apparatus for use in blood transfusion comprising a chamber, an inlet for blood and an outlet for blood, the apparatus being arranged in use with the inlet higher than the outlet, the chamber being divided into two parts separated
55 from one another by at least one mesh filter element arranged in pleated form with folds of the pleats arranged to extend substantially from a top region to a bottom region of the chamber, the inlet communicating with one part of the chamber and
60 the outlet communicating with the other part of

the chamber.
   2. Apparatus as claimed in Claim 1, in which a single mesh filter element is provided, the mesh size lying in the range 15—70 μm.
65   3. Apparatus as claimed in Claim 2, in which the mesh size of the filter element is 40 μm.
   4. Apparatus as claimed in Claim 1, in which two filter elements are provided, the mesh size of the first being in the range 50—70 μm and that of
70 the second being less than 25 μm.
   5. Apparatus as claimed in Claim 4, in which the mesh size of the first element is 60 μm.
   6. Apparatus as claimed in Claim 4 or Claim 5, in which the mesh size of the second element is
75 23 μm.
   7. Apparatus as claimed in any of Claims 4 to 6, in which the first filter element is on the blood inlet side of the apparatus and the second filter element on the blood outlet side.
80   8. Apparatus as claimed in any preceding Claim, in which the or each filter element is provided with at least one monofilament supporting mesh of large mesh size.
   9. Apparatus as claimed in Claim 8, in which
85 the or each filter element is sandwiched between two monofilament supporting meshes of large mesh size.
   10. Apparatus as claimed in Claim 8 or Claim 9, in which the mesh size of the supporting mesh is
90 of the order of 1 millimetre.
   11. Apparatus as claimed in Claim 8, 9 or 10, in which the monofilament supporting mesh is made from polyester material.
   12. Apparatus as claimed in any preceding
95 claim, in which the apparatus comprises a dome part having one of said blood inlet or blood outlet therein and a base part having the other of said blood inlet or blood outlet therein.
   13. Apparatus as claimed in Claim 12, in which
100 the dome part is made of transparent material.
   14. Apparatus as claimed in Claim 12 or Claim 13, in which the or each filter element is sealingly secured at one of its open ends to the base part, and at another open end to a cap member secured
105 in spaced relation to said base part within the dome part.
   15. Apparatus as claimed in Claim 13 in which said base part has the blood outlet therein, said blood outlet communicating with a space in the
110 interior of the or each tubular filter element.
   16. Apparatus as claimed in any preceding claim, in which the apparatus is made of plastics material.
   17. Apparatus as claimed in any preceding
115 claim, in which the or each filter element is made from polyester monofilament.
   18. Apparatus as claimed in any preceding claim, in which the or each filter element is made from nylon monofilament.
120   19. Apparatus substantially as hereinbefore described with reference to the accompanying drawings.

Figure 8.3   Example of a UK Patent Application (reproduced by permission of the inventor and the Comptroller of Her Majesty's Stationery Office, © Crown Copyright)

United States Patent

Pinto

[11]   **4,213,954**

[45]   **Jul. 22, 1980**

[54] **AMMONIA PRODUCTION PROCESS**

[75] Inventor: **Alwyn Pinto**, Stockton-on-Tees, England

[73] Assignee: **Imperial Chemical Industries Limited**, London, England

[21] Appl. No.: **939,916**

[22] Filed: **Sep. 5, 1978**

[30]   **Foreign Application Priority Data**

   Sep. 16, 1977 [GB]   United Kingdom 38713/77

[51] **Int. Cl.²** **C01C 1/04**
[52] **U.S. Cl.** **423/359; 423/361; 422/148**
[58] **Field of Search** 423/359-363; 252/374-377; 55/75; 422/148

[56]   **References Cited**

U.S. PATENT DOCUMENTS

3,441,393	4/1969	Finneran et al.	423/359
3,705,009	12/1972	Dougherty	423/361

FOREIGN PATENT DOCUMENTS*

1156002	6/1969	United Kingdom	423/359
1186939	4/1970	United Kingdom	423/359

OTHER PUBLICATIONS

Olsen et al., Unit Processes & Principles of Chem. Eng., 1932, pp. 1-3.

Primary Examiner—O. R. Vertiz
Assistant Examiner—Thomas W. Roy
Attorney, Agent, or Firm—Cushman, Darby & Cushman

[57]   **ABSTRACT**

An ammonia production process comprising synthesis gas generation and ammonia synthesis with heat recovery in the generation section by steam raising and steam superheating and heat recovery in the synthesis section by steam raising or boiler feed water heating is made more controllable by effecting part of the superheating in the synthesis section, so that a shut-down of the synthesis section results not only in a decrease in steam flow through the superheaters, but also in a decrease in the temperature of the steam fed to them. Thus overheating of the superheaters, a defect of some earlier processes, is avoided. The process is operated with a synthesis pressure under 150, preferably 40–80 bar abs. and preferably synthesis gas generation is based on primary hydrocarbon steam reforming at such a pressure that synthesis gas compression is by under 50%.

7 Claims, 1 Drawing Figure

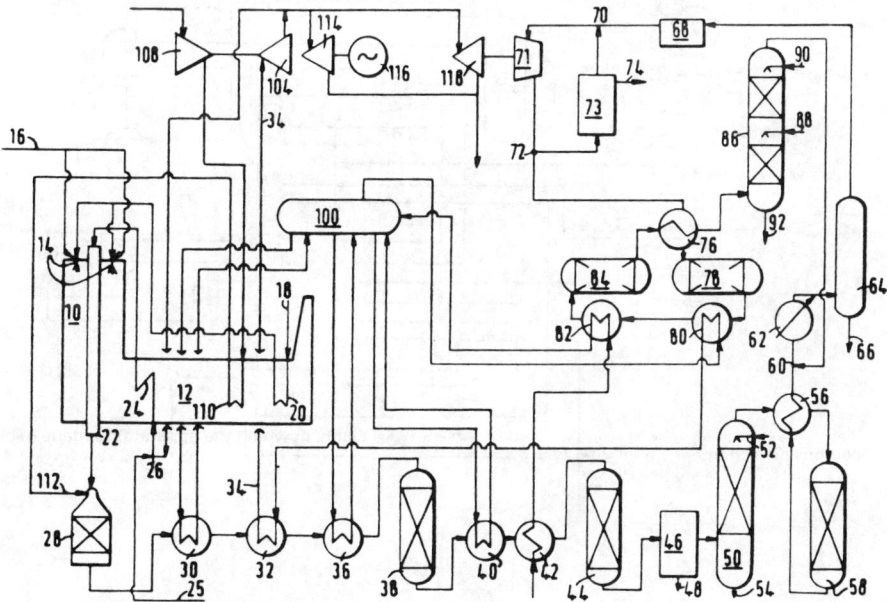

U.S. Patent

Jul. 22, 1980

4,213,954

4,213,954

1

AMMONIA PRODUCTION PROCESS

PROCESS OF PRODUCING AMMONIA

This invention relates to a steam system and in partic- 5
ular to a process of producing ammonia including the
system.

Processes for producing ammonia comprise the succession of three main chemical reactions, namely

reaction of carbon with steam and/or oxygen, for 10
example:

$$CH_4 + H_2O \rightarrow CO + 3H_2 \qquad\qquad Ia$$

$$C + H_2O \rightarrow CO + H_2 \qquad\qquad Ib \quad 15$$

$$C + O \rightarrow CO \qquad\qquad Ic$$

$$CH_4 + O \rightarrow CO + 2H_2 \qquad\qquad Id;$$

20

shift reaction

$$CO + H_2O \rightarrow CO_2 + H_2 \qquad\qquad II;$$
and

25

ammonia synthesis

$$N_2 + 3H_2 \rightarrow 2NH_3 \qquad\qquad III.$$

Each of these reactions produces a gas at over 400° C.,
from which useful heat can be recovered as steam or 30
boiler feed water. In the last decade processes have
become established in which steam has been generated
at high pressure, usually in the range 60–200 bar abs.,
superheated, let down in turbines exhausting at the
pressure just over that at which reaction Ia is carried 35
out, partly used in reaction Ia and for the rest let down
in condensing turbines or turbines exhausting at low
pressure. Various ways of integrating the heat recoveries from the hot gases have been proposed. In one system, described in U.S. Pat. No. 3,441,393, steam is raised 40
in boilers heated by process gas from reactions I and II
and by flue gas from the furnace in which reaction Ia is
carried out, the steam is superheated by the flue gas
before being let down in the turbines, and the heat
evolved in reaction III is taken up by heating feed water 45
for the boilers. There have been many proposals to raise
steam in a boiler heated by reacted ammonia synthesis
gas (reaction III), and such steam can be superheated by
heat exchange with the flue gas derived from reaction I.
A process in which steam is superheated by heat ex- 50
change with process gas is described in Chemiker-
Zeitung 1974, 98(9), 438-445.

These prior systems are subject to a control defect,
namely that a shut-down of the ammonia synthesis section (steps c and d below) of the ammonia production 55
process while the synthesis gas generation section (steps
a and b below) continues in operation results in a decrease in the supply of boiler feed water and/or steam
from the synthesis section and thus to a decrease in the
flow of steam through the superheaters. Since it is im- 60
practicable to shut down the generation section quickly,
the superheaters become overheated and may suffer
damage.

We have now realised that if the ammonia synthesis
pressure is less than 150 bar abs., for example in the 65
range 40–80, especially under 60 bar abs., it becomes
possible to avoid this control defect. This is because it is
practicable to construct a heat exchanger having steam

2

at 60–150 bar abs., on its cold side and ammonia synthesis gas at less than 150 bar abs., on its hot side, and thus
to superheat steam in the ammonia synthesis section. By
contrast, in synthesis processes at higher pressure it was
practicable only to have water on the cold side of such
a heat exchanger, in order to keep down its temperature. Thus in our process, whereas a shut-down of ammonia synthesis decreases the flow of steam in the superheaters in the generation section, it also decreases
the total superheating capacity of the plant and leaves
those superheaters adequately loaded.

According to the invention an ammonia production
process comprises the steps of

(a) reacting a carbonaceous feedstock with steam and-
/or oxygen to give a gas containing carbon monoxide;

(b) reacting the product of step (a) over a shift catalyst
to convert carbon monoxide and steam to carbon
dioxide and hydrogen and removing carbon oxides
and steam from the resulting gas;

(c) reacting the hydrogen with nitrogen over an ammonia synthesis catalyst, cooling the reacted synthesis
gas and recovering ammonia from it;

(d) heat exchanging hot gases produced in at least step
(c) with water in a boiler producing steam at a pressure in the range 60–200 bar abs. and/or in a water
heater feeding such a boiler;

(e) superheating steam from such a boiler by heat exchange with hot gases produced in step (a) or step (b);
and

(f) letting down such steam in one or more expansion
engines and thereby recovering useful power; and is
characterised by

in step (c) (i) reacting and cooling the synthesis gas at a
a pressure under 150 bar abs., and

(ii) cooling the reacted synthesis gas at least partly by
superheating steam produced in step (d) and/or re-heat-
ing steam within or from one or more of the engines of
step (f).

In the succeeding description the term "stage" will be
used to denote an operation forming part of the above
defined steps.

Step (a) can be any suitable gasification process, for
example non-catalytic partial oxidation of coal, residual
hydrocarbon or vaporisable hydrocarbon, catalytic
partial oxidation of vaporisable hydrocarbon or catalytic steam reforming. Catalytic steam reforming is
conveniently carried out in two stages:

primary reforming a hydrocarbon feedstock with
steam to give a gas containing carbon oxides, hydrogen
and methane. This stage is carried out with external
heating with the aid of a furnace, the flue gas of which
is one of the hot gases produced in step (a); and

secondary reforming the gas from primary reforming
by introducing air and bringing the mixture to equilibrium, whereby to produce a hot gas containing nitrogen, carbon oxides, hydrogen and a decreased quantity
of methane.

The hydrocarbon feedstock is preferably methane or
other steam reformable hydrocarbon such as a normally
gaseous or liquid hydrocarbon boiling at up to about
220° C. Primary reforming can be in one stage over a
catalyst with external heating or, when the feedstock is
of a higher molecular weight than methane and especially when it is normally liquid, can be in two stages, in
the first of which the feedstock is converted to a gas of
high methane content at an outlet temperature under

3

4,213,954

4

650° C. and in the second of which that gas is reacted in the externally heated process. Various types of supported nickel catalyst are available for these hydrocarbon-steam reactions. The feedstock preferably should be substantially sulphur-free (under 0.5 ppm S) and may 5 have been subjected to a preliminary catalytic hydrodesulphurisation and H_2S-removal treatment.

External heating in primary reforming can be effected by having the catalyst in tubes surrounded by a furnace or in adiabatic beds preceded by re-heat zones. 10 Before entering the catalyst, when present in heated tubes, the hydrocarbon steam mixture is preheated, preferably to a temperature in the range 600°-700° C. When the hydrocarbon is methane or ethane it can be used directly at such a high preheat temperature. When 15 it contains 3 or more carbon atoms in the molecule or includes unsaturated hydrocarbons, in preliminary conversion to gas of high methane content is useful in permitting the preferred high preheat temperature.

For a high level of fuel economy the pressure at the 20 outlet of primary reforming is preferably in the range 30–120, especially 40–80 bar abs. and the temperature is in the range 750°-850° C. The steam ratio is preferably in the range 2.5–3.5 molecules of total steam per atom of carbon in the hydrocarbon if all the steam and hydro- 25 carbon are fed to this stage. The methane content of the product gas is typically in the range 10 to 20% by volume on a dry basis and this is preferred, although it is considerably higher than is normally thought suitable in the primary reformer gas of an ammonia synthesis plant. 30

The feed to secondary reforming includes the primary reforming gas, which may if desired be further heated before entering secondary reforming, and an oxygen-containing gas, which is conveniently air and preferably is preheated to a temperature in the range 35 400°-700° C. Further steam can be fed or further hydrocarbon feedstock if it is desired to minimise the total steam ratio without having too low a steam ratio in primary reforming but usually all the hydrocarbon and steam are fed to primary reforming. The outlet pressure 40 is conveniently about the same as at the outlet of primary reforming apart from the pressure drop through the secondary reforming catalyst. The outlet temperature is preferably in the range 950°-1050° C. and the outlet methane content in the range 0.2 to 10% v/v on 45 a dry basis. The excess of air results in a gas containing 2.0 to 2.9, especially 2.2 to 2.7 molecules of hydrogen equivalent (i.e. total of H_2 and CO) per molecule of nitrogen. The catalyst can be for example a supported nickel catalyst or chromium oxide catalyst or a combi- 50 nation thereof.

The process in which there is used a quantity of air in excess of what would introduce 1 molecule of nitrogen per 3 molecules of hydrogen preferably combined, as described in our copending UK application taking pri- 55 ority from UK application 35096/77, 44766/77 and 44996/77, with the subsequent feature, as part of step (c), of treating synthesis gas after reaction to synthesise ammonia to separate a stream enriched in hydrogen and returning the enriched stream to the synthesis. 60

In the steps so far described, heat is recovered from the reforming furnace flue gas, if steam reforming is used. The flue gas is typically at 900°-1000° C. after it has left the radiative zone of the furnace in which the catalyst is heated. Suitably in the flue gas duct (also 65 referred to as the "convective zone") it is heat-exchanged first with the steam-hydrocarbon mixture about to be fed to the catalyst, then possibly though not

preferably with high pressure steam, then possibly with the tubes of a boiler raising steam at 60–200 bar abs., then possibly with a boiler feed water heater, heaters for natural gas and secondary reformer air and finally a low grade heat recovery such as a furnace combustion air direct preheater or a water heater as described in our UK application 8732-14553/75 (published as German application 2608486) supplying hot water as a source of heat for furnace combustion air or of part of the process steam. The steam at 60–200 bar abs. is to be used, after superheating, as the working fluid in an expansion engine such as a turbine, the exhaust of which is partly used as the steam supply to step (a) and partly, after reheating, in one or more further turbines exhausting to condenser or at a few bar pressure. Steam that has been or is to be superheated or re-heated in heat exchange with reacted ammonia synthesis gas and possibly also by process gas can advantageously be superheated or reheated in part by furnace flue gas.

Whether or not step (a) involves primary steam reforming, heat is recovered from the hot process gas when cooling it to the inlet temperature of step (b). The heat recoveries are similar to those of furnace flue gas, except that preheating of the steam-hydrocarbon mixture or of secondary reformer air is not usually effected by cooling process gas. Superheating or reheating by process gas can be advantageously applied to steam that has been or is to be superheated or reheated in heat exchange with reacted ammonia synthesis gas.

For step (b) the gas from step (a) is cooled with the recovery of useful heat to the inlet temperature of the "shift" catalyst over which the reaction of carbon monoxide with steam occurs. Usually this temperature is in the range 300°-400° C., especially 320°-350° C., appropriate to iron-chrome shift catalysts. The reaction over the iron-chrome catalyst is exothermic (outlet temperature 400°-450° C.) and the outlet gas is again cooled with recovery of useful heat in a similar manner to the cooling of process gas from step (a), including superheating or reheating of steam that has been or is to be superheated or reheated in heat exchange with reacted ammonia synthesis gas.

In addition to the superheating and/or reheating by one or more of flue gas, reformer gas or shifted gas, it is preferred to use an independently fired heater to provide 20° to 50° C. of the steam temperature increase. Such a heater can be rapidly shut down if the steam flow to the other superheaters or reheaters decreases. Suitably the heat taken from reacted synthesis gas and the fired heater amounts to 40–60% of the total superheating.

The heat recoveries after high temperature shift cool the gas sufficiently to permit further shift reaction, and this is carried out preferably over a copper-containing catalyst, for which the inlet temperature is suitably 200°-240° C. and the outlet temperature 240°-270° C. Such a low-temperature shift stage produces a gas containing usually 0.2 to 0.6% v/v of carbon monoxide on a dry basis. Since the pressure is higher than has been generally used in low temperature shift, the steam to dry gas ratio over the low temperature shift catalyst is kept down to a level that avoids damage to the catalyst, preferably in the range 0.1 to 0.3. This means that the steam to carbon ratio in primary and secondary reforming should not be too high, but it can be readily attained using ratios in the range 2.5 to 3.5 (methane feedstock) or 2.4 to 3.2 (feedstock of empirical formula CH_2) or in intermediate ranges for hydrocarbons of intermediate

4,213,954

5

composition. The higher nitrogen content due to excess air helps to keep down the steam to dry gas ratio. If it is desired to use a higher steam to carbon ratio, or to add extra steam in or after secondary reforming, the steam to dry gas ratio can be kept down by recycling synthesis gas after carbon oxides removal.

The outlet temperature of the low temperature shift stage is too low to produce high pressure steam, but useful heat can be recovered from the shifted gas in lower grade heat recoveries such as boiler feed water heating and feedstock heating in series or parallel with the recoveries from the hotter gases already described and also such as carbon dioxide removal solution regeneration and aqueous ammonia distillation.

If other shift conversion systems such as iron chrome/CO_2 removal/iron chrome or systems based on catalysts containing other Group VI and Group VIII metals and oxides or an alkali metal compounds of weak acids are used, analogous heat recoveries are effected.

Removal of carbon oxides is usually carried out in a first stage in which carbon dioxide is substantially removed, and in a second leaving carbon oxides at such a low level that the synthesis catalyst is not significantly poisoned by them. If desired, any carbon monoxide remaining after shift can be selectively oxidised to carbon dioxide.

The shifted gas is cooled to below the dewpoint of steam, water is separated, and then the first stage can be carried out using any liquid absorbent. Chemical systems such as Benfield's potassium carbonate or diethanolamine potassium carbonate, "Vetrocoke", "Catacarb" or amine systems such as monoethanolamine can be used. These have, however, the disadvantage of consuming a substantial quantity of steam in the regeneration of the liquid absorbent, a requirement that is especially inconvenient when the shift steam to dry gas ratio is at the preferred low levels. A 2-stage carbon dioxide removal, in the first stage of which most of the carbon dioxide is absorbed in an amine, such as triethanolamine, that is regenerable substantially without heating, and in the second stage of which an absorbent is used that does require heat regeration, is therefore preferable. The preferred high pressure in step (a), which is substantially maintained apart from inevitable pressure drops in subsequent steps, makes possible the use of "physical" absorbents, the preferred examples of which can be regenerated merely by lowering pressure. Suitable absorbents used in industrially developed processes are tetramethylene sulfone ("Sulfinol"), propylene carbonate (Fluor), N-methyl-2-pyrrolidone ("Purisol"), methanol ("Rectisol") and the dimethyl ether of polyethyleneglycol ("Selexol").

If desired, part or all the carbon dioxide can be removed by absorption in anhydrous or aqueous ammonia. Such a procedure is especially useful if the ammonia to be produced by the process of the invention is to be used for urea synthesis or for making ammonium sulphate by the calcium sulphate process. In a convenient form of the process, applicable especially when it is desired to operate at pressures less than optimal for using physical absorbents, the bulk of the carbon dioxide can be removed in a physical absorbent and the remainder in a chemical solvent as mentioned above or in ammonia. The latter procedure can be designed to suit any desired relative outputs of ammonia, carbon dioxide and urea.

The second stage of carbon oxides removal can be carried out by cryogenic separation or by contacting

6

the gas with a carbon oxides absorbent such as copper liquor, but is most conveniently effected by catalytic methanation, for example over a supported nickel catalyst at an outlet temperature in the range 250°–400° C. This decreases the carbon oxides content to a few parts per million by volume but produces water, which can be removed by cooling, separation and passage over a water-absorbent such as alumina or a molecular sieve. If nitrogen is not already present it can be added by means of liquid nitrogen washing of the gas after the first stage of carbon oxides removal.

The dried gas from step (b) contains nitrogen, hydrogen preferably in less than the stoichiometric ratio for ammonia synthesis, a small quantity (usually under 1% v/v) of methane and fractional percentages of noble gases introduced with the secondary reformer air, and is thus ready for use as an ammonia synthesis gas. It may be compressed to any convenient synthesis pressure up to 150 bar abs., but at the high pressure preferred for steps (a) and (b) it is suitable for use in the synthesis with less than 50% compression and preferably no more than the increase in pressure (for example up to 20%) encountered in a synthesis gas circulator. As an alternative a rather greater degree of compression can be used such as can be provided ay a single-barrel compressor circulator, conveniently by 20–80 bar. As a result of operating with such low degrees of compression, the requirement for high pressure steam is considerably less than in most conventional ammonia plants and consequently the superheating that can be conveniently provided in the synthesis loop is a greater fraction of the total superheat requirement.

The "fresh" synthesis gas from step (b) can be fed through a succession of catalytic stages and ammonia removal stages but, as in most ammonia synthesis processes, is preferably mixed with synthesis gas recycled from an ammonia removal stage. At the preferred synthesis pressures the attainable pass conversion over the synthesis catalyst is relatively low, giving an ammonia outlet concentration in the range 8 to 12% v/v. The ratio of recycled gas to fresh gas is suitable in the range 4 to 6.

The catalyst used in the ammonia synthesis can be of the usual composition, namely iron with promoting quantities of non-reducible oxides such as those of potassium, calcium, aluminium and others such as of beryllium, cerium or silicon. In order to afford maximum activity and thus to compensate for the lower rate of reaction due to low pressure, the iron catalyst preferably contains also cobalt, suitably to the extent of 1–20% w/w calculated as Co_3O_4 on the total oxidic composition from which the catalyst is made by reduction and in which the iron oxide is assumed to be all Fe_3O_4. The catalyst can be in the form of particles in the sieve range 18 to 4 ASTM (1–4.7 mm) especially 10 to 5 (2–4 mm), if it is desired to maximise their available contact surface or larger, for example up to 20 mm; the arrangement of the catalyst in the synthesis reactor preferably therefore may afford short gas flow paths, such as by radial or secantial flow in a cylindrical reactor. The outlet temperature of the synthesis catalyst is preferably in the range up to 500° C., especially 350°–430° C. This is lower than has been usual, in order to obtain a more favourable synthesis equilibrium. The catalyst volume is suitable in the range 100–200 m^3 per 100 metric tons per day output; this is higher than has been usual but can be tolerated because at the low operating temperature and

4,213,954

7

pressure the reactor can be of simple construction for example of the hot-wall type.

Reacted gas can be cooled during, between stages of or after the synthesis by any convenient means, but according to the invention the hot gas at some point is 5 heat exchanged with high pressure steam (60–200 bar abs.) generated by cooling furnace flue gas or step (a) process gas or high temperature shift outlet gas, or with intermediate pressure steam (30–120, especially 40–80 bar abs.) exhausted from a pass-out turbine or generated 10 independently. Preferably this heat exchange of steam precedes the other superheating or reheating heat exchanges. Suitably it raises the temperature of high pressure steam by 20°–60° C., and effects 15 to 30% of the total superheating. Preferably the steam is heat ex- 15 changed with reacted synthesis gas leaving one, preferably the first, of a succession of catalyst beds, and preferably before the reacted gas has been cooled in any other heat exchange. The heat exchange with steam is preferably followed by heat exchange with feed water 20 for the boilers of step (a) or step (b) and these two heat exchangers are operated preferably so as to cool the reacted synthesis gas from one catalyst bed to the temperature at which they are to enter the next bed. After the gas has left all the catalyst beds or parts thereof it is 25 cooled by heat exchange with incoming unreacted synthesis gas and conveniently also with boiler feed water, each in one or more stages and in any convenient order and then finally cooled to ammonia separation temperature. 30

The recovery of ammonia from reacted synthesis gas can be carried out by ordinary air-cooling or water-cooling if the pressure is high enough, but at preferred pressures in the range 40–80 bar. abs. is best carried out by absorption in water. Absorption in an acid or on a 35 solid such as zinc chloride can be used if convenient. Absorption in water is conveniently carried out in two or more stages, for example in the first of which the gas contacts a relatively strong ammonia solution (for example 15 to 30% w/w) and in the last pure water or a 40 weak ammonia solution (for example up to 10% w/w). After the absorption stage the gas is dried in order to prevent too-rapid deactivation of the catalyst be water. The aqueous ammonia product can be used as such or distilled to recover anhydrous ammonia from it. 45

When the fresh synthesis gas contains nitrogen in excess of the stoichiometric proportion, noble gases and also methane to an extent dependent on the incompleteness of the secondary reforming reaction and of the shift reaction, the continued removal of ammonia from it, 50 especially in a recycle process, results in a substantial concentration of non-reacting gases. It is preferred to treat the gas mixture so as to keep the concentration of such gases below 10% v/v especially below 5.0% v/v. This treatment could be applied to the fresh or mixed synthe- 55 sis gas before entering the synthesis or to the whole of the reacted gas after removal of ammonia, but it is preferred to apply it only to a side stream, because then any failure of the treatment plant does not cause a shutdown of the whole production process. The side stream 60 can conveniently be taken from the gas downstream of the ammonia separation and treated for hydrogen separation, whereafter the hydrogen is returned to the circulating synthesis gas. It could in principle be taken before ammonia separation but the treatment would then in- 65 volve also ammonia recovery. The hydrogen separation treatment involves a pressure-drop and may involve also a pressure let-down through an expansion engine in

8

order to decrease the gas temperature for cryogenic separation; consequently the hydrogen stream has to be compressed in order to return it to the synthesis. Preferably therefore the side stream is taken from the effluent of the circulator, where the gas pressure in the system is highest, and the separated hydrogen stream is returned to the inlet of the circulator, where the gas pressure is lowest. Part or all of the separated hydrogen stream can be recycled to the low temperature shift inlet.

The hydrogen separation treatment can be by any suitable means, for example by cryogenic fractionation, molecular sieve adsorption of gases other than hydrogen or palladium membrane diffusion. The hydrogen stream returned to the synthesis can be substantially (over 90% v/v) pure but in any event should contain at least 3 molecules of hydrogen per nitrogen molecule. The non-reactive gases discarded from the hydrogen separation treatment should of course be substantially free of hydrogen, since any discarded hydrogen represents wasted energy. If the side stream contains methane, the separation treatment can be designed and operated to separate a methane-rich stream and that stream can be used as process feed or furnace fuel for primary reforming or fed to secondary reforming. It will be appreciated that the magnitude of the side stream and the purity of the returned hydrogen are the subject of design optimisation. A typical side stream flow rate is in the range 15 to 30% of total gas flow.

The plant in which the process of the invention takes place is a new combination and constitutes a further feature of the invention.

The drawing, a flowsheet of one preferred form of the invention, shows the major processing steps and heat exchanges.

The process is based on steam-natural gas reforming. The reforming furnace has a radiative zone 10 and a convective zone 12. Zone 10 is heated by burners 14 fed with preheated natural gas from 16 and air fed in at 18 and preheated at 20 at the low temperature end of convective zone 12. The catalyst tubes 22 heated in radiative zone 10 are fed with a steam/natural gas mixture strongly preheated at 24 in the hottest part of convection zone 12. The mixture is formed at 26 from warm natural gas fed at 25 from a desulphurisation plant (not shown) and medium pressure steam the source of which will be described below. Gas leaving tube 22 consists of carbon oxides, hydrogen and methane, and is partly burnt with air supplied at 112 (after preheating at 110 at an intermediate level in convective zone 12) in secondary reformer 28 and then brought to equilibrium at a lower methane content over the secondary reformer catalyst. The gas leaving secondary reformer 28 is cooled in boiler 30 feeding high pressure steam drum 100, then cooled further in the heat exchangers 32 (a steam superheater feeding steam through line 34 to turbine 104) and boiler 36 also feeding stream drum 100. At high temperature shift inlet temperature it now enters reactor 38 in which carbon monoxide reacts with steam to give carbon dioxide and hydrogen. This reaction is exothermic and the hot reacted gas is cooled in heat exchanger 40 (a boiler feeding steam drum 100) and 42 (a boiler feed water heater) to the inlet temperature of the low temperature shift catalyst in reactor 44. In 44 the carbon monoxide/steam reaction goes almost to completion. The reacted gas is treated in 46, which indicates generally low grade heat recovery by for example heating boiler feed water and heating a carbon dioxide absorbent regeneration, followed by a cooler

4,213,954

9

and condensate separator from which water is taken at 48. It then passes into packed carbon dioxide absorber 50 in which it contacts absorbent liquid fed in at 52. Charged absorbent leaves the absorber at 54 and is passed through a regenerator (not shown) and then returned to feed point 52. Gas leaving the top of absorber 50 is heated in feed/effluent heat exchanger 56 to methanation inlet temperature and passed over a supported nickel catalyst in reactor 58 to convert residual carbon oxides to methane. Methanated gas is united at 60 with a recycle stream to be described and the mixture is cooled at 62 to below the dewpoint of steam and passed into separator 64, from which condensate is removed at 66. Gas leaving separator 64 is dried at 68, mixed at 70 with a recycle hydrogen stream to be described and fed to circulator 71 in which its pressure is increased sufficiently to maintain circulation. A part stream of gas is diverted at 72 into hydrogen separation unit 73 which typically includes a cryogenic fractionation plant in which the gas is cooled to about minus 188° C. to condense out nitrogen, methane and noble gases, whereafter the uncondensed fraction is fed back to point 70 as the recycle hydrogen stream already mentioned. In 73 the condensed fraction is re-evaporated in order to cool gas entering the unit, whereafter it leaves at 74. If the condensed fraction contains sufficient methane it can be fed to 16 as part of the fuel in furnace 14: alternatively it can be fractionated during evaporation in order to produce a methane stream to be used as furnace fuel or process feed. The stream recycled to 70 can be a nitrogen-hydrogen mixture if the circulating gas is approximately stoichiometric in composition or can be over-rich in hydrogen if the circulating gas contains excess nitrogen. The remainder of the gas passes from point 72 to feed/effluent heat exchanger 76, in which it is heated to catalyst inlet temperature, and thence to the inlet of the catalyst bed in synthesis reactor 78. Ammonia synthesis takes place and the resulting hot reacted gas is cooled in steam superheater 80 and boiler feed water heater 82 and passed through the catalyst bed in reactor 84 where it undergoes further ammonia synthesis. The resulting hot reacted gas passes out through the hot side of feed/effluent heat exchanger 74 and further heat exchanger and cooling stages (not shown) and enters two-stage ammonia absorber 86 in which the bulk of its ammonia content is removed by contact with lean aqueous ammonia fed in at 88 and its residual ammonia content is removed by cold water fed in at 90. A strong aqueous ammonia solution is taken off at 92 and fed to a distillation plant (not shown) from which anhydrous liquid ammonia is removed by condensation overhead and from which streams 88 and 90 are recovered as bottoms. Gas leaving the top of absorber 86 is recycled to point 60, at which it is united with fresh synthesis gas.

The steam system that characterises the process according to the invention is based, in this preferred flowsheet, on common steam drum 100. Drum 100 is fed with treated boiler water heated in exchanger 82 by reacted ammonia synthesis gas: the feed to exchanger 82 is already warm as a result of low grade heat recoveries such as 42 and other positions not shown such as in coolers in item 46, in cooler 62, between items 84 and 86 and in convective zone 12 of the steam reforming furnace. Water circulates from drum 100 through boilers 30, 36 and 40 already described. Steam passing overhead from drum 100 is superheated in heat exchanger 89 by reacted ammonia synthesis gas, superheated further

10

by hot raw synthesis gas at 32 and let-down in high pressure passout turbine 104 driving process air compressor 108, which supplies hot air as a result of limited inter-stage cooling and feeds secondary reformer 28 at point 112 by way of preheater 110. Exhaust steam from turbine 104 is at medium pressure. A part-stream of it is mixed with hot natural gas at 26, strongly preheated at 24 and fed to the catalyst in tubes 22, as already described. A further part stream is fed to turbine 114 driving alternator 116 supplying the electricity to be used in drives other than process air compressor 108 and circulator 68. A third part stream is fed to turbine 118 driving circulating compressor 68. Turbines 114 and 118 are both of the pass-out type. Their exhaust steam, led out at 120, is used in low-grade heating duties, principally the distillation of aqueous ammonia taken off at 92. In such duties it is condensed and the condensate may be recovered and after de-aeration used as boiler feed water.

In this process it is evident that if the ammonia synthesis section represented by numerals 60 to 90 is shut down, for example as a result of operation of a control device protecting circulator 71, there will be no exothermic heat of synthesis recovered in boiler feed water heater 82, so that cooler water will be fed to drum 100 and its steam output will be decreased. At the same time, however, the smaller flow of steam will be passed to superheater 32 at a lower temperature because, in the absence of exothermic heat of synthesis, it has not been superheated at 80. Consequently the heat load in superheater 32 is maintained at approximately its normal level. The same would apply to a superheater in flue gas duct 12 instead of or in addition to superheater 32, if one were used.

In a typical ammonia synthesis process producing 1500 metric tons (te) per day of ammonia 210 te/h of steam are generated in drum 100 at 340° C., 146 bar abs. pressure. This steam is superheated to 395° C. by heat exchange with reacted synthesis gas at 80, to 480° C. in heat exchange with process gas at 32 and to 510° C. in a fired heater (not shown) independent of the primary reformer. It is then let down in turbine 104 to 58 bar abs. and the exhaust is in part (100 te/h) fed to the process at 26 and for the rest let down in turbines 114 and 118.

The Table shows the high pressure steam output rates and heat loads for a process according to the invention in which the primary reformer outlet pressure at 22 and the ammonia synthesis pressure (outlet of 71) are both 50 bar abs., in comparison with a conventional process in which those pressures are respectively 30 and 220 bar abs. and compression is by a compressor driven by a turbine in which high pressure steam is let down.

TABLE

	Conventional	This invention
Steam output te h^{-1}	262.5	210
Total heat load in saturated steam, te cal^{-1}	145975	116780
Total steam superheat load, te cal h^{-1}	65687	35550
Heat recovered from synthesis as boiler feed water (hence as steam), te cal h^{-1}	40000	23000
as steam superheat, te cal h^{-1}	—	17000
Controlled superheat in fired heater, te cal h^{-1}	—	4150
Synthesis steam ÷ total steam	27.4%	19.7%
Synthesis and controlled		

4,213,954

| 11 | 12 |

TABLE-continued

	Conventional	This invention
superheat ÷ total superheat	—	59.5%

It is evident that, by using less of the heat recovered from synthesis to raise steam via boiler feed water heating, the effect of the loss of this steam in a shut-down is less, since it is only 19.7% of the total instead of 27.4% as in a conventional process. Since the superheat contribution of the synthesis heat recovery and the small fired heater is 59.5%, loss of the synthesis heat leaves the process gas superheater 32 more than fully loaded. If, however, the total synthesis heat recovery (40000 te cal h^{-1}) had been as boiler feed water in the 50 bar abs. process, the decrease in steam flow resulting from shut down of ammonia synthesis would have been 34.2% and thus superheater 32 would have been subject to overheating.

I claim:

1. In an ammonia production process which comprises the steps of

(a) reacting a carbonaceous feedstock with steam and/or oxygen to give a gas containing carbon monoxide;

(b) reacting the product of step (a) over a shift catalyst to convert carbon monoxide and steam to carbon dioxide and hydrogen and removing carbon oxides and steam from the resulting gas;

(c) reacting the hydrogen with nitrogen over an ammonia synthesis catalyst, cooling the reacted synthesis gas and recovering ammonia from it;

(d) heat exchanging hot gases produced in at least step (c) with water in a boiler producing steam at a pressure in the range 60–200 bar abs. and/or in a water heater feeding such a boiler;

(e) superheating steam from such a boiler by heat exchange with hot gases produced in step (a) or step (b); and

(f) letting down such steam in one or more expansion engines and thereby recovering useful power,

said process being previously subject to the defect that an accidental shut down of ammonia synthesis in step (c) decreases the supply of steam from

step (d) and thus causes overheating of the heat exchangers providing superheating in step (e);

in step (c) (i) reacting and cooling the synthesis gas at a pressure under 150 bar abs., and

(ii) cooling the reacted synthesis gas at least partly by superheating steam produced in step (d) and-/or reheating steam within or from one or more of the engines of step (f) whereby such a decrease in said supply of steam is accompanied by a compensating decrease in superheating capacity.

2. A process according to claim 1 in which step (a) comprises primary and secondary catalytic steam reforming of a hydrocarbon feedstock and in step (ii) high pressure steam is superheated in heat exchange with reacted ammonia synthesis gas and further superheated in heat exchange with the flue gases of the furnace used in primary reforming and/or with the hot gas produced by secondary reforming.

3. A process according to claim 1 in which ammonia synthesis is at a pressure in the range 40–80 bar abs., the outlet pressure of primary steam reforming is also at a pressure in that range and the gas after step (b) and before step (c) is compressed by less than 50%.

4. A process according to claim 1 in which 15–30% of the total steam superheating is by heat exchange with reacted ammonia synthesis gas.

5. A process according to claim 1 in which part of the steam superheating is by an independently fired heater and the heat taken from reacted synthesis gas and the fired heater amounts to 40–60% of the total superheating.

6. A process according to claim 1 in which ammonia synthesis is carried out in a succession of catalyst beds and the high pressure steam is heat exchanged with reacted synthesis gas leaving the first of such beds before that gas has been cooled in any other heat exchange.

7. A process according to claim 6 in which the heat exchange with steam is followed by heat exchange with feed water for the boilers of step (a) or step (b) and these two heat exchangers are operated so as to cool the synthesis gas to the inlet temperature of the next catalyst bed.

•   •   •   •   •

Figure 8.4   Example of a US Patent (reproduced by permission of Imperial Chemical Industries PLC and the US Patent and Trademark Office)

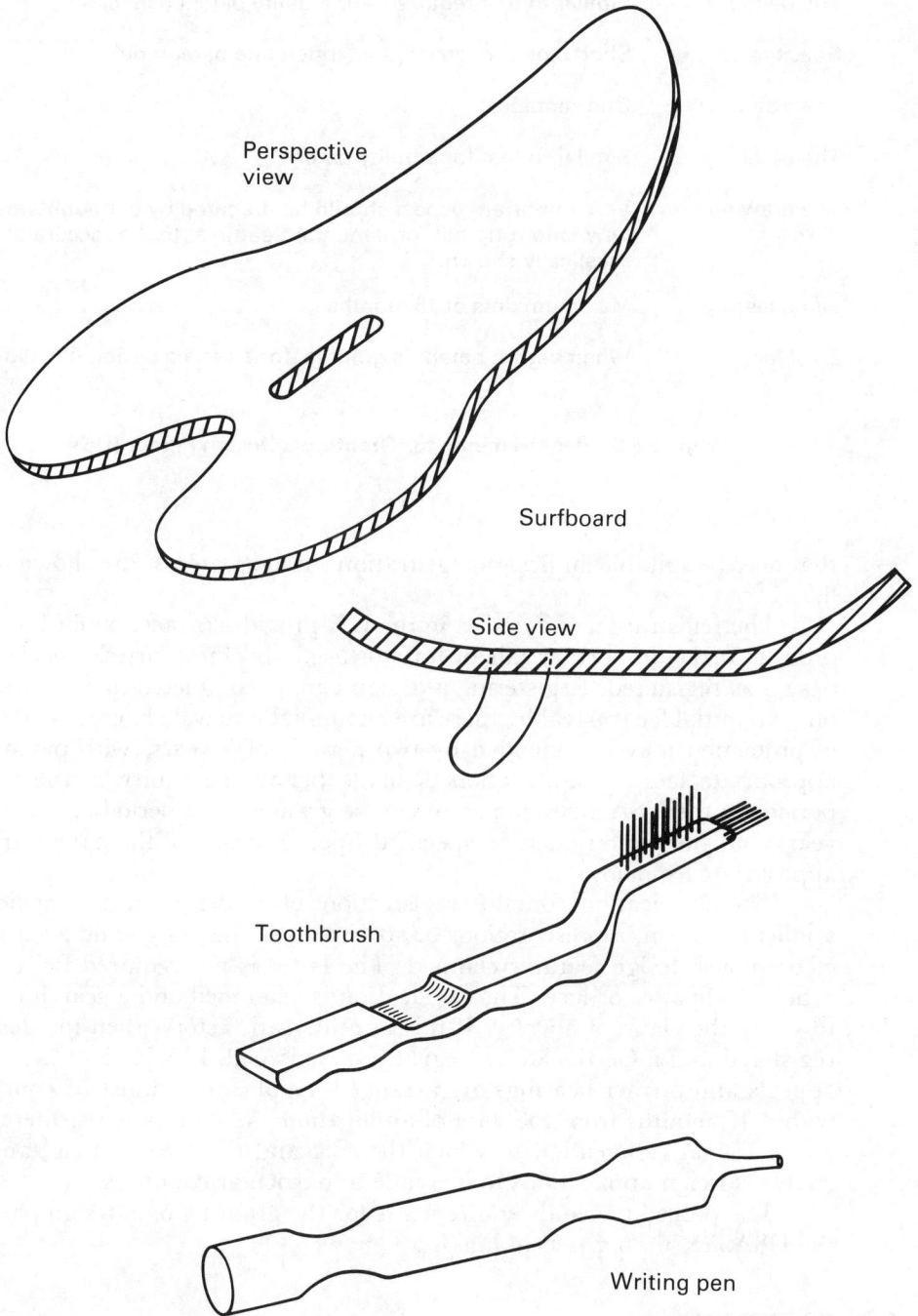

Figure 8.5 Typical (Not Actual) Examples of Design Registrations (UK) or Design Patents (USA)

The petition:	Similar to that required for a utility patent (Fig. 8.2).
Specification:	Short, concise, descriptive (often one paragraph).
The claim:	One sentence.
The oath:	Similar to that for a utility patent.
The drawing:	Very important aspect should be prepared by a draughtsman. All new and original ornamental features to be accurately and artistically shown.
Filing fee:	Maximum time of 15 months
Final fee:	When design patent is granted, for a certain period of validity.

Figure 8.6   Requirements for Granting a Design Patent (USA)

that may be suitable for design registration (design patents) are shown in Fig. 8.5.

The registration of a design in the UK provides protection for 5 years in order to make, import for sale or hire, or use in business, articles bearing the design as registered. Registering a design can be completed in 3 months and only an initial fee is payable; there are no annual renewal charges. The period of protection may be extended by two periods of 5 years (with payment of appropriate fees), if application is made before the expiry of the current period. In the USA a design patent can be granted for a period of 3½, 7 or 14 years, but the period must be specified upon granting of the patent and the appropriate fee paid.

The application for the registration of a design must include an application form, representations or specimens of the design and a statement of the novel design features claimed. The latter is not required for a textile article, wallpaper or lace. The application is examined and a search is made to verify the claim of novelty. If this appears satisfactory, then the design is registered and a Certificate of Registration is issued. If objections are raised, appeals and further hearings are arranged. Applications must be completed within 15 months from the date of application. As with patents, there is an International Convention to which the UK and the USA belong, and this enables foreign applications to be made to particular countries.

The procedures and requirements for the granting of a design patent in the USA are summarised in Fig. 8.6.

8.3 COPYRIGHT

New industrial designs are also protected under the provisions of the Copyright Act 1956 as amended by the Design Copyright Act 1968 in the UK. This protection is afforded without any form of registration and is valid for 15 years from the date on which the articles are manufactured in quantity, or first marketed in the UK or elsewhere. Protection takes effect immediately upon production of a design. Any document, drawing or sketch that is produced for a design, automatically acquires copyright without the need for registration. This is different from the situation in most countries. Protection under the Registered Designs Acts 1949–61 (if an application is made and accepted) and the Design Copyright Act 1968 thus offers two defences against infringement.

Unlike patents and designs where protection is given to the first article filed for application, copyright is only infringed by actual copying. Thus if two similar articles are shown to have originated from independent sources, then the article produced later is not deemed to have infringed the protection of the first.

In the USA the relevant copyright law is Public Law 94-553 which became effective on 1 January 1978. This law grants protection to the original author(s) of the exclusive right to prevent copying of the copyrighted work. A list of categories of items that can be copyrighted and an application form can be obtained from the Register of Copyrights at the Library of Congress (Washington, DC). The application, a copy of the work to be protected and the fee should be submitted in order to obtain granting of copyright. A copyright notice consists of the word 'copyright', the year and the name of the owner. For example, Copyright 1969, by Martyn Ray, or in shortened form as © 1969, Martyn Ray.

8.4 TRADE MARKS

A trade mark is a word, letter or symbol that identifies certain products or the origin of the products. It is used in the course of trade so that goods to which it is applied may be readily distinguished from other similar goods. It should be distinctive, not merely descriptive, and as unique as possible in its use. The use of a trade mark has certain distinct advantages. It can create an image for a product and create goodwill and consumer preference, e.g. associating vacuum cleaners with the name 'Hoover'. It has an infinite life, provided that it is used and the appropriate renewal fees are paid. Also it need not be restricted to one article but may be used with a wide range of products, e.g. '57' for various Heinz foods.

A trade mark should be easy to pronounce, easy to remember, easy to spell and write, attractive in sound, easy to reproduce, suggestive of good qualities, simple and short. It should not be offensive to the eye or ear and should not be deceptive or misleading. Misspellings or phonetic renderings of known words will be judged for registration as though they consisted of known words!

* * *

Exercises

Examine some common items and identify various trade marks. Compare and contrast their features including uniqueness, distinctiveness, non-descriptive nature and the other requirements listed above.

* * *

The important legislation in the UK governing the registration of trade marks is the Trade Marks Act 1938, and the Trade Marks Rules 1938, as amended. Registration of a trade mark confers a statutory monopoly on its use in relation to the goods for which it is registered. In the UK a trade mark can be registered in Part A or Part B of the Register depending upon the nature of the particular goods or classes of goods, as classified in Schedule IV of the Trade Marks Rules 1938, as amended. Part A provides more comprehensive protection to the owner and a stricter standard is required for registration. Section 9 of the Trade Marks Act 1938 sets out the criteria for distinctiveness by which trade marks are to be judged for registration in Part A of the Register; for example, it may contain or consist of an invented word or words.

The requirements for registration in Part B are set out in Section 10 of the Act. Restrictions on registration are set out in Sections 11 and 12 of the Act. Section 15 of the Act says that the rights of a registered word mark may be lost if the word becomes established as a general description of an article (i.e. a generic name); for example, the words aspirin or linoleum. There is also an extensive list of things that may not appear on a trade mark. These include certain devices such as royal crests, certain words such as common surnames, colour names, 'patent' and representations of members of the royal family. Careful consideration and research to devise a new trade mark will be time well spent.

Applications to register trade marks in the UK are examined in the Trade Marks Registry of the Patent Office. Details are published in the *Trade Marks Journal,* if considered to be *prima facie* acceptable. If no objections are raised within 1 month, then a Certificate of Registration is issued for an initial period of 7 years. After this time the mark may be renewed for further periods of 14 years, if the appropriate fee is paid. If objections to the application do arise, the case can be appealed to a higher level.

There are many trade marks in everyday use that are not registered, since this is not compulsory in the UK. Sometimes the user cannot be bothered to apply for registration or it may be difficult to establish ownership of a mark. Some examples of common trade marks are shown in Fig. 8.7.

In the USA an application must be made to the US Patent and Trademark Office to register a trade mark according to the US Trademark Act of 5 July 1946. The Act provides for principal registration and supplemental registration (or register). The Supplemental Register enables trade marks to be registered that lack the distinctiveness required by the Principal Register. The trade mark may be transferred to the Principal Register if it becomes distinctive through prolonged use. The Act also provides for registration either to be restricted to one state or to be valid

Figure 8.7 Examples of Registered Trade Marks

throughout the USA (and is then subject to federal control). A request for trade mark registration must include an application, specification, a drawing, a claim, five specimens of the mark, name and address of the applicant and a filing fee. If granted, the registration lasts for 20 years and may be renewed prior to expiry.

8.5 STANDARDS AND CODES

The engineer will frequently refer to, and make substantial use of, a wide range of standards and codes of practice, particularly in the detailed design stage. They provide a readily acceptable set of guidelines. Many have no legal requirement enforcing their use, although they are often adopted by government, local authorities or customers as the basis of industrial contracts. Their use enables standardisation of components, thus providing interchangeability, standardised materials' specification and (for codes) a common standard of good design practice.

Standards may result from influences that are international, national or local, or even from within particular industries or companies. National and international standards may originate from standards employed by particular industries or companies, e.g. automobile manufacturers and petroleum companies in the USA.

The widespread use of standards has benefited companies by reducing the number of products, materials or components that need to be manufactured or held in stock. The consumer has also found this procedure advantageous. An International Standards Organisation (ISO) exists in order to promote international acceptance of particular standards. There is usually collaboration between the ISO and the standards organisations which exist in most countries, e.g. the British Standards Institution (UK) and the National Bureau of Standards (USA). The development of international standards has been hampered by the variety of measurement systems used in different countries. Within any country, standards usually exist that provide a complete specification for machine components, electrical appliances, materials with industrial applications and aspects of machine performance.

Codes of practice describe recommended procedures for the design of pressure vessels, welded structures, etc., also for applications of concrete technology, low-temperature environments, etc. They provide guidelines for the designer in common situations and are frequently updated to incorporate new ideas and developments. Extracts from a British Standard Code of Practice relating to Sound Insulation and Noise Reduction are reproduced in Fig. 8.8.

In the USA standards and codes of practice can originate from the US Department of Commerce, National Bureau of Standards, or from specialist organisations such as the American Society for Testing and Materials (ASTM), Philadelphia, Pennsylvania; the American Society of Mechanical Engineers (ASME), New York; the National Fire Protection Association,

BRITISH STANDARD CODE OF PRACTICE
CP 3 : Chapter III : 1972

CODE OF BASIC DATA FOR THE DESIGN OF BUILDINGS

(formerly Code of Functional Requirements of Buildings)

CHAPTER III

SOUND INSULATION AND NOISE REDUCTION

Incorporating amendment issued March 1974 (AMD 1385).

THE COUNCIL FOR CODES OF PRACTICE
BRITISH STANDARDS INSTITUTION
2 PARK STREET, LONDON WIA 2BS

UDC 699.844 : 534.83

CP 3 : Chap. III : 1972

CONTENTS

TABLES

FIGURES

CP 3 : Chap. III : 1972

This Code of Practice represents a standard of good practice and, therefore, takes the form of recommendations. Compliance with it does not confer immunity from relevant statutory and legal requirements.

CP 3 : Chap. III : 1972

BRITISH STANDARD CODE OF PRACTICE CP 3

CHAPTER III. SOUND INSULATION AND NOISE REDUCTION

INTRODUCTION

This Chapter should be considered in conjunction with the other chapters, as certain recommendations in them may be incompatible, or may be reconciled only with difficulty, with recommendations in this Chapter. The designer should therefore consider the functional requirements of a building as a whole, and determine which recommendations should have precedence. Attention is particularly drawn to the possibility of increasing the fire hazard in a building by the introduction of acoustic treatments which may assist flame spread.

It cannot be too strongly emphasized that the best defence against noise, both outdoor and indoor, lies in intelligent planning precautions taken before building development begins. If such precautions are not taken, the solution to sound insulation or noise reduction problems by structural or technical measures may, in some circumstances, be attainable only at very great cost, if at all. For this reason the section on ' Planning against Noise ' is placed first.

In each section the problem has been considered under the general headings of (a) Outdoor noise and (b) Indoor noise. Where possible, recommendations have been made in respect of both. It has not been practicable to be entirely consistent, owing partly to the different nature of the problem in varying classes of building, and partly to the comparative lack of data available for a scientific study of certain building types (e.g. hospitals), as compared with others (e.g. houses). Much more research needs to be undertaken before a proper scientific assessment of the problem can be made in regard to certain building types, such as hospitals and office buildings.

Finally, a general understanding of the behaviour of sound is necessary in designing for the control of noise. Methods of construction for sound insulation, even the commonest and simplest, may, when conditions depart from normal, be misapplied if this understanding is lacking. For this reason brief explanations and general definitions are given in Appendix A of the properties and behaviour of sound, and how it is propagated and controlled. Reference should also be made to BS 661, ' Glossary of acoustical terms ', in which formal definitions of most of the acoustical terms and quantities mentioned in this Code will be found.

In accordance with the change to the metric system in the construction

CP 3 : Chap. III : 1972

industry this part of the Code has been metricated giving values in terms of SI units. For further information on SI units reference should be made to BS 3763, ' The International System of units (SI) '.

As amended Aug. 1975   The values in this Code represent the equivalents of the values in imperial units in the 1960 edition rounded to convenient numbers. Although the values are not exact equivalents of the imperial ones, this Code is not a technical revision but it will supersede the 1960 edition of Chapter III.

Certain documents referred to in Part 1 have been revised and in these cases a footnote has been added giving the present position, and where necessary minor alterations have been made to the text.

SECTION ONE: PLANNING AGAINST NOISE

101. General. Planning against noise should be an integral part of Town and Country Planning proposals, ranging from regional proposals to detailed zoning and three-dimensional layouts and road design within built-up areas. Noise nuisance should be fully recognized in zoning regulations.

Noise is either generated by traffic (airborne, road, rail and underground) or it arises from zones and buildings within built-up areas (industry, commerce, offices, public and residential buildings). Planning surveys should examine all the possible causes of noise and consider the various factors making for actual nuisance.

Noise by night, causing disturbance of sleep, is more of a nuisance than noise by day. For this reason factories that work by night are liable to cause serious complaints if housing estates adjoin them.

There are two aspects of defence by planning. The first is to plan so as to keep the noise at a distance. Under this aspect comes the separation of housing from traffic noise by interposing buffer zones, and the protection of schools and hospitals by green belts, public gardens, golf courses, etc. The second is the principle of shading or screening. This consists of deliberately interposing a less vulnerable building to screen a more vulnerable. It is well illustrated by planning a factory so that the offices and canteen blocks are interposed between the workshops and adjoining houses or schools. Typical noise levels from various sources are shown in Table 1*.

102. Planning against traffic noise. *a. Airborne traffic.* Consultations should be held between the Local Planning Authorities and the Airport Authorities, to ascertain what extensions of the aerodrome are anticipated in the future, and what areas are likely to be allocated for maintenance testing.

* From ' Acoustics, Noise and Buildings ', by Parkin and Humphreys (Faber, London, 1969).

TABLE 1. OCTAVE ANALYSES OF SOME COMMON NOISES

Noise	Distance	Octave bands (Hz)								Remarks
	m	37–75	75–150	150–300	300–600	600–1 200	1 200–2 400	2 400–4 800	4 800–9 600	
		Sound-pressure level (dB)								
Large (4 engines) jet air-liner	38	112	121	123	124	123	120	117	109	Maximum values when passing overhead at take-off power. No mufflers.
Single-engined jet fighter	38	102	114	116	116	117	115	111	102	Maximum values when passing overhead at take-off power.
Large (4 engines) piston-engined air-liner	38	111	117	114	108	107	108	106	97	Maximum values when passing overhead at take-off power.
Electric trains over steel bridge	6	94	93	99	99	95	84	81	73	
Kerb side, main road in London at rush hour	5	78	81	81	79	72	67	63	55	
	(average)									
Electric trains	30	77	77	76	74	73	67	59	54	In open air.
Pneumatic drills	38	75	72	72	66	69	71	67	65	
Riveting on large (6 m by 4 m) steel plate	2	88	96	105	106	111	109	113	110	
Nylon factory	Reverberant sound	87	86	92	93	97	97	96	87	Up-twisting process.
Weaving shed	Reverberant sound	78	71	77	81	86	86	84	78	
Canteen (hard ceiling)	Reverberant sound	52	54	59	67	67	61	55	49	Average levels. Peak values up to 20 dB higher.
Typing office with acoustic ceiling	Reverberant sound	68	64	60	56	55	55	53	50	Ten typewriters, one tele-type machine.
Male speech	1	52	55	59	66	65	60	52	40	Average values.

TABLE 8. EFFECT OF WINDOW SIZE ON THE INSULATION OF WALLS

Sound insulation values of a 225 mm brick wall with different percentage areas of window; the figures are for closed windows, single and double, taken from Table 7.

Percentage of glazing	Single window	Double window
	dB	dB
100	20	40
75	21	41
50	23	43
33⅓	25	44
25	26	45
10	30	47
Nil	Value of a 225 mm brick wall=50 dB	

NOTE. If 10% or more of the wall is open window, the insulation will not be more than 10 dB. It is seen from the Table that the range of percentages of glazing from 10% to 100% produces a range of variation of the insulation of 10 dB for single windows and 7 dB for double windows. The more important glazing range from 25% to 75% produces an insulation variation range of 5 dB for single windows and 4 dB for double windows.

5. Doors and lobbies. The main factors determining the insulation of single doors are the mass of the door (i.e. of the panels) and the air leakage round the edges; usually it is the latter that is decisive. For example, if a normal room door has a panel weight of 5 kg/m² (6 mm plywood panels or hardboard faced flush skeleton door) its inherent insulation is 22 dB; but if it has edge gaps amounting to 15 000–20 000 mm², which is quite common, the net insulation of the doorway falls to 17–18 dB; increasing the panel weight to 25 kg/m² (45 mm solid cored door) without decreasing the gap, raises the net insulation by only 1 dB. If, however, the gaps are reduced to such an extent as would make the net insulation of the lighter door 21 dB, then the substitution of the heavier door would raise the insulation to 25–26 dB, an increase of 4 or 5 dB. Higher insulation than this cannot be obtained for a single door unless a special ' refrigerator ' method of closure, with heavier construction if necessary, is employed.

Higher insulation can however be obtained by the use of double doors, i.e. two doors separated by an air-space or lobby. As with double windows, the air-space should be as wide as possible; it should never be less than 100 mm, and should be increased to form a lobby or corridor whenever possible.

CP 3 : Chap. III : 1972   **App, F**

$$x = \frac{2}{\lambda}[D_\mathrm{S}(\sqrt{1 + H^2/D_\mathrm{S}^2} - 1) + D_\mathrm{L}(\sqrt{1 + H^2/D_\mathrm{L}^2} - 1)] \quad . \quad . \quad . \quad \text{(i)}$$

where λ is the wavelength of the sound and other symbols as diagram below (all dimensions in same units)

OR

when D_L is much greater than D_S and D_S is not less than H, then

$$x = \frac{H^2}{\lambda D_\mathrm{S}} \text{ (approximately)} \quad . \quad . \quad . \quad . \quad . \quad . \quad \text{(ii)}$$

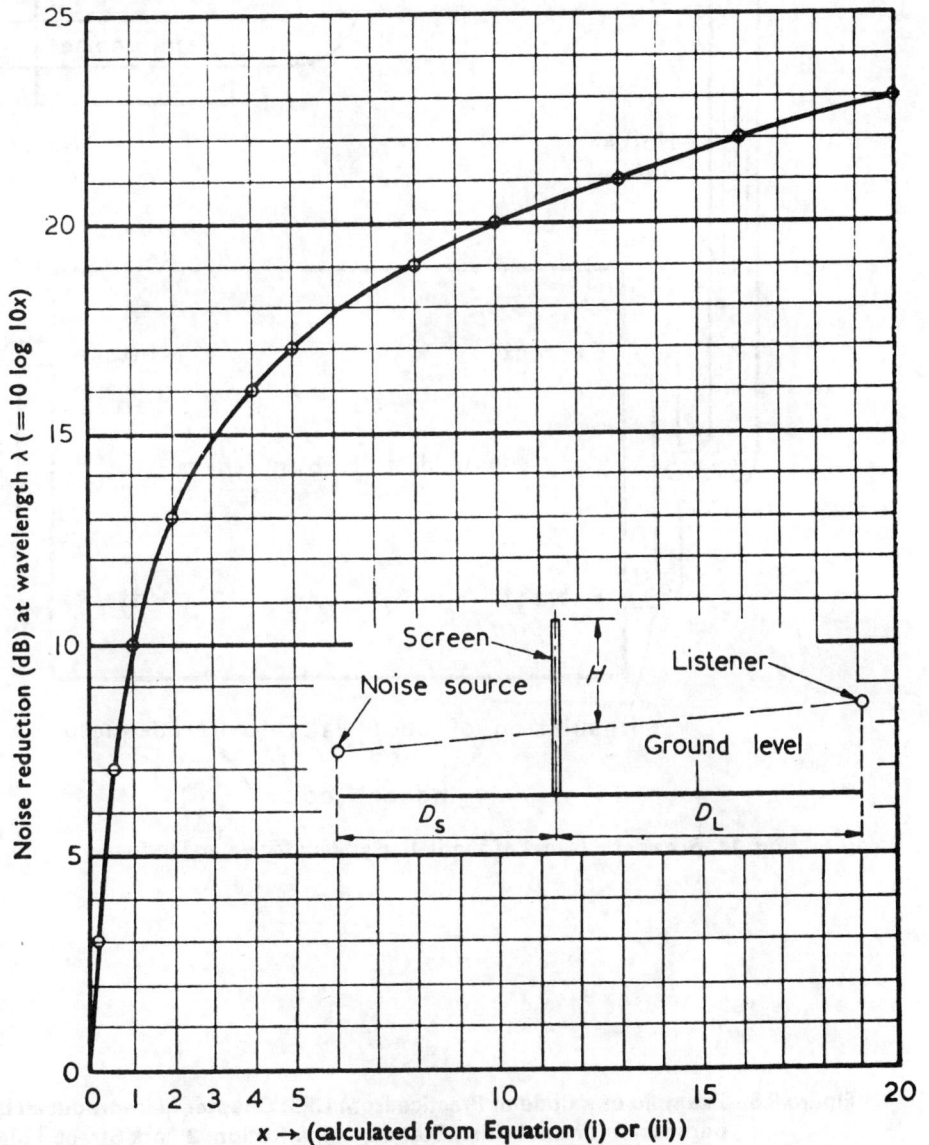

Fig. 10. Noise reduction by screens in open air

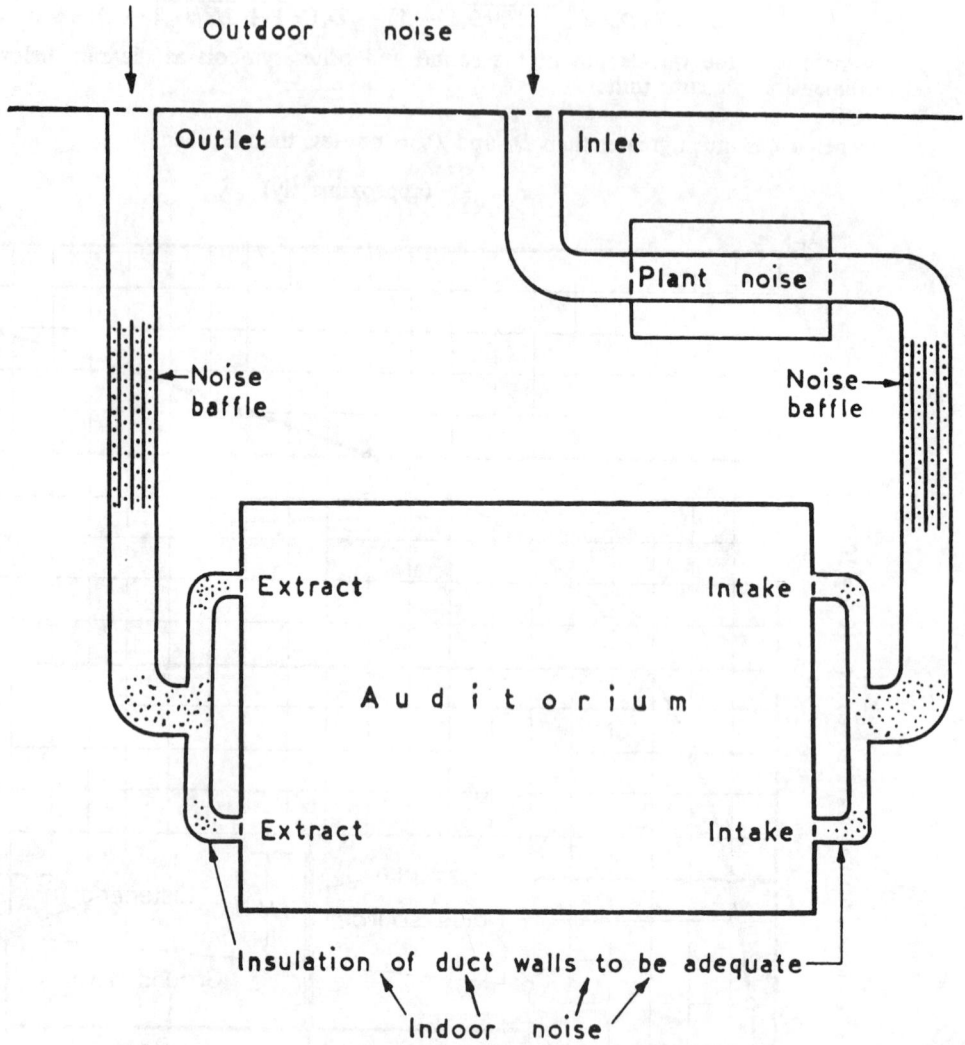

Fig. 11. Schematic layout of ventilation system for an auditorium

Figure 8.8   Example of a Code of Practice from CP3: Chapter III (reproduced by permission of the British Standards Institution, 2 Park Street, London W1A 2BS, from whom complete copies of the document can be obtained)

an agreement

A publication agreement made between _____ *(author)*

and _____ *(publisher) on* _____ *,19___*

THE AUTHOR AND PUBLISHER AGREE THAT:

1. The Author will write for publication a work on _____

ELEMENTS OF ENGINEERING DESIGN: AN INTEGRATED APPROACH

Grant of Rights

The Author grants this work to the Publisher with the exclusive right to publish and sell the work, under its own name and under other imprints or trade names, during the full term of copyright and all renewals thereof, and to copyright it in the Publisher's name or any other name in all countries; also the exclusive rights listed in paragraph *1* below; with exclusive authority to dispose of said rights in all countries and in all languages.

Delivery of Manuscript

2. The manuscript, containing about _____ words or their equivalent,

will be delivered by the Author by_____ ,19___

Royalties

3. When the manuscript is ready for publication, it will be published at the Publisher's own expense. The Publisher will pay the Author a royalty, based on the actual cash received by the Publisher, of 10% on the first 3000 copies sold and 15% on all over 3000 copies sold.

Payments

4. The Publisher will report on the sale of the work in March and September of each year for the six-month period ending the prior December 31 and June 30, respectively. With each report of sales, the Publisher will make settlement for any balance shown to be due.

5. Paragraphs *A-N* inclusive, on pages 2, 3 and 4 following, are parts of this agreement as though placed before the signatures.

_____

Author

By _____

Publisher

A. The Author will deliver the manuscript in typewritten form (or, in the case of anthologies and revisions, in typewritten and printed form) double-spaced on 8½ " × 11 " sheets on one side only. The manuscript will be submitted in duplicate and a third copy will be retained by the Author. It will be in proper form for use as copy by the printer, and the content will be such as the Author and Publisher are willing to have appear in print. If the Author fails to deliver a satisfactory manuscript on time, the Publisher will have the right to terminate this agreement and to recover from the Author any sums advanced in connection with the work. Until this agreement has been terminated and until such sums have been repaid, the Author may not have the work published elsewhere. The Author will read the proofs, correct them in duplicate, and promptly return one set to the Publisher. The Author will be responsible for the completeness and accuracy of such corrections and will bear all costs of alterations in the proofs (other than those resulting from printer's errors) exceeding 10% of the cost of typesetting. These costs will be deducted from the first payments otherwise due the Author.

Submission of Manuscript

Changes in Proofs

B. The Author will furnish the following items along with the manuscript: title page; preface; foreword (if any); table of contents; index; teacher's manual or key (if requested by the Publisher).

Items Furnished by Author

C. The Author hereby warrants:-

(a) that the work is an original work (and does not infringe on any other copyright) and that the Author is the sole owner of the work and that the Author has not made any assignment of or granted any licence in respect of any of the rights the subject of this agreement and that the work has not been published;

(b) that the work does not contain any scandalous libellous obscene or unlawful matter and that the Author further agrees to indemnify the Publisher against all claims costs damages and expenses which the Publisher may incur by reason of any breach or alleged breach of any of the above warranties and until such indemnity has been fulfilled the Publisher may at its discretion withhold any sums due to the Author under the terms of this Agreement.

Author's Warranty

D. The work will contain no material from other copyrighted works without the Publisher's consent and the written consent of the owner of such copyrighted material. The Author will be responsible for obtaining such consents and for the costs of such consents, and will file them with the Publisher.

Quoted Material

E. The Publisher will have the right to edit the work for the original printing and for any reprinting, provided that the meaning of the text is not materially altered.

Editing

F. The Publisher will have the right: (1) to publish the work in suitable style as to paper, printing and binding; (2) to fix or alter the title and price; (3) to use all customary means to market the work.

Publishing Details

G. The Publisher may distribute such copies as it thinks fit for the purposes of review or criticism or advertisement and shall furnish six copies of the work to the Author without charge and additional copies shall be supplied at the Author's request at a discount of twenty-five per cent from the Publisher's list price.

Author's Copies

H. The Author agrees to revise the work if the Publisher considers it necessary in the best interests of the work. The provisions of this agreement shall apply to each revision of the work by the Author as though that revision were the work being published for the first time under this agreement. Should the Author be unable or unwilling to provide a revision within a reasonable time after the Publisher has requested it, or should the Author be deceased, the Publisher may have the revision prepared and charge the cost including, without limitation, fees or royalties, against the Author's royalties, and may display in the revised work, and in advertising, the name of the person, or persons, who revise the work.

Revisions

I. The Publisher may permit others to publish, broadcast by radio, make recordings, mechanical or electromagnetic renditions, publish book club and micro-film editions, make translations and other versions, show by motion pictures or by television, syndicate, quote, and otherwise utilize this work, and material based on this work. The net amount of any royalty compensation received from such use shall be divided equally between the Publisher and the Author. The Publisher may authorize such use by others without compensation, if, in the Publisher's judgement, such use may benefit the sale of the work. If the Publisher itself uses the work for any of the foregoing purposes, the Publisher will pay the Author a royalty of 5% of the cash received from such use. On copies of the work or sheets sold outside continental United States, or sold by any means at a discount of 50% or more, the Author will be paid a royalty of 10% of the cash received from such sales. On copies of the work sold through any of the Publisher's book club divisions or institutes, under government adoption contracts, or by radio, television, mail order or coupon advertising direct to the consumer or through any of its subsidiaries, or to elementary and secondary schools, the Publisher shall pay to the Author a royalty of 5% of the cash received from such sales. If the Publisher sells any stock of the work at a price below the manufacturing costs of the book plus royalties, no royalties shall be paid. All copies of the work sold and all compensation from sales of the work under this paragraph shall be excluded in computing the royalties payable under paragraph 3 above and shall be computed and shown separately in reports to the Author.

Subsidiary Rights

J. If the balance due to the Author for any settlement period is less than ten dollars, the Publisher will make no accounting or payment until the next settlement period at the end of which the cumulative balance has reached ten dollars. If, after the expiration of two years following publication, the sales in any twelve-month period ending December 31 do not exceed a total of 500 copies, the royalty on such sales shall be one-half the stipulated royalty, this reduction in royalty being agreed upon to enable the Publisher to keep the work in print as long as possible. The Publisher may deduct from any funds due the Author, under this or any other agreement between the Author and the Publisher, any sum that the Author may owe the Publisher. When the Publisher decides that the public demand for this work no longer warrants its continued manufacture, the Publisher may discontinue manufacture and destroy any or all plates, books, and sheets without liability to the Author.

Discontinuing Manufacture

Competing Publications	*K*. The Author agrees that during the term of this agreement the Author will not agree to publish or furnish to any other publisher any work on the same subject that will conflict with the sale of this work.

L. This agreement may not be changed unless the parties to it agree in writing.

Heirs	*M*. This agreement shall be binding upon the parties hereto, their heirs, successors, assigns and personal representatives, and references to the Author and to the Publisher shall include their heirs, successors, assigns and personal representatives.

Construction	*N*. This agreement shall be construed in accordance with the Laws of England and the parties agree to submit to the jurisdiction of the English Courts.

Figure 8.9 Example of a Typical Contract or Agreement

Boston, Massachusetts; and the Tubular Exchanger Manufacturers' Association (TEMA), New York.

8.6 CONTRACTS

A contract is a formal written agreement between two parties, the supplier and the customer. Both parties agree to abide by the conditions that are laid down, in all respects. However, certain clauses ('let-outs') may be included in order to accommodate unforeseen difficulties that may arise.

The widespread use of contracts in business ventures is an attempt to achieve harmony and agreement between the parties involved. It is not always successful because there is always the possibility of interpreting a contract in a different way and also some people are unscrupulous in their dealings; ultimately, there are the vagaries of human nature.

A contract related to an engineering design will be concerned with three main areas of the work. These are the price that will be paid for the product, the time that will elapse before it is completed and the final operating ability or standard of the product. With large or complex projects, it is normal to impose conditions to be achieved at different stages in the progress of the project for all three aspects mentioned. This will provide a periodic check on the design being implemented, and also on the supplier's performance.

An industrial contract will contain many clauses, each concerned with a separate aspect of the work, e.g. terms of payment, certificates, drawings, defects after delivery, and so forth. The actual contents and the format will depend upon the nature of the agreement, e.g. supply of machinery, installation of factory services or purchase of materials from a foreign country. An example of a typical contract or agreement between an author and a publisher for the production of a textbook is shown in Fig. 8.9.

Design often involves new technology, new applications and new ideas, and so the designer is dealing partly with the unknown, although hopefully with reliable estimates. Care must be taken when preparing or signing a contract that the claims or requirements are not too ambitious, and, if modifications become necessary, that the financial penalties are not too harsh.

* * *

Exercise

Try to obtain contracts relating to some of the following transactions:

(a) car purchase or car hire;

(b) house purchase or leasing;

(c) household or luxury item purchased by credit or hire purchase;

(d) a stereo warranty;

(e) holiday health insurance policy;

(f) contract of temporary or permanent employment;

(g) any other form of contract or agreement.

Assess the strengths and weaknesses, protection and liability, 'loopholes', etc., in these contracts.

Decide which features you would include (and omit!) when preparing a contract for a particular situation.

*   *   *

8.7 SUMMARY

The legal factors involved in an engineering design may be formal or informal requirements. The protection of property by the use of patents, design registration, copyright or trade mark is well documented and the legal aspects are clearly defined. This does not mean that they are not open to interpretation, or misinterpretation! Contracts between suppliers and customers are more flexible in their content and application—until they are signed! They must then be considered binding and formal. Standard specifications and codes of practice are informal aspects of a design. Although they may be specified by a customer, they may not necessarily command any legal observance.

Some of the consequences that arise from the use of a design should be accepted by the designer as his responsibility. This is mainly a moral responsibility, but in some cases legal aspects will also be involved, e.g. if an engineer is shown to be negligent in his duties or if he has been influenced by corrupt practices. Morals, ethics, personal principles and professionalism are all bound together. The engineer should attempt to produce the best design he can, with the most economical (optimum) use of all available resources. He should be honest and he should only use his engineering skills in a way that produces benefits to mankind. 'Do of your best, and do as you would be done by' is a good motto for the design engineer. The role of engineering in society will be discussed in detail in Chapter 11.

KEYWORDS

legal	patent agent (or attorney)
patent	Patents Act 1977
invention	patent application
monopoly	embodiment
specification	priority date
novelty	Abstract of the Invention
licence	examiner

search

claim

substantive examination

amendment

patent life

renewal fee

International Convention for the
    Protection of Industrial Property

royalties

patent of addition

infringement

design registration

design patent

Registered Designs Acts 1949–61

Design Copyright Act 1968

industrial design

registration

representations

Certificate of Registration

copyright

Copyright Act 1956

trade mark

distinctive

unique

Trade Marks Act 1938

Trade Marks Rules 1938, as
    amended

Register—Parts A and B

Schedule IV

generic name

Trade Marks Registry

standards

International Standards
    Organisation (ISO)

British Standards Institution (BSI)

codes of practice

contract

morals

ethics

professionalism

BIBLIOGRAPHY

Brichta, A.M., and Sharp, P.E.M., *From Project to Production*, Pergamon Press Ltd., Oxford (1970).
  See Chapter 6—Inventions, Patents and Design Registration; before UK Patents Act 1977.

Capsey, S.R., *Patents. An Introduction for Engineers and Scientists*, Newnes-Butterworths, London (1973).
  An excellent 'easy to read' introduction. Before UK Patents Act 1977.

Carter, E.F. (Ed.), *Dictionary of Inventions and Discoveries*, Crane, Russak and Co. Ltd, New York (1977).
  A reference book.

Chuse, R., *Pressure Vessels*, 5th Ed., McGraw-Hill Book Co., New York (1977).
  Examples of codes and standards.

Dunham, C., Young, R., and Bockrath, J., *Contracts, Specifications and Law for Engineers*, 3rd Ed., McGraw-Hill Book Co., New York (1979).
  Presents the basic rules of law for engineers.

Hajeck, V.G., *Management of Engineering Projects*, 2nd Ed., McGraw-Hill Book Co., New York (1977).
  Includes contracts, clauses and the law, and many engineering topics.

Horgan, M.O'C., and Roulston, F.R., *The Elements of Engineering Contracts*, E. and F.N. Spon Ltd, London (1977).
  A basic guide (100 pages), as per title.

Kock, W.E., *The Creative Engineer: The Art of Inventing*, Plenum Press, New York (1978).
Examples of inventions; interesting material.

Liebesny, F., *Mainly on Patents. The Use of Industrial Property and its Literature*, Butterworth and Co. Ltd, London (1972).
Collection of papers on many topics, excellent coverage—before the UK Patents Act 1977.

Muncheryan, H.M., *Patent it Yourself*, Tab Books Inc., Blue Ridge Summit, Pennsylvania (1982).
An excellent guide to the patent system in the USA.

Newby, F., *How to Find Out about Patents*, Pergamon Press Ltd., Oxford (1967).
Good background reading despite being dated in parts.

Pike, A., *Engineering Tenders, Sales and Contracts*, E. and F.N. Spon Ltd, London (1982).
A handbook of typical contracts and agreements; a useful reference source.

Taylor, C.T., and Silberston, Z.A., *The Economic Impact of the Patent System. A Study of the British Experience*, Cambridge University Press, Cambridge (1973).
A detailed survey as title—only for an in-depth study of this topic!

Thring, M.W., and Laithwaite, E.R., *How to Invent*, Macmillan Press Ltd, London (1977).
A general interest book.

A Guide to the Preparation of Engineering Specifications, The Design Council, London (1980).
A useful, short (30 pages) guide to this subject.

Publications Available from the Patent Office, London

Patents. A Source of Technical Information (free booklet).
Patents Act 1977.
Patents Rules 1977, and amendments.
How to Prepare a UK Patent Application and then Apply for a Patent (free booklet).
Instructions for the Preparation of Specification Drawings (free pamphlet).
Manual of Office Practice (Patents).
Structure of the Classification Key (free pamphlet).
Classification Key to United Kingdom Patent Specifications.
Reference Index to the Classification Key.
Classified Abstract Pamphlets.
Classified Abridgment Volumes, in respect of patents granted under the 1949 and previous acts.
Index of Names of Applicants, in respect of published patent applications.
Reports of Patent, Design and Trade Mark Cases (RPC).
Classification and Information Retrieval Services Bulletin.
Protection of Industrial Designs (June 1981) (free pamphlet).
Registered Designs Act 1949.
Patents and Designs (Renewals, Extensions and Fees) Act 1961.

The Designs Rules 1949, and amendments.

Copyright Act 1956.

Design Copyright Act 1968.

The Registered Designs Appeal Tribunal Rules 1950, with amendment 1970.

The Registered Designs Appeal Tribunal (Fees) Order 1973.

Applying for a Trade Mark (September 1978) (free pamphlet).

Trade Marks Act 1938, and amendments.

International Classification of Goods and Services to which Trade Marks are applied, and the Supplements thereto.

Guide to Schedule III of the Classification of Goods.

Various publications concerning patents are available from the Superintendent of Documents, US Printing Office, Washington, DC 20402, including:

Annual indexes of patents, trade marks and the applicants.

Annual volumes of decisions in patent and trade mark cases.

Pamphlets concerning patent laws, rules of patent practice, trade mark laws, and others.

Copies of patents and trade mark registrations that are granted (a fee is charged for these items).

Also available from the addresses given:

International Patent Classification, 3rd Ed. (January 1980). Carl Heymanns Verlag KG, Munich.

Help for the Inventor, and other booklets, British Technology Group (BTG), 101 Newington Causeway, London.

List of Patent Agents, Chartered Institute of Patent Agents, Staple Inn Buildings, London.

Guide to the Science Reference Library and *Literature in the British Patents Section (Aids to Readers No. 10)*, Science Reference Library, London.

Journals

The Official Journal (Patents). Published weekly in the UK, contains details of all applications filed, specifications published and patents granted. Also contains a list of designs registered (except 'textile articles'), comprising alphabetical list of names of registered owners, registration numbers, dates of registration and brief particulars of the articles in respect of which the designs are registered.

Trade Marks Journal. Published weekly in the UK.

Official Gazette. Published weekly in the USA, contains US patents and trade marks that have been issued.

Communications

Chapter Objectives

To encourage the reader to consider:

1 the ways in which information can be presented;

2 the factors that influence the communication of ideas;

3 methods of presentation and communication strategies, which will improve the effectiveness of information transfer.

* * *

Exercises

1 You are required to attend an employment interview. List the personal communication factors that will influence your performance and success.

2 Discuss the factors that will influence your success when obtaining information by telephone concerning:
   (a) an air-conditioning unit;
   (b) travel arrangements between London and San Francisco, including a stop-over in New York.

3 You are about to launch a new type of 'do-it-yourself' double glazing unit. Consider details of the requirements for a promotional video or television film.

4 What are the important factors when advertising a radio telephone in a national newspaper or a business magazine?

5 You have completed the 'on-paper' design of a wheelchair. Now consider possible uses and advantages of scale models to assist an evaluation of the ideas proposed.

6 You have designed a domestic cooker. Make recommendations for building and testing a prototype unit.

7 What documents should be available when a design project is completed?

8 What recommendations would you make for the presentation of reports? How would these recommendations differ for different types of material to be recorded and presented? Compare your ideas with the requirements for laboratory and project reports in your course.

9 What are the differences between an abstract and a summary? Prepare an abstract. Let a colleague criticise it and then discuss it together.

10 Recommend the packaging to be used for a 20-piece construction toy, suitable for children aged 9 to 12 years.

11 You manufacture a multipurpose aerosol lubricant. Prepare technical literature for purposes of advertising, marketing and sales.

12 Make recommendations for the contents of a washing machine's maintenance and repair manual. What communications relate to the servicing of the machine by an engineer?

13 Assess the effectiveness of this book, and any associated lectures and tutorials, for the communication of particular information.

*   *   *

9.1 COMMENTS

Communication is nearly always symbolic: the common symbols are words (written and spoken) and pictures (graphs, diagrams, etc.). The symbols are then combined to convey information, that is, to formulate a message. Information science is concerned with studying the message used in a communication. However, if the message is viewed in isolation without consideration of both the sender and the receiver, the result may be a good message but poor communication.

The designer will need to be aware of all aspects of the communication system and to view this as an activity which involves people. Although machines can record, relay and act upon messages, they must originally be programmed by humans. It is the human transfer of information and the response which are important in design and engineering. Surprisingly, many messages are prepared and sent without any thought at all for the recipient. The sender often prepares the information in a form that he considers to be appropriate, i.e. as *he* would prefer to receive it, and this is the reason why so many communications are at best inefficient and at worst completely useless. Even some lecturers make the mistake of ignoring the requirements and capabilities of their audience, when a part of their job is supposed to be information transfer!

Always prepare information bearing in mind the person who will act upon it. Both students and lecturers would benefit if some time were devoted to 'effective' communications! Communication is the *efficient* transfer of information. If it is not efficient, then it is not communication.

There are many guidelines that could be proposed and entire books have been written describing just one aspect of communication, e.g. report writing or effective public speaking. Some suggestions will be discussed in this chapter. However, the best way to improve your communicating ability is to be widely read, generally observant and interested in a wide range of topics. In this way, you will improve your knowledge and awareness of the alternative means of presenting information, and be able to appreciate examples of good and bad communications.

9.2 ORAL COMMUNICATION

Purely oral communications occur if two people are working in the dark or
with an intervening partition; the most common example is talking on the
telephone. The efficient transfer of information then depends upon both
speech and hearing, as well as the way in which the message is constructed.

Oral communication can also occur by personal contact, but it is then
influenced by facial expressions, tone of voice and mannerisms. Transfer of
information using spoken words can be in the form of questions and answers,
or as direct statements. For example:

'*Can* I have component X?'

'Not until next week.'

'I *want* six of these.'

'They are out of stock.'

Problems can arise if the two people who are conversing have different
native tongues or different accents or dialects. The manner in which the words
are spoken will also affect their understanding, as will any background noise.
Ultimately, if the person receiving the message does not possess satisfactory
hearing or if he is not concentrating, then the interpretation of the message
will be incorrect.

An oral presentation can be used to inform, advise or instruct; the first
of these is the most common use. Attending oral presentations requires a
greater contribution of time and effort for the audience than is required when
reading a written report on the same topic. If oral communication is to be
effective then the speaker must also prepare his material and approach
thoroughly; the time spent should be at least equal to that required for a
written report. Oral presentation should not be considered a quick, easy,
'off-the-cuff' way of communicating information. The audience will expect
additional benefits by attending an oral presentation, otherwise they would
merely ask for a written summary.

A reader who becomes bored with a written report can put it aside and
yet retain the option of giving it further attention later. An audience that
becomes bored with an oral report loses both the content and the opportunity
for a second attempt. Audiences are generally polite but they are by definition
human, and if a message is not conveyed in an appropriate and satisfactory
manner, then they will probably show their dissatisfaction. Ultimately they
may 'vote with their feet', i.e. leave.

There are several disadvantages that may arise with oral reporting. Less
material can be presented, although additional printed material can be
distributed; this, however, may act as a distraction for the audience. The
listener cannot easily refer back to material presented earlier, and first
impressions are particularly important, especially as corrections and changes
reduce the effectiveness of the material. The ability to revise written material
makes it effective in its final form; revision is not possible once an oral report
has begun. If the listener loses interest, it will be difficult to recapture. If the
oral report is part of a series, it may be difficult to maintain 100% attendance
by the audience.

However, there are also advantages to be gained. The main advantage is the opportunity to influence the audience by the speaker's delivery, charisma, intonations, gestures, personality, etc. If this does not work, the advantage rapidly becomes a disadvantage! A more thorough coverage of a particular topic and increased audience participation can be achieved by the use of questions and answers, tutorials and problem-solving and group discussion sessions. Oral reporting *should* encourage the speaker to concentrate only upon the essential details, and it may also be more economical than report preparation and distribution. The speaker will have the opportunity to involve the audience directly and to use their abilities and experience in the reporting session. He will also have the opportunity to improve and enhance his material and its presentation by using appropriate audio and visual aids.

One final important point: both oral and written reports should be prepared for different audiences and different situations. The oral report has the added dimensions of vocal and visual expression, but it is also normally necessary to present it as a continuous single unit. Written reports prepared for readers should *never* be read aloud to an audience: they do not produce good communications. Consider this point in relation to your university lectures!

9.3 WRITTEN COMMUNICATION

An opinion that is sometimes expressed in schools, universities and industry is that engineering students need to be taught how to 'write', or in particular how to express themselves in writing. There are no reliable statistics available to confirm or refute this opinion, and anyway it is a subjective view. However, it is important to realise that there are two aspects of writing. These are literacy and competence. Literacy involves correct spelling, grammar and a reasonable vocabulary. The associated skills are not difficult, and therefore poor performance is often severely criticised. Competence is the ability to express thoughts, ideas, logical arguments, etc., and depends upon the subject-matter being presented, whereas literacy is independent of the subject. A competent writer uses language in a clear, concise and easily understandable way when communicating with others. Someone who is literate will not necessarily be competent, and vice versa, although competence requires some grammatical ability.

The skills of literacy should be acquired mainly in the schools, whereas competence will be developed and refined at the college level and thereafter. Those who are critical of the written work of engineers (and students) will need to decide which aspects are unsatisfactory, before they can begin to identify the causes and solutions to the problem.

Lecturers should attempt to make the work performed by the students typical of the real world. Maximum limits should be stated for written work, and any 'excess' included in appendices. Peters and Waterman conducted a detailed study of American companies and found that Proctor and Gamble and United Technologies, two of the most successful companies, had introduced the 'mini-memo' requiring that all written communications be summarised in one page. The efficient communication of information was identified as a common feature of other successful American companies.

The remainder of this section will consider different types of written material which the designer will be required to prepare or evaluate. Ideas and suggestions will be mentioned briefly as the student will only become more adept in written communications if he is required to produce his own material, and to evaluate its effectiveness.

The designer will be required to record all data and decisions associated with a project as the work proceeds. This will provide the full product specification and documentation and will consist mainly of data and records, including relevant standards, calculations, quotations, technical literature, patents, etc., but will also contain details of why and how certain decisions and choices were made. The documentation will contain full sets of detailed drawings associated with the product and its subsequent manufacture. This will include drawings of all components, sub-assemblies and the general overall layout or assembly. There will be an extensive amount of cross-referencing, and an appropriate system of identification and drawing number sequencing should be adopted as early as possible. An efficient filing system should also be implemented which includes all aspects of the work, all references including those that are not readily available as copies, suppliers' names, addresses and telephone numbers, and all relevant cost and quotation information and dates of acquisition and/or use. This should not be left until the project documentation has grown significantly, otherwise it will be an onerous, if not impossible, task. A system should be implemented that is easy to understand, use and cross-reference, so that new staff can easily appreciate its functions.

Report writing is one of the most important tasks that the design engineer performs. The report may be a policy document, a technical document, results of a feasibility study, recommendations for a strategy to be adopted, or the record and conclusions of a particular stage of a project. Company reports are usually confidential and are intended to provide a permanent record of a project. Report writing is a necessary aspect of the designer's work. It is important that a report is presented in the correct format and that the information it contains is both relevant and complete. The following guidelines will be useful in many situations.

Questions to be answered
(A) Who are the readers, e.g. engineers, salesmen, customers?
(B) What is the purpose of the report, e.g. information, use (a servicing manual), instruction (to be acted upon)?
(C) Is it confidential?
(D) What type of material does it contain, e.g. technical or non-technical, facts or conjecture?

Features of the presentation

(a) Assign a brief but meaningful title.

(b) Include an abstract or summary, or both (what is the difference between them?).

(c) Include only necessary background information with a complete reference list.

(d) Describe material in a depth suitable for the particular reader.

(e) Describe the background before the details.

(f) Plan the contents of the report as a logical sequence.

(g) Emphasise essential points.

(h) Distinguish clearly between facts and opinions.

(i) Use clear, concise language.

(j) Quantify statements wherever possible, e.g. mild steel is cheaper than aluminium, should be followed by — and the prices are £X and £Y per kg, respectively.

(k) Plan the report so that it is easy to read and can be read in stages, e.g. appropriate headings, graphical presentation of data.

(l) Choose the appropriate type of report. This may be a memo, letter, informal short report, instruction, specification, formal report, proposal, article or technical paper. What particular features or functions relate to each of these types of report? These aspects of the presentation have been expanded in more detail by Rathbone (see Rosenstein *et al.* in the Bibliography).

Preparation of the report

(1) Prepare an outline.

(2) Write a draft copy.

(3) Set the draft aside for a while.

(4) Then revise the draft by observing the questions to be answered (A to D above) and the required features of the presentation (a to l above).

(5) Repeat these preparation stages 3 and 4 as necessary.

Data presentation

• Only include *necessary* data in the body of the report, include all other data in appendices or as a separate volume.

• Present the data in a form that emphasises the points to be made in the text.

• In general, graphs, charts and diagrams are preferable, but there are situations where tables are more appropriate.

• If using tables only include relevant information, but present it in a clear, complete and unambiguous manner. Refer to Woodson (pp. 338 – 42) and Vidosic (pp. 145, 148 – 51) for examples of the alternative ways in which data can be presented.

• Remember that data can be presented in many forms, not just as graphs and tables. Alternatives are charts, block diagrams, flowcharts, schematics, nomographs, curves, sketches, drawings, maps and photographs.

• Only include visual information which is necessary and relevant.

Avoid
- Mistakes in wording and spelling.
- New, unusual or undefined symbols or abbreviations.
- Incomplete diagrams or tables.
- Incorrect references.
- Ambiguous or unfamiliar words.
- Personal or biased opinions and comments.
- Disordered or wrongly numbered pages.
- Marked or damaged paper.
- Poor reproduction.
- Typing errors and untidy corrections.
- Inadequate binding.
- Damage during postage.

Report presentation
  (i) Contents list with page references.
 (ii) Abstract and/or summary.
(iii) Data presented in appropriate sections of the report.
 (iv) Conclusions — brief, concise statements.
  (v) Recommendations.
 (vi) Actions required — identify persons to be responsible and the time-scale.
(vii) Full and complete reference list.
(viii) Nomenclature.
 (ix) Appendices for any information not specifically required in the main body of the report, e.g. tables of data presented graphically in the text, sample calculations, standards, specifications and suppliers' quotations.

9.4 MARKETING, ADVERTISING, SALES AND SERVICING

There are two other categories of written communication that concern the design engineer. These are aspects of communication that are intended to induce sales of the product and those that are related to the after-sales services.

In order to achieve sales of a product, a company will require expertise in marketing, advertising and sales. These may be provided by departments within the company or by outside specialist firms. All three aspects involve communication, in this case between the company and prospective customers.

The marketing department should identify who will buy the product, at what price, and what features of the final article are considered positive or negative. Market research and marketing of a product are specialised functions, and some consideration of this aspect of the design should be included as early as possible, and later as a continuous activity. Marketing involves communication with the customers and with the design team, ensuring that the customer will want what is produced, and that it is possible to produce the required item. It will be necessary to print introductory leaflets, sales brochures, technical sales literature, press releases and technical articles. As with all forms of communication it is essential to produce written information that is appropriate to the particular reader, and to ensure that this person receives the information.

Advertising involves selecting the best method of making the customer aware that the product exists, and also inducing the customer to buy this particular product rather than an alternative, or none at all! Our society is now bombarded by advertising from many sources, including television, radio, newspapers, journals, handouts, hoardings, stickers, word of mouth, packaging, etc. The job of the advertising department will be to select the most effective type of advertising medium, the nature of the message to be conveyed, the format and the promotional campaign.

Finally, the sales department should be kept adequately informed of all relevant details of the product, so that potential customers can be persuaded to buy it as soon as it becomes available. It is important to brief the sales department thoroughly so that enquiries relating to non-standardised uses of the product can be answered efficiently and accurately. There should always be liaison (feedback and feed-forward) between sales, marketing and design personnel, so that the advantages of the product can be exploited and any disadvantages minimised or quickly eliminated.

The after-sales services for a product are concerned with packaging, installation, maintenance and servicing. Communications in these areas are mainly written instructions, and as previously emphasised, they should be prepared for a particular reader. The instructions should be clear, simple and unambiguous, but also as brief as possible and yet adequate for most situations. They should clearly state the 'dos' and 'don'ts' and any emergency procedures.

The preparation of maintenance and servicing information will depend upon whether these tasks are to be performed by the supplier (specialist personnel) or the customer (general expertise). Some instructions may be oral, either from a supervisor to a workman or by telephone between the supplier and the customer.

9.5 VISUAL COMMUNICATION

The construction of models and prototypes is part of the design and development stage of the project. Model-making is really an extension of the work of the drawing office, and building a prototype is a part of the project engineering function. However, both of these representations of the final product provide a useful form of visual communication which should be exploited to its full advantage.

Models can vary from simple cardboard and glue components to accurately machined scale models that include many of the features of the final (envisaged) product. Models may provide only size and layout representation, e.g. a model of a housing estate or a chemical plant model to assist in piping layout. Alternatively, models may have moving parts, e.g. opening doors on a model car, or they may be illustrative working models, e.g. water flowing through a model of a dam.

A prototype is usually a full-scale, fully operational unit built to conform to the detailed design specifications. It is used to test the design data and assumptions, to obtain experimental results, to provide an accurate product specification and to test the operation of the product to determine unsatisfactory features. It can be used to provide an accurate visual representation of the final product, which will be useful for the designer, the production engineer and the potential customer. It will also enable the marketing and advertising departments to obtain realistic photographs of the product for use in their promotional literature.

9.6 ORAL AND VISUAL COMMUNICATION

There are many visual aids that can be used to supplement an oral presentation (see Section 9.2). In some cases the sound may be an integral part of the visual material, e.g. a video film, or it may be provided separately by the speaker.

There are also many audio-visual aids available, and so it will be important to ascertain the costs of alternative systems. The following list is not exhaustive but will provide the reader with an appreciation of what is available.

(a) Overhead projector transparencies (acetate sheets) and felt-point pens.
(b) Slides (35 mm).
(c) Slides (35 mm) and tape commentary.
(d) Boards for magnetic and stick-on materials.
(e) Cutaway model sections.
(f) Visual-cast reflection projector (epidiascope).
(g) Film strip projector, with or without sound.
(h) Video films.
(f) Closed-circuit television.

It will be necessary, before deciding upon a particular form of audio-visual presentation, to determine the time required for preparation and the facilities required for its use.

9.7 SUMMARY

Having studied this chapter, there are two important points that the reader should appreciate. First, prepare material for a specific audience and decide upon the most efficient and effective presentation. Second, prepare audio and visual material *thoroughly*.

Finally, some rules for effective communication. Hopefully the reader will already have thought of these.

(a) Decide the purpose of the communication.
(b) Make an appropriate presentation.
(c) Make the message short, sharp and to the point.
(d) Continually re-evaluate your communications.
(e) Practice listening skills.

Evaluate the effectiveness of the communications that are used in your work and non-work situations.

KEYWORDS

communication	project documentation
words	filing system
pictures	report writing
message	marketing
information science	advertising
sender	sales
receiver	servicing
lecturer	packaging
oral communication	installation
audience	maintenance
speaker	visual communication
listener	models
written communication	prototypes
literacy	oral and visual communication
competence	audio-visual aids

BIBLIOGRAPHY

Barrass, R., *Scientists Must Write. A guide to better writing for scientists, engineers and students,* Chapman and Hall Ltd., London (1978).
The title says it all, easy reading, highly recommended.

Beard, R.M., *Research into Teaching Methods in Higher Education Mainly in British Universities,* 2nd Ed., Society for Research into Higher Education Ltd, London (1968).

Beard, R.M., *Course Design, Teaching Methods and Departmental Decisions,* University of London, Institute of Education (1974).

Beard, R.M., *Teaching and Learning in Higher Education*, 3rd Ed., Penguin Books Ltd, London (1979).
All three publications by Beard are concerned with 'effective' teaching methods and contain much practical advice concerning all forms of communication.

Bragg, G.M., *Principles of Experimentation and Measurement*, Prentice-Hall Inc., New Jersey (1974).
As title, including reporting results and report writing (with examples).

Brichta, A.M. and Sharp, P.E.M., *From Project to Production*, Pergamon Press Ltd, Oxford (1970).
See Chapter 10, pp. 250–61, Models and Prototypes. Also Chapter 11, pp.274–82, concerning literature associated with a product.

Brinkworth, B.J., *An Introduction to Experimentation*, The English Universities Press Ltd, London (1971).
As title, including presentation and analysis of results (Chapter 6) and reporting the work (Chapter 7).

Garvey, W.D., *Communication: The Essence of Science*, Pergamon Press Ltd, Oxford (1979).
Sub-title: 'Facilitating information exchange among librarians, scientists, engineers and students' — good as background reading.

Hall, W.C., and Cannon, R., *University Teaching*, Advisory Centre for University Education, University of Adelaide, Adelaide (1975).
Good advice and analysis of communication using lectures, tutorials and audio-visual aids.

Herbert, A.J., *The Structure of Technical English*, Longman Group Ltd, London (1977).
Intended to teach language, not engineering, but uses engineering examples. A workbook or programmed text of how to write.

Peters, T.J., and Waterman Jr, R.H., *In Search of Excellence (Lessons from America's Best Run Companies)*, Harper and Row, New York (1981).
One of these lessons was short, concise memos.

Rosenstein, A.B., Rathbone, R.R., and Schneerer, W.F., *Engineering Communications*, Prentice-Hall, Inc., Englewood Cliffs, New Jersey (1964).

Scharf, B., *Engineering and its Language*, Frederick Muller Ltd, London (1971).
A reference book describing terms used in engineering. Headings such as plastics and mechanical testing.

Ulman Jr, J.N., and Gould, J.R., *Technical Reporting*, 3rd Ed., Holt, Reinhart and Winston, Inc., New York (1972).
A comprehensive text dealing with written communications.

Vidosic, J.P., *Elements of Design Engineering*, Ronald Press Co., New York (1969).
See Chapter 8.

Woodford, F.P. (Ed.), *Scientific Writing for Graduate Students. A Manual on the Teaching of Scientific Writing*, Rockefeller University Press, New York (1968).
Recommended, easy to read, good notes on tables and figures, and also oral presentation.

Woodson, T.T., *Introduction to Engineering Design,* McGraw-Hill Book Co., New York (1966).

See Chapter 17, p. 321, Communication and the Engineer, and also Chapter 18, p. 347, Oral Reporting.

Yaggy, E., *How to Write Your Term Paper,* 4th Ed., Harper and Row, New York (1980).

Recommended — a student's manual.

Management

Chapter Objectives

To understand:

1 the meaning of management;

2 the role of a manager;

3 the interactions between management, engineering and design.

---

* * *

Exercises

1 For a small design and development section or team, list the main responsibilities of the senior engineer and the projects manager.

2 List in order of importance the main factors that influence the job satisfaction, motivation and standard of work performed, for:
 (a) a graduate engineer;
 (b) an engineering technician;
 (c) the technical sales and marketing manager;
 (d) the Director of Engineering in a large company.

3 What salary and benefits policy would you recommend for the engineers and managers in an engineering company?

4 Identify any management tasks that are the responsibility of the young graduate engineer.

5 Although it is not possible to be experienced in every aspect of business, what qualifications and experience would you expect the senior staff of a manufacturing company to possess?.

6 You are required to improve the productivity of a small component assembly unit. You identify one of the problems to be the inefficiency of a member of the team. The man is 54 years old and has worked for the company for 23 years. What are your possible courses of action, and which one would you recommend?.

7 The paint spraying unit of a car assembly plant has a poor work record due to many instances of disputes and 'down tools'. The foreman of the group is an active union member. Discuss the merits and drawbacks of attempting to implement one of the following strategies:
 (a) wait for an excuse to sack the foreman;
 (b) promote the foreman;
 (c) transfer the foreman;
 (d) impose a time and motion study;
 (e) introduce a productivity and wages incentive scheme;
 (f) eliminate manual work by introducing automatic spraying machines.

8 Can undergraduates or graduate engineers be taught the skills of management, or can they only be acquired 'on the job'?

9 What changes would you suggest in the contents of an engineering degree scheme, in order to prepare graduates for the management roles they will be required to assume when they join a company?

10 What management skills do you exhibit in your day-to-day life?

11 What are the responsibilities of managers? What are the skills required by managers? What are the personality traits possessed by 'good' managers?

12 What experience do you think should be gained by a graduate engineer in his first 5 years of work?

13 You have been awarded a contract to produce a one-off, non-standard, mechanical plant assembly. Strict deadlines have been imposed and accepted. The design specification is complete and all the necessary components and materials are available. The unit must be fabricated and assembled within 7 days, and it is intended that this work should be carried out in the workshop of the Engineering Department of your company. Three more days are allocated for commissioning and modifications. Workshop staff must be available during this period to perform any immediate practical work which becomes necessary.

The shop steward in the workshop has indicated that completion of the work on time can only be 'guaranteed' if bonus payments are agreed. This is not normal company policy and may set a precedent.

What course of action would you take?

What strategy would you devise for a similar situation in the future?

*   *   *

10.1 DEFINING MANAGEMENT

Consider the following dictionary definitions:

(a)   manage — handle, conduct, control;
(b)   manager — person organising a business;
(c)   the management — governing body;
(d)   management — verbal senses of (a); also trickery, deceitfulness, contrivance.

Nothing very surprising except perhaps the last item, which implies that management may include actions that are not wholly ethical or moral, or perhaps that are even dishonest. Certainly in recent years, the activities of both private and public institutions have come under close scrutiny, and some disquieting practices have been revealed. The managers will jump to their own defence and point out that the company has a responsibility to pay dividends to the shareholders and adequate wages to their employees. This is a debate that has occurred many times. However, ideas taken out of context may present a false impression. Advertisers would probably think twice about describing products as 'sophisticated' if the public knew that one dictionary definition of the word was 'artificial or false'!

The definitions above do show that management is concerned with power, control over people and things, organisation, authority and leadership. What is not stated directly but is implied by these features, is that management must be based upon reliable decisions. Decision-making will be one of the primary functions of any manager.

10.2 TYPES OF MANAGEMENT

An engineer working in a company will be in contact with two types of management, or rather two different but related managerial roles. The first is technical management and this has the most influence upon the day-to-day work of the engineer. It involves co-ordinating the work of a design team, production department, etc., and assigning particular tasks to individuals or groups. The technical manager will be the chief engineer, senior designer or the projects manager. The successful engineer will assume the responsibilities of technical management quite naturally as his career advances, unless he decides to specialise in a particular, narrow area of engineering expertise. In this case, he may become the company specialist in that area or an independent consultant.

The second type of management within a company is executive management, which involves very different activities and responsibilities from those undertaken by the technical managers. The executives of a company are responsible for an overall view of its activities, and their decisions are often based upon long-term predictions and plans. The executive will require annual budgets, and details of whether these are actually achieved. He will also be responsible for preparing a long-term plan, say 5 or 10 years, for a particular aspect of a company's operations. In an automobile company there will probably be different senior executives with responsibilities for cars, trucks, buses, etc., and also for overseas sales, new developments and ventures, etc. Within a large chemical company there will be an executive manager responsible for specific types of chemical production, e.g. plastics, dyes and pigments, and fibres. A company will also appoint managers with specific responsibilities for corporate planning, personnel, legal affairs, foreign operations, etc.

The tasks to be performed by the technical manager should be fairly easily and naturally assumed by the experienced and successful engineer. They will represent a move away from detail to a wider view of the design, development and production process. Some extra expertise or training may be required in certain areas, but this is not usually a problem. However, the skills and experience required by a company executive are very different from those of the engineer or technical manager.

The executive should be qualified or have some experience in legal affairs, accountancy, financial planning, personnel management, etc. These are not normal areas of expertise for the engineer, and additional qualifications and experience will be required for the engineer who wishes to become an executive manager. In a large company there may only be one or two senior executives specifically qualified and experienced as engineers, whereas there will be several executives responsible for the 'business' affairs of the company. This is one reason why there are few engineers who are directors or managing directors of large companies, even engineering companies. Perhaps another reason is that an engineer's advancement is usually based upon his successes and achievements, and lawyers and administrators are more adept at strategies and manipulation!

10.3 FUNCTIONS OF MANAGEMENT

There are three main functions that managers should perform. These are:

(a) decision-making;
(b) control;
(c) authority.

Decision-making and planning should go hand in hand. If decisions are made without considering the possible consequences, then the company's performance will be impaired. Decisions are taken at two levels in the company hierarchy. Decisions at the senior executive level are broad-based policy decisions. For example, a decision to abandon the car model range currently produced owing to falling sales, and to introduce a new style of car with particular marketable features. This decision will be based upon sales figures, competitors' products, the general economic situation, future forecasts of petrol prices, government legislation for safety and pollution, etc. In arriving at the decision, the executive responsible will have ascertained that it is technically possible to produce the required models, and will have obtained estimates of the costs involved and the time required. The decision has implications for the company far into the future, say 5 or 10 years ahead: it requires long-term planning. It is also a broad-based policy, the overall objective being to improve or maintain the company's profitability.

However, this decision has been made without considering in detail the 'nuts and bolts' issues involved in producing a new car. Aspects such as a large research and development expenditure, necessary to overcome a particular technical problem, will have been assessed. The engineering aspects will be left to the managers further down the company management structure. Once higher-level decisions are made and passed down the ladder of command they become another manager's responsibility. The senior executive will only want to know whether a decision has been implemented, and if not, the reasons why and the alternatives that are available. He will have imposed certain constraints, such as time and cost, but again the answer required will be whether these have been achieved or not.

The major difference between executive management and technical management is that the former makes decisions and then communicates them to the appropriate staff to be implemented, largely relinquishing the responsibility for the methods by which they are achieved. Technical management makes decisions but must then ensure that they are implemented, and must be kept aware of all developments and problems which arise in all departments.

At both management levels, decision-making and planning are interdependent. A plan is evolved and decisions are made, then more planning occurs and further decision-making, etc. However, the process will only be effective if the decisions are implemented or if necessary changes or modifications are communicated to the manager responsible. Management is also required to ensure that adequate communications exist within the company.

Control is related to the way in which the decisions are implemented. An easily identifiable company structure and efficient organisation will be required. Appropriate control over projects and particular courses of action will be necessary to ensure that materials, resources and time are used in the most efficient and profitable manner. Management control should also mean that authority can be delegated and that there is sufficient flexibility within the system to ensure that appropriate action can be taken, if and when necessary.

Authority is the power and the right to enforce a particular course of action. By virtue of the company's infrastructure, managers at various levels have the right to impose and enforce certain decisions. The organisation of a company should be such that the extent of any individual's control is well known. However, the ability to exercise authority also depends upon the personality of the individual. A manager must be in control of the people working for him and he must be sure that his instructions will be carried out. The pattern of leadership within a company will emanate from above, and an ineffective manager within the chain of command will undermine the efforts and authority of those who respond to him. It is therefore essential that managers are appointed because they can perform the tasks required, and not merely because of the time they have served the company. An engineer who is technically well qualified and experienced will not necessarily possess the personal qualities needed to become a good manager.

Although a manager must possess authority in order to implement decisions, he should also remember that the aim of a business is to produce items and profits. If management adopts an attitude of strength and intransigence, then confrontation will result. This will not be good for morale or motivation, and if the employees stop working, the company's profits will be reduced. Thus management should maintain a balance between control and authority and satisfaction and motivation within the company. Good managers inspire others to work by their example and leadership. A manager who is not *seen* to be hardworking, keen and dedicated cannot expect these traits in others.

A good manager should also realise that the objectives determined for the success of the company will not necessarily be embraced by the employees. Motivation, job satisfaction and personal satisfaction are as important as the economic considerations of wages, benefits, security, etc.

10.4 SUMMARY

Managers in a company perform various functions including planning, decision-making, organisational control and delegation of authority. They must also possess particular personal qualities of authority and leadership, and still maintain the ability to be conciliatory and to provide motivation for their subordinates.

KEYWORDS

management	annual budget
organisation	corporate planning
control	accounting
authority	personnel management
leadership	communication
decision-making	delegation of authority
technical management	motivation
executive management	job satisfaction
long-term plan	

BIBLIOGRAPHY

Archibald, R.D.,*Managing High-Technology Programs and Projects*, John Wiley and Sons Inc., New York (1976).
   Intended for managers but gives a good insight for students.
Semler, E.G. (Ed.), *Management for Engineers*, Institute of Mechanical Engineers, London (1972).
   Selected papers.
Shannon, R.E., *Engineering Management*, John Wiley and Sons Inc., New York (1980).
   As title.
Wortman, L.A., *Effective Management for Engineers and Scientists*, John Wiley and Sons Inc., New York (1981).
   An easy-to-read book containing good ideas.

Engineering and Society

Chapter Objectives

1 To make the student think about his future role in society.

2 To make the engineer question the value of his work.

3 To make the engineer assess the good and bad uses, the advantages and disadvantages, and the exploitation and consequences of engineering developments and technological advances.

4 To make students and engineers think about the needs of society, the resources available and the possible solutions.

* * *

Exercises

1 Identify the major world problems that need to be solved. Make proposals that may help to alleviate these problems.

2 For two specific problem situations identify the causes, the effects (present and future) and the alternative strategies.
   What are the possible disadvantages or objections to the 'best solution' available?

3 Which major problems are most likely to be overcome (at least partially) in the near future? Why and how?

4 Identify the main reasons why certain problems are unlikely to be solved.

5 What influences should government have upon the developments and applications of technology?
   What influences should society have?

6 Assess the importance, the achievements and the possible future role in world affairs of international organisations such as the United Nations and the World Health Organisation.

7 What are the personal and professional responsibilities of an engineer?

8 Identify possible areas of major technological advance in the near future.
   What are the possible consequences of such progress?

9 Evaluate the usefulness/effectiveness to society of such major areas of expenditure as space exploration, military defence spending and nuclear power generation.

10 Devise your own 'master plan' for the organisation of a new society consisting of:
   (a) settlers on a distant planet;
   (b) the survivors of nuclear devastation.

* * *

11.1 COMMENTS

This chapter is deliberately brief. Those who are content to be ostrich-like will be unaffected whatever the length and depth of these arguments. The concerned and creative engineers will only need a stimulus, since they will always be questioning and evaluating their work and searching for new, improved solutions.

The subject-matter of this chapter is invariably found (as here) either as the introduction or the conclusion of a textbook of this type. I feel that these chapters are invariably brushed aside in the search for the 'real' contents of the book, and I would have liked to insert this chapter in the middle of the book. The aim would have been to jolt the reader out of the confines of the narrow thought processes defined by the syllabus, and into real and important problems. Perhaps it would have been too different, for both the reader and the publisher.

11.2 SOCIETY AND GOVERNMENT

A society is an organisation of people possessing some common traits and living within certain boundaries. Government makes and implements decisions that affect the people within that society. In a democracy, the government is periodically elected by the people. We often talk about the needs of society, implying the entire world population. However, what is beneficial to one society may be harmful to another, e.g. alien foods or vaccines for diseases not encountered by particular races. Also our social, cultural and political experiences may mean we are unable to understand the way another society functions. This may influence our ability to implement solutions to the problems of other societies.

11.3 THE PROBLEMS

The major problems that exist can be categorised as:

   (a) those present or imminent in poor and developing countries, e.g. food shortages, population growth;

(b) those present in most countries but on different scales, e.g urban development problems, overloaded transportation systems;

(c) those that are world-wide, e.g. pollution such as 'acid rain', increased carbon dioxide in the ionosphere, effects of deforestation.

Woodson (see Chapter 9, bibliography) has listed nine topics that represent challenges to engineers. These are energy, raw materials, urban environment, transportation, information handling, biotechnology, health and medicine, education and developing nations. He also considers in more detail the needs and problems that these subjects represent. The topics themselves are not particularly surprising, but the fact that the book was published in 1966 and the list originated from the Engineer's Joint Council in 1962, should cause some comment. If I had asked the reader to prepare his own list first, it would probably have been very similar, even now! Some solutions will have been implemented, perhaps related to biotechnology or health, but I wonder to what extent. Information handling is now easier owing to cable communication and the micro 'chip', but energy was recognised as a problem even in 1962, *before* the oil 'crises' of the 1970s. The problems of the developing nations listed by Woodson, e.g. population growth and food, are as real today as they were then!

The montage of newspaper extracts in Fig. 11.1 were all obtained at approximately the same time. These headlines indicate some of the problems facing society and the human race today, but similar items could have been obtained many times from newspapers since the Second World War, and could probably be obtained again in the future. These problems, e.g. wars, famine and social unrest, have not been solved, but only almost forgotten or camouflaged, to re-emerge in different locations. There are also many 'happy' headlines in the newspapers but the media have a vested interest in publishing disaster stories: the indications are that many serious problems exist and they are not being solved by the implementation of long-term, reliable strategies.

11.4 GOVERNMENT PRIORITIES

The reasons why so many world problems exist, and why effective solutions are not being implemented on a larger scale, are complex and dependent upon many factors. However, it is interesting to consider government priorities that exist now in the developed world. The major area of spending, development and rhetoric is in military defence. The arms race, defence and nuclear weapons accounts for billions of dollars worth of investment in non-productive (in terms of society's advancement) technology. Similarly, the space programme, which began in the 1960s, continues to be supported by a wide range of scientific arguments. It would be interesting to study a list of the positive advantages to society that are attributable to space exploration.

Unemployment jump dampens optimism

Unemployment rose by 120,300 to 3,199,878 in January, dampening some of this week's Government and business optimism of a sustained improvement in the economy.

The race against the clock and thermometer

After these initial changes, carbon dioxide rises would bring more pronounced effects. General warming would become much more noticeable, and the world's climatic regions would change. For instance, arable wheatlands would be pushed northwards — to the benefit of Russia and Canada and to the detriment of the United States. Eventually — when temperature rises reached 10 deg. Centigrade — the Antarctic ice cap would start to melt and ocean levels would rise, flooding low level regions, such as London, the Fens and Holland. The apocalypse would truly have arrived.

These are alarming predictions — and are certainly not accepted by all scientists. Nevertheless most now accept that some global warming is inevitable — given that every year we pour five billions tons of carbon, in the form of invisible carbon dioxide, into the atmosphere as a result of burning fossil fuels.

Normally, sunlight is reflected back into space as infra-red radiation, or heat. However, carbon dioxide absorbs infra-red radiation, so the more there is, the more Earth's heat is blocked from escaping. This is known as the 'greenhouse effect' in which carbon dioxide acts like a global pane of glass.

Male pill poses hairy dilemma

WOMEN grew moustaches after their lovers started using an experimental new method of male contraception, doctors report today. Seventeen men were given a daily pill containing the female hormone progesterone to suppress their sperm production.

Saudis tell US marines must go

Britain rejects proposal for UN peace force in Falklands

£60m satellite lost in space

A WORLDWIDE radar search continued yesterday for the world's most heavily insured satellite, which was lost in space after its launch from the Shuttle Challenger.

The giant Westar 6 communication satellite — insured through Lloyd's for £60 million — mysteriously vanished on Friday evening as it was blasted into high orbit after being positioned in space near the shuttle.

American radar stations round the globe hunted desperately for the missing craft.

Britain leads rescue of aid to third world

BRITAIN is to lead an international rescue mission to save the world's most important aid programme from disaster

Hi-tech to stop Olympic terror

LOS ANGELES police last week unveiled for the Press an ominous-looking 'hi-tech' arsenal of weaponry and surveillance devices that, they hope, will deter international terrorists who may be planning to air their political grievances at the 1984 Olympic games, before a global TV audience of 2.5 billion.

There were guns galore, British-made 'night vision' spying gadgets, a bomb-busting robot named 'Fearless Felix.'

Nuclear power station dumps will store waste from weapons research

Radioactive waste fron weapons research will be buried with nuclear power station waste in the proposed dumps under Billingham, Teesside, and Elstow, Bedfordshire.

The nuclear industry Reactive Waste Executive (Nirex) confirmed yesterday that the repositories would be used for the small proportion of waste from Ministry of Defence contracts including material from weapons research establishments like Aldermaston.

Figure 11.1 Problems Facing Society

A feature of the world at present is the number of wars and conflicts in progress. I will not risk being out of date before this book is even published by listing these wars; they are reported daily by the media for us all to see. Many of the countries involved in wars are poor — often they cannot even feed their own population adequately.

While government priorities continue in the areas mentioned, significant improvements in the situation of Third World countries is unlikely. If the time, effort and money spent by the USA in the long conflict in Vietnam had been directed instead towards solving the problems of agricultural development in the Third World, then this would probably not still be a major problem today.

11.5 SOLUTIONS

Upon first consideration, the solutions to some world problems may appear straightforward. Provide money, equipment, expertise to poor countries, legislate against pollution, provide education for birth control, create jobs for ethnic minorities, etc. However, apart from the obvious restriction that money is often not available, the solutions to particular problems are influenced by underlying and less obvious causes. Birth control is not only dependent upon education and contraceptive devices, it is also influenced by cultural and religious factors. Providing resources and equipment is not sufficient in itself: local people must be trained to take over the responsibility for a project, and be capable of solving problems as they arise. A course of action does not represent a solution if it merely shifts the burden of responsibility and the cost from a poor government to a developed nation.

What is required is people of ability who are concerned about these major problems, people with enthusiasm, compassion and a desire to effect improvements. Many of the problems have no simple solution, otherwise such solutions would already have been implemented. What is required is engineering skill and innovation, a creative approach and the ability to use engineering fundamentals and technological advances in old situations but in new ways. Engineers can and will provide the answers to many of today's problems. The main question is, When? However, when solutions are found, the governments concerned must be fully prepared to make the efforts necessary for effective implementation.

The human race must also learn to be more adaptive and creative in its thinking. The world has many natural resources such as the atmosphere, the oceans, deserts and polar ice caps, as well as such natural forces as the sea, the wind and the sun. It will become necessary to find ways of harnessing these resources of our planet and then using them for the good of society.

Problem solutions and technical advances often arise out of necessity and urgency. If the price of oil had continued to rise by as much as it did in the early and mid-1970s, more effort would have been made to find viable energy alternatives. What actually happened was that conservation was seen as a short-term solution, but as the price and supply of oil fluctuated, conservation became a permanent 'stop–go' measure. Hence no significant alternative energy source has been developed. If the food supply of Canada fell to the levels of India or of some African countries, then productive efforts would be made immediately. Perhaps, when social unrest becomes too great to ignore, efforts will be made to redress some of the imbalance between the poor countries and the rich nations.

11.6 MORALS, ETHICS AND RESPONSIBILITIES

Engineers have two types of responsibilities in the work they perform. These are the legal and professional obligations which are laid down, and which define the standard of the work and the nature of business agreements. Any manufactured item should be designed and produced so that it will operate as required and will be safe to use; also contracts should not be obtained by the use of bribery or other inducements. There are also social responsibilities which the engineer can legally choose to ignore. They include the consideration of those aspects of a project that determine whether the final item will be used for the benefit of society or whether it may be misused, whether it will cause undesirable or irreversible environmental damage, whether it will cause social, racial or religious divisions within society and whether it can be used as a destructive force. The answers to these questions are not always obvious. They should be easier for the engineer to answer, as he usually works towards a definite goal, rather than for the scientist who often does not know the final outcome or applications of his work.

Each engineer should set his own moral standards, and decide which courses of action he considers ethical and moral. He should do this as early as possible in his career, so that he knows the position he will wish to take before particular events occur. In this way he will avoid succumbing to arguments and pressures that he will later regret. The supply of goods or services to governments that create and maintain social, religious or racial divisions within their country is a case in point. The engineer should recognise the personal limitations and economic constraints that may require compromises to be made on matters of principle. It will be better to admit to human weakness than to follow the dictates of others unthinkingly.

If the engineers working in society are people of principle who believe in, and adhere to, particular moral values and ethical codes, then less problems will be created in the future and there will be a better chance of solving the problems that already exist.

11.7 CONCLUSIONS

Probably the major factor preventing the solution of world problems is government inflexibility. This may be because of social, cultural or religious pressures, lack of experience or the corrupting influence of power.

The most effective resources available to overcome these problems are a combination of creative thought and engineering skill.

BIBLIOGRAPHY

The evidence, the literature, the examples, the problems are everywhere, to read about and to see. The students who will become the 'best' engineers will already be familiar with the sources of written material. They will always be observing and learning.

The answers, the ideas and the solutions will be more difficult to find. To start with, look no further than your own mind.

Index

Nutrition and Mental Performance

Nutrition and Mental Performance

A Lifespan Perspective

Edited by
Leigh Riby, Michael Smith and
Jonathan Foster

palgrave
macmillan

First published 2012 by
PALGRAVE MACMILLAN

Palgrave Macmillan in the UK is an imprint of Macmillan Publishers Limited,
registered in England, company number 785998, of Houndmills, Basingstoke,
Hampshire RG21 6XS.

Palgrave Macmillan in the US is a division of St Martin's Press LLC,
175 Fifth Avenue, New York, NY 10010.

Palgrave Macmillan is the global academic imprint of the above companies
and has companies and representatives throughout the world.

Palgrave® and Macmillan® are registered trademarks in the United States,
the United Kingdom, Europe and other countries.

ISBN 978–0–230–29989–4 hardback
ISBN 978–0–230–29990–0 paperback

This book is printed on paper suitable for recycling and made from fully
managed and sustained forest sources. Logging, pulping and manufacturing
processes are expected to conform to the environmental regulations of the
country of origin.

A catalogue record for this book is available from the British Library.

A catalog record for this book is available from the Library of Congress.

10   9   8   7   6   5   4   3   2   1
21   20   19   18   17   16   15   14   13   12

Printed and bound in Great Britain by
CPI Antony Rowe, Chippenham and Eastbourne

Contents

Contents

Figures and Tables

Figures

Tables

The Editors

Leigh M. Riby, PhD
Department of Psychology,
Cognition and Communication Research Centre,
Brain, Performance and Nutrition Research Centre,
Northumbria University,
Northumberland Building,
Newcastle-upon-Tyne
UK

Web: diabetesbrain.com
Tel: 01912 273571
Fax: 01912 274515
Email: Leigh.Riby@northumbria.ac.uk

Michael A. Smith, Ph.D.
Department of Psychology,
Northumbria University,
Northumberland Building,
Newcastle-upon-Tyne
UK

Tel: 01912 437169
Fax: 01912 274515
Email: michael4.smith@northumbria.ac.uk

Jonathan K. Foster, FRMS DPhil
School of Psychology and Speech Pathology
Curtin University of Technology
GPO Box U1987
Perth, Western Australia, 6845

Tel: +61 8 9266 9157
Fax: +61 8 9266 2464
Email: j.foster@curtin.edu.au

Acknowledgements

We would like to thank Valerie Gunn for useful comments on the early drafts of the individual chapters. Dr Riby would like to thank his daughter Jessica and wife Debbie for continued support for his academic endeavours.

Notes on Contributors

Dr Stephane Bastianetto is a neuroscientist at the Douglas Mental Health University Institute at Montréal (Canada). He has expertise in the use of plant extracts (e.g. *Ginkgo biloba* extract) and polyphenols (e.g. resveratrol, catechins) for their possible beneficial effects on brain ageing and age-related neurological disorders, such as Alzheimer's disease. In 2002 he created a website (www.neuromedia.ca) about brain ageing for seniors and their caregivers.

Dr Louise A. Brown is a lecturer in psychology at the School of Social Sciences, Nottingham Trent University (UK). She is a cognitive psychologist who is primarily interested in issues of cognitive ageing. Her research focuses on short-term memory and attention processes.

Dr David Camfield is a postdoctoral fellow in the Centre for Human Psychopharmacology at Swinburne University, Melbourne (Australia). Since obtaining his PhD he has been involved in a number of publications investigating nutraceutical and dietary interventions for enhancing cognition and countering age-related cognitive decline. His research interests include the neuroimaging of cognitive enhancement as well as the identification of biomarkers that may be associated with cognitive ageing.

Dr Paul Cherniack carried out a study that was recently published in the *Journal of the American Geriatrics Society* assessing the effects of supplementation of vitamin D on calciotropic hormones and safety. He has published several book chapters and numerous journal articles about the health effects of vitamin D in the elderly. He is a co-investigator on a project to determine the effects of vitamin D supplementation on the physical performance of the elderly. He is Director of the Geriatrics Clinic at the Miami VA Medical Center, FL (USA), and an associate professor at the University of Miami Miller School of Medicine, FL (USA).

Kristen E. D'Anci is an assistant professor in psychology at Salem State University in Salem, MA (USA). Her research interests include the effects of nutritional factors on animal and human behaviour. Recent research has focused on the effects of dietary macronutrients on human cognition, and creatine and cognitive behaviour in rats. She has published extensively in the areas of drug abuse, nutrition and behaviour, including several book chapters.

She is an associate editor for the journal *Nutrition Reviews*. She is a member of the American Society for Nutrition, the Society for Neuroscience, the International Behavioral Neuroscience Society and the Society for the Study of Ingestive Behavior. She received a BA in psychology from the University of Southern Maine in Portland, ME (USA), and an MS and PhD in psychology from Tufts University in Medford, MA (USA).

Dr Kathy DeBarr is an associate professor in the Department of Public Health at the University of Illinois at Springfield, IL (USA). Her research interests include health education and technology. She teaches courses on research methods, evaluation and community health education.

Magnus Domellöf is an associate professor and Head of Pediatrics in the Department of Clinical Sciences, Umeå University (Sweden), as well as a senior consultant neonatologist at Umeå University Hospital, Sweden. His research focuses on paediatric nutrition, especially with regard to iron, low-birth-weight infants and the importance of early nutrition for later health. He has carried out a study that was recently published in the *Journal of the American Geriatrics Society* assessing the effects of supplementation of vitamin D on calciotropic hormones and safety, and he has written several book chapters and numerous journal articles about the health effects of vitamin D in the elderly. He is a co-investigator on a project to determine the effects of supplementation of vitamin D on the physical performance of the elderly. He is Director of the Geriatrics Clinic at the Miami VA Medical Center, FL (USA) and an associate professor at the University of Miami Miller School of Medicine, FL (USA).

Dr Caroline J. Edmonds is a senior lecturer in psychology at the University of East London (UK) and an honorary research fellow at the MRC Childhood Nutrition Research Centre, Institute of Child Health, University College London (UK). She obtained a PhD in psychology from University College London. She then undertook postdoctoral work at the MRC Childhood Nutrition Research Centre, examining the effect of early nutrition and nutritional interventions on long-term cognitive outcome. More recently she has published research papers on the effects of hydration on cognitive performance.

Prof. Jonathan Foster obtained a DPhil in neuropsychology from the University of Oxford (UK) in 1989 before undertaking a clinical postdoctoral fellowship at the Rotman Research Institute, Baycrest Centre, University of Toronto (Canada). He is a senior research fellow at Curtin University (Australia) and a clinical professor in the Neurosciences Unit, Health Department of Western Australia. He also holds honorary positions at the Telethon Institute for Child Health Research and the University of Western Australia. Prof. Foster has written more than 100 peer-reviewed journal articles relating to the mechanisms underlying cognitive functioning across the lifespan. He has published widely on the influence of glucose ingestion on cognitive performance.

Stine Hasselholt is a doctor of veterinary medicine and did her thesis on vitamin C and cognition. She is currently a PhD student at the University of Copenhagen (Denmark).

Dr Jeanet Ingwersen completed her PhD in psychology in 2011 at Northumbria University (UK) and currently works as a senior lecturer in psychology at Teesside University (UK). Her research interests include the relationship between food intake and children's cognition and behaviour, in particular the cognitive and behavioural effects of breakfast and snack consumption in children and how these can be related to school performance.

Prof. Jack E. James has a degree in psychology and a master's in clinical psychology, both from the University of New South Wales (Australia), and he did his PhD at the University of Western Australia, Perth (Australia). After working in clinical and community settings as a clinical psychologist, he has pursued an academic career in clinical psychology, behavioural medicine and health psychology. In 1991 he was appointed Foundation Professor of Behavioural Health Sciences at La Trobe University, Melbourne (Australia). He was later elected Founding National Chair of the College of Health Psychologists (a College of the Australian Psychological Society). In 1998 he relocated to the National University of Ireland, Galway (Ireland), then in 2012 he moved to Reykjavik University (Iceland). He has a particular interest in the implications of dietary caffeine for human health and well-being, and he is Editor-in-Chief of the *Journal of Caffeine Research*. He also has research interests in cardiovascular behavioural health, with particular reference to the psychophysiological correlates of stress, and in applied behaviour analysis.

Dr Emma Jones completed her PhD in psychology and nutrition in 2008 at Lancaster University (UK). She is a research associate in the Psychology and Communication Technologies Laboratory, Northumbria University (UK). Her research interests include the effects of nutrition on brain function and cognition across the lifespan as well as health and lifestyle predictors of healthy ageing.

Jens Lykkesfeldt holds an MSc in chemistry, a PhD in biochemistry and a DSc in medicine. He is a professor and chair in experimental pharmacology and toxicology at the University of Copenhagen (Denmark). His research interests include the roles of oxidative stress and antioxidants in neurogenesis, chronic diseases and ageing, in particular those of vitamin C.

Dr Therese O'Sullivan is an accredited practising dietician in Australia. Her PhD was in the area of dietary glycaemic intake and insulin resistance. She is a senior lecturer in the masters of nutrition and dietetics course at Edith Cowan University (Western Australia) and is involved in nutrition research through the Telethon Institute for Child Health Research (Australia). Her research interests include investigating the roles of dietary carbohydrate and fats on cardiometabolic health.

Dr Lauren Owen carried out her PhD at Lancaster University (UK) and she has since worked as a postdoctoral fellow at the Centre for Human Psychopharmacology at Swinburne University, Melbourne (Australia), where she is researching the effects of nutrition/pharmacology and metabolic activity on brain functioning. She has recently accepted a Marie Curie Fellowship at the University of Keele (UK) to continue her work in the area of psychopharmacology.

Dr Michele L. Pettit is an assistant professor in the Department of Health Education and Health Promotion at the University of Wisconsin-La Crosse, WI (USA). Her research interests include mental and emotional health, health risk behaviours, health disparities and health education. She teaches courses in mental and emotional health, sexual health promotion, foundations of health education and emerging public health issues.

Dr Andrew Pipingas is a cognitive neuroscientist at the Centre for Human Psychopharmacology, Swinburne University of Technology, Melbourne (Australia). He has expertise in the measurement of neurocognitive ageing and risk factors for cognitive decline. He has conducted a number of randomised clinical trials assessing effects of dietary interventions, such as pine bark extract, multivitamins and fish oil on neurocognitive function.

Prof. Remi Quirion, OC, PhD, CQ, FRSC became Québec's first chief scientist in September 2011 and has held numerous prestigious positions in the areas of Alzheimer's disease, neuropeptides, neuroprotection, cell death and psychiatric illnesses. He has over 650 publications in prominent scientific journals and is one of the most extensively cited neuroscientists in the world. He has received several awards and honours, including the Ordre national du Québec (Chevalier du Québec, CQ) in 2003, the Prix Wilder-Penfield (Prix du Québec) in 2004 and the Order of Canada (OC) in 2007. Prof. Quirion is also a member of the Royal Society of Canada.

Dr Jonathon Lee Reay is a principal lecturer in the Department of Psychology at Northumbria University (UK). His research investigates the physiological and psychological effects of phytochemicals and nutrients in healthy humans.

Dr Leigh Riby earned a BSc (Hons) in psychology from the University of Lincolnshire (UK). He then gained a PhD in experimental psychology in the area of cognitive ageing and frontal lobe deficits in the Department of Experimental Psychology, Bristol University (UK). During postdoctoral work at the University of Stirling (UK), he gained expertise in multi-modal brain imaging (EEG and fMRI). More recently, he has published a catalogue of research papers on the topic of glycaemic modulation of cognitive processes using behavioural and neuroimaging techniques in younger and older adults. Dr Riby is a senior lecturer at Northumbria University where he is the programme leader for the MSc in nutrition and psychological science. He is also a member of the Brain, Performance and Nutrition Centre and the Cognition and Communication Research Centre at Northumbria University. He holds an honorary contract with the National Health Service related to his more applied work in diabetes and brain performance.

Prof. Andrew Scholey is Co-Director of the Centre for Human Psychopharmacology (www.humanpsych.com) at Swinburne University, Melbourne (Australia). He has 15 years' experience in researching the human neurocognitive effects of natural products, including 'metabolic' interventions (e.g. oxygen and glucose) and a range of herbal extracts and natural supplements (e.g. ginkgo, ginseng, sage, cocoa polyphenols, green tea catechins and vitamins) across the lifespan. More recent work is aimed at uncovering the mechanisms underlying these effects with appropriate biomarkers and neuroimaging.

Dr Michael Smith holds a BA (Hons) in psychology and a PhD in cognitive neuroscience from the University of Western Australia. His PhD research investigated the influence of glucose ingestion on cognitive performance in adolescents. After undertaking postdoctoral work at the Telethon Institute for Child Health Research (Australia), he took up his current post as a senior lecturer at Northumbria University (UK). Previous and current research interests include the influence of ginseng, carbohydrates and breakfast on cognitive performance across the lifespan, and the role of omega-3 fatty acids in neurocognitive development and mood.

Con Stough is Professor of Cognitive Neuroscience and Co-Director of the Centre for Human Psychopharmacology at Swinburne University, Melbourne (Australia). He has over 15 years' experience in research examining the pharmacological basis of cognition and intelligence. He is lead investigator on a major Australian Government-funded trial on pharmacological interventions for cognitive ageing (Australian Research Council Longevity Intervention) and is working with several organisations to develop novel pharmacological interventions for cognition.

Dr Ewa Szymlek-Gay is a senior research fellow in the Department of Clinical Sciences, Umeå University Hospital (Sweden). Her primary research interest is concerned with nutrition in toddlers, and she has published in the areas of vitamin D, zinc and milk intake. Other interests include physical activity among cancer survivors.

Pernille Tveden-Nyborg is a doctor of veterinary medicine and did her PhD in embryology. She is currently an assistant professor in veterinary pharmacology at the University of Copenhagen (Denmark) and studies the role of vitamin C in early brain development.

CHAPTER 1

Nutrition and Mental Performance: A Lifespan Perspective

Leigh M. Riby, Michael A. Smith and Jonathan K. Foster

Introduction

In recent years the desire to understand the relationship between nutrition and behaviour has grown substantially. Indeed, the impact of nutrients and other food substances on brain function, cognition and mental performance has received rigorous scientific research attention as well as media interest. Sensational headlines such as 'Good nutrition for babies critical to IQ' (*Independent* online, 27 November 1998), '[children's] diet does boost your intelligence' (*Telegraph* online, 8 February 2011), 'Eat berries to keep the brain young' (*Telegraph* online, 23 August 2010), 'Dehydration makes young brains inefficient' (*Telegraph* online, 16 May 2010) and '...fish oil may stop dementia' (*Sunday Times* online, 18 May 2003) have caught the imagination of the public. Importantly, these issues are relevant to a number of professionals, including psychologists, medical practitioners, nutritionists, dietitians, food scientists, health advisers, educators and policy makers, showing an interest in healthy nutrition generally and, particularly relevant here, brain function. For example, food scientists and manufacturers making claims regarding the health benefits (including the improvement of brain function) of foods, drinks or supplements need to meet European Union legislative guidelines. For this reason, any claims of cognitive enhancement need to be based on empirical evidence. Here, we target both the popular science reader and the abovementioned (and other) academic and clinical disciplines.

Across the chapters that follow, we have purposefully selected a range of different nutrients to probe the nature of the relationship between nutrition and mental performance during one's life. We combine insights from micronutrients, macronutrients and other key substances (e.g. mild stimulants, such as caffeine), emphasising the role that these substances play in general health, brain function and cognitive performance. This is achieved by

1

examining works on nutrient deficiency and studies considering the usefulness of supplementation in enhancing brain function and mental performance. A critical aspect of the text is the developmental lifespan perspective that is adopted throughout. It is important to note that some of the nutrients presented may be differentially important/influential for cognitive function at different stages of development. Contributors are mindful of this consideration throughout the text.

To illustrate the lifespan perspective, among the chapters that follow we consider how nutrient supply during early development is important for optimal health and neurodevelopment. Alongside folate, which is advocated as an important nutrient during pregnancy (e.g. for neural tube protective effects; Daly et al. 1995), other B vitamins, iron and n-3 fatty acids are among the key nutrients reported to be beneficial during this phase of life, particularly when consumed in combination (for a discussion, see Leung et al. 2011). The period between the 24th and 42nd weeks of gestation is considered particularly critical. During this time frame, deficiency in any of these key nutrients may impact significantly on a number of neurodevelopmental processes, including synaptogenesis and myelination (Georgieff 2007). Deficits or abnormalities in these neuronal processes during early brain development have significant implications for cognitive function throughout subsequent development; Georgieff (see influential review paper, 2007) outlined the key nutrients central to typical brain development and the candidate brain areas affected by nutrient deficiency. For instance, iron deficiency can impact on the integrity of the hippocampus (adversely affecting memory and learning) and frontal cortex (impairing higher cognitive functions, such as planning and decision making), while other nutrients may have a more global impact on cognition. One of the aims of this book is to outline evidence for global (e.g. speed of processing) versus selective (i.e. cognitive domain specific) impacts of different nutrients on cognition. Although it is difficult to disentangle the effect(s) of a single nutrient on specific cognitive processes, great strides have been made in recent research (as outlined in this volume), particularly when using innovative methods of cognitive assessment and brain-imaging techniques. Through such approaches, it may well be feasible to parse out specific elements of cognition that are influenced by dietary input.

In terms of post-natal development, research considers the period between one and five years of age to be critical, due to brain plasticity and the development of key cognitive skills. The concepts of *critical periods* and *windows of sensitivity* are particularly relevant in terms of nutritional deficiency; these issues will be considered throughout this book. The importance of early nutritional wellbeing is further emphasised by the impact, more widely, of malnutrition or undernutrition throughout childhood. The prevalence of maternal and child undernutrition is particularly problematic in low- to middle-income countries (providing a cross-cultural perspective to the issue of nutrition and cognitive function). This topic was noted as a priority research area in a recent review highlighting the significant impact of malnutrition on brain development and cognitive ability (Black et al. 2008). Black and colleagues contemplate the cognitive consequences of undernutrition alongside the implications of other important health issues; for example, how nutritional deficiency impacts on the immune system. Indeed, cognitive consequences of nutrition should be considered in the context of other, more wide-ranging, health complications.

In this introductory section of this chapter, we provide a flavour of the topics that relate to the link between nutrition and early human development; for example, the importance of key micronutrients. To provide a consolidated view of nutritional impacts, coverage of macronutrients is also central to the volume. Macronutrients, including carbohydrates, protein and fat, are the nutrients that we consume in large amounts in our daily diet. From a developmental perspective, there is evidence that at certain points in development the intake of sufficient macronutrients is essential. During childhood and adolescence, a great deal of research has linked nutritional composition to academic performance. For example, skipping breakfast has been cited as an influential factor contributing to poor academic performance. In addition, the make-up of the breakfast is critical, with carbohydrates and – in particular – low glycaemic index (GI) foods being reported to be beneficial. This is hardly surprising given that glucose is the primary source of fuel for the brain, and given that slow-release foods, such as oatmeal, have been found to be beneficial to academic and cognitive performance in children (e.g. 9–11 year olds; Mahoney et al. 2005). In summary, this book will address both micronutrients and macronutrients in our consideration of the impact of nutrition on cognition.

However, these two categories of micro and macronutrients do not fully encapsulate the range of foods that can impact on cognition, such that other foodstuffs that have attracted recent research and/or media attention with respect to their potential impact on cognition and the brain will also be considered. For example, polyphenols (such as those in red wine or green tea) have received a great deal of media interest in terms of their putative ability to prevent or decelerate age-related cognitive decline. This last example underscores a point made earlier in this introductory section; namely, supplementation or intervention may be more beneficial and/or appropriate in one population (or age group) than another. Intake of these potentially protective substances may therefore be especially important during ageing, thereby re-emphasising the developmental lifespan approach taken throughout this volume.

To provide a balance, adult ageing is also given priority here in addition to the earlier periods of human development (e.g. prenatal and early post-natal stages). Successful ageing is considered as the ability to capitalise on intact functioning and use environmental support (including nutritional intervention) to minimise cognitive decline. This is especially important with respect to the increasing number of older adults in the population. Simple lifestyle changes, including changes in nutritional intake, represent a significant factor contributing to successful ageing. Related to health more generally, the World Health Organization has emphasised an urgent need to consider malnutrition among the elderly and to consider recommended daily allowances of certain nutrients (particularly given the need for increases in the intake of some types of nutrient and a reduction in others; see World Health Organization 2011). Changes in eating habits which accompany major lifestyle changes (e.g. loss of independence and difficulties with activities of daily living, including shopping and meal preparation) among the elderly make this demographic group particularly vulnerable to nutritional deficiency. And with respect to 'pathological ageing', a sufficient supply of key nutrients (e.g. omega-3 fatty acid and B vitamins) may be able to prevent, or at least delay, the onset of dementia.

In summary, it is highly likely that some nutrients impact upon the developing brain, while others are more critical in maintaining (or may enhance) mental performance in adolescence, adulthood or dementia when key cognitive faculties (e.g. memory and attention) become compromised. These issues are addressed from a multidisciplinary perspective throughout the volume, with chapters focusing on a range of key nutrients. We will now review the central issues covered by each of the contributors.

Organisation of this Book

This book brings together the work of prominent researchers within the field of nutrition and cognitive performance, taking the approach outlined in the introductory section of this chapter. In each chapter, the authors explore the mechanisms by which a nutrient or substance leads to enhanced or impaired cognitive and mental function at a specific point in development. Part I focuses on vitamins and minerals that are consumed in relatively small quantities and are classified as micronutrients. These substances have been studied largely in terms of their ability to enhance and protect brain function during childhood and adulthood, and in patient populations where cognitive deficiency is a characteristic of the condition. Part II focuses on macronutrients; that is, those substances required in larger amounts for normal function and development, such as proteins and carbohydrates. Finally, Part III focuses on other substances that have attracted media and research attention due to the possibility of their enhancing and protective properties; these do not easily fall under the categories of micronutrient or macronutrient.

Part I begins with Chapter 2, in which Magnus Domellöf and Ewa Szymlek-Gay examine iron-deficiency anaemia, which is considered one of the most common micronutrient deficiencies in the general population. Young children represent a particular risk group due to the rapid physiological changes that occur during childhood, and the importance of iron for developing neurones and glia. Selective effects of iron deficiency have been observed in brain regions subserving learning and memory (e.g. the hippocampus), and also in the dopamine system, which has been associated with executive function abilities. As well as considering the impact of nutritional deficiency, it is noted by these authors that iron supplementation may also have a positive effect on psychomotor development. They highlight that, even in affluent societies, the necessary iron requirements are difficult to achieve without iron-fortified foods or supplements. Therefore, supplementation has been recommended to prevent impaired neurodevelopment, although there are important caveats because excessive iron intake can impact negatively in iron-sufficient individuals.

In Chapter 3, Stine Hasselholt and colleagues examine the role of another micronutrient, vitamin C, in human health and cognition, with a focus on early development. They point out that ever since the identification of a link between the deadly disease scurvy and vitamin C deficiency, no one has questioned the importance of vitamin C to human health. Although deficiency leading to scurvy is rare, hypovitaminosis is reported by the authors to be problematic in some populations and associated with oxidative brain damage. Indeed, people of low socio-economic status, smokers and children may be particularly vulnerable to this deficiency and its consequences. The authors provide details of key studies describing the biology of the central nervous system in relation to the role of vitamin C (e.g. cofactor and antioxidant functions). Overall, the link between vitamin C deficiency and cognition remains unclear; the authors point to an urgent need for well-designed clinical studies to address the functional significance of this deficiency.

Therese O'Sullivan also writes about childhood in Chapter 4, where she reviews the different B vitamins (e.g. thiamine (B1), riboflavin (B2), niacin (B3), pantothenic acid (B5), pyridoxine (B6), biotin (B7), folic acid (B9) and cobamides (B12)), their primary functions, recommended intake level, food sources, and symptoms of deficiency and toxicity. Health implications are related to the water-soluble nature of B vitamins, such that deficiency can only be avoided by regular consumption. The chapter evaluates empirical evidence of the association between B vitamins and cognitive abilities in childhood, before moving through the developmental lifespan to contemplate adolescence and adulthood (e.g. considering aspects of psychopathology through the topics of mood, depression and alcoholism during adulthood). The beneficial impact of B vitamin (B1) supplementation is emphasised by minimising symptoms of memory impairment in alcoholics who are at high risk of developing Wernicke–Korsakoff syndrome. Therefore, Chapter 4 considers a range of issues related to deficiency and supplementation of B vitamins.

Chapter 5 is firmly rooted in the adult range of development and considers vitamin D. Paul Cherniack is a leader in research concerned with vitamin D, energy regulation and mental health. He notes that older adults are particularly vulnerable to vitamin D deficiency. In fact, between 40 and 90 per cent of older adults are reported to be vitamin D insufficient due to both inadequate amounts of dietary vitamin D and being exposed to less sunlight because they tend to cover up with clothing more than their younger counterparts. The author points out that, although so named, vitamin D is actually a steroid hormone that is important for regulating energy within cells. Through its anti-inflammatory actions, it may prevent or reverse the impact of vascular disease on dementia and depression. This chapter considers the evidence for the possible role of vitamin D in cognition, mental health and brain function. Preliminary evidence suggests global cognitive function and perhaps more specific domains of cognition, such as executive functions, are related to deficiency. The author makes recommendations for the optimal supplementation doses of vitamin D in older adults.

For the final chapter of Part I on micronutrients (Chapter 6), we return to B vitamins, moving from childhood/adulthood to older adulthood in a consideration of impacts of this group of nutrients on ageing and dementia. Here, Jonathan Reay critically discusses both epidemiological and experimental research exploring the psychological processes associated with B vitamins. He provides suggestions with respect to the underlying biological mechanisms associated with supplementation, which has been targeted towards the older end of the age spectrum. We have already outlined how B vitamins provide protection from neural tube deficits during development, but importantly supplementation may contribute to functional capacity in later adulthood by ameliorating cognitive decline associated with both normal and pathological ageing. This chapter brings to an end our consideration of the impact of micronutrients on cognition and health across the lifespan.

Part II focuses on macronutrients (e.g. carbohydrates, protein, fats) and again contemplates their role(s) at different points within the developmental spectrum. In Chapter 7, we begin in early development with the role of protein. Emma Jones highlights the fact that dietary protein has an important role in the development of brain structure and neurotransmitter synthesis. Proteins are particularly important during prenatal and early neonatal development, when there is rapid neural growth leading to the formation of the central nervous system and the brain. Meat is a key source of protein, such that meat intake during childhood has been reported to significantly impact on cognitive abilities later in life. For example, adults who self-report high meat intake during childhood have been found to exhibit enhanced performance in tasks that assess episodic memory ability. The literature concerning protein supplementation during pregnancy that is discussed here is, however, mixed with reliable health effects only found on tasks tapping motor capacity rather than central cognitive performance. The author suggests possible mechanisms underlying the link between intake of sufficient protein during gestation and infancy, and later healthy brain function and cognitive abilities.

In Chapter 8, the editors examine carbohydrates and the often reported 'glucose cognitive facilitation effect'. The majority of studies in this area have reported a beneficial effect of oral glucose ingestion on mental performance; most notably in the domain of episodic memory or in tasks which are particularly cognitively demanding. The chapter covers studies across the lifespan (with adolescents, younger and older adults) and examines possible mechanisms responsible for cognitive facilitation following glucose ingestion.

Importantly, it also provides a foundation for the subsequent two chapters, which focus on other macronutrients.

In Chapter 9, Jeanet Ingwersen examines the impact of breakfast on cognitive performance in adolescents and adults. The majority of such research has examined the effects of breakfast consumption compared with breakfast omission, and suggests that skipping breakfast can have adverse effects on cognitive performance. Recently there has been increased interest in the cognitive consequences of different breakfast types and the relevance of the GI to performance. Research has suggested that performance is differentially affected by GI, depending on the complexity of the carbohydrate. In particular, it has been suggested that breakfast foods consisting of slow-release glucose may be beneficial to cognitive function. The author also debates whether children's attendance at breakfast clubs has a positive effect on cognitive function and academic performance. This work has recently received significant press attention, with researchers recommending that a nutritionally balanced breakfast may be beneficial to cognitive and social development (Defeyter et al. 2010).

In Chapter 10, Kristen D'Anci addresses the contentious issue that reduced calorie diets have a negative impact on mood and cognition. From a psychological perspective, it is proposed that reduced calorie diets negatively affect mood and mental performance via intrusions of food-related thoughts and via unpleasant physical sensations of hunger. Interestingly, he discusses this work in terms of a well-recognised cognitive model (i.e. Working Memory; Baddeley and Hitch 1974) and suggests that the demands of dieting and the pre-occupation with food-related thoughts require cognitive resources that are diverted away from the task at hand. In addition, the author argues that weight loss diets restrict energy intake from key nutrients, and that reduced energy intake in the short term may disrupt mental function. Indeed, he reports that severely restricting carbohydrate intake impacts, acutely, on memory ability and, chronically, by adversely affecting an individual's mood. The author suggests that foods with a lower GI or with greater proportions of complex carbohydrates or protein may be more beneficial in improving cognitive ability and alertness than foods with a greater proportion of simple sugars.

In Chapter 11, Caroline Edmonds discusses water, dehydration and cognitive performance. The negative effect of dehydration is discussed in terms of health implications in children and adults. Dehydration in adults has been associated with poor performance in a number of domains of cognition, including verbal memory, mathematical ability and visuospatial skills. In children, evidence points to possible deficits in verbal short-term memory. She clearly outlines both behavioural and recent imaging work (magnetic resonance imaging (MRI)) which has linked hydration status (via studies examining both dehydration and water consumption) to poor health and cognitive performance. The recent use of MRI and functional MRI techniques has been particularly informative, with one study observing increased activity in frontoparietal regions in individuals with dehydration (explained in terms of greater neurocognitive resources required to achieve the same level of task performance). Further studies using such innovative imaging techniques are clearly warranted.

Louise Brown considers the consumption of fish and omega-3 fatty acids in Chapter 12. It has been suggested that the intake of omega-3 fatty acids (typically found in oily fish, nuts and seeds) has decreased across millennia (Sanders 2000). This could pose a range of age-related health risks, including cognitive dysfunction, since the brain relies on these compounds for healthy neuronal function. The author outlines and evaluates the role of omega-3 in protection from age-related cognitive decline and dementia. Based on the evidence acquired to date, she suggests we should be aiming to consume the recommended quantities of oily fish as one aspect of our healthy diet, although there is an urgent need for randomised controlled trials.

Finally, Part III focuses on a group of substances that have attracted media and research attention over the last decade due to their possible enhancing and protecting properties (but which do not easily fall into the micro and macronutrient categories).

The consumption of energy drinks represents an emerging trend that is considered by Michele Pettit and Kathy DeBarr in the first chapter of this section (Chapter 13). Increased availability of these drinks has sparked the attention of the media, the public and researchers interested in public health. The primary ingredients in energy drinks include carbohydrates, taurine, glucoronolactone and caffeine (Ivy et al. 2009). This chapter explores energy drink consumption in relation to perceived stress, cognitive performance and overall health, especially for the high intake group of adolescents and younger adults. Although the combination of caffeine and sugar has been reported to improve alertness and cognition, the risks of consuming energy drinks with high doses of caffeine may outweigh the benefits. Potential risks may include cardiovascular difficulties. The authors critically evaluate the research on energy drinks (often mistaken for sports drinks) in relation to marketing and regulatory issues.

Caffeine is a primary ingredient of energy drinks. In Chapter 14, Jack James highlights the effects of this most widely used psychoactive substance. He details in depth the main mechanisms of action related to the effects of caffeine on cognitive performance and mood. A central question is whether enhanced performance and mood traditionally attributed to the stimulant properties of caffeine represent net benefits, or whether differences between caffeine and control conditions are better characterised as being due to reversal of adverse withdrawal effects. Related to the lifespan aspect of this book, it is noted by the author that positive effects of caffeine have been reported in Alzheimer's disease, and possible methodological concerns with this work are identified. Similar claims of neuroprotective properties of caffeine have been made for Parkinson's disease. The author suggests that it is premature to recommend caffeine as a prophylactic against age-related cognitive decline, given the methodological issues which he highlights. However, more work is warranted in this area.

Certain herbal extracts may be effective in maintaining psychological health or 'wellbeing', particularly in ageing. In Chapter 15, David Camfield and colleagues evaluate evidence that herbal extracts have the capacity to improve aspects of mental function. This work has received a great deal of media attention over recent years. Although the focus of this chapter is on cognitive performance, the authors include relevant material on the modulation of mood. There is evidence that extracts of ginkgo, ginseng, salvia, guarana, lemon balm, bacopa and polyphenols have cognitive-enhancing properties. In some cases, the mechanisms of action are reasonably clear (e.g. the pro-cholinergic effect of sage), while the mechanisms of action for other agents are less certain. As well as considering mechanisms, the chapter also considers emerging technologies and how they can contribute to our understanding of the relationship between herbal extracts and cognition.

In the final chapter (Chapter 16), Stephane Bastianetto and Remi Quirion consider the preventive effects of resveratrol on neurological disorders associated with ageing. A growing number of epidemiological studies indicate that older adults who consume red wine (in moderation), green tea, fruits and vegetables have a lower risk of developing age-related neurological disorders, including cognitive deficits associated with ageing and dementia. The chapter reviews the most recent findings, and considers the mechanisms of action which are likely to mediate the impact of resveratrol. However, more clinical trials are desirable to assess its impact on various age-related disorders. Positive health effects of resveratrol are suggested for age-related neurological disorders, including macular degeneration, stroke and cognitive deficits with and without dementia.

In this introductory chapter, it has been possible to provide an overview of the nutrients that will be considered in more depth in the rest of this book, while also outlining how the volume will provide insights into cognitive and health implications of nutrient intake/supplementation across the developmental spectrum. The multidisciplinary approach adopted here will provide the reader with a thorough coverage of the relationships between nutrition and cognitive performance throughout human development.

References

Adam, S. (2010). Eat berries to keep the brain young. *Telegraph* online, 23 August 2010.

Alleyne, R. (2011). Food for thought – diet does boost your intelligence. *Telegraph* online, 8 February 2011.

Baddeley, A. D. & Hitch, G. (1974). Working memory. In G. H. Bower (ed.), *The Psychology of Learning and Motivation: Advances in Research and Theory*, vol. 8, pp. 47–89. New York: Academic Press.

Black, R., Allen, L., Bhutta, Z. & Caulfield, L. (2008). Maternal and child undernutrition: global and regional exposures and health consequences. *The Lancet, 371*, 243–260.

Burns, J. (2003). Irish scientists find fish oil may stop dementia. *Sunday Times* online, 18 May 2003.

Cooper, G. (1998). Good nutrition for babies critical to IQ. *Independent* online, 27 November 1998.

Daly, L., Kirke, P., Molloy, A., Weir, D. & Scott, J. (1995). Folate levels and neural tube defects: implications for prevention. *Journal of the American Medical Association, 274*, 1698–1702.

Defeyter, M. A., Graham, P.L., Walton, J. & Apicella, T. (2010). Breakfast clubs: availability for British schoolchildren and the nutritional, social and academic benefits. *Nutrition Bulletin, 35*, 245–253.

Georgieff, M. (2007). Nutrition and the developing brain: nutrient priorities and measurement. *American Journal of Clinical Nutrition, 85*, 614S–620S.

Gray, R. (2010). Dehydration makes young brains inefficient. *Telegraph* online, 16 May 2010.

Ivy, J. L., Kammer, L., Ding, Z., Wang, B., Bernard, J. R., & Liao, Y. (2009). Improved cycling time – Trial performance after consumption of a caffeine energy drink. *International Journal of Sport Nutrition and Exercise Metabolism, 19*, 61–78.

Leung, B., Wiens, K. & Kaplan, B. (2011). Does prenatal micronutrient supplementation improve children's mental development? A systematic review. *BMC Pregnancy and Childbirth, 11*, 1–12.

Mahoney, C., Taylor, H. & Kanarek, R. (2005). Effect of breakfast composition on cognitive processes in elementary school children. *Physiology and Behavior, 85*, 635–645.

Sanders, T. A. B. (2000). Polyunsaturated fatty acids in the food chain in Europe. *American Journal of Clinical Nutrition, 71*, 176S–178S.

World Health Organization (2011) Nutrition for older persons. Retrieved 29 June 2011, from http://www.who.int/nutrition/topics/ageing/en/index1.html

PART I

Micronutrients

CHAPTER 2

Iron Nutrition and Neurodevelopment in Young Children

Magnus Domellöf and Ewa A. Szymlek-Gay

Chapter Overview

- Iron-deficiency anaemia is the most common micronutrient deficiency in the world, affecting an estimated 600 million individuals.
- Young children are a particular risk group due to the rapid changes that occur during childhood and the requirement for iron during those processes.
- Iron is essential for early brain development and development of the central nervous system.
- Empirical evidence suggests that iron is important for cognitive performance. Several well-controlled studies have shown an association between iron-deficiency anaemia and poor cognition.
- In this chapter we will focus on early childhood as a key developmental period that is central to exploring the relationship between iron intake and neurodevelopment.

Introduction

Iron deficiency is the most common single-nutrient deficiency in the world, affecting over a billion individuals in both industrialised and economically developing countries (McLean et al. 2009). Three successive stages of iron deficiency are generally recognised: iron depletion; iron-deficient erythropoiesis and iron-deficiency anaemia. The first stage is characterised by a gradual depletion of iron stores, but the body still has enough iron to sustain the production of red blood cells. If depletion of the iron stores continues, the second stage of iron deficiency – iron-deficient erythropoiesis – takes place, during which iron stores become exhausted and the production of red blood cells is reduced. In the third and final stage of iron deficiency, iron stores are

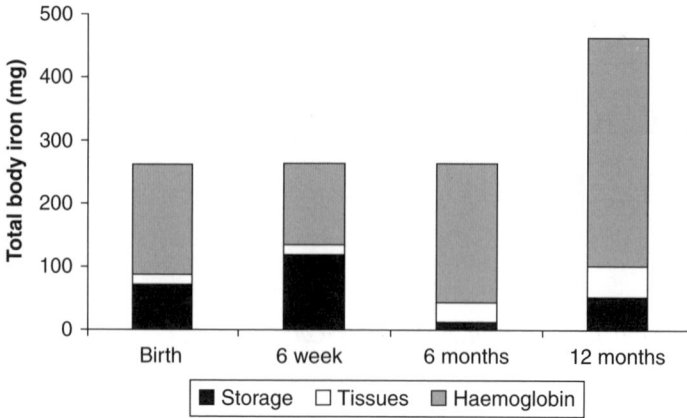

Figure 2.1   Changes in body iron during the first 12 months of life
Source: Modified from Domellöf (2007).

exhausted, levels of circulating iron are low, the production of red blood cells is severely compromised and anaemia develops.

Young children are especially vulnerable to nutritional iron deficiency, primarily due to increased physiological requirements during rapid growth that exceed their dietary iron intakes. At birth, the total body iron content averages about 270 mg (75 mg/kg) in healthy, full-term, normal birthweight infants. In newborns, more than 65 per cent of total body iron is in the haemoglobin of red blood cells and approximately 25 per cent is in the form of iron stores (Figure 2.1). The remaining iron is found in other tissues, mainly in myoglobin and enzymes. A very small amount of iron (transport iron) circulates in the plasma bound to the transport protein, transferrin. After birth, increased oxygen saturation leads to an effective down-regulation of haemoglobin synthesis, which results in a rapid fall in the concentration of blood haemoglobin during the first six weeks of life, accompanied by a corresponding increase of iron stores. Due to this redistribution between iron compartments, the infant is virtually self-sufficient with regard to iron during the first six months of life, and iron deficiency is very unusual during this period in healthy, full-term, normal birthweight infants. However, at six months of age, iron stores are almost depleted and between 6 and 24 months of age, infants need to increase their total body iron by 77 per cent, resulting in very high dietary requirements, estimated to be approximately 1 mg/kg/day (Domellöf 2007). Iron requirements per kilogram of body weight per day also remain relatively high during toddlerhood (Table 2.1). Low birthweight infants (< 2500 g) rapidly deplete their iron stores and need iron supplements during the first six months of life to prevent iron-deficiency anaemia (Berglund et al. 2010).

Even in affluent societies, the estimated required intake of iron at 6–24 months of age is difficult to achieve without regular consumption of

Magnus Domellöf and Ewa A. Szymlek-Gay   15

Table 2.1 Recommended daily iron intakes for pre-adolescent
children

Age	Recommended intake	
	mg/day	mg/kg/day
6–12 months	7.8–11.0	0.9–1.3
1–3 years	5.8–9.0	0.5–0.8
4–8 years	6.1–10.0	0.3–0.5
9–13 years	8.0–11.0	0.2–0.3

Sources: Institute of Medicine (2001), World Health Organization & Food and Agriculture Organization of the United Nations (2004), Department of Health (1991), German Nutrition Society (2002), Australian Government Department of Health and Aging, New Zealand Ministry of Health & National Health and Medical Research Council (2006) and Nordic Council of Ministers (2004). The mg/kg/day values are based on average body weights (World Health Organization 2006, 2007).

foods fortified with iron or separate iron supplements. As a result, the prevalence of iron-deficiency anaemia, the most severe stage of iron-deficiency, is higher in infancy and early childhood than at any other stage of life. It has been estimated that approximately 50 per cent of the world's pre-school children are anaemic, with half of all cases being due to iron deficiency (McLean et al. 2009; World Health Organization 2001). In industrialised countries, where iron-rich complementary foods are generally available, the prevalence of iron-deficiency anaemia has been estimated at 2–3 per cent in 12-month-old infants (Male et al. 2001) and 3–4 per cent in children aged 1–3 years (Looker et al. 1997; Thane et al. 2000). Risk factors for iron-deficiency anaemia among children in high-income countries include low birthweight, high intakes of cow's milk (especially > 500 ml/day), low intakes of iron-rich foods and low socio-economic status (Gunnarsson et al. 2004; Sutcliffe et al. 2006).

A number of physiological manifestations have been attributed to iron deficiency in children, including growth retardation, impaired immune responses and poor temperature regulation. However, the high prevalence of iron-deficiency anaemia in early childhood is of particular concern because there is an established association between iron-deficiency anaemia and poor neurodevelopment. Growth and development of the central nervous system is rapid during the first years of life, and iron is critical for this process because it is essential for myelination, monoamine synthesis and energy metabolism in neurons and glial cells (Beard 2007). It is therefore important to prevent iron deficiency in young children. To achieve this, foods fortified with iron or iron supplements are often recommended for this at-risk group. However, in contrast to most other nutrients, excess iron cannot be excreted by the human body, and it has recently been suggested that excessive iron supplementation of infants may have adverse effects on growth, the risk of infections and possibly even cognitive development (Domellöf 2007;

Lozoff et al. 2008). Thus, recommendations regarding iron intake must not only prevent iron deficiency but also must avoid excessive iron intakes in iron-sufficient infants.

Mechanisms Linking Iron-Deficiency Anaemia to Altered Cognitive Performance

The effects of iron deficiency on brain and behavioural development depend on the time of onset, duration and severity of iron deficiency. Iron deficiency is most common in young children, with the highest prevalence among infants and toddlers. The first two years of life coincide with a brain growth spurt, during which the brain attains 80 per cent of its adult size and hippocampal and cortical regional development; as well as myelinogenesis, dendritogenesis, and synaptogenesis reach their peaks. Animal studies show that a lack of iron available to oligodendrocytes – the glial cells responsible for the production of myelin – results in a decreased synthesis of myelin lipids and phospho-lipids in the brain. The synthesis of myelin proteins also declines, leading to hypomyelination of nerve cells. Hypomyelination reduces central nervous system conduction velocities and is associated with impaired brain function. The altered composition and amount of myelin in the brain persisted into adulthood in some of these animal experiments, which is a potential mech-anism for the suggested long-lasting adverse effects of severe iron deficiency (Lozoff et al. 2006a).

Iron is essential for a number of enzymes involved in the synthesis of monoamine neurotransmitters such as serotonin, norepinephrine and dopamine. Studies in animals show that iron deficiency appears to alter syn-thesis and metabolism of monoamine neurotransmitters, especially dopamine. These changes are proportional to the degree of iron deficiency and may per-sist into adulthood, despite iron therapy, if iron deficiency occurs during the neonatal period (Beard & Connor 2003). Dopamine is important for many brain functions, including behaviour, emotion, cognition and movement, and dysmaturation of dopamine systems has been associated with, for example, attention deficit hyperactivity disorder in children (Nieoullon & Coquerel 2003).

Iron is also a cofactor for enzymes involved in neuronal and glial energy metabolism, for example, cytochrome *c* oxidase, suggesting that iron defi-ciency may predispose the brain tissue to damage in situations with limited energy supply. Indeed, it has been shown that iron deficiency is associated with hippocampal injury in rats subjected to mild/moderate perinatal hypoxia-ischaemia (Rao et al. 2007). The hippocampus seems to be the brain region which is most affected by iron deficiency. In rodents, perinatal and early post-natal iron deficiency irreversibly alters dendritic structure and reduces synaptic connections in the hippocampal area CA1 (Carlson et al. 2009; Jorgenson et al. 2005). Given that the hippocampus is a primary region of the brain, controlling the formation of memories and learnt behaviours, it is plausible that the changes in the hippocampus resulting from iron deficiency may lead to deficits in learning and impaired spatial memory (Bird & Burgess 2008).

Effects of Inadequate Intakes of Iron on Cognitive Performance

Numerous studies have investigated the association between iron deficiency and neurodevelopment. Animal studies have been used to show the effects of iron deficiency on brain development and function. Case-control studies have established the association between iron-deficiency anaemia and poor neurodevelopment in young children. Finally, randomised controlled trials have been used to establish the causal relationship between iron intake and neurodevelopment in children.

Animal Studies

Animal models permit the study of the effects of iron deficiency while controlling environmental factors, which is not possible in human studies. The vast majority of animal studies in this area have been based on rodent models. In rats, as in humans, myelination and dendritic arborisation occur primarily in the post-natal period. In rat pups, post-natal day 10 is equivalent to full-term human birth with regard to brain development, and post-natal day 25 corresponds to a stage reached by the human brain by 3–4 years (Dobbing & Sands 1973). Mouse and non-human primate models have also been used to study the effects of iron deficiency on neurodevelopment (Carlson et al. 2009; Golub et al. 2006; Lubach & Coe 2008).

In animal models of iron deficiency, reduced motor activity has been the most consistent observation, although this has been found to resolve itself with iron repletion (McCann & Ames 2007). Negative effects on cognitive and behavioural functions have also been observed in some studies of animals exposed to pre and post-natal iron deficiency (Beard 2007; Mohamed et al. 2011). Compared with iron-sufficient animals, those that were iron deficient showed deficits in spatial memory, learning, fear conditioning and attention (Beard 2007; Mohamed et al. 2011). These effects appeared to be long-lasting because neurological function was not fully restored after iron repletion in most of these animal studies (McCann & Ames 2007). Recently, however, Mohamed et al. (2011) showed that treatment with methylphenidate improved cognitive deficits in iron-sufficient rats that had iron deficiency in early life. These preliminary findings suggest that pharmacotherapy may be used in the future for treating persistent cognitive difficulties in children exposed to severe iron deficiency in infancy (Mohamed et al. 2011).

A limitation in most of these studies is that the animals have been severely iron deficient, with severe anaemia and growth failure, making it difficult to apply the findings to milder forms of iron deficiency.

Case-Control Studies

Lozoff et al. (2006a) and McCann & Ames (2007) examined evidence from studies that evaluated the relationship between child development and iron deficiency. Several well-performed case-control studies in children have shown a consistent association between iron-deficiency anaemia and poor cognitive and behavioural performance, even though these observations may have been confounded by other nutritional deficiencies and socio-economic factors (Lozoff et al. 2006a; McCann & Ames 2007).

In scientific studies of neurodevelopment in infants and young children, the Bayley Scales of Infant Development is the most commonly used test. This test evaluates an individual's mental (cognitive), motor (psychomotor) and behavioural development.

An important case-control study was performed by Lozoff et al. (1987) in a cohort of 191 Costa Rican 12–23-month-old children who had varying degrees of iron deficiency. Lozoff et al. (1987) observed that infants with iron-deficiency anaemia attained poorer Bayley mental development scores by an average of eight points and poorer Bayley psychomotor development scores by an average of ten points compared with non-anaemic children. In this cohort, children who had severe iron deficiency in infancy (haemoglobin concentration ≤ 100 g/l) continued to have lower cognitive scores compared with controls at age 5 (Lozoff et al. 1991), 11–14 (Lozoff et al. 2000) and up to 19 years of age (Lozoff et al. 2006b) compared with non-anaemic children, despite iron therapy that corrected iron-deficiency anaemia in infancy. Because the anaemic children came from more disadvantaged environments than did the non-anaemic children, Lozoff et al. (1987, 2000, 2006b, 1991) adjusted their findings for many environmental and family factors.

Recent reports continue to provide support for earlier findings. Carter et al. (2010) observed that nine-month-old infants with iron-deficiency anaemia were less likely to exhibit object permanence compared with non-anaemic iron-deficient and iron-sufficient infants. Moreover, the authors found a dose-response effect of iron status on object permanence, with the lowest proportion of infants passing this test in the iron-deficient anaemic group and the highest proportion in the iron-sufficient group. Recognition memory for photographs (preference for novel stimulus) was also affected by iron status as infants with more severe iron-deficiency anaemia (as assessed by haemoglobin concentration) performed more poorly in this subtest in the Fagan Test of Infant Intelligence compared with infants without iron-deficiency anaemia. Poorer orientation/engagement (more shy behaviour) in the infants with iron-deficiency anaemia may partially explain these cognitive deficits. However, the differences in cognitive performance were no longer evident following daily supplementation with 22 mg of elemental iron for three months (Carter et al. 2010).

A series of studies have been performed using newborn infants of diabetic mothers as a model of prenatal iron deficiency. These studies show that prenatal iron deficiency may lead to lasting impairments in recall memory during childhood, observable already in the neonatal period (Riggins et al. 2009). However, a weakness of these studies is that infants of diabetic mothers also experience prenatal hypoxia (Taricco et al. 2009), which itself may lead to poor neurodevelopment (Ornoy 2005).

The auditory and visual pathways of the central nervous system also appear to be negatively affected by iron-deficiency anaemia. Roncagliolo et al. (1998) observed that infants who had iron-deficiency anaemia at six months of age had a trend towards less mature auditory brainstem responses as shown by longer central conduction time (suggesting poor central nervous system myelination) compared with non-anaemic infants. These differences increased somewhat and became statistically significant at 12 and 18 months of age despite effective iron therapy (Roncagliolo et al. 1998). Likewise, significantly longer latencies of electroencephalogram (EEG) visual evoked potentials were observed in 6–24-month-old children with iron-deficiency anaemia compared with iron-sufficient infants (Monga et al. 2010). Furthermore, in that study, neural conduction times were negatively associated with the severity of iron-deficiency anaemia (as assessed by haemoglobin concentration) in infants with iron-deficiency anaemia (Monga et al. 2010).

Lozoff et al. (2010) demonstrated that iron-deficiency anaemia was associated with the rate of spontaneous eye-blinking, assumed to be a marker of brain dopamine activity (Colzato et al. 2009), which was lower in 9–10-month-old infants who had iron-deficiency anaemia compared with controls. This deficit was responsive to iron therapy as after three months of daily supplementation with 22 mg of elemental iron, the rate of eye-blinking increased in infants with iron-deficiency anaemia and was no longer different from that of non-anaemic infants (Lozoff et al. 2010). Lukowski et al. (2010) observed that chronic and persistent iron-deficiency anaemia in early infancy was associated with impaired executive function and recognition memory in early adulthood, suggesting that deficits in the dopamine system, as well as the hippocampus, may be long-lasting (Lukowski et al. 2010).

Iron also appears to play an important role in infant motor development, and iron-deficiency anaemia has consistently been shown to be associated with poor gross and fine motor development in infants (Angulo-Barroso et al. 2011; Lozoff et al. 2006a; McCann & Ames 2007; Pala et al. 2010; Shafir et al. 2009).

Iron-deficiency anaemia in infancy has been negatively associated with social-emotional development and mother–infant interaction. Infants affected by iron-deficiency anaemia have been shown to exhibit behaviours that could contribute to functional isolation by appearing to maintain closer contact with their caregiver, show less pleasure, be more wary, hesitant, less playful, more easily tired and less attentive (Lozoff et al. 1998). In those who are affected by chronic iron-deficiency anaemia in infancy, social-emotional deficits may be long-lasting, with externalising (aggressive and delinquent behaviour) and internalising (social withdrawal, somatic complaints, anxiety and depression) problems persisting well into adolescence. These problems, however, may resolve by early adulthood (Corapci et al. 2010).

Mothers of children affected by iron-deficiency anaemia appear to exhibit less developmentally facilitative behaviour by showing less pleasure in the child, less affection towards the child and less eye contact, and they try less frequently to encourage their infant to perform various tasks (Corapci et al. 2006; Lozoff et al. 1998). Mother–infant interaction during infant feeding has also been negatively correlated with iron-deficiency anaemia in infants. Mothers of infants affected by iron-deficiency anaemia appear to respond with less sensitivity to the behavioural cues of their infants. This may in turn lead to poorer feeding practices, resulting in iron-deficiency anaemia in the infant. Furthermore, mothers of infants with iron-deficiency anaemia exhibit poorer social-emotional and cognitive growth fostering behaviour compared with mothers of non-anaemic infants, which may indicate poorer parenting, leading in turn to poorer infant development (Armony-Sivan et al. 2010). The associations between infant iron-deficiency anaemia and poorer parenting behaviour in the mother may be caused by less behavioural cues from the infant but may also be a confounding effect of the disadvantaged environment.

A limitation of all these case-control studies is that infants with iron-deficiency anaemia, compared with controls, are more likely to have other socio-economic and familial risk factors, as well as other nutritional deficiencies. Even though confounders have been statistically adjusted for in most studies, randomised intervention trials are needed to prove causality in the relation between iron-deficiency anaemia and cognitive/behavioural development in humans.

Randomised Controlled Trials

To date, there have been very few randomised controlled trials of iron supplementation in young children in which neurodevelopmental outcomes have been assessed. Those trials that have assessed neurodevelopment have shown that in infants, preventive daily supplementation with 7.5–10 mg of iron given as iron drops from 1 to 5 (Friel et al. 2003), 6 to 9 (Yalçin et al. 2000), or 6 to12 months of age (Lind et al. 2004) had no effect on infant cognitive development. In the two larger and longer trials, however, Friel et al. (2003) and Lind et al. (2004) showed a positive effect of iron supplementation on the Bayley psychomotor development indices, which at the completion of treatment were higher by an average of 3–7 points in the iron-supplemented infants compared with the non-supplemented infants. None of the infants in Yalçin et al.'s (2000) or Friel et al.'s (2003) studies had iron-deficiency anaemia at baseline. Lind et al.'s trial (2004) was not purely preventive because 8 per cent of the infants had iron-deficiency anaemia at baseline (haemoglobin concentration < 110 g/l).

The daily use of iron-fortified infant formula for prevention of iron deficiency may have similar effects on child development as iron drops. Moffatt et al. (1994) showed that Canadian infants from very low-income families consuming formula containing 12.8 mg of iron per litre from 0–2 months to 15 months of age attained higher Bayley psychomotor development indices by 4 points at 9 months, and by 6 points at 12 months of age compared with infants consuming low-iron formula (1.1 mg of iron per litre). Infants' mental development and behaviour were not affected in this trial (Moffatt et al. 1994) and psychomotor development indices were no longer different between the two groups at the study endpoint (15 months), suggesting that possible gains in psychomotor development achieved with iron supplementation or iron-fortified formula in infancy may be transient. Likewise, daily supplementation with 10 mg of iron for six months during infancy had no effect on cognitive performance at nine years of age in Thai children. However, the prevalence of iron-deficiency anaemia in infancy was rare (≤ 3 per cent) in this sample (Pongcharoen et al. 2011).

Therapeutic trials of iron supplementation in young children affected by iron-deficiency anaemia have shown inconsistent effects of iron supplementation on cognitive and psychomotor development. Idjradinata and Pollitt (1993) showed that daily supplementation with 3 mg of iron per kilogram of body weight in 12–18-month-old infants with iron-deficiency anaemia resulted in successful resolution of iron-deficiency anaemia and improved Bayley mental and psychomotor development indices by 19–23 points. Aukett et al. (1986) observed that 24 mg of iron given daily to 17–19-month-old children for two months did not result in an overall effect on the mean developmental score, even though a higher proportion of iron-supplemented infants acquired at least six new skills. In the largest trial of iron-fortified formula to date, Morley et al. (1999) randomised 493 nine-month-old cow's milk-fed UK infants to receive an iron-fortified formula (12 mg iron per litre), low-iron formula (0.9 mg iron per litre) or cow's milk from 9 to 18 months of age. In this trial, consumption of iron-fortified formula improved iron status but had no significant effect on Bayley mental or psychomotor development indices at 18 months (Morley et al. 1999). Likewise, Lozoff et al. (1982) found no effect of iron supplementation on reversing developmental delays in 6–24-month-old children affected by iron-deficiency anaemia. This trial, however, was only of one-week's duration (Lozoff et al. 1982).

A few randomised controlled trials of iron have investigated other neurodevelopmental effects than cognitive and psychomotor development. In a secondary analysis of data from two large randomised controlled trials in Nepal and Zanzibar, 12-month supplementation with iron in infants at high risk of iron deficiency resulted in longer sleep duration as reported by mothers

(Kordas et al. 2009). In a randomised controlled trial of iron supplementation (0, 1 or 2 mg/kg/day) to marginally low birthweight Swedish infants, significant effects on iron status at six months were observed, while no effects on conduction velocities were found as measured by auditory brain stem response (Berglund et al. 2010).

A meta-analysis of 17 randomised clinical trials in children that included cognitive outcomes showed that iron supplementation had a positive effect on mental development indices, although this improvement in mental development scores was modest and equivalent to 1.5–2 points of 100 (Sachdev et al. 2005). This effect was more apparent for children who were initially anaemic or who had iron-deficiency anaemia, suggesting that iron supplements have positive cognitive effects in iron-deficient children. This meta-analysis showed a more pronounced effect in children aged seven years or older and no convincing evidence for an effect of iron supplements on neurodevelopmental outcomes in children below two years of age. This lack of effect in the youngest infants may be due to irreversible effects of iron deficiency on the developing brain or the fact that cognitive development and behaviour are more difficult to measure in young children.

Similar to the results of Sachdev et al. (2005), a recent systematic review and meta-analysis of 14 randomised controlled trials showed that iron supplementation improved attention and concentration in individuals aged six years or older (including four studies in adult women), irrespective of their baseline iron status (Falkingham et al. 2010). Moreover, iron supplementation improved the intelligence quotient by 2.5 points in anaemic groups, but had no effect on non-anaemic individuals, or on memory psychomotor skills or scholastic achievements (Falkingham et al. 2010). In this meta-analysis it was concluded that most of the studies were small and methodologically weak and that there was a possible modest publication bias towards studies which demonstrated significant outcomes.

A recent meta-analysis concluded that iron supplements in infancy may have a positive effect on motor development (Szajewska et al. 2010), although this conclusion was reached based on only three randomised controlled trials described earlier (Friel et al. 2003; Lind et al. 2004; Moffatt et al. 1994). In line with results reported in the original studies (Friel et al. 2003; Lind et al. 2004; Moffatt et al. 1994), Szajewska et al. (2010) found no effect of iron supplementation on infant behaviour or mental development.

Effects of High Intakes of Iron on Cognitive Performance

Recent studies suggest that excessive iron intakes can have negative effects on brain development. In a mouse model, Parkinson's-like progressive mid-brain neurodegeneration was seen after a period of high dietary iron intake (Kaur et al. 2007). These findings are supported by data from a randomised controlled trial in which healthy Chilean infants with a birthweight of $\geq$ 3 kg and without iron-deficiency anaemia at six months of age were randomised to receive formula with either a high (12 mg iron per litre) or low (2.3 mg iron per litre) iron content from 6 to 12 months of age (Lozoff et al. 2008). Motor development, cognitive development, spatial memory, reading and arithmetic and visual-motor integration were assessed at ten years of age. The high-iron group had lower scores on all of these outcomes, significantly so for spatial memory and visual-motor integration scores. Effects depended on initial iron status: high-iron formula had a greater negative effect on the outcome measures in children who initially had higher haemoglobin concentrations, while the opposite was true in infants with an initially lower haemoglobin concentration. The effect size in visual-motor integration was 2 standard deviations, which corresponds to a score difference of 15 points out of 100. The physiological mechanisms behind this possible negative effect of excessive iron intake on cognitive development are unknown, but iron-mediated oxidative stress has been suggested (Kaur et al. 2007).

In future intervention trials in children, it is therefore important to assess not only positive effects of iron but also possible adverse effects of iron on growth, infections and neurodevelopment in iron-sufficient individuals.

Summary

Iron deficiency is the most common nutritional deficiency in the world. Young children are at particular risk of iron deficiency due to high iron requirements during rapid growth and inadequate iron intakes. Globally, approximately a quarter of pre-school children are affected by iron-deficiency anaemia, the most severe stage of iron deficiency. The proportion of young children with iron-deficiency anaemia from economically developed countries is about 2–4 per cent, although this prevalence is higher in certain at-risk groups with low socio-economic status.

Animal studies have clearly shown that iron is essential for several aspects of brain development. Case-control studies have shown that iron-deficiency anaemia in young children is associated with long-lasting cognitive and behavioural deficits. A few randomised controlled trials in young children suggest that iron may improve neurodevelopment, at least psychomotor development, particularly in those initially anaemic or at risk of iron deficiency. There is a clear need for additional randomised controlled trials in order to determine the effects of different iron intakes on different aspects of neurodevelopment at various ages, and to better establish iron requirements in young children. Such trials must focus on the prevention of iron-deficiency anaemia or the study of different doses of iron because it is no longer ethically acceptable to perform placebo-controlled intervention trials in anaemic children. The available evidence suggests that it is important to prevent iron-deficiency anaemia in young children to avoid impaired brain development. This can be achieved by using iron supplements, consuming iron-rich foods (meat products and iron-fortified foods) and avoiding high intakes of cow's milk in infants and toddlers. However, because excessive iron intakes have been suggested to have adverse effects in iron-sufficient children, unnecessary iron supplementation or fortification should be avoided.

References

Angulo-Barroso, R. M., Schapiro, L., Liang, W., Rodrigues, O., Shafir, T., Kaciroti, N., et al. (2011). Motor development in 9-month-old infants in relation to cultural differences and iron status. *Developmental Psychobiology, 53*(2), 196–210.

Armony-Sivan, R., Kaplan-Estrin, M., Jacobson, S. W., & Lozoff, B. (2010). Iron-deficiency anemia in infancy and mother-infant interaction during feeding. *Journal of Developmental & Behavioral Pediatrics, 31*(4), 326–332.

Aukett, M. A., Parks, Y. A., Scott, P. H., & Wharton, B. A. (1986). Treatment with iron increases weight gain and psychomotor development. *Archives of Disease in Childhood, 61*(9), 849–857.

Australian Government Department of Health and Aging, New Zealand Ministry of Health, & National Health and Medical Research Council. (2006). *Nutrient reference values for Australia and New Zealand.* Canberra: National Health and Medical Research Council.

Beard, J. (2007). Recent evidence from human and animal studies regarding iron status and infant development. *Journal of Nutrition, 137*(2), 524S–530S.

Beard, J. L., & Connor, J. R. (2003). Iron status and neural functioning. *Annual Review of Nutrition, 23*, 41–58.

Berglund, S., Westrup, B., & Domellöf, M. (2010). Iron supplements reduce the risk of iron deficiency anemia in marginally low birth weight infants. *Pediatrics, 126*, e874–e883.

Bird, C. M., & Burgess, N. (2008). The hippocampus and memory: insights from spatial processing. *Nature Reviews Neuroscience, 9*(3), 182–194.

Carlson, E. S., Tkac, I., Magid, R., O'Connor, M. B., Andrews, N. C., Schallert, T., et al. (2009). Iron is essential for neuron development and memory function in mouse hippocampus. *Journal of Nutrition, 139*(4), 672–679.

Carter, R. C., Jacobson, J. L., Burden, M. J., Armony-Sivan, R., Dodge, N. C., Angelilli, M. L., et al. (2010). Iron deficiency anemia and cognitive function in infancy. *Pediatrics, 126*(2), e427–e434.

Colzato, L. S., van den Wildenberg, W. P., van Wouwe, N. C., Pannebakker, M. M., & Hommel, B. (2009). Dopamine and inhibitory action control: evidence from spontaneous eye blink rates. *Experimental Brain Research, 196*(3), 467–474.

Corapci, F., Calatroni, A., Kaciroti, N., Jimenez, E., & Lozoff, B. (2010). Longitudinal evaluation of externalizing and internalizing behavior problems following iron deficiency in infancy. *Journal of Pediatric Psychology, 35*(3), 296–305.

Corapci, F., Radan, A. E., & Lozoff, B. (2006). Iron deficiency in infancy and mother-child interaction at 5 years. *Journal of Developmental & Behavioral Pediatrics, 27*(5), 371–378.

Department of Health. (1991). *Dietary reference values for food energy and nutrients for the United Kingdom: report of the Panel on Dietary Reference Values of the Committee on Medical Aspects of Food Policy.* London: HMSO.

Dobbing, J., & Sands, J. (1973). Quantitative growth and development of human brain. *Archives of Disease in Childhood, 48*(10), 757–767.

Domellöf, M. (2007). Iron requirements, absorption and metabolism in infancy and childhood. *Current Opinion in Clinical Nutrition and Metabolic Care, 10*(3), 329–335.

Falkingham, M., Abdelhamid, A., Curtis, P., Fairweather-Tait, S., Dye, L., & Hooper, L. (2010). The effects of oral iron supplementation on cognition in older children and adults: a systematic review and meta-analysis. *Nutrition Journal, 9*, 4.

Friel, J. K., Aziz, K., Andrews, W. L., Harding, S. V., Courage, M. L., & Adams, R. J. (2003). A double-masked, randomized control trial of iron supplementation in early infancy in healthy term breast-fed infants. *Journal of Pediatrics, 143*(5), 582–586.

German Nutrition Society. (2002). *Reference values for nutrient intakes* (1st ed.). Bonn, Germany: Umschau Braus GmbH.

Golub, M. S., Hogrefe, C. E., Germann, S. L., Capitanio, J. P., & Lozoff, B. (2006). Behavioral consequences of developmental iron deficiency in infant rhesus monkeys. *Neurotoxicology and Teratology, 28*(1), 3–17.

Gunnarsson, B. S., Thorsdottir, I., & Palsson, G. (2004). Iron status in 2-year-old Icelandic children and associations with dietary intake and growth. *European Journal of Clinical Nutrition, 58*(6), 901–906.

Idjradinata, P., & Pollitt, E. (1993). Reversal of developmental delays in iron-deficient anaemic infants treated with iron. *Lancet, 341*(8836), 1–4.

Institute of Medicine. (2001). *Dietary reference intakes for vitamin A, vitamin K, arsenic, boron, chromium, copper, iodine, iron, manganese, molybdenum, nickel, silicon, vanadium, and zinc.* Washington, DC: National Academy Press.

Jorgenson, L. A., Sun, M., O'Connor, M., & Georgieff, M. K. (2005). Fetal iron deficiency disrupts the maturation of synaptic function and efficacy in area CA1 of the developing rat hippocampus. *Hippocampus, 15*(8), 1094–1102.

Kaur, D., Peng, J., Chinta, S. J., Rajagopalan, S., Di Monte, D. A., Cherny, R. A., et al. (2007). Increased murine neonatal iron intake results in Parkinson-like neurodegeneration with age. *Neurobiology of Aging, 28*(6), 907–913.

Kordas, K., Siegel, E. H., Olney, D. K., Katz, J., Tielsch, J. M., Kariger, P. K., et al. (2009). The effects of iron and/or zinc supplementation on maternal reports of sleep in infants from Nepal and Zanzibar. *Journal of Developmental & Behavioral Pediatrics, 30*(2), 131–139.

Lind, T., Lönnerdal, B., Stenlund, H., Gamayanti, I. L., Ismail, D., Seswandhana, R., et al. (2004). A community-based randomized controlled trial of iron and zinc supplementation in Indonesian infants: effects on growth and development. *American Journal of Clinical Nutrition, 80*(3), 729–736.

Looker, A. C., Dallman, P. R., Carroll, M. D., Gunter, E. W., & Johnson, C. L. (1997). Prevalence of iron deficiency in the United States. *Journal of the American Medical Association, 277*(12), 973–976.

Lozoff, B., Armony-Sivan, R., Kaciroti, N., Jing, Y., Golub, M., & Jacobson, S. W. (2010). Eye-blinking rates are slower in infants with iron-deficiency anemia than in nonanemic iron-deficient or iron-sufficient infants. *Journal of Nutrition, 140*(5), 1057–1061.

Lozoff, B., Beard, J., Connor, J., Barbara, F., Georgieff, M., & Schallert, T. (2006a). Long-lasting neural and behavioral effects of iron deficiency in infancy. *Nutrition Reviews, 64*(5 Pt 2), S34–S43.

Lozoff, B., Brittenham, G. M., Viteri, F. E., Wolf, A. W., & Urrutia, J. J. (1982). The effects of short-term oral iron therapy on developmental deficits in iron-deficient anemic infants. *Journal of Pediatrics, 100*(3), 351–357.

Lozoff, B., Brittenham, G. M., Wolf, A. W., McClish, D. K., Kuhnert, P. M., Jimenez, E., et al. (1987). Iron deficiency anemia and iron therapy effects on infant developmental test performance. *Pediatrics, 79*(6), 981–995.

Lozoff, B., Castillo, M., Clark, KM, Smith, JB. "Iron-fortified vs low-iron infant formula: Developmental outcome at 10 years." Arch Pediatr Adolesc Med 2011 Nov 7 [Epub ahead of print].

Lozoff, B., Jimenez, E., Hagen, J., Mollen, E., & Wolf, A. W. (2000). Poorer behavioral and developmental outcome more than 10 years after treatment for iron deficiency in infancy. *Pediatrics, 105*, e51. Available at: http://pediatrics.aappublications.org/cgi/content/full/105/104/e151. Accessed December 110, 2005.

Lozoff, B., Jimenez, E., & Smith, J. B. (2006b). Double burden of iron deficiency in infancy and low socioeconomic status: a longitudinal analysis of cognitive test scores to age 19 years. *Archives of Pediatrics & Adolescent Medicine, 160*(11), 1108–1113.

Lozoff, B., Jimenez, E., & Wolf, A. W. (1991). Long-term developmental outcome of infants with iron deficiency. *New England Journal of Medicine, 325*(10), 687–694.

Lozoff, B., Klein, N. K., Nelson, E. C., McClish, D. K., Manuel, M., & Chacon, M. E. (1998). Behavior of infants with iron-deficiency anemia. *Child Development, 69*(1), 24–36.

Lubach, G. R., & Coe, C. L. (2008). Selective impairment of cognitive performance in the young monkey following recovery from iron deficiency. *Journal of Developmental & Behavioral Pediatrics, 29*(1), 11–17.

Lukowski, A. F., Koss, M., Burden, M. J., Jonides, J., Nelson, C. A., Kaciroti, N., et al. (2010). Iron deficiency in infancy and neurocognitive functioning at 19 years: evidence of long-term deficits in executive function and recognition memory. *Nutritional Neuroscience, 13*(2), 54–70.

Male, C., Persson, L. A., Freeman, V., Guerra, A., van't Hof, M. A., Haschke, F., et al. (2001). Prevalence of iron deficiency in 12-mo-old infants from 11 European areas and influence of dietary factors on iron status (Euro-Growth Study). *Acta Paediatrica, 90*(5), 492–498.

McCann, J. C., & Ames, B. N. (2007). An overview of evidence for a causal relation between iron deficiency during development and deficits in cognitive or behavioral function. *American Journal of Clinical Nutrition, 85*(4), 931–945.

McLean, E., Cogswell, M., Egli, I., Wojdyla, D., & de Benoist, B. (2009). Worldwide prevalence of anaemia, WHO Vitamin and Mineral Nutrition Information System, 1993–2005. *Public Health Nutrition, 12*(4), 444–454.

Moffatt, M. E., Longstaffe, S., Besant, J., & Dureski, C. (1994). Prevention of iron deficiency and psychomotor decline in high-risk infants through use of iron-fortified infant formula: a randomized clinical trial. *Journal of Pediatrics, 125*(4), 527–534.

Mohamed, W. M. Y., Unger, E. L., Kambhampati, S. K., & Jones, B. C. (2011). Methylphenidate improves cognitive deficits produced by infantile iron deficiency in rats. *Behavioural Brain Research, 216*(1), 146–152.

Monga, M., Walia, V., Gandhi, A., Chandra, J., & Sharma, S. (2010). Effect of iron deficiency anemia on visual evoked potential of growing children. *Brain and Development, 32*(3), 213–216.

Morley, R., Abbott, R., Fairweather-Tait, S., MacFadyen, U., Stephenson, T., & Lucas, A. (1999). Iron fortified follow on formula from 9 to 18 months improves iron status but not development or growth: a randomised trial. *Archives of Disease in Childhood, 81*(3), 247–252.

Nieoullon, A., & Coquerel, A. (2003). Dopamine: a key regulator to adapt action, emotion, motivation and cognition. *Current Opinion in Neurology, 16*(Suppl. 2), S3–S9.

Nordic Council of Ministers. (2004). *Nordic Nutrition Recommendations 2004: Integrating nutrition and physical activity* (4th ed.). Copenhagen, Denmark: Nordic Council of Ministers.

Ornoy, A. (2005). Growth and neuro developmental outcome of children born to mothers with pregestational and gestational diabetes. *Pediatric Endocrinology Reviews, 3*(2), 104–113.

Pala, E., Erguven, M., Guven, S., Erdogan, M., & Balta, T. (2010). Psychomotor development in children with iron deficiency and iron-deficiency anemia. *Food and Nutrition Bulletin, 31*(3), 431–435.

Pongcharoen, T., DiGirolamo, A. M., Ramakrishnan, U., Winichagoon, P., Flores, R., & Martorell, R. (2011). Long-term effects of iron and zinc supplementation during infancy on cognitive function at 9 y of age in northeast Thai children: a follow-up study. *American Journal of Clinical Nutrition, 93*(3), 636–643.

Rao, R., Tkac, I., Townsend, E. L., Ennis, K., Gruetter, R., & Georgieff, M. K. (2007). Perinatal iron deficiency predisposes the developing rat hippocampus to greater injury from mild to moderate hypoxia-ischemia. *Journal of Cerebral Blood Flow & Metabolism, 27*(4), 729–740.

Riggins, T., Miller, N. C., Bauer, P. J., Georgieff, M. K., & Nelson, C. A. (2009). Consequences of low neonatal iron status due to maternal diabetes mellitus on explicit memory performance in childhood. *Developmental Neuropsychology, 34*(6), 762–779.

Roncagliolo, M., Garrido, M., Walter, T., Peirano, P., & Lozoff, B. (1998). Evidence of altered central nervous system development in infants with iron deficiency anemia at 6 mo: delayed maturation of auditory brainstem responses. *American Journal of Clinical Nutrition, 68*(3), 683–690.

Sachdev, H., Gera, T., & Nestel, P. (2005). Effect of iron supplementation on mental and motor development in children: systematic review of randomised controlled trials. *Public Health Nutrition, 8*(2), 117–132.

Shafir, T., Angulo-Barroso, R., Su, J., Jacobson, S. W., & Lozoff, B. (2009). Iron deficiency anemia in infancy and reach and grasp development. *Infant Behavior and Development, 32*(4), 366–375.

Sutcliffe, T. L., Khambalia, A., Westergard, S., Jacobson, S., Peer, M., & Parkin, P. C. (2006). Iron depletion is associated with daytime bottle-feeding in the second and third years of life. *Archives of Pediatrics & Adolescent Medicine, 160*(11), 1114–1120.

Szajewska, H., Ruszczynski, M., & Chmielewska, A. (2010). Effects of iron supplementation in nonanemic pregnant women, infants, and young children on the mental performance and psychomotor development of children: a systematic review of randomized controlled trials. *American Journal of Clinical Nutrition, 91*(6), 1684–1690.

Taricco, E., Radaelli, T., Rossi, G., Nobile de Santis, M. S., Bulfamante, G. P., Avagliano, L., et al. (2009). Effects of gestational diabetes on fetal oxygen and glucose levels in vivo. *International Journal of Obstetrics and Gynaecology, 116*(13), 1729–1735.

Thane, C. W., Walmsley, C. M., Bates, C. J., Prentice, A., & Cole, T. J. (2000). Risk factors for poor iron status in British toddlers: further analysis of data from the

National Diet and Nutrition Survey of children aged 1.5–4.5 years. *Public Health Nutrition, 3*(4), 433–440.

World Health Organization. (2001). *Iron deficiency anaemia: assessment, prevention and control. A guide for programme managers. WHO/NHD/01.3.* Geneva: World Health Organization.

World Health Organization. (2006). *WHO child growth standards.* Geneva, Switzerland: World Health Organization.

World Health Organization. (2007). *WHO reference 2007.* Geneva, Switzerland: World Health Organization.

World Health Organization & Food and Agriculture Organization of the United Nations. (2004). *Vitamin and mineral requirements in human nutrition* (2nd ed.). Geneva: World Health Organization.

Yalçin, S. S., Yurdakök, K., Açikgöz, D., & Özmert, E. (2000). Short-term developmental outcome of iron prophylaxis in infants. *Pediatrics International, 42*(6), 625–630.

CHAPTER 3

Vitamin C and Its Role in Brain Development and Cognition

Stine Hasselholt, Pernille Tveden-Nyborg and Jens Lykkesfeldt

Chapter Overview

- So far, vitamin C deficiency has been mostly associated with its potentially mortal manifestation, scurvy.
- Vitamin C is the most abundant antioxidant in the brain, reaching levels of up to 10 mmol/l in neurons. It apparently plays a pivotal role in the redox homeostasis of the brain.
- Recent experimental evidence suggests that inadequate vitamin C levels may impair neuronal and cognitive development.
- This chapter discusses the importance of vitamin C in brain development and the implications for human nutrition.

Introduction

Most animal species are capable of synthesising L-ascorbic acid or vitamin C on their own. Humans, non-human primates, flying mammals and guinea pigs, however, do not have this ability and therefore completely rely on a sufficient dietary supply of this micronutrient. Prolonged insufficient intake of vitamin C leads to a deficiency state that may ultimately result in the potentially fatal condition, scurvy. The risk of developing scurvy in humans increases with severe depletion, defined as a plasma concentration of vitamin C < 11 µmol/l. While scurvy is rarely observed in the western world, hypovitaminosis C, defined as a plasma concentration of vitamin C < 23 µmol/l, is surprisingly common, affecting large sub-populations (Hampl et al. 2004; Johnston & Thompson 1998; Madruga de et al. 2004; Jacob et al. 1987). This includes pregnant and breastfeeding women. Since foetuses are completely dependent on the transportation of vitamin C across the placenta (Pate et al. 1996; Streeter & Rosso 1981) and vitamin C is transported in milk to newborns, there is a risk that women with hypovitaminosis C convey their AA status to their offspring.

Pre- and post-natal malnutrition can have serious consequences for maturation and functional development of the brain (Dunn et al. 1980; Morgane
et al. 1993). Although the precise mechanisms are still to be fully elucidated,
studies in humans and animal models have linked malnutrition to memory
impairment, reduction in cognitive abilities and reduced hippocampal volume
(Isaacs et al. 2000; Ranade et al. 2008). In guinea pigs, intrauterine malnutrition (i.e. foetal malnutrition) has been shown to result in decreased neuronal
numbers and reduced growth of neuropil in the hippocampus, a central structure in cognition (Mallard et al. 2000). Here, vitamin C could potentially
be of importance as growth retardation has been associated with a decrease
in the vitamin C content of human foetal brain tissue (Zalani et al. 1989).
Moreover, the immature brain has a high cellular growth rate and a moderate
amount of enzymatic antioxidants that renders it susceptible to accumulation
of reactive oxygen species (ROS). In guinea pigs, it has been shown that vitamin C-deficient pups are unable to prevent oxidative damage, despite selective
preservation of the micronutrient in the brain and induction of deoxyribonucleic acid repair (Lykkesfeldt et al. 2007). Also, chronic post-natal vitamin C
deficiency has been shown to impair spatial memory and reduce the number
of neurons in the hippocampus of guinea pigs (Tveden-Nyborg et al. 2009).
These results suggest that an adequate supply of vitamin C could be essential
for the normal development of the brain.

Vitamin C Homeostasis in the Brain

Uptake and Recycling

Due to a defect in the gene coding for L-gulono-γ-lactone oxidase, the
enzyme responsible for the last step in the conversion of D-glucose to vitamin
C, uptake of dietary vitamin C over the intestinal epithelium via the Sodium-
dependent Vitamin C Transporter 1 and facilitated diffusion is essential to
humans (Chatterjee 1973; Tsukaguchi et al. 1999) and other species lacking
biosynthetic capacity.

Vitamin C reaches the central nervous system (CNS) via the bloodstream,
primarily in the monovalent anion form, ascorbate (ASC) (Dhariwal et al.
1991), and is transported to the cerebrospinal fluid (CSF) across the epithelium of the choroid plexus via the Sodium-dependent Vitamin C Transporter 2
(SVCT2) (Tsukaguchi et al. 1999; Angelow et al. 2003). Vitamin C is also able
to cross the blood–brain barrier in the two-electron oxidation product form
dehydroascorbic acid (DHA) by facilitated diffusion on glucose transporters
(GLUTs) (Agus et al. 1997; Vera et al. 1993). However, this mechanism is
probably of minor importance (Qiu et al. 2007; Harrison & May 2009).

Once in the brain, ASC and DHA diffuse from the CSF to the extracellular
fluid and enter neurons by high affinity transport via SVCT2 and GLUTs,
respectively (see Figure 3.1). For glia cells, which apparently lack the SVCT2

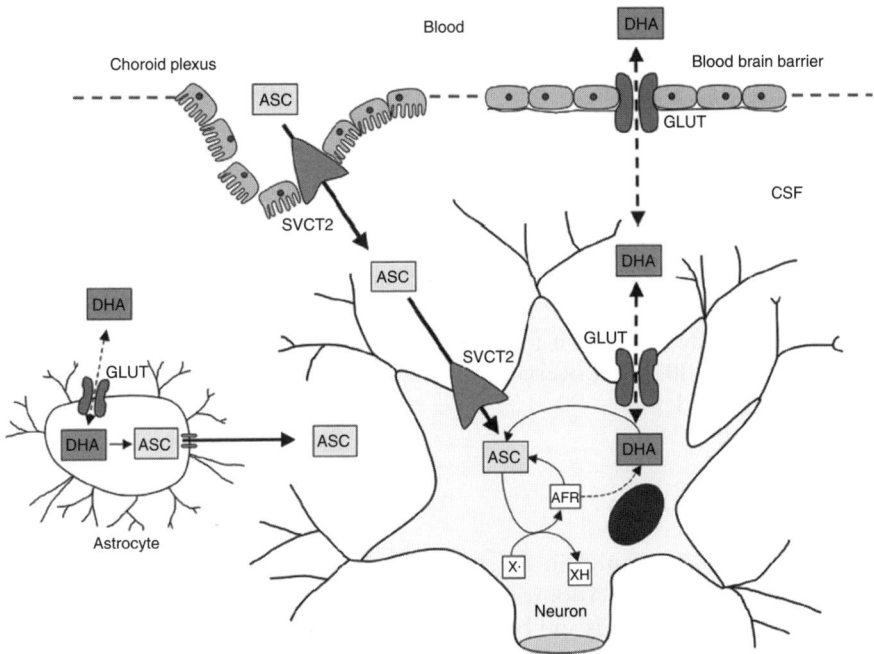

Figure 3.1   Ascorbate uptake and metabolism in the CNS

ASC, ascorbate; AFR, ascorbyl free radical; DHA, dehydroascorbic acid; CSF, cerebrospinal fluid; X, oxidizing free radical species.
Ascorbate enters the CSF either directly through the choroid plexus via the SVCT2 or possibly as DHA via GLUTs across the blood–brain barrier. Similarly, ascorbate enters the neuron through the SVCT2 or as DHA via the GLUTs. AFR generated by X· dismutates to form DHA and ascorbate. Both the AFR and the DHA are recycled back to ascorbate by the cellular metabolism. Glia cells obtain ascorbate from the recycling of DHA that enters via GLUTs, whereas neurons also have the SVCT2 for direct acquisition of ascorbate.
Source: Reprinted from *Free Radical Biology & Medicine, 46* (2009), Harrison & May, Vitamin C function in the brain: vital role of the ascorbate transporter SVCT2, 719–730, © 2009, with permission from Elsevier.

transporter (Berger & Hediger 2000; Mun et al. 2006; Astuya et al. 2005), uptake of vitamin C as DHA via GLUTs and subsequent intracellular reduction is the only possibility.

ASC remains within the cell because of its negative charge at physiological pH (Qiu et al. 2007). It can, however, be released from both neurons and glia in order to maintain the extracellular ASC homeostasis (Schenk et al. 1982; Miele & Fillenz 1996). Alternatively, vitamin C can be transported across the cellular wall to the extracellular space as DHA via GLUTs.

Both the one-electron oxidation product of ASC, semidehydroascorbate or ascorbyl free radical (AFR), and DHA can be reverted to ASC within the cell. AFR is reduced to ASC by NADH- and NADPH-dependent reductases (Kobayashi et al. 1991; May et al. 1998; Nishino & Ito 1986) or two molecules of AFR can disproportionate via dimer formation to one molecule of ASC and one of DHA (Bielski et al. 1971). DHA can, in turn, be reduced by NADPH- or glutathione-dependent reductases (Fornai et al. 1999; Rose 1993; Del et al. 1994; May et al. 1997). Furthermore, glutathione can chemically convert DHA to ASC (May et al. 1997; May & Asard 2004). By means of these recycling processes, vitamin C is able to engage in cellular redox reactions more than once. The recycling of DHA and AFR preserves ASC and is particularly important during deficiency where the amount of vitamin C is limited. It has been proposed that glia cells may play a role in maintaining sufficient concentrations of ASC for neuronal uptake by reducing DHA to ASC and subsequently releasing it following a physiological or pathological stimulus (Qiu et al. 2007). An overview of vitamin C transport in the CNS can be seen in Figure 3.1.

Distribution

ASC is widely distributed in the body, as shown in Figure 3.2. However, concentrations of the micronutrient are tissue dependent, with high concentrations in the adrenal glands and brain compared with, for example, erythrocytes and muscle (Chinoy 1972; Hornig 1975). In plasma, the ASC concentration is regulated by intestinal absorption and renal excretion (Levine et al. 1996). In humans a daily intake of up to 200 mg of vitamin C is almost completely absorbed, whereas doses of more than 500 mg will be largely excreted (Levine et al. 1996). In contrast, a nutritional deficiency of vitamin C will result in release of ASC from the adrenal glands and an almost quantitative reabsorption of vitamin C from the kidneys to ensure a maximum use of the available amount of the compound and support a continuous supply to the brain (Lee et al. 2006; Corpe et al. 2010; Rebec & Pierce 1994).

The high concentration of ASC in the brain and selective retention of ASC in this organ during states of deficiency suggest that vitamin C has a central role in the CNS (Hughes et al. 1971; Lykkesfeldt et al. 2007). In a study of adult guinea pigs subjected to long-term marginal vitamin C deficiency by Frikke-Schmidt et al. (2011), brain levels of ASC decreased over time during the first 90 days of the study to a tissue concentration of around 40 per cent compared with controls kept on a diet sufficient in vitamin C. In contrast, the plasma concentration of ASC dropped to approximately 15 per cent of the initial level within two weeks on a non-scorbutic, low-vitamin C diet. The comparatively high level of vitamin C in the deficient brain was still evident after 180 days on the diet, whereas the concentration of ASC in the liver of deficient animals was 25 per cent compared with controls (Frikke-Schmidt et al. 2011). This demonstrates that even during prolonged states of deficiency the brain is able to maintain a relatively high concentration of vitamin C. The retention of ASC in the brain – despite severe depletion in other parts of the body – is achieved through the active transport by the SVCT2.

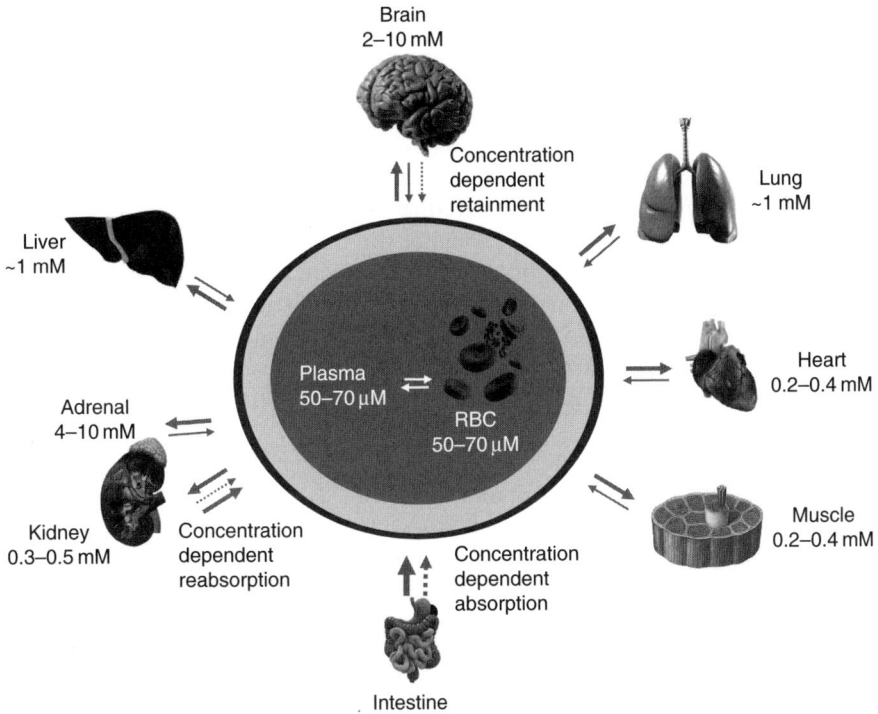

Figure 3.2 Distribution of vitamin C

Vitamin C is taken up from the intestine and distributed with the blood to the tissues. Active transport through SVCTs is essential for the vitamin C homeostasis; and particularly the strongly concentration-dependent intestinal absorption and renal reabsorption and the ability of the brain to selectively retain vitamin C are vital safeguards in the protection against vitamin C deficiency in the CNS.

Within the brain, the ASC distribution is region- and cell type-specific. Neurons have an intracellular ASC content ten times that of glia cells (Rice & Russo-Menna 1998), possibly reflecting the higher oxidative metabolism in neurons (Siesjö 1978). Also, ASC content decreases in the anterior-posterior direction, consistent with the increasing white matter content of more caudal regions of the brain (Rice et al. 1995; Milby et al. 1982). The hippocampus, cortex cerebri, hypothalamus and amygdala all have high ASC concentrations in both humans and animals (Mefford et al. 1981; Milby et al. 1982; Harrison et al. 2010b). This correlates well with a high expression of the SVCT2 in the hippocampus, cortex cerebri and amygdala, but not with a comparably low expression of the SVCT2 in the hypothalamus (Astuya et al. 2005; Mun et al. 2006). In spite of these regional differences, it should be recognised that there is a generally high concentration of ASC in the CNS (Mefford et al. 1981; Milby et al. 1982).

Vitamin C Function in the Brain

Vitamin C as an Antioxidant

ASC has the properties of a unique reducing agent. Thus, in all known bio-logical reactions in which ASC takes part, its molecular mechanism is that of an antioxidant, that is it donates reducing equivalents to the reaction and is itself oxidised in the process. Likewise, the functions of ASC in the brain are all a result of the compound's ability to donate electrons. Both ASC and AFR have remarkably low one-electron reduction potentials, enabling them to reduce virtually all physiologically relevant oxidants and also regenerate other antioxidants – such as vitamin E and glutathione to their active reduced states once they have been oxidised (Buettner & Schafer 2004; Witting & Stocker 2004).

ASC and AFR act directly to scavenge ROS generated both during nor-mal cellular metabolism and neuronal activity as well as under conditions of increased oxidative stress. The DHA generated hereby can subsequently be reverted to the effective reduced antioxidant form in glia and neurons (Figure 3.1). Another important feature is that at physiological pH, the AFR does not induce free radical damage but primarily disproportionates (Buettner & Schafer 2004).

In agreement with the physiochemical properties, several studies have shown ASC to play a role in the antioxidant defence of the brain. In vitro, the SVCT2 transporter protects hippocampal neurons from direct oxidant-induced cell death, and Qiu et al. (2007) attribute this to the ability of the cells to retain ASC (Qiu et al. 2007). This is in agreement with results from Grant et al. (2005), who found a decreased formation of protein carbonyls (a marker of oxidative damage to proteins) in ASC-treated neuroblastoma cells subjected to peroxide insult (Grant et al. 2005). Furthermore, the addition of ASC to rat brain microsomes or brain slices inhibits lipid oxidation induced by various agents in vitro (Seregi et al. 1978; Kovachich & Mishra 1983). In vivo, ASC is selectively retained in the brain of guinea pigs during vitamin C deficiency and keeps other antioxidants – including vitamin E and glutathione – reduced dur-ing increased oxidative stress (Lykkesfeldt et al. 2007). Thus, both in vivo and in vitro data support a role for ASC as an important antioxidant in the brain.

Vitamin C as a Cofactor and Neuromodulator

One of the best characterised functions of vitamin C is its role in collagen synthesis. ASC acts as co-substrate for the Fe^{2+}-2-oxoglutarate-dependent dioxygenases in the hydroxylation of unfolded collagen polypeptide chains, resulting in the formation of triple-helix collagen. If ASC levels are decreased, the collagen polypeptide chain residues are insufficiently hydroxylated, preventing the assembly of the stable triple-helical structure and hence the release of procollagen from the cell (see Figure 2.1) (Hara & Akiyama 2009; Telang et al. 2007; Yoshikawa et al. 2001; De Tullio 2004; Walmsley et al. 1999).

In mice lacking the SVCT2, the principal transport mechanism for vitamin C is unavailable and transportation of ASC to the brain is drastically reduced. Pups display cerebral haemorrhage that has been attributed to a decreased amount of mature type IV collagen in the basement membranes of brain micro vessels (Harrison et al. 2010a). This links ASC to the formation of brain blood vessels and it appears that at least minimal amounts are important for the formation of both brain vasculature and neural sheets (Harrison & May 2009; Sotiriou et al. 2002; Eldridge et al. 1987).

ASC is also a cofactor for Fe^{2+}-2-oxoglutarate-dependent dioxygenases involved in hydroxylation of hypoxia-inducible transcription factors (HIFs) (Flashman et al. 2010; Bruick & McKnight 2001). These HIFs are important for the cellular responses to low oxygen conditions and target genes which modulate cell survival, vascular development and metabolic homeostasis during the development of the brain (Lee et al. 2001; Iyer et al. 1998). Mutant mice with brain-specific deletion of the HIF-1α gene were shown to have impaired development of cortex cerebri and reduced vessel density in the same area, along with deficits in cognitive abilities (Tomita et al. 2003), demonstrating the importance of the transcription factors for normal brain development. Acting as a central electron donor, it seems likely that ASC deficiency could affect the rate of HIF hydroxylation, thereby interfering with HIF-mediated gene regulation and ultimately influencing the normal development of the brain (Figure 3.3).

In neural tissues, ASC serves as a specific cofactor in the biosynthesis of catecholamine, providing the reducing equivalents for dopamine β-hydroxylase in the conversion of dopamine (DA) to norepinephrine (NE) (Diliberto, Jr. & Allen 1980, 1981; Levine et al. 1985). In guinea pigs, vitamin C deficiency was associated with an increased level of DA in the brain accompanied by a decrease in NE. Catecholamine levels returned to within normal ranges following repletion of vitamin C (Hoehn & Kanfer 1980), linking the compound with the ability to convert DA into NE.

As shown in Figure 3.3, DA-containing neurons are prone to oxidative stress and ROS-mediated degeneration due to the conversion of DA to ROS and quinines, both of which can cause cellular damage (Rabinovic et al. 2000; Berman et al. 1996; Halliwell & Gutteridge 2007). Quinones can also deplete glutathione (Spencer et al. 1998) and in this way diminish the recycling of DHA to ASC and thus reduce the cellular defence against oxidative stress. There seems to be a well-regulated interaction between DA and ASC function (Desole et al. 1991; Saponjic et al. 1994). A high level of DA in the brain has been shown to be cytotoxic in neurons, a cytotoxicity that can be diminished by concurrent ASC administration (Hastings et al. 1996). In chromaffin cells

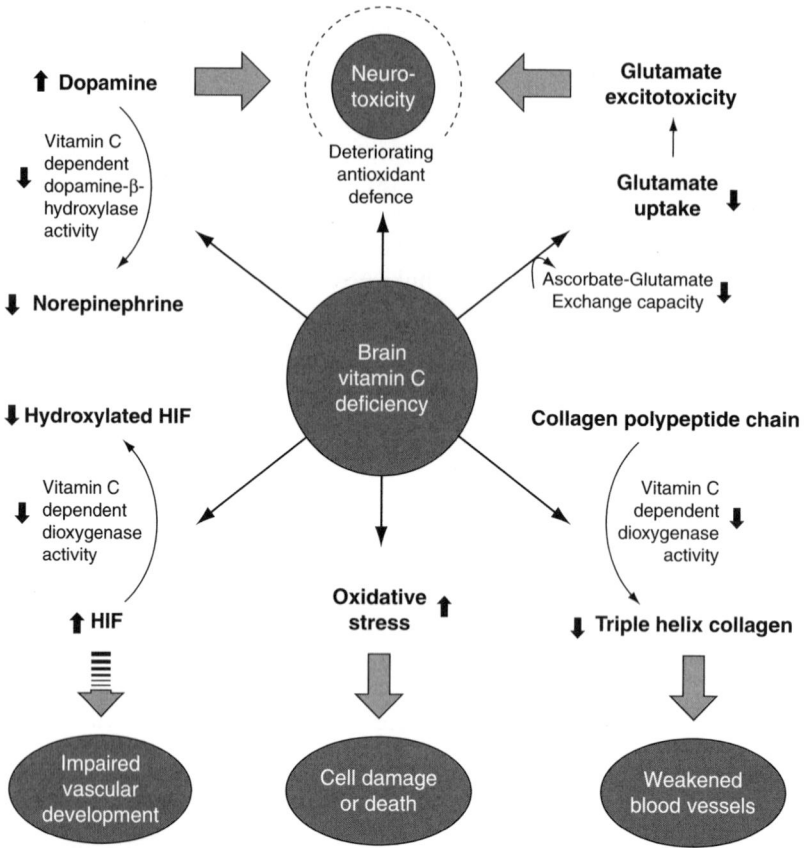

Figure 3.3　Possible consequences of vitamin C deficiency in the brain

Vitamin C helps to maintain CNS function through various molecular pathways. In vitamin C deficiency, the conversion of dopamine to norepinephrine catalysed by dopamine-β-hydroxylase is impaired, resulting in increased concentrations of dopamine. Excess dopamine can result in the generation of reactive oxygen species and quinones which can normally be removed by antioxidant action; however, in the case of vitamin C deficiency, the defence is unlikely to function optimally and prevent neurotoxicity. The ascorbate–glutamate exchange system is also likely to be affected by deficiency leading to decreased glutamate uptake and allowing for extracellular accumulation and subsequent excitotoxic neural injury. Vitamin C deficiency can also result in cell damage and death through increased oxidative stress in general. Moreover, vitamin C deficiency can affect collagen assembly and hydroxylation of hypoxia-inducible transcription factors (HIFs), thus influencing vascular stability and development and brain growth regulation.

of the adrenal gland, a concomitant secretion of ASC and catecholamines has been observed in vitro (Levine et al. 1983) and a low concentration of ASC has been associated with increased apoptosis in these cells despite normal levels of DA and NE (Bornstein et al. 2003). The same may be true for chromaffin cells in the brain, thereby providing a role for ASC in sustaining the viability of catecholaminergic cells. There is also evidence that ASC could be involved in the regulation of acetylcholine release from synaptic vesicles, further substantiating the role of vitamin C in neuronal function (Kuo et al. 1979; Harrison & May 2009).

The excitatory neurotransmitter glutamate is released to the extracellular environment in substantial amounts during neuronal death or breakdown of normal ion gradients. This results in an excessive and prolonged increase of the cation concentration in the neurons (Halliwell & Gutteridge 2007) of which particularly high amounts of intracellular Ca^{2+} can be cytotoxic (Halliwell & Gutteridge 2007; Majewska et al. 1990).

Cellular uptake of glutamate or activation of this compound's ionotropic receptors has been coupled to ASC release in several studies (Cammack et al. 1991; Sandstrom & Rebec 2007; Miele et al. 1994, 2000). Surplus extracellular glutamate in the CNS can result in neurotoxic excessive excitation and so the glutamate-ASC exchange may provide protection against glutamate-mediated excitotoxicity by decreasing the available amount of extracellular glutamate. Furthermore, ASC inhibits the binding of glutamate to the N-methyl-D-aspartate receptor (Majewska et al. 1990) and is able to function as a scavenger of glutamate-generated ROS both intra and extracellularly (Rebec & Pierce 1994; Miele et al. 1994). The possible consequences of vitamin C deficiency in the brain can be seen in Figure 3.3.

Vitamin C in the Developing Brain

Viewing the specific characteristics of the brain there are certain features which renders it particularly sensitive to oxidative stress. The brain has a high content of polyunsaturated fatty acids that are prone to oxidation, which leads to the production of free radicals and in turn may start a chain reaction ultimately resulting in reduced cellular membrane integrity and impaired synaptic function (Hong et al. 2004; Omoi et al. 2006). Also, a large oxygen turnover is necessary for the brain to sustain neuronal intracellular ion homeostasis.

The developing brain can be regarded as even more susceptible to oxidative damage for several reasons. During brain development, the energy demand is enhanced – mainly for the generation and maintenance of ionic equilibria (Erecinska et al. 2004) – and leakage of high-energy electrons derived by neuronal oxidative phosphorylation in the electron transport chain is a noteworthy source to the generation of ROS (Halliwell & Gutteridge 2007; Loschen et al. 1971; Drechsel & Patel 2010). In the adult brain, the amount of enzymatic antioxidants is only moderate and particularly the catalase activity is low (Drechsel & Patel 2010; Sinet et al. 1980). However, in the immature brain, the antioxidant defence as a whole is not fully developed and thus renders this tissue even more exposed. Maintaining the homeostasis between oxidants and antioxidants during brain development is essential as imbalances in ROS levels can affect vital cellular regulatory mechanisms (Nose & Ohba 1996; Sen

& Packer 1996; Graziewicz et al. 2002). This includes the rate of cell division (Yoneyama et al. 2010) and apoptosis (Maycotte et al. 2010), both of which hold central roles in normal brain growth regulation (Ikonomidou & Kaindl 2011).

As a possible precaution to redox imbalance, the ASC content of the brain increases during gestation and is higher in the perinatal brain compared with the adult (Zalani et al. 1989; Adlard et al. 1973, 1974; Kratzing et al. 1985). The increased content of ASC in the immature brain may have an anatomical basis since the decline in ASC content correlates with the brain growth spurt, a period of glia proliferation, myelination, an increase in dendritic complexity and establishment of synaptic connections (Dobbing 1974). ASC has, however, also been shown to enhance the differentiation of precursor cells to neurons and astrocytes and to enhance the functional maturation of differentiated neurons in vitro (Lee et al. 2003; Shin et al. 2004).

Vitamin C and Cognition

Experimental Cognitive Tests and Brain Damage

Several neurodegenerative disorders and direct brain damage are associated with cognitive deficits. Of these, some of the most extensive studies have been carried out to evaluate the relationship between disorders associated with ageing and an impairment of brain function. Although symptoms, aetiology and specific location of lesions in the brain vary considerably, oxidative stress and antioxidant imbalances have consistently been shown to be implicated in diseases such as Alzheimer's (AD), Parkinson's (PD) and Huntington's disease (HD) (Pratico & Sung 2004; Pratico et al. 2001; Halliwell & Gutteridge 2007; Bogdanov et al. 2001). Animal models are often applied in the study of these disorders and subjected to a variety of behavioural tasks in an attempt to assess cognition by analysing animal learning and memory performance.

A challenge to be addressed when applying and evaluating behavioural tasks is the difficulty in determining single affected components resulting in observed functional deficits, due to the numerous interactions between different brain areas. Most often, an observed behaviour is a result of interaction between various centres in the brain. Therefore, impairment can result in varying degrees of behavioural change when tested in different maze constellations, depending on which brain areas are activated during the execution of the specific trials. Furthermore, it is important to clearly define the objective of the study, since procedures in the behavioural tasks vary accordingly. Thus, caution should be taken when comparing findings from behavioural studies.

The Morris Water Maze (MWM) test (Morris 1984) is one of the most frequently used behavioural tests in neuroscience (D'Hooge & De Deyn 2001;

Dringenberg et al. 2001; Vorhees & Williams 2006). It has been used for sophisticated studies of neurobiology and neuropharmacology involved in spatial learning and memory, and in the validation of rodent models of cognitive disorders (D'Hooge & De Deyn 2001). The test can be used to assess spatial and related forms of learning and memory depending on different procedural modifications (Morris 1984; Vorhees & Williams 2006). MWM is a key technique to examine the hippocampal circuitry and is particular sensitive to hippocampal damage. Involvement of various parts of the neocortex in MWM performance is also well established (Vorhees & Williams 2006; Squire et al. 2004; D'Hooge & De Deyn 2001).

The radial arm maze (RAM) can, like the MWM, be used to examine spatial memory in rodents (Olton & Samuelson 1976). This test has become standard protocol for assessing memory processing in a variety of research areas (Dubreuil et al. 2003). The experimental set-up determines the behaviour studied. In order to examine spatial abilities, inter-trial confinement of the animals is necessary in the full-baited RAM task (Dubreuil et al. 2003). This version of the RAM is used to test spatial working memory and is sensitive to hippocampal damage (Liu et al. 2002b). In other spatial versions of the RAM, only some of the arms are baited, testing the animals' ability to navigate to and remember specific bait positions from trial to trial or from training to test phase. These versions can be used to assess spatial working as well as reference memory, and performance is affected by hippocampal function (Crusio & Schwegler 2005; Floresco et al. 1997).

The more simply applicable T-maze is generally used to measure exploratory behaviour. It can, however, also be used to detect cognitive dysfunction if a discrete trial procedure is applied. By lengthening the retention interval between successive trials, it becomes possible to assess spatial memory. The behaviour studied is dependent on a functioning septo-hippocampal system, but other brain areas are also involved (Deacon & Rawlins 2006; Lalonde 2002).

A substantial proportion of the available data concerning the impact of oxidative stress on brain damage, cognition and spatial learning have been reported from studies of AD, PD, HD and ageing. Here, several authors have shown a decline in trial performance in the maze systems associated with deviations of the hippocampus, striatum, cortex cerebri and/or striato-cortical connections (Ardayfio et al. 2008; Fukui et al. 2001, 2002; Furtado & Mazurek 1996; Leon et al. 2010; Liu et al. 2002b). In an animal model of global ischaemia, reductions in hippocampal neuron number were seen and associated with functional impairments (Langdon et al. 2008). Likewise, studies of prenatal exposure to ethanol in guinea pigs and post-natal exposure in rats have revealed cellular deviations in the hippocampus (CA1 area) as well as a poorer performance in the MWM (Richardson et al. 2002; Pauli et al. 1995; McGoey et al. 2003). These results demonstrate that an association between neuronal damage of different origins and a subsequent functional impairment of the hippocampus can be demonstrated by a behavioural test system in old, adult and early life models.

Importance of Vitamin C in the Brain

Studies in Animal Models

The possible involvement of oxidative damage in neurodegenerative disorders and the sensibility of the immature brain to ROS combined with the unique antioxidant properties of ASC have generated an interest in vitamin C as a neuroprotective agent.

Deficiency of the SVCT2 transporter has been found to be lethal in new-born mice, suggesting that vitamin C is of vital importance during brain development. The mice die at birth with respiratory failure, and cortical and brain stem haemorrhage. Barely detectable levels of vitamin C, increased levels of biomarkers of oxidative damage to lipids (F_2-isoprostanes and F_4-neuroprostanes) and apoptosis of cortical neurons were present in the brain cortex of SVCT2 (-/-) mice. The loss of neurons and supporting cells in crucial brain areas such as the brain stem – responsible for the autonomic regulation of respiration – due to haemorrhage and oxidative stress, could well be the cause of death in the SVCT2 (-/-) mice (Harrison et al. 2010a; Sotiriou et al. 2002).

The hypothesis that adequate levels of vitamin C are crucial during brain development is supported by findings in Gulo (-/-) mice unable to synthesise vitamin C. In a study by Harrison et al. (2010c), foetuses had increased levels of vitamin C in the cortex and cerebellum compared with dams. Furthermore, post-natal F_2-isoprostane levels in the cortex and glutathione levels in the cortex and cerebellum were elevated in Gulo (-/-) mice compared with wild-type littermates, probably as a response to oxidative stress (Harrison et al. 2010c). Elevated levels of F_4-neuroprostanes in the cortex and cerebellum compared with wild-type controls were also seen in adult Gulo (-/-) mice. Trials in the MWM system did not reveal an effect on cognitive performance of Gulo (-/-) mice compared with wild-type controls (Harrison et al. 2008).

However, a functional consequence of vitamin C deficiency has been reported in newborn guinea pigs. Post-natal guinea pigs subjected to chronic vitamin C deficiency displayed a significantly decreased spatial memory in the MWM and an approximately 30 per cent reduction in the number of neurons in the hippocampus compared with non-deficient counterparts (Tveden-Nyborg et al. 2009). This study revealed a direct effect of vitamin C deficiency on brain function in the developing hippocampus. Furthermore, the reduction in neuronal numbers is quantitatively comparable to reported reductions in models following neonatal hypoxia or foetal alcohol exposure, proposed to lead to cognitive disabilities in children (Gibson et al. 2000; Livy et al. 2003; Nunez et al. 2003).

In a double transgenic mouse model of Alzheimers Disease (AD) with elevated brain levels of F_4-neuroprostanes and malondialdehyde (MDA: another biomarker of oxidative damage to lipids), a vitamin C-supplemented diet resulted in a decrease in the concentration of these biomarkers. No main effect of treatment with vitamin C was seen on T-maze or MWM performance. However, AD mice on the control diet did not exhibit a spatial memory equivalent to that of vitamin C-supplemented and wild-type counterparts (Harrison et al. 2009a). In middle-aged (12 months) and very old (24 months) AD mice, ASC had a positive effect on performance in the probe trial of MWM in the middle-aged mice, but this was not seen in the very old mice. In this study, neither genotype nor ASC treatment had a significant effect on MDA and ASC levels in brain cortex, but the very old mice had higher levels than the middle-aged. The greatest benefit of ASC treatment was observed in very old mice that displayed the poorest performance in MWM, indicating that ASC treatment may be most beneficial in systems that are heavily compromised (Harrison et al. 2009b).

The results from the limited number of studies performed to assess the putative association between vitamin C and neuronal development are conflicting. Though the level of vitamin C in the brain can influence the extent of oxidative damage, estimated by measuring biomarkers in models of ageing and neurodegeneration, the impact on animal performance in the behavioural tasks is inconsistent. One explanation could be that the damage in the brain needs to be extensive before a functional deficit can be detected; that is, it needs to exceed a level above which the brain is unable to further compensate. Another possible implicated factor could be that vitamin C has an effect in different brain areas than the ones under investigation. Also, the relationship between vitamin C levels and oxidative stress varies depending on the brain area under investigation (Harrison et al. 2010b). Finally, it may be that the compound just does not have a generalized effect on cognition.

Human Studies

Studies investigating the relationship between vitamin C, neurodevelopment and cognitive function in humans are sparse. This is surprising, as hypovitaminosis C is more prevalent in the western world than generally presumed, but it is probably related to the difficulty in performing cognitive tests in general, combined with problems in isolating an effect of vitamin C deficiency from other deficiencies in high-risk individuals. Sub-populations at risk of deficiency include people of low socio-economic status and smokers (Mosdol et al. 2008; Wrieden et al. 2000). The highest prevalence of vitamin C deficiency was observed in a study by Hampl et al. (2004) observed among fertile women (Hampl et al. 2004); and in a healthy outpatient population from Arizona, USA, 31.6 per cent of the participating pregnant women were found to be vitamin C deficient (Johnston & Thompson 1998). In a study with pregnant women from Cuenca, Spain, the vitamin C-deficient group included 43.5 per cent of non-smokers and 56.3 per cent of smokers (Ortega et al. 1998). Since there is a correlation between the plasma concentration of vitamin C in pregnant women and their newborns, this could be a potential problem (Madruga de et al. 2004, 2009; Scaife et al. 2006).

A central issue in this context is the precise requirements of vitamin C of the foetus. In case of maternal deficiency during pregnancy, vitamin C concentrations in the blood from the umbilical cord have been shown to be several times higher than the maternal blood concentration, thus demonstrating that vitamin C concentration in foetal plasma is maintained at an increased level compared with the maternal vitamin C status (Madruga de et al. 2004, 2009; Scaife et al. 2006). Madruga de et al. (2004) found a prevalence of hypovitaminosis C in newborns from smoking and non-smoking women (São Paulo, Brazil) of 2.5 and 1.1 per cent, respectively (Madruga de et al. 2004); in a study of Mexican children by Villalpando et al. (2003), 30.3 per cent of the children 0–2 years of age had blood levels of vitamin C ≤ 0.2 mg/dl (corresponding to 11.4 μmol/l) and 23 per cent of all participating children younger than 12 years of age fell into this category (Villalpando et al. 2003). Thus, despite the aforementioned retention of vitamin C in the developing foetus, low levels of vitamin C after birth and during early life are not uncommon. This is particularly worrying in light of the considerable growth and development that the human brain undergoes during the first years of life (Dobbing 1974) and the proposed need for a high level of antioxidants to maintain the redox balance.

Results from studies in subjects with neurodegenerative disorders may provide some information regarding the role of vitamin C as a neuroprotectant. In AD patients, Riviere et al. (1998) found a gradual, significant decrease in plasma vitamin C as a function of cognitive impairment (Riviere et al. 1998). Quinn et al. (2003) found a significantly higher mean CSF: plasma ratio of vitamin C in AD than in control subjects, and hippocampal volume was correlated with CSF and plasma vitamin C and inversely correlated with the CSF: plasma vitamin C ratio. The group hypothesise that the oxidatively stressed AD brain has a greater requirement for vitamin C and depletes the peripheral circulation in an attempt to maintain brain vitamin C homeostasis (Quinn et al. 2003). This is in agreement with a study by Pratico & Sung (2004), who found an increased level of lipid oxidation and a lower level of ASC in the brain in AD patients compared with matched controls (Pratico & Sung 2004). In contrast, Bowman et al. (2009) found a slower rate of cognitive decline in subjects with higher CSF : plasma vitamin C ratio. However, they also found evidence of a dysfunctional blood–brain barrier in subjects with lower CSF: plasma vitamin C ratios and suggest that this impairs the ability of the brain to maintain high CSF: plasma AA ratios, as AA may then diffuse out of the brain according to its concentration gradient (Bowman et al. 2009).

A frequently overlooked problem is that vitamin C uptake is highly dose dependent. This means that individuals saturated from their daily diet are unlikely to benefit from further supplementation with vitamin C as they will excrete any surplus of the compound (Lykkesfeldt & Poulsen 2010; Levine et al. 1996; Terpstra et al. 2010). Thus, an effect of vitamin C cannot be expected if the included individuals are all vitamin C saturated. In many of the previous published human studies, the initial vitamin C status of included individuals is unclear, making these studies less useful for evaluation of the possible effects of vitamin C supplementation on human health (Lykkesfeldt & Poulsen 2010).

Conclusion

In the last decade, numerous experimental studies have shown vitamin C to be an important cofactor in the brain, particularly during development. Moreover, functional brain damage and inferior performance in behavioural tasks have been observed following vitamin C deficiency. However, despite an increased focus on this issue, the importance of vitamin C in cognition is not clearly defined and, most importantly, the clinical significance of deficiency, beyond that of scurvy, remains to be established in humans.

The increased prevalence of subclinical vitamin C deficiency among smokers, groups of fertile women and young children demonstrates that this condition is indeed common. Moreover, the brain's unique ability to retain vitamin C compared with other tissues, and the fact that oxidative damage is consistently observed in the brain following states of deficiency, indicates a central role for this vital micronutrient in the CNS.

Regardless of the considerable biochemical and experimental evidence, the profound lack of well-designed clinical studies in the available literature prevents conclusions on the possible functional importance of vitamin C deficiency in human brain and cognitive development but rather suggests that such clinical studies are required to adequately address this issue.

In future studies, clearly defined inclusion and exclusion criteria and increased study specificity are required in order to allow proper evaluation of the effect of vitamin C deficiency or supplementation (Lykkesfeldt & Poulsen 2010). Knowledge acquired in this way may in time form the basis for new dietary recommendations for vitamin C with specific reference to supplementation of high-risk sub-populations such as pregnant women and children.

References

Adlard, B. P., de Souza, S. W., & Moon, S. (1973). The effect of age, growth retardation and asphyxia on ascorbic acid concentrations in developing brain. *Journal of Neurochemistry, 21*, 877–881.

Adlard, B. P., de Souza, S. W., & Moon, S. (1974). Ascorbic acid in fetal human brain. *Archives of Disease in Childhood, 49*, 278–282.

Agus, D. B., Gambhir, S. S., Pardridge, W. M., Spielholz, C., Baselga, J., Vera, J. C. et al. (1997). Vitamin C crosses the blood-brain barrier in the oxidized form through the glucose transporters. *Journal of Clinical Investigation, 100*, 2842–2848.

Angelow, S., Haselbach, M., & Galla, H. J. (2003). Functional characterisation of the active ascorbic acid transport into cerebrospinal fluid using primary cultured choroid plexus cells. *Brain Research, 988,* 105–113.

Ardayfio, P., Moon, J., Leung, K. K., Youn-Hwang, D., & Kim, K. S. (2008). Impaired learning and memory in Pitx3 deficient aphakia mice: a genetic model for striatum-dependent cognitive symptoms in Parkinson's disease. *Neurobiology of Disease, 31,* 406–412.

Astuya, A., Caprile, T., Castro, M., Salazar, K., Garcia, M. L., Reinicke, K., et al. (2005). Vitamin C uptake and recycling among normal and tumor cells from the central nervous system. *Journal of Neuroscience Research, 79,* 146–156.

Berger, U. V. & Hediger, M. A. (2000). The vitamin C transporter SVCT2 is expressed by astrocytes in culture but not in situ. *Neuroreport, 11,* 1395–1399.

Berman, S. B., Zigmond, M. J., & Hastings, T. G. (1996). Modification of dopamine transporter function: effect of reactive oxygen species and dopamine. *Journal of Neurochemistry, 67,* 593–600.

Bielski, B. H. J., Comstock, D. A., & Bowen, R. A. (1971). Ascorbic acid free radicals. I. pulse radiolysis study of optical absorption and kinetic properties. *Journal of the American Chemical Society, 93,* 5624–5629.

Bogdanov, M. B., Andreassen, O. A., Dedeoglu, A., Ferrante, R. J., & Beal, M. F. (2001). Increased oxidative damage to DNA in a transgenic mouse model of Huntington's disease. *Journal of Neurochemistry, 79,* 1246–1249.

Bornstein, S. R., Yoshida-Hiroi, M., Sotiriou, S., Levine, M., Hartwig, H. G., Nussbaum, R. L. et al. (2003). Impaired adrenal catecholamine system function in mice with deficiency of the ascorbic acid transporter (SVCT2). *FASEB Journal, 17,* 1928–1930.

Bowman, G. L., Dodge, H., Frei, B., Calabrese, C., Oken, B. S., Kaye, J. A. et al. (2009). Ascorbic acid and rates of cognitive decline in Alzheimer's disease. *Journal of Alzheimers Disease, 16,* 93–98.

Bruick, R. K. & McKnight, S. L. (2001). A conserved family of prolyl-4-hydroxylases that modify HIF. *Science, 294,* 1337–1340.

Buettner, G. R. & Schafer, F. Q. (2004). Ascorbate as an antioxidant. In H. Asard, J. M. May, & N. Smirnoff (Eds.), *Vitamin C functions and biochemistry in animals and plants* (pp. 173–188). New York: BIOS Scientific Publishers.

Cammack, J., Ghasemzadeh, B., & Adams, R. N. (1991). The pharmacological profile of glutamate-evoked ascorbic acid efflux measured by in vivo electrochemistry. *Brain Research, 565,* 17–22.

Chatterjee, I. B. (1973). Evolution and the biosynthesis of ascorbic acid. *Science, 182,* 1271–1272.

Chinoy, N. J. (1972). Ascorbic acid levels in mammalian tissues and its metabolic significance. *Comparative Biochemistry and Physiology. A, Comparative Physiology, 42,* 945–952.

Corpe, C. P., Tu, H., Eck, P., Wang, J., Faulhaber-Walter, R., Schnermann, J. et al. (2010). Vitamin C transporter Slc23a1 links renal reabsorption, vitamin C tissue accumulation, and perinatal survival in mice. *Journal of Clinical Investigation, 120,* 1069–1083.

Crusio, W. E. & Schwegler, H. (2005). Learning spatial orientation tasks in the radial-maze and structural variation in the hippocampus in inbred mice. *Behavioral and Brain Functions, 1,* doi: 10.1186/1744-9081-1-3.

D'Hooge, R. & De Deyn, P. P. (2001). Applications of the Morris water maze in the study of learning and memory. *Brain Research Reviews, 36,* 60–90.

De Tullio, M. C. (2004). How does ascorbic acid prevent scurvy? A survey of the nonantioxidant functions of vitamin C. In H. Asard, J. M. May, & N. Smirnoff (Eds.), *Vitamin C functions and biochemistry in animals and plants* (pp. 159–171). New York: BIOS Scientific Publishers.

Deacon, R. M. & Rawlins, J. N. (2006). T-maze alternation in the rodent. *Nature Protocols, 1,* 7–12.

Del, B. B., Maellaro, E., Sugherini, L., Santucci, A., Comporti, M., & Casini, A. F. (1994). Purification of NADPH-dependent dehydroascorbate reductase from rat liver and its identification with 3 alpha-hydroxysteroid dehydrogenase. *Biochemical Journal, 304*(Pt. 2), 385–390.

Desole, M. S., Miele, M., Enrico, P., Esposito, G., Fresu, L., De, N. G. et al. (1991). Investigations into the relationship between the dopaminergic system and ascorbic acid in rat striatum. *Neuroscience Letters, 127,* 34–38.

Dhariwal, K. R., Hartzell, W. O., & Levine, M. (1991). Ascorbic acid and dehydroascorbic acid measurements in human plasma and serum. *American Journal of Clinical Nutrition, 54,* 712–716.

Diliberto, E. J., Jr. & Allen, P. L. (1980). Semidehydroascorbate as a product of the enzymic conversion of dopamine to norepinephrine. Coupling of semidehydroascorbate reductase to dopamine-beta-hydroxylase. *Molecular Pharmacology, 17,* 421–426.

Diliberto, E. J., Jr. & Allen, P. L. (1981). Mechanism of dopamine-beta-hydroxylation. Semidehydroascorbate as the enzyme oxidation product of ascorbate. *Journal of Biological Chemistry, 256,* 3385–3393.

Dobbing, J. (1974). The later growth of the brain and its vulnerability. *Pediatrics, 53,* 2–6.

Drechsel, D. A., & Patel, M. (2010). Respiration-dependent H2O2 removal in brain mitochondria via the thioredoxin/peroxiredoxin system. *Journal of Biological Chemistry, 285,* 27850–27858.

Dringenberg, H. C., Richardson, D. P., Brien, J. F., & Reynolds, J. N. (2001). Spatial learning in the guinea pig: cued versus non-cued learning, sex differences, and comparison with rats. *Behavioural Brain Research, 124,* 97–101.

Dubreuil, D., Tixier, C., Dutrieux, G., & Edeline, J. M. (2003). Does the radial arm maze necessarily test spatial memory? *Neurobiology of Learning and Memory, 79,* 109–117.

Dunn, H. G., Crichton, J. U., Grunau, R. V., McBurney, A. K., McCormick, A. Q., Robertson, A. M. et al. (1980). Neurological, psychological and educational sequelae of low birth weight. *Brain and Development, 2,* 57–67.

Eldridge, C. F., Bunge, M. B., Bunge, R. P., & Wood, P. M. (1987). Differentiation of axon-related Schwann cells in vitro. I. Ascorbic acid regulates basal lamina assembly and myelin formation. *Journal of Cell Biology, 105,* 1023–1034.

Erecinska, M., Cherian, S., & Silver, I. A. (2004). Energy metabolism in mammalian brain during development. *Progress in Neurobiology, 73,* 397–445.

Flashman, E., Davies, S. L., Yeoh, K. K., & Schofield, C. J. (2010). Investigating the dependence of the hypoxia-inducible factor hydroxylases (factor inhibiting HIF and prolyl hydroxylase domain 2) on ascorbate and other reducing agents. *Biochemical Journal, 427,* 135–142.

Floresco, S. B., Seamans, J. K., & Phillips, A. G. (1997). Selective roles for hippocampal, prefrontal cortical, and ventral striatal circuits in radial-arm maze tasks with or without a delay. *Journal of Neuroscience, 17*, 1880–1890.

Fornai, F., Saviozzi, M., Piaggi, S., Gesi, M., Corsini, G. U., Malvaldi, G. et al. (1999). Localization of a glutathione-dependent dehydroascorbate reductase within the central nervous system of the rat. *Neuroscience, 94*, 937–948.

Frikke-Schmidt, H., Tveden-Nyborg, P., Birck, M. M., & Lykkesfeldt, J. (2011). High dietary fat and cholesterol exacerbates chronic vitamin C deficiency in guinea pigs. *British Journal of Nutrition, 105*, 54–61.

Fukui, K., Omoi, N. O., Hayasaka, T., Shinnkai, T., Suzuki, S., Abe, K. et al. (2002). Cognitive impairment of rats caused by oxidative stress and aging, and its prevention by vitamin E. *Annals of the New York Academy of Sciences, 959*, 275–284.

Fukui, K., Onodera, K., Shinkai, T., Suzuki, S., & Urano, S. (2001). Impairment of learning and memory in rats caused by oxidative stress and aging, and changes in antioxidative defense systems. *Annals of the New York Academy of Sciences, 928*, 168–175.

Furtado, J. C. & Mazurek, M. F. (1996). Behavioral characterization of quinolinate-induced lesions of the medial striatum: relevance for Huntington's disease. *Experimental Neurology, 138*, 158–168.

Gibson, M. A., Butters, N. S., Reynolds, J. N., & Brien, J. F. (2000). Effects of chronic prenatal ethanol exposure on locomotor activity, and hippocampal weight, neurons, and nitric oxide synthase activity of the young postnatal guinea pig. *Neurotoxicology and Teratology, 22*, 183–192.

Grant, M. M., Barber, V. S., & Griffiths, H. R. (2005). The presence of ascorbate induces expression of brain derived neurotrophic factor in SH-SY5Y neuroblastoma cells after peroxide insult, which is associated with increased survival. *Proteomics, 5*, 534–540.

Graziewicz, M. A., Day, B. J., & Copeland, W. C. (2002). The mitochondrial DNA polymerase as a target of oxidative damage. *Nucleic Acids Research, 30*, 2817–2824.

Halliwell, B. & Gutteridge, J. (2007). *Free radicals in biology and medicine* (4th ed.). New York: Oxford university press.

Hampl, J. S., Taylor, C. A., & Johnston, C. S. (2004). Vitamin C deficiency and depletion in the United States: the Third National Health and Nutrition Examination Survey, 1988 to 1994. *American Journal of Public Health, 94*, 870–875.

Hara, K. & Akiyama, Y. (2009). Collagen-related abnormalities, reduction in bone quality, and effects of menatetrenone in rats with a congenital ascorbic acid deficiency. *Journal of Bone and Mineral Metabolism, 27*, 324–332.

Harrison, F. E., Allard, J., Bixler, R., Usoh, C., Li, L., May, J. M. et al. (2009a). Antioxidants and cognitive training interact to affect oxidative stress and memory in APP/PSEN1 mice. *Nutritional Neuroscience, 12*, 203–218.

Harrison, F. E., Dawes, S. M., Meredith, M. E., Babaev, V. R., Li, L., & May, J. M. (2010a). Low vitamin C and increased oxidative stress and cell death in mice that lack the sodium-dependent vitamin C transporter SVCT2. *Free Radical Biology and Medicine, 49*, 821–829.

Harrison, F. E., Green, R. J., Dawes, S. M., & May, J. M. (2010b). Vitamin C distribution and retention in the mouse brain. *Brain Research, 1348*, 181–186.

Harrison, F. E., Hosseini, A. H., McDonald, M. P., & May, J. M. (2009b). Vitamin C reduces spatial learning deficits in middle-aged and very old APP/PSEN1 transgenic and wild-type mice. *Pharmacology, Biochemistry and Behavior, 93*, 443–450.

Harrison, F. E. & May, J. M. (2009). Vitamin C function in the brain: vital role of the ascorbate transporter SVCT2. *Free Radical Biology and Medicine, 46*, 719–730.

Harrison, F. E., Meredith, M. E., Dawes, S. M., Saskowski, J. L., & May, J. M. (2010c). Low ascorbic acid and increased oxidative stress in gulo(-/-) mice during development. *Brain Research, 1349*, 143–152.

Harrison, F. E., Yu, S. S., Van Den Bossche, K. L., Li, L., May, J. M., & McDonald, M. P. (2008). Elevated oxidative stress and sensorimotor deficits but normal cognition in mice that cannot synthesize ascorbic acid. *Journal of Neurochemistry, 106*, 1198–1208.

Hastings, T. G., Lewis, D. A., & Zigmond, M. J. (1996). Role of oxidation in the neurotoxic effects of intrastriatal dopamine injections. *Proceedings of the National Academy of Sciences of the United States of America, 93*, 1956–1961.

Hoehn, S. K. & Kanfer, J. N. (1980). Effects of chronic ascorbic acid deficiency on guinea pig lysosomal hydrolase activities. *Journal of Nutrition, 110*, 2085–2094.

Hong, J. H., Kim, M. J., Park, M. R., Kwag, O. G., Lee, I. S., Byun, B. H. et al. (2004). Effects of vitamin E on oxidative stress and membrane fluidity in brain of streptozotocin-induced diabetic rats. *Clinica Chimica Acta, 340*, 107–115.

Hornig, D. (1975). Distribution of ascorbic acid, metabolites and analogues in man and animals. *Annals of the New York Academy of Sciences, 258*, 103–118.

Hughes, R. E., Hurley, R. J., & Jones, P. R. (1971). The retention of ascorbic acid by guinea-pig tissues. *British Journal of Nutrition, 26*, 433–438.

Ikonomidou, C. & Kaindl, A. M. (2011). Neuronal death and oxidative stress in the developing brain. *Antioxidants and Redox Signaling, 14*, 1535–1550.

Isaacs, E. B., Lucas, A., Chong, W. K., Wood, S. J., Johnson, C. L., Marshall, C. et al. (2000). Hippocampal volume and everyday memory in children of very low birth weight. *Pediatric Research, 47*, 713–720.

Iyer, N. V., Kotch, L. E., Agani, F., Leung, S. W., Laughner, E., Wenger, R. H. et al. (1998). Cellular and developmental control of O2 homeostasis by hypoxia-inducible factor 1 alpha. *Genes and Development, 12*, 149–162.

Jacob, R. A., Skala, J. H., & Omaye, S. T. (1987). Biochemical indices of human vitamin C status. *American Journal of Clinical Nutrition, 46*, 818–826.

Johnston, C. S. & Thompson, L. L. (1998). Vitamin C status of an outpatient population. *Journal of the American College of Nutrition, 17*, 366–370.

Kobayashi, K., Harada, Y., & Hayashi, K. (1991). Kinetic behavior of the monodehydroascorbate radical studied by pulse radiolysis. *Biochemistry, 30*, 8310–8315.

Kovachich, G. B. & Mishra, O. P. (1983). The effect of ascorbic acid on malonaldehyde formation, K+, Na+ and water content of brain slices. *Experimental Brain Research, 50*, 62–68.

Kratzing, C. C., Kelly, J. D., & Kratzing, J. E. (1985). Ascorbic acid in fetal rat brain. *Journal of Neurochemistry, 44*, 1623–1624.

Kuo, C. H., Hata, F., Yoshida, H., Yamatodani, A., & Wada, H. (1979). Effect of ascorbic acid on release of acetylcholine from synaptic vesicles prepared from different species of animals and release of noradrenaline from synaptic vesicles of rat brain. *Life Sciences, 24*, 911–915.

Lalonde, R. (2002). The neurobiological basis of spontaneous alternation. *Neuroscience and Biobehavioral Reviews, 26,* 91–104.

Langdon, K. D., Granter-Button, S., & Corbett, D. (2008). Persistent behavioral impairments and neuroinflammation following global ischemia in the rat. *European Journal of Neuroscience, 28,* 2310–2318.

Lee, J. H., Oh, C. S., Mun, G. H., Kim, J. H., Chung, Y. H., Hwang, Y. I. et al. (2006). Immunohistochemical localization of sodium-dependent L-ascorbic acid transporter 1 protein in rat kidney. *Histochemistry and Cell Biology, 126,* 491–494.

Lee, J. Y., Chang, M. Y., Park, C. H., Kim, H. Y., Kim, J. H., Son, H. et al. (2003). Ascorbate-induced differentiation of embryonic cortical precursors into neurons and astrocytes. *Journal of Neuroscience Research, 73,* 156–165.

Lee, Y. M., Jeong, C. H., Koo, S. Y., Son, M. J., Song, H. S., Bae, S. K. et al. (2001). Determination of hypoxic region by hypoxia marker in developing mouse embryos in vivo: a possible signal for vessel development. *Developmental Dynamics, 220,* 175–186.

Leon, W. C., Canneva, F., Partridge, V., Allard, S., Ferretti, M. T., DeWilde, A. et al. (2010). A novel transgenic rat model with a full Alzheimer's-like amyloid pathology displays pre-plaque intracellular amyloid-beta-associated cognitive impairment. *Journal of Alzheimers Disease, 20,* 113–126.

Levine, M., Asher, A., Pollard, H., & Zinder, O. (1983). Ascorbic acid and catecholamine secretion from cultured chromaffin cells. *Journal of Biological Chemistry, 258,* 13111–13115.

Levine, M., Conry-Cantilena, C., Wang, Y., Welch, R. W., Washko, P. W., Dhariwal, K. R. et al. (1996). Vitamin C pharmacokinetics in healthy volunteers: evidence for a recommended dietary allowance. *Proceedings of the National Academy of Sciences of the United States of America, 93,* 3704–3709.

Levine, M., Morita, K., Heldman, E., & Pollard, H. B. (1985). Ascorbic acid regulation of norepinephrine biosynthesis in isolated chromaffin granules from bovine adrenal medulla. *Journal of Biological Chemistry, 260,* 15598–15603.

Liu, J., Head, E., Gharib, A. M., Yuan, W., Ingersoll, R. T., Hagen, T. M. et al. (2002a). Memory loss in old rats is associated with brain mitochondrial decay and RNA/DNA oxidation: partial reversal by feeding acetyl-L-carnitine and/or R-alpha -lipoic acid. *Proceedings of the National Academy of Sciences of the United States of America, 99,* 2356–2361.

Liu, L., Ikonen, S., Heikkinen, T., Heikkila, M., Puolivali, J., van, G. T. et al. (2002b). Effects of fimbria-fornix lesion and amyloid pathology on spatial learning and memory in transgenic APP+PS1 mice. *Behavioural Brain Research, 134,* 433–445.

Livy, D. J., Miller, E. K., Maier, S. E., & West, J. R. (2003). Fetal alcohol exposure and temporal vulnerability: effects of binge-like alcohol exposure on the developing rat hippocampus. *Neurotoxicology and Teratology, 25,* 447–458.

Loschen, G., Flohe, L., & Chance, B. (1971). Respiratory chain linked H(2)O(2) production in pigeon heart mitochondria. *FEBS Letters, 18,* 261–264.

Lykkesfeldt, J. & Poulsen, H. E. (2010). Is vitamin C supplementation beneficial? Lessons learned from randomised controlled trials. *British Journal of Nutrition, 103,* 1251–1259.

Lykkesfeldt, J., Trueba, G. P., Poulsen, H. E., & Christen, S. (2007). Vitamin C deficiency in weanling guinea pigs: differential expression of oxidative stress and DNA repair in liver and brain. *British Journal of Nutrition, 98,* 1116–1119.

Madruga de, O. A., Rondo, P. H., & Barros, S. B. (2004). Concentrations of ascorbic acid in the plasma of pregnant smokers and nonsmokers and their newborns. *International Journal for Vitamin and Nutrition Research, 74*, 193–198.

Madruga de, O. A., Rondo, P. H., & Oliveira, J. M. (2009). Maternal alcohol consumption may influence cord blood ascorbic acid concentration: findings from a study of Brazilian mothers and their newborns. *British Journal of Nutrition, 102*, 895–898.

Majewska, M. D., Bell, J. A., & London, E. D. (1990). Regulation of the NMDA receptor by redox phenomena: inhibitory role of ascorbate. *Brain Research, 537*, 328–332.

Mallard, C., Loeliger, M., Copolov, D., & Rees, S. (2000). Reduced number of neurons in the hippocampus and the cerebellum in the postnatal guinea-pig following intrauterine growth-restriction. *Neuroscience, 100*, 327–333.

May, J. & Asard, H. (2004). Ascorbate recycling. In H. Asard, J. M. May, & N. Smirnoff (Eds.), *Vitamin C functions and biochemistry in animals and plants* (pp. 139–157). New York: BIOS Scientific Publishers.

May, J. M., Cobb, C. E., Mendiratta, S., Hill, K. E., & Burk, R. F. (1998). Reduction of the ascorbyl free radical to ascorbate by thioredoxin reductase. *Journal of Biological Chemistry, 273*, 23039–23045.

May, J. M., Mendiratta, S., Hill, K. E., & Burk, R. F. (1997). Reduction of dehydroascorbate to ascorbate by the selenoenzyme thioredoxin reductase. *Journal of Biological Chemistry, 272*, 22607–22610.

Maycotte, P., Guemez-Gamboa, A., & Moran, J. (2010). Apoptosis and autophagy in rat cerebellar granule neuron death: role of reactive oxygen species. *Journal of Neuroscience Research, 88*, 73–85.

McGoey, T. N., Reynolds, J. N., & Brien, J. F. (2003). Chronic prenatal ethanol exposure induced decrease of guinea pig hippocampal CA1 pyramidal cell and cerebellar Purkinje cell density. *Canadian Journal of Physiology and Pharmacology, 81*, 476–484.

Mefford, I. N., Oke, A. F., & Adams, R. N. (1981). Regional distribution of ascorbate in human brain. *Brain Research, 212*, 223–226.

Miele, M., Boutelle, M. G., & Fillenz, M. (1994). The physiologically induced release of ascorbate in rat brain is dependent on impulse traffic, calcium influx and glutamate uptake. *Neuroscience, 62*, 87–91.

Miele, M. & Fillenz, M. (1996). In vivo determination of extracellular brain ascorbate. *Journal of Neuroscience Methods, 70*, 15–19.

Miele, M., Mura, M. A., Enrico, P., Esposito, G., Serra, P. A., Migheli, R. et al. (2000). On the mechanism of d-amphetamine-induced changes in glutamate, ascorbic acid and uric acid release in the striatum of freely moving rats. *British Journal of Pharmacology, 129*, 582–588.

Milby, K., Oke, A., & Adams, R. N. (1982). Detailed mapping of ascorbate distribution in rat brain. *Neuroscience Letters, 28*, 169–174.

Morgane, P. J., Austin-LaFrance, R., Bronzino, J., Tonkiss, J., Diaz-Cintra, S., Cintra, L. et al. (1993). Prenatal malnutrition and development of the brain. *Neuroscience and Biobehavioral Reviews, 17*, 91–128.

Morris, R. (1984). Developments of a water-maze procedure for studying spatial learning in the rat. *Journal of Neuroscience Methods, 11*, 47–60.

Mosdol, A., Erens, B., & Brunner, E. J. (2008). Estimated prevalence and predictors of vitamin C deficiency within UK's low-income population. *Journal of Public Health, 30*, 456–460.

Mun, G. H., Kim, M. J., Lee, J. H., Kim, H. J., Chung, Y. H., Chung, Y. B. et al. (2006). Immunohistochemical study of the distribution of sodium-dependent vitamin C transporters in adult rat brain. *Journal of Neuroscience Research, 83,* 919–928.

Nishino, H. & Ito, A. (1986). Subcellular distribution of OM cytochrome b-mediated NADH-semidehydroascorbate reductase activity in rat liver. *Journal of Biochemistry, 100,* 1523–1531.

Nose, K. & Ohba, M. (1996). Functional activation of the egr-1 (early growth response-1) gene by hydrogen peroxide. *Biochemical Journal, 316* (Pt 2), 381–383.

Nunez, J. L., Alt, J. J., & McCarthy, M. M. (2003). A novel model for prenatal brain damage. II. Long-term deficits in hippocampal cell number and hippocampal-dependent behavior following neonatal GABAA receptor activation. *Experimental Neurology, 181,* 270–280.

Olton, S. & Samuelson, J. (1976). Remembrance of places passed: spatial memory in rats. *Journal of Experimental Psychology: Animal Behavior Processes, 2,* 97–116.

Omoi, N. O., Arai, M., Saito, M., Takatsu, H., Shibata, A., Fukuzawa, K. et al. (2006). Influence of oxidative stress on fusion of pre-synaptic plasma membranes of the rat brain with phosphatidyl choline liposomes, and protective effect of vitamin E. *Journal of Nutritional Science and Vitaminology, 52,* 248–255.

Ortega, R. M., Lopez-Sobaler, A. M., Quintas, M. E., Martinez, R. M., & Andres, P. (1998). The influence of smoking on vitamin C status during the third trimester of pregnancy and on vitamin C levels in maternal milk. *Journal of the American College of Nutrition, 17,* 379–384.

Pate, S. K., Lukert, B. P., & Kipp, D. E. (1996). Tissue vitamin C levels of guinea pig offspring are influenced by maternal vitamin C intake during pregnancy. *Journal of Nutritional Biochemistry, 7,* 524–528.

Pauli, J., Wilce, P., & Bedi, K. S. (1995). Spatial learning ability of rats following acute exposure to alcohol during early postnatal life. *Physiology and Behavior, 58,* 1013–1020.

Pratico, D. & Sung, S. (2004). Lipid peroxidation and oxidative imbalance: early functional events in Alzheimer's disease. *Journal of Alzheimers Disease, 6,* 171–175.

Pratico, D., Uryu, K., Leight, S., Trojanoswki, J. Q., & Lee, V. M. (2001). Increased lipid peroxidation precedes amyloid plaque formation in an animal model of Alzheimer amyloidosis. *Journal of Neuroscience, 21,* 4183–4187.

Qiu, S., Li, L., Weeber, E. J., & May, J. M. (2007). Ascorbate transport by primary cultured neurons and its role in neuronal function and protection against excitotoxicity. *Journal of Neuroscience Research, 85,* 1046–1056.

Quinn, J., Suh, J., Moore, M. M., Kaye, J., & Frei, B. (2003). Antioxidants in Alzheimer's disease-vitamin C delivery to a demanding brain. *Journal of Alzheimers Disease, 5,* 309–313.

Rabinovic, A. D., Lewis, D. A., & Hastings, T. G. (2000). Role of oxidative changes in the degeneration of dopamine terminals after injection of neurotoxic levels of dopamine. *Neuroscience, 101,* 67–76.

Ranade, S. C., Rose, A., Rao, M., Gallego, J., Gressens, P., & Mani, S. (2008). Different types of nutritional deficiencies affect different domains of spatial memory function checked in a radial arm maze. *Neuroscience, 152,* 859–866.

Rebec, G. V. & Pierce, R. C. (1994). A vitamin as neuromodulator: ascorbate release into the extracellular fluid of the brain regulates dopaminergic and glutamatergic transmission. *Progress in Neurobiology, 43,* 537–565.

Rice, M. E., Lee, E. J., & Choy, Y. (1995). High levels of ascorbic acid, not glutathione, in the CNS of anoxia-tolerant reptiles contrasted with levels in anoxia-intolerant species. *Journal of Neurochemistry, 64,* 1790–1799.

Rice, M. E. & Russo-Menna, I. (1998). Differential compartmentalization of brain ascorbate and glutathione between neurons and glia. *Neuroscience, 82,* 1213–1223.

Richardson, D. P., Byrnes, M. L., Brien, J. F., Reynolds, J. N., & Dringenberg, H. C. (2002). Impaired acquisition in the water maze and hippocampal long-term potentiation after chronic prenatal ethanol exposure in the guinea-pig. *European Journal of Neuroscience, 16,* 1593–1598.

Riviere, S., Birlouez-Aragon, I., Nourhashemi, F., & Vellas, B. (1998). Low plasma vitamin C in Alzheimer patients despite an adequate diet. *International Journal of Geriatric Psychiatry, 13,* 749–754.

Rose, R. C. (1993). Cerebral metabolism of oxidized ascorbate. *Brain Research, 628,* 49–55.

Sandstrom, M. I. & Rebec, G. V. (2007). Extracellular ascorbate modulates glutamate dynamics: role of behavioral activation. *BMC Neuroscience, 8,* doi: 10.1186/1471-2202-8-32

Saponjic, R. M., Mueller, K., Krug, D., & Kunko, P. M. (1994). The effects of haloperidol, scopolamine, and MK-801 on amphetamine-induced increases in ascorbic and uric acid as determined by voltammetry in vivo. *Pharmacology, Biochemistry and Behavior, 48,* 161–168.

Scaife, A. R., McNeill, G., Campbell, D. M., Martindale, S., Devereux, G., & Seaton, A. (2006). Maternal intake of antioxidant vitamins in pregnancy in relation to maternal and fetal plasma levels at delivery. *British Journal of Nutrition, 95,* 771–778.

Schenk, J. O., Miller, E., Gaddis, R., & Adams, R. N. (1982). Homeostatic control of ascorbate concentration in CNS extracellular fluid. *Brain Research, 253,* 353–356.

Sen, C. K. & Packer, L. (1996). Antioxidant and redox regulation of gene transcription. *FASEB Journal, 10,* 709–720.

Seregi, A., Schaefer, A., & Komlos, M. (1978). Protective role of brain ascorbic acid content against lipid peroxidation. *Experientia, 34,* 1056–1057.

Shin, D. M., Ahn, J. I., Lee, K. H., Lee, Y. S., & Lee, Y. S. (2004). Ascorbic acid responsive genes during neuronal differentiation of embryonic stem cells. *Neuroreport, 15,* 1959–1963.

Siesjö, B. K. (1978). Regional Metabolic Rates in the Brain. In *Brain energy metabolism* (pp. 131–150). Chichester: Wiley-Interscience.

Sinet, P. M., Heikkila, R. E., & Cohen, G. (1980). Hydrogen peroxide production by rat brain in vivo. *Journal of Neurochemistry, 34,* 1421–1428.

Sotiriou, S., Gispert, S., Cheng, J., Wang, Y., Chen, A., Hoogstraten-Miller, S. et al. (2002). Ascorbic-acid transporter Slc23a1 is essential for vitamin C transport into the brain and for perinatal survival. *Nature Medicine, 8,* 514–517.

Spencer, J. P., Jenner, P., Daniel, S. E., Lees, A. J., Marsden, D. C., & Halliwell, B. (1998). Conjugates of catecholamines with cysteine and GSH in Parkinson's disease: possible mechanisms of formation involving reactive oxygen species. *Journal of Neurochemistry, 71,* 2112–2122.

Squire, L. R., Stark, C. E., & Clark, R. E. (2004). The medial temporal lobe. *Annual Review of Neuroscience, 27,* 279–306.

Streeter, M. L. & Rosso, P. (1981). Transport mechanisms for ascorbic acid in the human placenta. *American Journal of Clinical Nutrition, 34,* 1706–1711.

Telang, S., Clem, A. L., Eaton, J. W., & Chesney, J. (2007). Depletion of ascorbic acid restricts angiogenesis and retards tumor growth in a mouse model. *Neoplasia, 9,* 47–56.

Terpstra, M., Torkelson, C., Emir, U., Hodges, J. S., & Raatz, S. (2010). Noninvasive quantification of human brain antioxidant concentrations after an intravenous bolus of vitamin C. *NMR in Biomedicine, 24,* doi: 10.1002/nbm.1619

Tomita, S., Ueno, M., Sakamoto, M., Kitahama, Y., Ueki, M., Maekawa, N. et al. (2003). Defective brain development in mice lacking the Hif-1alpha gene in neural cells. *Molecular and Cellular Biology, 23,* 6739–6749.

Tsukaguchi, H., Tokui, T., Mackenzie, B., Berger, U. V., Chen, X. Z., Wang, Y. et al. (1999). A family of mammalian Na+-dependent L-ascorbic acid transporters. *Nature, 399,* 70–75.

Tveden-Nyborg, P., Johansen, L. K., Raida, Z., Villumsen, C. K., Larsen, J. O., & Lykkesfeldt, J. (2009). Vitamin C deficiency in early postnatal life impairs spatial memory and reduces the number of hippocampal neurons in guinea pigs. *American Journal of Clinical Nutrition, 90,* 540–546.

Vera, J. C., Rivas, C. I., Fischbarg, J., & Golde, D. W. (1993). Mammalian facilitative hexose transporters mediate the transport of dehydroascorbic acid. *Nature, 364,* 79–82.

Villalpando, S., Montalvo-Velarde, I., Zambrano, N., Garcia-Guerra, A., Ramirez-Silva, C. I., Shamah-Levy, T. et al. (2003). Vitamins A, and C and folate status in Mexican children under 12 years and women 12–49 years: a probabilistic national survey. *Salud Publica de Mexico, 45* (Suppl. 4), S508–S519.

Vorhees, C. V. & Williams, M. T. (2006). Morris water maze: procedures for assessing spatial and related forms of learning and memory. *Nature Protocols, 1,* 848–858.

Walmsley, A. R., Batten, M. R., Lad, U., & Bulleid, N. J. (1999). Intracellular retention of procollagen within the endoplasmic reticulum is mediated by prolyl 4-hydroxylase. *Journal of Biological Chemistry, 274,* 14884–14892.

Witting, P. K., & Stocker, R. (2004). Ascorbic acid in atherosclerosis. In H.Asard, J. M. May, & N. Smirnoff (Eds.), *Vitamin C functions and biochemistry in animals and plants* (pp. 261–290). New York: BIOS Scientific Publishers.

Wrieden, W. L., Hannah, M. K., Bolton-Smith, C., Tavendale, R., Morrison, C., & Tunstall-Pedoe, H. (2000). Plasma vitamin C and food choice in the third Glasgow MONICA population survey. *Journal of Epidemiology and Community Health, 54,* 355–360.

Yoneyama, M., Kawada, K., Gotoh, Y., Shiba, T., & Ogita, K. (2010). Endogenous reactive oxygen species are essential for proliferation of neural stem/progenitor cells. *Neurochemistry International, 56,* 740–746.

Yoshikawa, K., Takahashi, S., Imamura, Y., Sado, Y., & Hayashi, T. (2001). Secretion of non-helical collagenous polypeptides of alpha1(IV) and alpha2(IV) chains upon depletion of ascorbate by cultured human cells. *Journal of Biochemistry, 129,* 929–936.

Zalani, S., Rajalakshmi, R., & Parekh, L. J. (1989). Ascorbic acid concentration of human fetal tissues in relation to fetal size and gestational age. *British Journal of Nutrition, 61,* 601–606.

CHAPTER 4

Exploring B Vitamins and Their Impact on Cognitive Function and Mood from Conception to Early Adulthood

Therese O'Sullivan

Chapter Overview

- This chapter details the sources, structure and function of vitamins B1, B3, B6, B9 and B12, as these B vitamins are suggested to have the strongest links with cognition and mood.
- The neurological functions of these B vitamins are discussed along with associated interrelationships.
- The role of B vitamins in cognition and mood is presented across the lifespan from conception to early adulthood.

Introduction

The water-soluble vitamins (vitamins B and C) were originally identified as a single factor, 'water soluble' B, to distinguish them from the 'fat soluble' A. Nowadays, the term *vitamin B complex* refers to a group of eight water-soluble vitamins: thiamine (B1); riboflavin (B2); niacin (B3); pantothenic acid (B5); pyridoxine (B6); biotin (B7); folic acid (B9) and cobamides (B12). These vitamins have many important roles in the body, including energy production from macronutrients, cell growth and nervous system function. Given the link with the nervous system, B vitamins are of particular interest in the field of neuroscience. This chapter will focus on vitamins B1, B3, B6, B9 and B12, as these are suggested to have the strongest links with cognition and mood. Application to the lifespan from conception to early adulthood will be examined, with Chapter 6 focusing on B vitamins in older adults and dementia.

Being water soluble, B vitamins are found in watery portions of foods and are absorbed directly into the bloodstream during digestion. Most are unable

Table 4.1 Recommended dietary intake of different B vitamins across the lifespan (per day)

Age range/status	Thiamine (mg)	Niacin (equiv mg)	Vitamin B6 (mg)	Folate (µg)	Vitamin B12 (µg)
0–6 months	0.2*	2*#	0.1*	65*	0.4*
7–12 months	0.3*	4*	0.3*	80*	0.5*
1–3 years	0.5	6	0.5	150	0.9
4–8 years	0.6	8	0.6	200	1.2
9–13 years	0.9	12	1.0	300	1.8
14–50 years (M/F)	1.2/1.1	16/14	1.3/1.2	400/400	2.4/2.4
51–70 years (M/F)	1.2/1.1	16/14	1.7/1.5	400/400	2.4/2.4
>70 years (M/F)	1.2/1.1	16/14	1.7/1.5	400/400	2.4/2.4
Pregnancy	1.4	18	1.9	600	2.6
Lactation	1.4	17	2.0	500	2.8

(M/F) = male/female, * = adequate intake, # = preformed niacin.
Source: Australian Government National Health and Medical Research Council: *Nutrient Reference Values for Australia and New Zealand*. Accessed March 2011 from http://www. nrv.gov.au/nutrients

to be used in the form in which they are absorbed and need to be converted or modified to an active form. As the body contains a high proportion of water, water-soluble vitamins circulate easily. Vitamin B6 is an exception as it is stored in muscle. Kidneys regulate levels of water-soluble vitamins and excrete excess through the urine.

The water-soluble nature of B vitamins means regular consumption is important to maintain optimal levels and avoid deficiency. Symptoms of deficiency may arise relatively quickly if intake is not sustained, particularly during the rapid growth phases of childhood. Generally, water-soluble vitamins need to be replenished every few days, although it is possible for some B vitamins to be stored in the body for longer periods of time. Folic acid stores can last longer than a few days, and the liver can hold several years' supply of B12 (Stampfer 2008). Requirements for B vitamins change across the lifespan (Table 4.1).

B vitamins are widely distributed across the different core food groups in relatively small amounts. Whole grains and metabolically active tissues such as meats, liver and yeast extracts are particularly good sources of B vitamins (Table 4.2).

Vitamin B1 – Thiamine

As the name suggests, B1 or thiamine was the first B vitamin to be identified. Lack of thiamine causes the deficiency disease beriberi, a complex illness with two forms: wet and dry. In wet beriberi, cardiovascular problems and swelling of the legs occur, while in dry beriberi, the peripheral nervous system is affected, particularly in the legs. If the person is alcoholic or has rapid weight

Table 4.2   Food sources of B vitamins

B vitamin	Rich or important dietary sources	Examples
Thiamine	Thiamine-enriched flour, liver and yeast	1 cup of fortified ready-to-eat breakfast cereal can provide 0.25 mg 2 slices of grain bread can provide 0.50 mg 50 g liver can provide 0.10 mg
Niacin	Wholegrain cereals, beef, pork, eggs and milk	100 g beef steak can provide 10 niacin equivalent mg 50 g liver can provide 10 niacin equivalent mg 2 slices of grain bread can provide 5 niacin equivalent mg
Vitamin B6	Organ meats (e.g. liver and kidney), muscle meats, breakfast cereals, nuts, whole grains, vegetables and fruits	85 g cooked chicken can provide 0.55 mg 1 baked potato with skin can provide 0.65 mg 1 banana can provide 0.40 mg 1 cup of baked beans can provide 0.35 mg
Folate	Fortified products (e.g. cereals and juice); vegetables, particularly green leafy vegetables; legumes; liver and oranges	50 g liver can provide 380 µg 1 cup of fortified ready-to-eat breakfast cereal can provide 120 µg 1 cup of cooked bok choy can provide 80 µg
Vitamin B12	Red meats, dairy products, certain algae and plants exposed to bacterial action or contaminated by soil or insects	50 g clams can provide 50 µg 50 g liver can provide 35 µg 6 raw oysters can provide 15 µg 1 cup of yoghurt can provide 1 µg 100 g pork chop can provide 0.9 µg

Sources: Xyris Software, *Foodworks 2009 Professional Edition* Version: 6.0.2539, using the AUSNUT 2007 Foods database, and Mahan and Escott-Stump (2008) *Krause's Food and Nutrition Therapy* 12th edn, Philadelphia: Elsevier.

loss, dry beriberi may also be accompanied by confusion, memory loss and brain damage. Early researchers noted that both forms of the disease were associated with rice-based diets. Brown rice is a source of thiamine through the thiamine-rich outer bran layer, but the content is greatly reduced when this layer is removed to create polished white rice (International Rice Research Institute 1979). Until the 1800s, rice was milled by hand. It was too difficult to remove all of the outer husk, meaning white as well as brown rice still contained the outer bran layer and, therefore, moderate amounts of thiamine. Milling was generally done on a daily basis as the intact grains were susceptible to rot and would attract insects and rodents in storage. Mechanised milling gave rice a longer shelf life, as removal of the outer skin and grain fats made the grain less likely to rot or become rancid. Interestingly, insects and rodents were not as keen on eating white rice, and this was seen as another bonus (Frankenburg 2009). Perhaps insects and rodents are smarter than we give them credit for?

As the industrial revolution improved access to white rice and grains, beriberi started to become a widespread problem. The connection between thiamine and health arose in the late 1800s, when Dutch military physician Christiaan Eijkman tried to create an animal model of the disease using chickens inoculated with material from beriberi patients. He became frustrated when both his control and test chickens became ill with beriberi. Then, to complicate things further, both groups suddenly became well again. After some investigation, Eijkman discovered that a new cook had changed the chicken feed from leftovers of the white rice from the officers' table to cheaper unmilled rice. Lucky for the chickens, the new cook had disapproved of the chickens' luxurious diet, believing that nice white military-grade rice should not be given to civilian chickens (Frankenburg 2009). Beriberi was eventually shown to be a deficiency disease which could be cured by consumption of thiamine-rich outer grain layers. Thiamine has since been added back in as a supplement to many processed grain food products.

Sources

Pork and yeast are concentrated sources of thiamine, but lower concentrations can be found in a wide variety of foods and can contribute significant amounts when consumed in large quantities. Fortified breads, cereals, pasta and whole grains tend to be good dietary sources.

Structure and function

Thiamine diphosphate is the predominant form of thiamine in living tissues. Also known as thiamine pyrophosphate or TPP, it acts as a coenzyme in the metabolism of carbohydrates and branched-chain amino acids. Thiamine triphosphate is involved in nerve transmission (Ball 2004). Although vitamin B1 is synthesised by the microflora that generally inhabit the large intestine, this exogenous vitamin does not appear to be available to the human host. Therefore, intake from the diet is required (Ball 2004).

Physiology

Dietary sources of thiamine, occurring mostly in the form of TPP, are hydrolysed to free thiamine in the intestinal lumen. Absorption of thiamine then takes place in the duodenum and proximal jejunum. After absorption, thiamine is transported through the portal blood to the liver. Both non-phosphorylated thiamine and thiamine monophosphate circulate in the bloodstream. Thiamine is then converted back to TPP, its coenzyme form, within the liver and in other body tissues (Ball 2004). If taken in high doses, only a small percentage of thiamine is absorbed, with elevated serum values resulting in prompt urinary excretion. Negligible amounts may be secreted in bile. Thiamine can also be lost through sweating (Bender & Bender 1997). Due to the role of the vitamin in metabolism, high calorie and high carbohydrate diets increase the demand for thiamine (Sechi & Serra 2007).

Vitamin B3 – Niacin

The introduction of corn from South America to the western world led to the discovery of niacin, vitamin B3. Along with a variety of treasures, Spanish conquistadors brought corn to Europe from the Aztecs and Incas in the 16th century. But the introduction of corn turned out to be both a blessing and a curse – it is a good source of energy and palatable, but is very low in available niacin compared with other staple grains. Niacin is present in corn, but not in a form that is available to the body. The Spanish didn't have the knowledge of the Aztecs and Incas, who grew corn with particular crops and prepared it with alkaline substances such as ashes or lime to free the niacin so it could be absorbed when eaten. This meant that although the Aztecs and Incas consumed a high corn diet, they were free of the niacin deficiency disease pellagra, which became a major problem in Europe and America when corn was introduced. Pellagra is characterised by photosensitive dermatitis, diarrhoea and dementia, and can ultimately lead to death. The pellagra epidemics in regions where corn was the dominant crop led to beliefs that it was caused by a germ or toxin in corn, and many thought it was an infectious disease like leprosy. It was also more common in poorer communities, where people ate poor diets high in corn but low in fresh produce. In the late 1920s, Dr Joseph Goldberger established that pellagra could be prevented by eating a balanced diet or a small amount of brewer's yeast, rich in vitamin B3. Niacin occurred naturally in yeast and seed husks but was also synthesised for use as a chemical in photography. Researcher Conrad Elvehjem found that pellagra could be cured through supplementing the diet with niacin, leading to the fortification of bread (Frankenburg 2009).

Vitamin B3 was originally termed nicotinic acid. After bread became fortified with nicotinic acid, the name was changed to niacin so that people would not think the bread contained nicotine (or perhaps that cigarettes contained vitamin B) (Frankenburg 2009).

Sources

Niacin differs from the other B vitamins in that humans can produce it from the essential amino acid tryptophan. The liver can synthesise 1 mg of niacin from 60 mg of tryptophan, although the efficiency varies depending on dietary and metabolic factors. The dietary requirement for niacin, therefore, tends to be lower with higher tryptophan intakes. However, if there is a shortage, use of tryptophan for protein synthesis takes priority over niacin formation. Niacin is still required in the diet but protein foods containing tryptophan can count towards 'niacin equivalents': for a given food, niacin equivalents are estimated as the amount of preformed niacin plus 1/60th of its tryptophan content. If the amount of tryptophan in a food is unknown, it can be assigned a value of 1 per cent of the protein content (Samman & Lyons-Wall 2003).

Due to the contribution of tryptophan to niacin formation, foods containing balanced protein such as red meat, poultry and liver are also important contributors to total niacin equivalent intake (Ball 2004). Cheese and eggs are not good sources of niacin but have a high niacin equivalent due to their tryptophan content. Legumes and peanut butter are good sources of niacin. Most fruit and vegetables can also be useful sources of niacin, along with wholegrain cereals, tea and coffee (Ball 2004). Corn, on the other hand, is not only low in available niacin, but also low in tryptophan. Most food composition tables give total niacin amounts, which includes niacin that is normally not bioavailable without hydrolysis treatment, such as preparation with lime as mentioned earlier. Therefore, values may be overestimated, for example with corn.

Structure and function

The term niacin is used to refer to nicotinic acid and nicotinamide. Niacin functions in many biological redox reactions in the form of coenzymes nicotinamide adenine dinucleotide (NAD) and nicotinamide adenine dinucleotide phosphate (NADP). NAD plays an important role in the transfer of electrons from metabolic intermediates into the electron-transport chain for the production of energy in the form of adenosine triphosphate (ATP). NADP in the reduced form is a reducing agent in biosynthetic reactions, protects against reactive oxygen species and is used in anabolic pathways, including synthesis of lipids and cholesterol. NAD also has a role in deoxyribonucleic acid (DNA) repair (Ball 2004).

Physiology

Most available niacin in the diet is in the form of NAD and NADP, with the digestive process resulting in nicotinamide. Nicotinic acid and nicotinamide are converted into NAD in the liver. NAD not required for use by the liver is then broken down to free nicotinamide and released in the circulation, along with unmetabolised nicotinic acid. These are then synthesised into NAD and NADP once taken into body cells. There is a continuous turnover of NADP and NAD in the body and little is stored. Any excess niacin is excreted in the urine, after being converted to methylated derivatives in the liver.

Vitamin B6 – Pyridoxine and Related Compounds

Hungarian physician Paul György named vitamin B6 in 1934, after discovering it was able to cure a type of scaly dermatitis (acrodynia) in rats. These rats were fed a diet without any of the B-vitamin complex but were supplemented with vitamins B1 and B2. This led to discovery of a 'rat acrodynia-preventive factor', designated as vitamin B6 (Ball 2004). In the mid-1900s, Esmond Snell discovered that vitamin B6 consisted of a group of three vitamers based on the pyridine ring – pyridoxine, pyridoxal and pyridoxamine – noting that each can occur in a phosphorylated form (PNP, PLP and PMP, respectively). All six types of vitamin B6 are considered to have approximately the same biological activity in humans.

Sources

Wheat bran, fish and sunflower seeds are good sources of vitamin B6. Apart from dietary sources, microflora in the large intestine synthesise vitamin B6. However, it is not known whether this is available to other body cells outside the intestine, and obtaining vitamin B6 from dietary sources is still thought to be required. In some plant foods, such as spinach, vitamin B6 exists as pyridoxine glucoside, which is less bioavailable than other forms (Lyons-Wall & Samman 2003).

Structure and function

Of the three vitamer forms, pyridoxine is the most stable. It is found mostly in plants, while pyridoxal and pyridoxamine are found mostly in animal tissue. Vitamin B6 plays an important role as a coenzyme in the metabolism of amino acids, glycogen and a class of lipids known as sphingolipids. Due to its importance in the metabolism of amino acids, the dietary requirement for vitamin B6 is linked to protein intake. PLP, synthesised in the liver, is the active form of vitamin B6, but all vitamers can be interconverted in the body. PLP plays an important role in the functioning of many enzymes, including as a coenzyme in haem biosynthesis, the synthesis of the neurotransmitter serotonin from tryptophan and for reactions in gluconeogenesis, where glucose is produced from amino acids. A deficiency of vitamin B6 can result in seborrhoeic dermatitis, microcytic anaemia, seizures, depression and confusion (National Academy of Sciences, Institute of Medicine & Food and Nutrition Board 1998), as well as an impaired immune system. However, a primary deficiency of this vitamin is rare in normal circumstances, as it is widely distributed in foods. Diets low in vitamin B6 are also likely to be of very poor quality and lack other B-group vitamins.

Physiology

Phosphatase-mediated hydrolysis of B6 occurs in the gut and the non-phosphorylated form of B6 is absorbed mainly in the jejunum. Vitamin B6 is generally well absorbed, even in very large doses. Pyridoxine glucoside is absorbed less effectively but can be absorbed intact and hydrolysed in a variety of different tissues. Absorbed B6 in the non-phosphorylated form quickly moves to the liver where it is metabolised. PN, PL and PM are converted back to the phosphorylated form. PLP is bound to proteins and stored in muscle tissue. This protects it from the action of phosphatases, and the capacity for protein binding limits accumulation of PLP by tissues, even at high doses. During periods of starvation, vitamin B6 in muscle glycogen phosphorylase is released to enable amino acid catabolising reactions for gluconeogenesis. This allows the release of glucose in a non-carbohydrate environment and helps maintain blood glucose concentrations within a normal range. This is particularly useful for the brain, which relies on glucose as a preferred source of fuel. In the brain, vitamin B6 concentrations are kept relatively constant, even in times of severe deficiency. This is due to regulated entry and exit across the blood–cerebrospinal fluid barrier (Ball 2004).

Products of vitamin B6 metabolism are excreted mostly through the urine, with a small amount of excretion through faeces.

Vitamin B9 – Folate

Vitamins B9 and B12 work together to produce red blood cells, and were both discovered through the study of haematology. English physician Lucy Wills was working in India with pregnant women suffering from megaloblastic anaemia when she suspected there may be a dietary cause. The women most likely to be anaemic were the poorest, who were very limited in their food choices. Megaloblastic anaemia is characterised by large immature red blood cells and can lead to heart failure and death. While doing laboratory work with anaemic monkeys, Lucy persuaded one monkey to eat Marmite, a yeast extract. The monkey's anaemia was cured, and she then went on to successfully treat many women with the same method. Her work contributed to the identification of folate as a vitamin needed to make red blood cells (Frankenburg 2009).

Sources

Folate takes its name from the Latin word for foliage, due to the high amounts found in green leafy vegetables. Other good sources of folate include liver, wheat bran and some processed foods fortified with folate, such as breakfast cereals.

Structure and function

Folate is a generic term referring to all derivatives of pteroic acid that show vitamin activity in humans. Folic acid is the most stable form of folate. Although this form occurs rarely in food in nature, it is used in food fortification and vitamin supplements. Folate can also be synthesised by bacteria in the large intestine. Folate functions as a coenzyme in the metabolism of nucleic and amino acids, and is involved in many reactions including DNA synthesis and remethylation of homocysteine. DNA synthesis is required for normal cell division and is therefore especially important during periods of rapid growth. Folate is particularly important for cells with high turnover, such as red blood cells. Deficiency hinders erythropoiesis and leads to megaloblastic anaemia. For the same reason, increased folate intake at the start of pregnancy can reduce the incidence of neural tube defects in the developing foetus by 75 per cent (Bower 1996). Folate remethylation of homocysteine recycles it into methionine. Elevated plasma homocysteine is known as an early risk factor for atherosclerosis.

The role of folate intake in healthy DNA, decreasing the risk of mutations and cancer, is supported by observational studies showing that increased intakes lead to decreases in colon cancer risk. However, folate supplements have also been associated with an increase in the recurrence of precursor cancer polyps, suggesting that although folate may decrease the risk of new cancers, it may promote growth of tumours already present (Stampfer 2008). The anti-cancer drug Methotrexate is a folate antagonist. In folate deficiency, plasma homocysteine increases and, because DNA cannot be synthesised, cells cannot divide (Christopherson 1996). This begs the question of whether folate fortification in foods is necessarily beneficial for everyone in the population.

Physiology

Food sources of folate provide the polyglutamate form of the vitamin, which first must be hydrolysed to free folic acid in the form of monoglutamate in the small intestine before absorption. This form is used for transport through the circulatory system, while the polyglutamate form is used in storage and as a coenzyme. The synthetic folic acid and folate synthesised by intestinal bacteria is in monoglutamate form. This type is absorbed more efficiently, leading to the concept of dietary folate equivalents (similar to niacin), where 1 microgram of food folate is equivalent to 0.6 micrograms from fortified food or supplement (Groff & Gropper 2000). Cellular transport of folate is controlled by membrane carriers or folate binding protein systems. If not bound to protein, circulating free folate is filtered through the kidneys and reabsorbed into the bloodstream. The body is proficient in conserving dietary folate, except in instances of malabsorption, and not much folate is lost through urinary or faecal excretion (Ball 2004).

Vitamin B12 – Cobamides

The discovery of vitamin B12 originated in studies of pernicious anaemia. Symptoms of pernicious anaemia are megaloblastic anaemia, similar to that seen in folate deficiency but, with the B12 deficiency, an irreversible degeneration of the nervous system could also observed. This degeneration includes symptoms such as fatigue, cognitive impairment and personality change. Both pernicious and megaloblastic anaemia result from impaired DNA synthesis causing abnormal maturation of cells. Pernicious anaemia was initially thought to be incurable. However, in 1926, Harvard physicians George Minot and William Murphy found that feeding patients large amounts of liver alleviated symptoms of the disease. The active substance in the liver was named vitamin B12, but it was present in such small amounts that full characterisation of the vitamin did not take place until the 1960s, with the Nobel-prize winning work of England's Dorothy Hodgkin and colleagues (Mathews et al. 2000).

Pernicious anaemia is not a deficiency disease like other illnesses previously mentioned in this chapter. It is caused by deficiency of a glycoprotein required for the absorption of vitamin B12 in the intestine, meaning that there can be enough vitamin B12 in the diet, but it is not absorbed properly. The uncomplexed vitamin can still be absorbed, but so weakly that massive doses (such as those supplied by large amounts of liver) are required to cure or prevent the disease (Mathews et al. 2000).

Sources

Vitamin B12 is found widely in animal foods, although absorption of vitamin B12 from eggs seems to be poor relative to other animal food products (Watanabe 2007). Vitamin B12 is synthesised almost exclusively by microorganisms. Dietary sources are ingested microorganisms (e.g. through oysters which are filter feeders, or meat of ruminant animals that obtain vitamin B12 from bacteria in the lumen) and animal products that acquire vitamin B12 from bacterial fermentation, such as cheese (Lyons-Wall & Samman 2003). Vitamin B12 can occur in plant foods, such as mushrooms, due to contamination, or potentially in legumes through nitrogen-fixing bacteria in root nodules (Lyons-Wall & Samman 2003). Bioavailability of the vitamin decreases with increasing intake of vitamin B12 in a meal, due to saturation of the intrinsic factor-mediated intestinal absorption system (Watanabe 2007).

There has long been speculation about whether or not mushrooms are a good source of vitamin B12, a vitamin normally associated with animal foods. Analyses showing appreciable amounts of vitamin B12 in mushrooms were initially used by the mushroom industry to promote mushrooms as 'meat for vegetarians'. This was disputed by some nutrition professionals who believed that only animal sources can provide vitamin B12 (Australian Mushroom Growers' Association 2009). It was thought that the vitamin B12 found in the initial analyses was coming from the manure the mushrooms were grown in, rather than from the mushrooms themselves. In 2009, researchers confirmed that mushrooms do have bioavailable B12 on both the surface and in the flesh (Koyyalamudi et al. 2009). As the majority of vitamin B12 is found on the surface of the cup, peeling the skin can result in a loss of this vitamin. This also suggests that the vitamin B12 is predominantly bacteria-derived.

Structure and function

Vitamin B12 compounds have a corrin ring with a central cobalt atom, hence the name 'cobalamins' or 'cobamides'. The colbalt gives the vitamin a bright red colour. Cyanocobalamin is the most air-stable form of vitamin B12 and is easy to crystallise and purify, making it a widely used form in the food industry. Cyanocobalamin is not directly used in the body but is converted to coenzymatically active forms of vitamin B12, including methylcobalamin. Vitamin B12 is involved in many processes, including neurological function, haemoglobin synthesis and DNA synthesis. In the form of methylcobalamin, vitamin B12 functions as a cofactor for methionine synthase. This enzyme helps convert homocysteine to methionine, which is required for the formation of S-adenosylmethionine. S-adenosylmethionine is a methyl donor for many different substrates, including hormones, proteins, lipids and DNA.

Physiology

Absorption of vitamin B12 is unique as it depends on intrinsic factor, a glycoprotein produced by the stomach. Both the stomach and the ileum must be functioning properly for effective absorption of the vitamin. Vitamin B12 is bound to protein in food, but is released by the activity of gastric acid and pepsin in the stomach. After vitamin B12 is released from protein in food through the stomach, the free form of the vitamin binds with intrinsic factor. This partnership helps vitamin B12 to escape hydrolytic attack from intestinal enzymes. It then travels to the ileum, where intrinsic factor receptors facilitate absorption of vitamin B12 (Lyons-Wall & Samman 2003). Absorption of vitamin B12 decreases drastically when the capacity of intrinsic factor is exceeded.

Homocysteine is an intermediate in the methylation cycle which uses vitamins B6, B9 and B12. The reaction where folate donates a methyl group to homocysteine to form methionine is dependent on vitamin B12, while vitamin B6 is required for an alternative reaction – changing homocysteine to cystathionine (Lyons-Wall & Samman 2003).

Absorption of vitamin B12 from food tends to decrease with age, due to a lack of stomach acid to release B12 from food into the free form. Use of certain drugs, such as antacids, can also reduce gastric secretions. Vitamin B12 found in supplements and fortified foods is already in the free form and therefore does not require stomach acid for digestion. Deficiencies can therefore be avoided in older people through sufficient intake from these sources (Stampfer 2008). Injections of vitamin B12 can also be administered, bypassing absorption barriers. Vitamin B12 is stored mainly in the liver, with additional storage in muscle. It is well retained in the body due to protein binding, and losses in urine are small. Good storage capacity and minimal excretion, combined with a long biological half life ($\sim$400 days), means that vitamin B12 deficiency takes years to develop (Lyons-Wall & Samman 2003). As a result, deficiency in developed countries is generally rare, although people who are vegans, pregnant women who are vegetarian (Koebnick et al. 2004) and those who have had gastrointestinal surgery or disorders are at particular risk.

Neurophysiological functions of B vitamins

B-vitamin intake is a modifiable factor that can have neurophysiological effects, particularly in the areas of neurotransmitter function (Gomez-Pinilla 2008) and plasma homocysteine levels (Troen et al. 2008). Effective cognitive function involves the participation of many different neurotransmitter systems across a variety of brain areas, including the frontal cortex, basal forebrain cholinergic and midbrain nuclei, the thalamus and parts of the limbic system, such as the hippocampus and amygdala (Levin 2006). Neurotransmitters are the chemicals used to communicate between these different brain areas. Neurotransmitters requiring B vitamins for biosynthesis include epinephrine, norepinephrine, serotonin and gamma-aminobutyric acid. Decreases or increases in the levels of neurotransmitters, or blockage or activation of their receptors, may alter cognitive functions such as learning and memory (Zarrindast 2006).

With the use of a mouse model, dietary deficiencies of folate, vitamin B12 and vitamin B6 intake have been shown to result in cognitive dysfunction related to high homocysteine (Troen et al. 2008). After ten weeks on a vitamin B-deficient diet, mice show significantly worse spatial learning and memory compared with control mice on a normal diet (no significant differences were observed in psychomotor performance). Vitamin B-deficient mice had higher levels of plasma homocysteine, as the conversion of homocysteine to methionine or cystathionine was limited without B vitamins, causing the homocysteine to accumulate. In the brains of vitamin B-deficient mice, brain capillaries were of a shorter length, particularly in the hippocampus. Despite these differences, the overall brain anatomy between the two groups was similar, without signs of cellular proliferation that would usually occur with neurological degeneration. This suggests that B vitamins may play a role in cognitive function through microvascular circulation.

Cognitive difficulties and psychological symptoms, such as depression, may appear earlier than symptoms of outright B-vitamin deficiency as described in the previous section (Benton & Donohoe 1999). This section gives more detail on the links between B vitamins and these neurophysiological functions.

Thiamine

The brain accounts for approximately 2 per cent of total body weight in the average adult human. But despite its relatively small size, the brain is very metabolically active. To perform its functions, the brain requires about 20 per cent of the body's oxygen, and therefore uses about 20 per cent of the energy obtained from food consumed (Clark & Sokoloff 1999). Despite widely varying brain activity levels, this high rate of metabolism stays relatively constant. No other organ is as sensitive to changes in its energy supply as the brain. But the high energy needs of the brain increases its vulnerability – if the brain's energy supply is cut off, even for a short time, permanent brain damage can result. Effective energy transfer from food to the brain is therefore of primary importance to brain function.

The preferred fuel for the brain is glucose, as an ATP substrate. As carbohydrate foods supply glucose, proper functioning of the brain and nervous systems therefore relies upon normal carbohydrate metabolism (see Chapter 8). This depends upon a dietary supply of vitamin B1 for conversion to TPP. TPP is required as a coenzyme in the main energy-yielding pathway of carbohydrate metabolism, the tricarboxylic acid (TCA) cycle. The TCA cycle converts glucose into carbon dioxide and water to generate energy as part of cellular respiration. However, the brain is not beaten without TPP – it finds alternative pathways to meet its energy requirements. But although the brain manages, nerve impulse transmission at synapses is impaired. Symptoms of subclinical thiamine deficiency can be difficult to diagnose as they may be quite general, such as headaches, fatigue and irritability (Sechi & Serra 2007).

Symptoms become more pronounced if the deficiency continues. Rats fed a vitamin B1-deficient diet for several weeks develop neurological abnormalities of posture and equilibrium, which can be reversed with administration of thiamine (Ball 2004). With severe deprivation of thiamine, mental changes occur. This includes the loss of emotional control, paranoid trends, manic or depressive episodes and confusion (Ball 2004), in addition to the deficiency diseases previously discussed. The neuropathy symptoms observed in dry beriberi are due to the thinning of myelin in peripheral nerves, progressing to fibre tract deterioration (Ball 2004).

Wernicke–Korsakoff syndrome is a term used to describe two related vitamin B1-deficiency disorders. The Wernicke's encephalopathy part of the syndrome is a degenerative brain disorder which has an acute onset and is characterised by mental status changes, vision impairment and motor problems, such as ataxia – similar to the signs of drunkenness, which can hamper diagnosis in an emergency setting. These symptoms predominantly result from an involvement of thalamic or mamillary bodies and range from a confused state to an impaired awareness of the current situation (Sechi & Serra 2007). Coma and death can occur if the condition is left untreated.

The Korsakoff part of the syndrome refers to a memory disorder leaving sufferers with an impaired ability to acquire new information or new memories. However, implicit learning is retained, so patients with this syndrome are able to learn new motor skills or develop conditioned reactions to stimuli (Sechi & Serra 2007). Other cognitive functions are also generally preserved in most patients. Although Wernicke's and Korsakoff's have different symptoms, they are generally considered to be different stages of the same disorder, with Wernicke's encephalopathy representing an acute phase and Korsakoff's syndrome representing the chronic phase.

As the body's stores of thiamine are only generally sufficient for up to 18 days, thiamine deficiency can lead to brain lesions in under a month, particularly in regions of the brain with a high thiamine content and turnover (Sechi & Serra 2007).

Niacin

As mentioned in the background to this vitamin, niacin deficiency causes pellagra, which includes the mental symptom of dementia. In the brain, a deficiency of niacin sees lower levels of NAD and NADP coenzymes, implicated in energy formation and DNA synthesis. Neuropathological changes seen in pellagra are restricted to neurons, and chromatolysis is the characteristic finding (Bémeur et al. 2011). Chromatolysis refers to the disintegration of the chromophil substance in a nerve cell (the histologic element that stains readily), which can occur after the cell becomes damaged or exhausted. The severe niacin deficiency associated with dementia may be also partly due to inadequate supply of tryptophan to the brain for serotonin synthesis (Samman & Lyons-Wall 2003).

Vitamin B6

Neurological deficits, such as depression and confusion, resulting from pyridoxine deficiency can largely be explained by decreased enzyme function affecting neurotransmitter function (Bémeur et al. 2011). Neurotransmitters affected include gamma-aminobutyric acid, the chief inhibitory neurotransmitter in the central nervous system which regulates the excitability of neurons, and serotonin, a neurotransmitter which contributes to feelings of wellbeing. Vitamin B6 deficiency in the rat results in decreased dendritic branching and reduced numbers of synapses and myelinated axons (Gerster 1996). In people, the effect of this can be seen with the use of some commonly used drugs which happen to be pyridoxine antagonists (such as the antibiotic Cycloserine, and vasodilator Hydralazine). Use of these drugs can result in neurological problems, including peripheral neuropathy and seizures, but co-administration of pyridoxine reverses these side effects without affecting the drug efficacy (Bémeur et al. 2011). Atrophy of the cerebral cortex and white matter positron has been observed in people with pyridoxine deficiency (Bémeur et al. 2011). On the other hand, high doses of vitamin B6 (150–300 mg/kg/day for up to 12 weeks) has been shown to cause pathology in the central nervous system of rats, with axonal atrophy and degeneration (Xu et al. 1989).

Folate

Adequate levels of folate are essential for brain function, and folate deficiency and resultant increased homocysteine levels have been linked experimentally and epidemiologically with neurodegenerative conditions, including dementia. The biological mechanisms by which folate deficiency influences brain function are not clearly understood. One theory is that increased concentration of homocysteine in the blood may act as a vascular toxin (Miller 2004). However, higher plasma and erythrocyte folate concentrations have been shown to be associated with better cognitive performance regardless of homocysteine concentration (de Lau et al. 2007; Durga et al. 2006). The mechanism of cognitive decline with low folate may occur through a homocysteine-independent mechanism, such as DNA infidelity and mitochondrial decay (Durga et al. 2006). Another theory is that decreased intracellular concentration of the methyl donor S-adenosylmethionine in the brain contributes to the decline (Miller 2004). In mice fed a folate-deficient diet, increased homocysteine and altered brain levels of neurotransmitters were observed (Kronenberg et al. 2008). The generation of neurons in the hippocampus was also impaired. This resulted in cognitive dysfunction and an anxious, despair-related phenotype, aggravated in animals lacking a DNA repair enzyme (Kronenberg et al. 2008). In humans, abnormal neurotransmitter metabolism has been observed in folate-deficient psychiatric patients, and low levels of serotonin metabolites have been reported in patients suffering from folate-responsive neuropsychiatric disease (Benton & Donohoe 1999).

Vitamin B12

Serious vitamin B12 deficiency is associated with neurological changes as well as pernicious anaemia. Although treatment with vitamin B12 successfully resolves the anaemia, neurological consequences may be permanent (Bémeur et al. 2011). To diagnose vitamin B12 deficiency, a blood test is used to identify large, immature and dysfunctional red blood cells associated with megaloblastic anaemia, connected with low serum B12. Unfortunately, in up to a quarter of patients, the neurological symptoms are the only signs of vitamin B12 deficiency or occur before the anaemia (Bémeur et al. 2011), meaning vitamin B12 deficiency is frequently missed. Vitamin B12 is involved in the metabolism of the fatty acids needed to produce myelin, which is an insulating sheath around the neuronal axon (Benton 2008). The neuro-logical effects of vitamin B12 deficiency are related to subacute combined degeneration of the spinal cord. Distension of myelin sheaths gives a spongy appearance to white matter that is affected. The distension is followed by disin-tegration of myelin and loss of axons. These changes can be reversed by vitamin B12 administration, but only if caught early (Agamanolis 2006). The disease process mechanism is yet to be clarified, but may be related to increased expres-sion of neurotoxic tumour necrosis factor-alpha (TNF-alpha) and decreased expression of neurotrophic epidermal growth factor in cerebrospinal fluid (Scalabrino 2009; Scalabrino & Peracchi 2006).

Interrelationships

Vitamin B6, folate and vitamin B12 may have interwoven neurochemical mechanisms by which these vitamins influence brain function through their role in methylation in the central nervous system (CNS). Folate, with vitamin B6 or vitamin B12 as a catalysing cofactor, may have a direct effect on the CNS via hypomethylation, inhibiting methionine synthesis and form-ing S-adenosylmethionine (SAM). SAM serves as the sole methyl donor for the CNS. Low SAM inhibits methylation reactions throughout the CNS, involving proteins, membrane phospholipids, DNA and the metabolism of neurotransmitters. Alternatively, there could be an indirect and possibly longer-term effect of vitamin B6, folate and vitamin B12 on the functioning of the brain via the cerebrovasculature through high levels of homocysteine, which is associated with detrimental effects on vascular tissue (Bryan et al. 2004).

The relationship between folate and vitamin B12 is an interesting one. In people with a normal vitamin B12 status, high serum folate has been shown to be associated with protection from cognitive impairment. But in people with low vitamin B12 status, high serum folate was associated with increased cognitive impairment (Morris et al. 2007). Further, an increase in the severity of vitamin B12-related neurologic impairment has been observed with rising serum folate concentrations (Savage et al. 1994). The idea that high folic acid intake exacerbates neurologic and neuropsychiatric effects of vitamin B12 deficiency is controversial (Smith 2007). While plausible mech-anisms could include effects of unmetabolised folic acid in the circulation or a reduction in the remethylation of homocysteine to methionine (leading to high homocysteine and low SAM), some researchers maintain that there is no reliable evidence that folic acid is a neurological poison (Dickinson 1995).

Application to cognition and mood

The information presented in this section is broken down across the lifespan. The period from gestation until the first two years of life is thought to be critical for rapid brain growth, and brain development at this time may be particularly sensitive to vitamin deficiency. But certain brain areas continue to develop past this period; for example, spurts of frontal lobe development occur from seven to nine years and in the mid-teenage years (Bryan et al. 2004). This chapter will not cover application to older adults, as this is covered in Chapter 6.

Pregnancy

The importance of folate during the period of early pregnancy was highlighted at the start of this chapter with respect to anaemic pregnant women in India. The importance of folate in early conception for prevention of neural tube defects is well known. Low maternal folate status during pregnancy and lactation is still a significant cause of morbidity for the mother in some areas. Although the folate status of the developing foetus tends to be protected at the expense of the mother, there is evidence to suggest that inadequate maternal folate status during pregnancy may lead to low infant birthweight and potential adverse developmental outcomes (Molloy et al. 2008).

Increasing evidence suggests that the beneficial effect of folate may be due to improved function of methionine synthase, the vitamin B12-dependent enzyme responsible for converting homocysteine to methionine (Refsum 2001). For women who follow vegetarian diets and are likely to be vitamin B12 deficient (Koebnick et al. 2004), increased folate intake without adjunct vitamin B12 supplementation may not offer much protection against birth defects and later developmental problems.

Infancy

Mothers who used pyridoxine supplements to counteract nausea during pregnancy have been shown to give birth to babies with higher pyridoxine requirements. As a result, pyridoxine dependency has been reported in some newborns with seizures. Pyridoxine treatment reverses the seizure activity in these infants (Bémeur et al. 2011). The importance of vitamin B6 was demonstrated in the 1950s, when an occurrence of seizures in infants was traced back to an unfortified liquid milk-based canned formula. The formula had undergone autoclaving during manufacturing and the high temperature destroyed the B vitamins (Coursin 1954). Work in animal models suggests that vitamin B6 deficiency during gestation and lactation also alters the functioning of the N-methyl-Daspartate receptor (NMDA), thought to play an important role in memory and learning (Guilarte 1993).

Vitamin B12 deficiency has been reported in infants born to vegetarian mothers. In the Netherlands, a mixed longitudinal cohort study investigated 50 infants who were being brought up with macrobiotic diets. A macrobiotic diet involves avoiding animal products and highly processed or refined foods, focusing instead on eating predominantly whole, unprocessed grains, with beans, vegetables, seeds and nuts. It is high in some B vitamins, but not vitamin B12, which is found in animal sources. While this type of diet can offer many health and environmental benefits, breast milk from macrobiotic mothers in the study was found to contain less vitamin B12 when compared with mothers on omnivorous diets (Dagnelie & van Staveren 1994). The infants in macrobiotic households showed significantly lower cobalamin concentrations (Dagnelie et al. 1989) and impaired psychomotor functioning (Dagnelie & van Staveren 1994) when compared with 57 matched omnivorous control infants. As a result of these and other findings, dietary recommendations were made and the macrobiotic families in the study gradually adopted lacto-vegetarian, lacto-ovo-vegetarian, or even omnivorous diets. Deficits in vitamin B12 (cobalamin) have negative consequences on the developing brain during infancy (Black 2008), thought to be due to disruptions to myelination and inflammatory processes. The long-term consequences of vitamin B12 deficiency in infancy were reviewed by Graham et al. (1992), who reported a consistent clinical pattern of failure to thrive and irritability, with marked developmental regression and poor brain growth.

Childhood and adolescence

A well-designed study by Ruth Harrell and colleagues (1946) examined the effects of thiamine supplementation in 120 children living in an orphanage. The diets of the children prior to intervention met the recommended intake for thiamine. The double-blind placebo-controlled study used matched pairs and involved the intervention group taking a 2 mg thiamine tablet daily for a year, in addition to their usual diet. Children who took the vitamin showed enhanced cognitive functioning compared with children who did not, performing better on average across all the tests administered. Tests included visual acuity, game skills, memory tasks and intelligence tests. These differences were significant ($P < 0.05$) for tasks involving memorising new material, remembering word-number pairs, remembering Morse code, code substitution and an intelligence test. This suggests that the thiamine had effects on working memory, encoding and cognitive skills. In the supplemented children, vision was also significantly better, and they were taller on average compared with the non-supplemented children. In support of thiamine impacting on cognition and mood, behavioural problems in adolescents consuming thiamine-deficient diets have been demonstrated to improve with treatment with thiamine alone, even in cases where drugs or psychotherapy did not help (Benton et al. 1997).

A relationship between vitamin B6 deficiency and autism in children (characterised by deficits in social interaction, communication and stereotypical behaviour patterns) has been suggested in some studies. However, a Cochrane review concluded that due to a lack of well-designed trials, the use of vitamin B6 for improving the behaviour of individuals with autism cannot currently be supported (Nye & Brice 2005). Due to large doses of vitamin B6 resulting in potential CNS pathology, there are insufficient data to recommend long-term use of pyridoxine in children at this stage (Bémeur et al. 2011).

Vitamin B deficiencies when young may result in cognitive problems in later life that are not reversed by subsequent adequate intake. The macrobiotic infants from the Netherlands study were reassessed in early adolescence to check whether a cognitive difference still existed compared with controls (Louwman et al. 2000). Following recommendations from the initial study, their diets had changed to include at least some animal products by a mean age of six years. Although the mean dietary intakes of the previously macrobiotic children were close to the recommended amount of vitamin B12 and the children were not showing any symptoms of B12 deficiency, their intake and B12 status was still lower on average than the control group in adolescence. As with the infant study, the control children performed better in most psychological tests than did the previously macrobiotic children with low or normal B12 status. Significant differences were observed in tests measuring fluid intelligence, spatial ability and short-term memory. The association between B12 status and fluid intelligence was stronger within the subgroup of macrobiotic children, suggesting that the observed relationship was not due to related characteristics within the group characteristics. Fluid intelligence involves the ability to reason and learn, and the capacity to think laterally and solve complex problems. Improvements in this area as a result of optimal vitamin B intake may therefore have important consequences for individual functioning (Louwman et al. 2000). Results of this follow-up study suggest that early vitamin B12 deficiency in infancy may lead to impaired cognitive performance in adolescence, even if normal B12 status is achieved over this period.

Depression in adolescence is associated with a variety of negative outcomes and considerable risk of morbidity and mortality, and this risk can carry through across the lifespan (Bamber et al. 2007). Depression during the adolescent period can manifest through irritability, a loss of interest in normal activities and mood shifts that can interrupt schoolwork or the enjoyment of time with friends (American Psychiatric Association 1994). The duration of a depressive episode can range from as short as a fortnight to as long as many years, with differences in severity from mild (only a modest deviation from normal functioning) to severe (requiring full-time care) (Bamber et al. 2007). A cross-sectional Japanese study suggests that a higher intake of dietary B vitamins, particularly folate and vitamin B6, is independently associated with a lower prevalence of depressive symptoms in early adolescence (Murakami et al. 2010). Low folate and low vitamin B12 status have been found in studies of depressive patients, and an association between depression and low levels of the two vitamins is found in studies of the general population (Coppen & Bolander-Gouaille 2005).

Adulthood

Thiamine is thought to be the vitamin with the most systematic data alluding to an association with mood, with increased irritability and depression among the earliest signs of deficiency (Benton & Donohoe 1999). It may take only a few weeks to become deficient in vitamin B1 because the body has only small stores. Benton and colleagues (Benton et al. 1997) investigated thiamine supplementation on mood in young, healthy and well-nourished women, who had adequate thiamine status according to traditional criterion. A placebo or 50 mg of thiamine was taken daily. Although mood did not differ before the intervention, the women who took thiamine reported a significant improvement in mood after two months. Similarly, in young adult males, poor thiamine status was associated with introversion, decreased self-confidence and poorer mood. Using a double-blind study procedure, supplementation with 3 mg thiamine resulted in increased sociability and sensitivity after two months (Heseker et al. 1990).

In a systematic review of randomised controlled trials investigating the effect of vitamin B6 in the treatment of premenstrual syndrome, Wyatt et al. (1999) found that doses of vitamin B6 up to 100 mg/day are likely to be of benefit in treating symptoms, including premenstrual depression. However, the authors noted that the studies were of insufficient quality to draw definitive conclusions. Williams and colleagues (2005) concurred with this finding in their subsequent review of vitamin B6 and depression – although a meaningful treatment effect of vitamin B6 for depression in general was not apparent, a consistent improvement in pre-menopausal syndrome or depression related to the contraceptive pill was observed. This suggests that vitamin B6 as a therapy may be effective for hormone-related depression in women.

It has been noted that even in studies where only a small percentage of subjects showed vitamin B12 or folate deficiencies, those with below-median levels still showed poorer scores on tests of cognitive functioning (Benton & Donohoe 1999). To test whether supplementation with vitamin B6, folate or vitamin B12 influences cognitive performance and mood among healthy women across the lifespan, Byran and colleagues (Bryan et al. 2002) investigated the effects of a five-week supplementation regime in 211 younger (20–30 years), middle-aged (45–55 years) and older women (65–92 years). The women took 75 mg of vitamin B6, 750 g of folate, 15 g of vitamin B12 or a placebo daily. Results showed few effects of B-vitamin supplementation on cognitive performance, mainly in the area of memory, and none on mood. The authors suggest that longer-term trials may be needed, because they did find associations between the usual dietary intake of B vitamins and cognitive performance, particularly for the younger group, of which almost a third of women were under the recommended intake for folate and 20 per cent did not meet requirements for vitamin B12. Folate intake was generally positively associated with performance on a variety of measures of cognition, whereas vitamin B12 and vitamin B6 intake was positively associated with memory performance. Interestingly, researcher Victor Herbert, who gave himself a self-inflicted folate deficiency, noted memory impairment as one of the prominent symptoms (Herbert 1962). This was resolved with folate supplementation. On the other hand, vitamin B6 was the only vitamin from B6, folate and B12 to be positively associated with cognitive performance after adjustment for cardiovascular disease (CVD) risk factors and CVD (Elias et al. 2006).

With less than a year of folate supplementation, Botez and colleagues (1984) demonstrated that depressed patients with initial low folate levels were less easily fatigued and less distracted. Intelligence scores, particularly in non-verbal skills, also improved.

Both low folate and low vitamin B12 status, along with increased plasma homocysteine, have been found in studies of depressive patients, while in the general population, an association between depression and low levels of the two vitamins has been observed (Coppen & Bolander-Gouaille 2005). A Cochrane review investigating folate for depressive disorders used two randomised trials to conclude that folate may have a potential role as a supplement to other treatment for depression (Taylor et al. 2009). In one trial of folate used alone as a treatment compared with an antidepressant drug, no significant difference was found. However, the study had limitations, including being underpowered. Through a review of folate and vitamin B12 in depression, Coppen & Bolander-Gouaille (2005) suggest that oral doses of both folic acid (800 µg daily) and vitamin B12 (1 mg daily) should be tried to improve treatment outcomes in depression. However, they add that larger double-blind randomised intervention studies are required to determine to what extent these vitamins can improve depression management.

Special Cases

Alcoholism

Alcohol displaces more nutritious foods from the diet, adds carbohydrates that require thiamine for metabolism, impairs the absorption of vitamins and lowers the capacity of the liver to store vitamins (Agamanolis 2006; Sechi & Serra 2007). This places alcoholics at high risk of vitamin B1 deficiency and subsequent Wernicke–Korsakoff syndrome. Long-term alcohol consumption does not result in Wernicke's encephalopathy if the dietary intake of thiamine is sufficient to avoid deficiency (Sechi & Serra 2007), so vitamin B1 supplementation is commonly considered to be mandatory for patients at risk. Wernicke–Korsakoff syndrome results in impaired memory, with additional symptoms of ataxia (inability to coordinate movements) and weakness of the eye muscles.

In rats, both long-term alcohol consumption and pyrithiamine-induced blockage of vitamin B1 uptake significantly impair the ability to learn, as tested by an active two-way avoidance task and a spatial reversal task (Irle & Markowitsch 1983). Compared with rats in the control group, rats in both the alcohol and blocked vitamin B1 groups displayed noticeable brain damage, particularly to hippocampal and cerebellar regions. This suggests that the cognitive changes seen with alcoholism are likely to be due (at least in part) to the resulting vitamin B1 deficiency. Vitamin B1 treatment has been shown to improve results on a test of working memory in subjects detoxifying from alcohol in a randomised, double-blind, multi-dose study (Ambrose et al. 2001). Results were evident after a test period of only two days, and a therapeutic relationship between dose and working memory performance was indicated.

Although Wernicke–Korsakoff syndrome is usually seen in people with alcoholism, it has also been observed in people consuming an inadequate diet and in people with conditions resulting in persistent vomiting. As thiamine is required for carbohydrate metabolism, the syndrome can also result from intravenous glucose feeding without a thiamine supplement. Acute symptoms are similar to that of drunkenness, which can make diagnosis difficult, particularly for people suffering from alcoholism. Given the severity of the syndrome, it is considered good medical practice to treat all patients presenting with coma, stuporous state, hypothermia or hyperthermia of unknown nature, or tachycardia and intractable hypotension of unknown cause, with parenteral thiamine (after determining thiamine status), to avoid permanent neurological damage or death (Sechi & Serra 2007).

Oral Contraceptive Users

Increased amounts of female hormones, resulting from taking oral contraceptive pills, can lead to changes in general health and nutritional requirements. Most research has focused on the levels of these vitamins or minerals in the blood, but it is difficult to draw definite conclusions because these analyses are not always the most accurate or sensitive measure of changes in nutritional status (Anderson 2010). Reports concerning the interaction between oral contraceptives and vitamins indicate that vitamin B2, vitamin B6 and vitamin B12 serum levels are likely to be reduced. Clinical effects due to vitamin deficiency have been described for vitamin B6, including depression (Wynn 1975), and depression is often cited as the reason for discontinuing use of oral contraceptives (Kulkarni 2007). Vitamin B6 metabolism appears to be different in women who take oral contraceptives, and the form of vitamin B6 in the blood is also changed, which also occurs during pregnancy. Oral contraceptives will aggravate an existing vitamin B2 deficiency (Anderson 2010). However, in terms of cognition, no significant difference in psychometric test performance was observed when women were taking oral contraceptives compared with when they were not, in a crossover study of 20 women examined twice at four-week intervals (Silber et al. 1987). Similarly, no significant difference in neurocognitive function (as assessed by standard concussion test battery, including computerised assessment of neurocognitive function) was observed between females using the oral contraceptive pill and eumenorrheic females not using the pill (Mihalik et al. 2009). In a study investigating the effects of menstrual cycle phase and oral contraceptives on verbal memory, naturally cycling women showed no changes in verbal memory across the cycle, whereas women on oral contraceptives showed enhanced memory during the active pill phase (Mordecai et al. 2008). It may be that the types of B vitamins affected by the oral contraceptive pill play a larger role in mood and depression than with aspects of thinking, processing and comprehension.

Bariatric Weight-Loss Surgery

Obesity is a major ongoing nutritional issue in the western world. In adults, surgical approaches to treat obesity are gaining popularity. Commonly performed weight-loss surgery techniques include adjustable gastric banding and stomach resectioning. After these operations, the decreased stomach and small intestine size means that it is difficult to eat large amounts of food, the risk of vomiting increases and less food is absorbed. This then increases the risk of vitamin B deficiencies. Thiamine deficiency is generally considered to be the most serious in these cases, because body stores are small and deficiency can result in permanent brain damage. Wernicke encephalopathy following bariatric surgery generally occurs between 4 and 12 weeks after the operation, and young women who have vomiting episodes are at particular risk (Singh & Kumar 2007). However, in terms of cognitive outcomes, bariatric surgery also has the potential to improve many co-morbid medical conditions that may be related to cognitive dysfunction (Gunstad et al. 2011), such as hypertension (Knecht et al. 2008).

Drug–Nutrient Interactions

Drugs may influence the nutritional status of vitamins by affecting food intake, absorption, use or excretion. Table 4.3 gives examples of potential drug–nutrient interactions.

Conclusions and Implications

This chapter has described the existing literature about the effects of B vitamins on neurophysiological function, mood and cognition. Deficiencies in B-group vitamins – thiamine (B1), niacin (B3), pyridoxine and related compounds (B6), folate (B9) or cobamides (B12) – have the potential to cause neuropathological changes impacting on cognitive performance or mood. Neuropathological changes include brain and nerve degeneration and impaired abilities of neurotransmitters, potentially affecting aspects of cognitive function, such as memory, learning, spatial ability and fluid intelligence, and aspects of mood, such as irritability and depression.

Interrelationships are also evident between the vitamins, such as with folate and B12, where folate can appear to be beneficial or detrimental depending on

Table 4.3 Examples of potential drug–vitamin B interactions

Drug type	B vitamin effect
Anti-infective agents	
Antibacterials (Neomycin)	Malabsorption of vitamin B12, folate
Sulphamethoxazole	Inactivates pyridoxine (B6), decreases folate utilisation, decreases B-vitamin synthesis
Pain relief	
Aspirin, codeine, endone and others	Long-term use can lead to folate depletion
Motility agents	
Calcium carbonate, sodium bicarbonate, losec, proton pump inhibitor and others	Destroys vitamin B1, malabsorption of vitamin B12
Anti-hypertensives	
ACE inhibitors	Vitamin B6 antagonist
Aldomet, methyldopa	Increases requirement for vitamin B12, folate
Nitrate	Decreases folic acid absorption
Anti-angina	
Cardizem, Dilzem, others	Decreases folic acid absorption
Cholesterol lowering	
Bile acid sequestrants	Decreases vitamin B12 absorption
Diuretic	
Diamox, frusemide, others	Depletes vitamin B1
Biguanide	
Metformin	Malabsorption of vitamin B12
Anti-gout	
Allopurinol, pro-gout, others	Malabsorption of vitamin B12
Anti-convulsants	
Diazepam, clonazepam, others	Risk of vitamin B12 and folate deficiency
Neurological system	
Monoamine oxidase inhibitors, nardil, parnate	Risk of vitamin B6 deficiency
Anti-metabolites	
Methotrexate, flurouracil, others	Folate antagonist, decreases vitamin B12 absorption

Adapted from Stewart (2009).

B12 status. Chapter 6 on older adults will go on to consider the role of these vitamins further in the area of dementia.

In practice, the neurocognitive benefits observed with B vitamins raise the question of whether to try to obtain these vitamins from food or rely on supplements. Some dietary supplements have been shown to be beneficial for certain health conditions. With B vitamins, the use of folic acid supplements by women of childbearing age who may become pregnant reduces the risk of some birth defects, as folate can be difficult to obtain in sufficient amounts from the diet. In addition, vitamin B12 in the crystalline form in supplements can assist people aged over 50 years who often have a reduced ability to absorb vitamin B12 from food (US Food and Drug Administration 2008). But just because B vitamins are water soluble doesn't mean that very high dosages from supplements are safe. For example, large amounts of vitamin B6 can cause nerve damage (Dalton & Dalton 1987). Supplements also come with potential risks, including drug interactions and side effects (US Food and Drug Administration 2008). Folate supplements can be considered risky in that they can mask vitamin B12 deficiency. The anaemia symptom improves, making a vitamin B12 deficiency diagnosis difficult, but the neurological symptoms can worsen (Frankenburg 2009). This can be of concern with the elderly as the body's ability to absorb naturally occurring vitamin B12 decreases.

Health policy has previously dealt with issues of the lack of B vitamins in the diet by supplementing food products. Rather than encourage people to eat brown rice and wholegrain breads, people continue to consume the more palatable white rice and bread, with the safeguard of having B vitamins artificially added to the food supply. This is certainly an easier option than changing to healthier eating habits, but is it the best option? Future research is still required regarding whether the total effect of the food surpasses artificial supplementation of individual aspects. As beneficial effects have been seen over the recommended daily intake levels of B vitamins, optimal intakes, particularly for areas such as cognitive function, are still under consideration. An improved understanding of the role of B vitamins on mood and cognition across the lifespan will help to identify cases where dietary intake can be improved to optimise neurological processes and mental health.

References

Agamanolis, D. P. (2006). Nutritional CNS disorders. Retrieved 10 January 2011, from http://www.neuropathologyweb.org/chapter8/chapter8Nutritional.html

Ambrose, M. L., Bowden, S. C., & Whelan, G. (2001). Thiamin treatment and working memory function of alcohol-dependent people: preliminary findings. *Alcoholism: Clinical and Experimental Research, 25*(1), 112–116, doi: 10.1111/j.1530-0277.2001.tb02134.x

American Psychiatric Association. (1994). *Diagnostic and Statistical Manual of Mental Disorders*, 4th ed. (DSM-IV). Washington, DC: American Psychiatric Association.

Anderson, J. E. (2010) Nutrition and Oral Contraceptives. *Food and Nutrition Series: Vol. no 9.323*: Colorado State University, U.S. Department of Agriculture and Colorado counties cooperating.

Australian Mushroom Growers' Association. (2009). Nutrition & health – surprises from the mushroom. Retrieved 3 January 2011, from http://www.mushroomsforlife.net/scientific_facts.html

Ball, G. (2004). *Vitamins: Their Role in the Human Body.* Oxford, UK: Blackwell Publishing Ltd.

Bamber, D. J., Stokes, C. S., & Stephen, A. M. (2007). The role of diet in the prevention and management of adolescent depression. *Nutrition Bulletin, 32*, 90–99, doi: 10.1111/j.1467-3010.2007.00608.x

Bémeur, C., Montgomery, J. A., & Butterworth, R. F. (2011). Vitamins deficiencies and brain function. In J. P. Blass (Ed.), *Neurochemical Mechanisms in Disease* (Vol. 1, pp. 103–124): Springer: New York.

Bender, D. A., & Bender, A. E. (1997). *Nutrition. A Reference Handbook.* Oxford: Oxford University Press.

Benton, D. (2008). Micronutrient status, cognition and behavioral problems in childhood. *European Journal of Nutrition, 47*(suppl 3), 38–50, doi: 10.1007/s00394-008-3004-9.

Benton, D., & Donohoe, R. T. (1999). The effects of nutrients on mood. *Public Health Nutrition, 2*(Suppl. 3a), 403–409, doi: doi:10.1017/S1368980099000555

Benton, D., Griffiths, R., & Haller, J. (1997). Thiamine supplementation, mood and cognitive functioning. *Psychopharmacology, 129*(1), 66–71, doi: 10.1007/s002130050163

Black, M. (2008). Effects of vitamin B12 and folate deficiency on brain development in children. *Food and Nutrition Bulletin, 29*(Suppl. 2), S126–S131.

Botez, M. I., Botez, T., & Maag, U. (1984). The Wechsler subtests in mild organic brain damage associated with folate deficiency. *Psychological Medicine, 14*(2), 431–437, doi: doi:10.1017/S0033291700003688

Bower, C. (1996). Folate and the prevention of birth defects. *Australian Journal of Nutrition and Dietetics, 53*(Suppl.), 5S–8S.

Bryan, J., Calvaresi, E., & Hughes, D. (2002). Short-term folate, vitamin B-12 or vitamin B-6 supplementation slightly affects memory performance but not mood in women of various ages. *The Journal of Nutrition, 132*(6), 1345–1356.

Bryan, J., Osendarp, S., Hughes, D., Calvaresi, E., Baghurst, K., & van Klinken, J.-W. (2004). Nutrients for cognitive development in school-aged children. *Nutrition Reviews, 62*(8), 295–306, doi: 10.1111/j.1753-4887.2004.tb00055.x

Christopherson, R. I. (1996). How folate functions inside the body. *Australian Journal of Nutrition and Dietetics, 53*(Suppl.), 8S–10S.

Clark, D. D., & Sokoloff, L. (1999). In G. J. Siegel, B. W. Agranoff, R. W. Albers, S. K. Fisher & M. D. Uhler (Eds.), *Basic Neurochemistry: Molecular, Cellular and Medical Aspects.* Philadelphia: Lippincott.

Coppen, A., & Bolander-Gouaille, C. (2005). Treatment of depression: time to consider folic acid and vitamin B12. *Journal of Psychopharmacology, 19*(1), 59–65. doi: 10.1177/0269881105048899

Coursin, D. B. (1954). Convulsive seizures in infants with pyridoxine-deficient diet. *Journal of the American Medical Association, 154*(5), 406–408, doi: 10.1001/jama.1954.02940390030009

Dagnelie, P., & van Staveren, W. (1994). Macrobiotic nutrition and child health: results of a population-based, mixed-longitudinal cohort study in The Netherlands. *The American Journal of Clinical Nutrition, 59*(5), 1187S–1196S.

Dagnelie, P., van Staveren, W., Vergote, F., Dingjan, P., van den Berg, H., & Hautvast, J. (1989). Increased risk of vitamin B-12 and iron deficiency in infants on macrobiotic diets. *The American Journal of Clinical Nutrition, 50*(4), 818–824.

Dalton, K., & Dalton, M. J. T. (1987). Characteristics of pyridoxine overdose neuropathy syndrome. *Acta Neurologica Scandinavica, 76*(1), 8–11, doi: 10.1111/j.1600-0404.1987.tb03536.x

de Lau, L. M., Refsum, H., Smith, A. D., Johnston, C., & Breteler, M. M. (2007). Plasma folate concentration and cognitive performance: Rotterdam Scan Study. *The American Journal of Clinical Nutrition, 86*(3), 728–734.

Dickinson, C. J. (1995). No reliable evidence that folate is harmful in B-12 deficiency. *British Medical Journal, 311*(7010), 949.

Durga, J., Boxtel, M. P. J. v., Schouten, E. G., Bots, M. L., Kok, F. J., & Verhoef, P. (2006). Folate and the methylenetetrahydrofolate reductase 677C→T mutation correlate with cognitive performance. *Neurobiology of Aging, 27*(2), 334–343.

Elias, M. F., Robbins, M. A., Budge, M. M., Elias, P. K., Brennan, S. L., Johnston, C., et al. (2006). Homocysteine, folate, and vitamins B6 and B12 blood levels in relation to cognitive performance: The Maine-Syracuse study. *Psychosomatic Medicine, 68*(4), 547–554, doi: 10.1097/01.psy.0000221380.92521.51

Frankenburg, F. R. (2009). *Vitamin Discoveries and Disasters: History, Science, and Controversies*. Santa Barbara: Praeger.

Gerster, H. (1996). The importance of vitamin B6 for development of the infant. Human medical and animal experimental studies. *Zeitschrift fur Ernahrungswiss, 35*(4), 309–317.

Gomez-Pinilla, F. (2008). Brain foods: the effects of nutrients on brain function. [10.1038/nrn2421]. *Nature Reviews Neuroscience, 9*(7), 568–578.

Graham, S. M., Arvela, O. M., & Wise, G. A. (1992). Long-term neurologic consequences of nutritional vitamin B12 deficiency in infants. *The Journal of Pediatrics, 121*(5), 710–714.

Groff, J., & Gropper, S. (2000). *Advanced Nutrition and Human Metabolism*. Belmont, CA: Wadsworth Thomson Learning.

Guilarte, T. R. (1993). Vitamin B6 and cognitive development: recent research findings from human and animal studies. *Nutrition Reviews, 51*(7), 193–198, doi: 10.1111/j.1753-4887.1993.tb03102.x

Gunstad, J., Strain, G., Devlin, M. J., Wing, R., Cohen, R. A., Paul, R. H., et al. (2011). Improved memory function 12 weeks after bariatric surgery. *Surgery for Obesity and Related Diseases, Corrected Proof*. doi: 10.1016/j.soard.2010.09.015

Harrell, R. F. (1946). Mental response to added thiamine. *The Journal of Nutrition, 31*(3), 283–298.

Herbert, V. (1962). Experimental nutritional folate deficiency in man. *Transactions of the Association of American Physicians, 75*, 307–320.

Heseker, H., Kubler, W., Westenhofer, J., & Pudel, V. (1990). Psychische Veranderungen als Fruhzeichen einer suboptimalen Vitaminversorgung. *Ernahrungs-Umschau, 37*, 87–94.

International Rice Research Institute. (1979). *Proceedings of the Workshop on Chemical Aspects of Rice Grain Quality*. Laguna, Philippines.

Irle, E., & Markowitsch, H. J. (1983). Widespread neuroanatomical damage and learning deficits following chronic alcohol consumption or vitamin-B1 (thiamine) deficiency in rats. *Behavioural Brain Research, 9*(3), 277–294, doi: 10.1016/0166-4328(83)90133-x

Knecht, S., Wersching, H., Lohmann, H., Bruchmann, M., Duning, T., Dziewas, R., et al. (2008). High-normal blood pressure is associated with poor cognitive performance. *Hypertension, 51*(3), 663–668, doi: 10.1161/hypertensionaha.107.105577

Koebnick, C., Hoffmann, I., Dagnelie, P. C., Heins, U. A., Wickramasinghe, S. N., Ratnayaka, I. D., et al. (2004). Long-term ovo-lacto vegetarian diet impairs vitamin B-12 status in pregnant women. *The Journal of Nutrition, 134*(12), 3319–3326.

Koyyalamudi, S. R., Jeong, S.-C., Cho, K. Y., & Pang, G. (2009). Vitamin B12 is the active corrinoid produced in cultivated white button mushrooms (Agaricus bisporus). *Journal of Agricultural and Food Chemistry, 57*(14), 6327–6333, doi: 10.1021/jf9010966

Kronenberg, G., Harms, C., Sobol, R. W., Cardozo-Pelaez, F., Linhart, H., Winter, B., et al. (2008). Folate deficiency induces neurodegeneration and brain dysfunction in mice lacking uracil DNA glycosylase. *The Journal of Neuroscience, 28*(28), 7219–7230, doi: 10.1523/jneurosci.0940-08.2008

Kulkarni, J. (2007). Depression as a side effect of the contraceptive pill. *Expert Opinion on Drug Safety, 6*, 371–374, doi: 10.1517/14740338.6.4.371

Levin, E. D. (2006). The rationale for studying transmitter interactions to understand the neural bases of cognitive function. In E. D. Levin (Ed.), *Neurotransmitter Interactions and Cognitive Function.* Switzerland: Birkhäuser Verlag.

Louwman, M. W., van Dusseldorp, M., van de Vijver, F. J., Thomas, C. M., Schneede, J., Ueland, P. M., et al. (2000). Signs of impaired cognitive function in adolescents with marginal cobalamin status. *The American Journal of Clinical Nutrition, 72*(3), 762–769.

Lyons-Wall, P., & Samman, S. (2003). Vitamins B6, B12 and folate. *Nutrition and Dietetics, 60*(1), 51–53.

Mahan. L. K. & Escott-Stump, S. (2008) *Krause's Food and Nutrition Therapy* (12th ed.). Philadelphia: Elsevier.

Mathews, C., van Holde, K., & Ahern, K. (2000). *Biochemistry* (3rd ed.). San Franciso: Addison Wesley Longman.

Mihalik, J. P., Ondrak, K. S., Guskiewicz, K. M., & McMurray, R. G. (2009). The effects of menstrual cycle phase on clinical measures of concussion in healthy college-aged females. *Journal of Science and Medicine in Sport, 12*(3), 383–387.

Miller, J. W. (2004). Folate, cognition, and depression in the era of folic acid fortification. *Journal of Food Science, 69*(1), SNQ61–SNQ65, doi: 10.1111/j.1365-2621.2004.tb17889.x

Molloy, A., Kirke, P., Brody, L., Scott, J., & Mills, J. (2008). Effects of folate and vitamin B12 deficiencies during pregnancy on fetal, infant, and child development. *Food and Nutrition Bulletin, 29*(Suppl. 2), S101–S115.

Mordecai, K. L., Rubin, L. H., & Maki, P. M. (2008). Effects of menstrual cycle phase and oral contraceptive use on verbal memory. *Hormones and Behavior, 54*(2), 286–293, doi: 10.1016/j.yhbeh.2008.03.006

Morris, M. S., Jacques, P. F., Rosenberg, I. H., & Selhub, J. (2007). Folate and vitamin B-12 status in relation to anemia, macrocytosis, and cognitive impairment in older Americans in the age of folic acid fortification. *The American Journal of Clinical Nutrition, 85*(1), 193–200.

Murakami, K., Miyake, Y., Sasaki, S., Tanaka, K., & Arakawa, M. (2010). Dietary folate, riboflavin, vitamin B-6, and vitamin B-12 and depressive symptoms in early adolescence: the Ryukyus Child Health Study. *Psychosomatic Medicine, 72*(8), 763–768, doi: 10.1097/PSY.0b013e3181f02f15

National Academy of Sciences, Institute of Medicine, & Food and Nutrition Board. (1998). *Dietary Reference Intakes for Thiamin, Riboflavin, Niacin, Vitamin B6, Folate, Vitamin B12, Pantothenic Acid, Biotin, and Choline: a report of the Standing*

Committee on the Scientific Evaluation of Dietary Reference Intakes and its Panel. Washington, DC.: National Academy Press.

Nye, C., & Brice, A. (2005). Combined vitamin B6-magnesium treatment in autism spectrum disorder. *Cochrane Database of Systematic Reviews*, (4): CD003497.

Refsum, H. (2001). Folate, vitamin B12 and homocysteine in relation to birth defects and pregnancy outcome. *British Journal of Nutrition, 85*(Suppl. S2), S109–S113, doi:10.1049/BJN2000302

Samman, S., & Lyons-Wall, P. (2003). Thiamin, riboflavin, niacin, pantothenic acid and biotin. *Nutrition and Dietetics, 60*(2), 131–133.

Savage, D., Gangaidzo, I., Lindenbaum, J., Kiire, C., Mukiibi, J. M., Moyo, A., et al. (1994). Vitamin B12 deficiency is the primary cause of megaloblastic anaemia in Zimbabwe. *British Journal of Haematology, 86*(4), 844–850, doi: 10.1111/j.1365-2141.1994.tb04840.x

Scalabrino, G. (2009). The multi-faceted basis of vitamin B12 (cobalamin) neurotrophism in adult central nervous system: lessons learned from its deficiency. *Progress in Neurobiology, 88*(3), 203–220, doi: 10.1016/j.pneurobio.2009.04.004

Scalabrino, G., & Peracchi, M. (2006). New insights into the pathophysiology of cobalamin deficiency. *Trends in Molecular Medicine, 12*(6), 247–254, doi: 10.1016/j.molmed.2006.04.008

Sechi, G., & Serra, A. (2007). Wernicke's encephalopathy: new clinical settings and recent advances in diagnosis and management. *The Lancet Neurology, 6*(5), 442–455.

Silber, M., Almkvist, O., Larsson, B., Stock, S., & Uvnäs-Moberg, K. (1987). The effect of oral contraceptive pills on levels of oxytocin in plasma and on cognitive functions. *Contraception, 36*(6), 641–650.

Singh, S., & Kumar, A. (2007). Wernicke encephalopathy after obesity surgery. *Neurology, 68*(11), 807–811, doi: 10.1212/01.wnl.0000256812.29648.86

Smith, A. D. (2007). Folic acid fortification: the good, the bad, and the puzzle of vitamin B-12. *The American Journal of Clinical Nutrition, 85*(1), 3–5.

Stampfer, M. J. (2008). *Vitamins and Minerals*. Boston: Harvard Health Publications.

Stewart, R. (Ed.). (2009). *Griffith Handbook of Clinical Nutrition and Dietetics* (3rd ed.). Southport, QLD: Griffith University.

Taylor, M. J., Carney, S. M., Geddes, J., & Goodwin, G. (2009). Folate for depressive disorders. *Cochrane Database of Systematic Reviews, 2003*(2). Art No.:CD003390.doi:10.1002/14651858.CD003390.

Troen, A. M., Shea-Budgell, M., Shukitt-Hale, B., Smith, D. E., Selhub, J., & Rosenberg, I. H. (2008). B-vitamin deficiency causes hyperhomocysteinemia and vascular cognitive impairment in mice. *Proceedings of the National Academy of Sciences, 105*(34), 12474–12479, doi: 10.1073/pnas.0805350105

U. S. Food and Drug Administration. (2008). FDA 101: dietary supplements. Retrieved 3 January 2011, from www.fda.gov/consumer/updates/supplements 080408.html

Watanabe, F. (2007). Vitamin B12 sources and bioavailability. *Experimental Biology and Medicine, 232*(10), 1266–1274, doi: 10.3181/0703-mr-67

Williams, A.-L., Cotter, A., Sabina, A., Girard, C., Goodman, J., & Katz, D. L. (2005). The role for vitamin B-6 as treatment for depression: a systematic review. *Family Practice, 22*(5), 532–537, doi: 10.1093/fampra/cmi040

Wyatt, K. M., Dimmock, P. W., Jones, P. W., & O'Brien, P. M. S. (1999). Efficacy of vitamin B-6 in the treatment of premenstrual syndrome: systematic review. *British Medical Journal, 318*(7195), 1375–1381.

Wynn, V. (1975). Vitamins and oral contraceptive use. *The Lancet, 305*(7906), 561–564.

Xu, Y., Sladky, J., & Brown, M. (1989). Dose-dependent expression of neuronopathy after experimental pyridoxine intoxication. *Neurology, 39*(8), 1077–1083.

Zarrindast, M. R. (2006). Neurotransmitters and cognition. In E. D. Levin (Ed.), *Neurotransmitter Interactions and Cognitive Function*. Switzerland: Birkhäuser Verlag.

CHAPTER 5

Vitamin D, Energy Regulation and Mental Health

Paul Cherniack

Chapter Overview

- Vitamin D insufficiency is a widespread problem among the elderly.
- Epidemiological studies have suggested a relationship between vitamin D insufficiency and cognitive impairment, depression and schizophrenia.
- No trials have yet been published showing that correction of vitamin D insufficiency prevents or reverses the symptoms of most mental or cognitive illnesses.
- A few small trials have used small doses of vitamin D to treat seasonal affective disorder without conclusive results.

Introduction

A growing body of evidence suggests a potential role for vitamin D in the pathogenesis of cognitive and mental illness. This chapter will describe the evidence and delineate the potential mechanisms for the relationship between health status and vitamin D.

A significant number of individuals globally lack sufficient vitamin D, particularly the elderly (Holick 2005). While the exact definition of optimal serum vitamin D concentrations is a matter of debate, most authorities believe that humans need at least 32 ng/ml for optimal health (Cherniack et al. 2008a). Hypovitaminosis D occurs along a spectrum, whereby 10–32 ng/ml vitamin D can be considered insufficient and < 10 ng/ml can be considered deficient. Vitamin D insufficiency ($<32- \geq 10$ ng/ml) heightens the prevalence of osteoporosis, falls, fractures and increased mortality (especially from heart disease). Possible progression to vitamin D deficiency (< 10 ng/ml) brings increased risk of rickets (see Figure 5.1; Cherniack et al. 2008b). Between 40 and 90 per cent of elderly persons are vitamin D insufficient, since they consume inadequate amounts of dietary vitamin D, spend less time in the sun and are more likely to cover up with clothing than younger persons (Cherniack

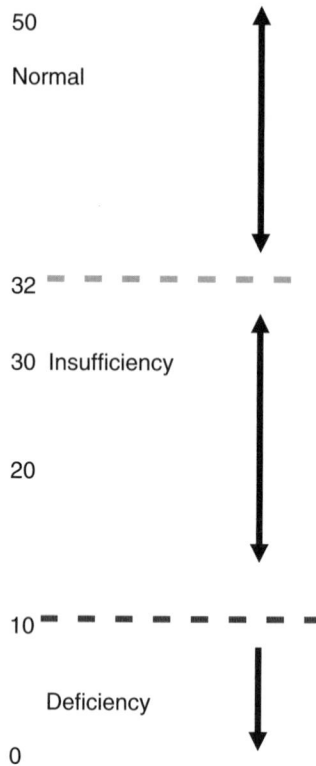

Figure 5.1   Spectrum of vitamin D deficiency (units ng/ml)
Source: Adapted from Cherniack et al. (2008).

et al. 2008b). Fish and milk have the largest amounts of dietary vitamin D of all foods (100–150 IU/serving). However, one must consume the equivalent of a 9 oz sockeye salmon fillet each day to meet the requirements for vitamin D from diet alone (US Department of Agriculture 2009). Fortunately, vitamin D can be simply and cheaply consumed as a capsule supplement (usually 1000–2000 IU/capsule) (Cherniack et al. 2008b).

Sunlight is another source of vitamin D (Cherniack et al. 2008b). Sunlight provides ultraviolet radiation which acts as a catalyst to convert 7-dehydrocholesterol to vitamin D3 (cholecalciferol). The latter is then converted by the liver and kidneys into 1,25 dihydroxycholecalciferol (calcitriol), which is metabolically active (Cherniack et al. 2008b). The skin will create vitamin D subject to the length of exposure to sunlight and the darkness of skin pigmentation (Hollis 2005). The further one travels from the equator, the greater the variability of sunlight, with lower intensity during the winter (Hollis 2005). Lighter skin absorbs ultraviolet light more readily than darker skin. A person with light skin outdoors in a bathing suit in the middle of the summer at a latitude of a city in northern USA would have to receive sun exposure for no less than ten minutes to receive enough ultraviolet radiation to produce enough vitamin D to meet daily requirements (Hollis 2005). In fact, this may be difficult to accomplish in much of the world, in which limited daylight hours in the winter months and culturally acceptable forms of dress prevent an individual from obtaining sufficient sunlight (Cherniack & Troen 2008). Indeed, in Lebanon, Turkey, Jordan and the USA, Arab women whose skin is more covered by clothing are more likely to be vitamin D deficient than those whose skin is more exposed (Batieha et al. 2011; Gannage-Yared et al. 2000; Hobbs et al. 2009; Mishal 2001).

Vitamin D and Cognitive Screening Tests

Several investigations have described an association between scores on screening evaluations for cognitive dysfunction and vitamin D levels (Annweiler et al. 2009a; Cherniack et al. 2009), although there have been no prospective trials of vitamin D supplementation on such tests. In one study, 32 individuals between the ages of 61 and 92, who were receiving care at a medical school outpatient clinic, were administered cognitive tests and had their vitamin D levels measured. There was a positive correlation between scores on the Mini-Mental Status Examination (MMSE) and vitamin D levels ($p = 0.006$) (Przybelski & Binkley 2007). In elderly attendees of a university hospital clinic in the Netherlands, higher MMSE scores were also associated with increased vitamin D levels ($p = 0.01$; Oudshoorn et al. 2008). This correlation was further defined in a population of 1766 older persons who were involved in the Health Survey for England 2000 (Llewellyn et al. 2009). Subjects in the lowest quartile group of vitamin D (3.2–12 ng/ml) were 2.3 times as likely to exhibit cognitive dysfunction than those in the highest quartile (26.4–68 ng/ml). In a further study involving 69 healthy older persons, oral intake of vitamin D was also positively correlated with MMSE scores (Rondanelli et al. 2007).

In another study, 40 'mildly demented' individuals were compared with 40 'normal' individuals, aged 60 years and above (Wilkins et al. 2006). A measure of vitamin D levels was obtained, and all participants were administered the Short Blessed Test (a six-item assessment of cognition similar in composition to the MMSE), a Clinical Dementia Rating test (a longer assessment) and the MMSE. A relationship was observed between vitamin D deficiency and performance on the Short Blessed Test and on the Clinical Dementia Rating score, but no relationship was observed between vitamin D deficiency and performance on the MMSE. In a further group of 60 individuals (vitamin D mean level 21.59 ng/ml, mean age 74.99) comprising both 'mild cognitively impaired' and healthy individuals, MMSE scores were not associated with deficient vitamin D levels (defined in this investigation as < 20 ng/ml), but the Short Blessed Test was (Wilkins et al. 2009). However, within this population, African-Americans (who manifested lower mean vitamin D scores than Caucasians) showed a positive association between vitamin D level and MMSE score.

Two other epidemiological studies have reported a relationship between vitamin D and cognition. In a nested cohort investigation, an association was found between vitamin D levels and scores on another cognitive assessment tool, the Short Portable Mental State Questionnaire, in women aged 57 years and over (taking part in an osteoporosis study; Annweiler et al. 2009b). In a further study, higher levels of parathyroid hormone (PTH) – which increases as vitamin D levels decrease – was associated with cognitive impairment in 514 individuals who participated in the Helsinki Ageing Study, evaluated at ages 75, 80 and 85 years (Bjorkman et al. 2009; Cherniack & Troen 2008). Vitamin D itself was not assessed in this investigation. Subjects whose PTH was > 62 ng/l at baseline had a 2.4 times increased chance of manifesting a reduction in their Clinical Dementia Rating or MMSE score on subsequent assessment, five years later.

The relationship between specific cognitive functions and vitamin D has not been as thoroughly investigated as that between vitamin D and performance on global cognitive screening instruments (Cherniack et al. 2009). The relationship has, however, been evaluated using assessment tools measuring speed of processing, memory and attention, and executive functions. These instruments, which assess more specific aspects of cognitive capacity, have yielded mixed results. In a study of 1080 older persons with a mean age of 75 years, who were receiving home care in Boston (USA), the relationship between vitamin D levels and performance in several tests of cognition was investigated (Buell et al. 2009, 2010). More than 65 per cent of subjects in this study were identified with deficient vitamin D levels (< 20 ng/ml). There was a positive association between vitamin D levels and performance on a global screening test (MMSE), on specific measures of cognitive functioning (attention and processing speed) and on assessments of executive functions (digit symbol coding, block design, matrix reasoning, Trails A and B). There was no relationship between vitamin D and memory. Buell et al. (2009) concluded from these findings that further work was warranted. An epidemiological investigation of international studies included assessments of vitamin D and specific measures of cognitive performance in 3369 men aged 40–79 years with a mean age of 60 years (Lee et al. 2009). These studies found correlations between vitamin D and one executive function test (the Digit Symbol Substitution Test), but no correlation was observed between vitamin D and other instruments (such as figure copying and memory tests). This work again suggests domain specificity when considering the relationship between vitamin D status and cognitive abilities.

A second large epidemiological study, the InCHIANTI trial of 858 older individuals (at least 65 of whom were assessed for three years), investigated the relationship between vitamin D deficiency and the prospective development of cognitive impairment (Llewellyn et al. 2010). Individuals with the most severe vitamin D deficiency (< 10 ng/ml) had an increased probability of losing three points on their MMSE (1.60; 1.19–2.00; 95 per cent confidence interval (CI)) or achieving lower scores on the Trail-Making Test B (1.31; 1.03–1.51; 95% CI), but not on the cognitively less-demanding Trail-Making Test A (which has a limited executive functioning component). Although those persons with insufficient vitamin D levels scored, on average, 0.3 points a year worse on the MMSE than those with adequate levels, this tells us little about the specific cognitive domains impaired in vitamin D-deficient individuals. The Trail-Making Test B performance decrements are perhaps the most informative in this study, suggesting that abilities involving cognitive flexibility and task switching are particularly affected if vitamin D levels are compromised.

In a final epidemiological study, the National Health Examination (which included 4809 participants aged at least 60 years) showed no association between cognition and vitamin D levels (McGrath et al. 2007). However, the cognitive assessment comprised only a single measure, in which subjects had to recall the details of a story read to them.

The Relationship between Vitamin D and Cognitive Disorders: Possible Mechanisms

While detailed discussion of the pathophysiology of dementing illness is beyond the scope of this chapter, it is important to consider the role of vascular disease and its relationship with vitamin D in cognitive disorders. Vascular pathology is a clinical feature of more than 80 per cent of all dementing illness (Dickstein et al. 2010), including Alzheimer's disease, multi-infarct dementia, and 'mixed dementia' (which embodies components of both Alzheimer's and multi-infarct disease; Dickstein et al. 2010). The most common form of dementia, Alzheimer's disease, has a complex pathology which includes genetic risk factors, hippocampal degeneration, mutations of the *tau* protein, deposition of amyloid, mitochondrial abnormalities and vascular changes (Querfurth & LaFerla 2010). Specifically, evidence of vascular abnormality has been observed in 60–90 per cent of the brains of patients with Alzheimer's disease on autopsy (Dickstein et al. 2010). A hallmark cellular pathologic feature of Alzheimer's disease is derangement in the processing of neuronal amyloid precursor protein, which is also an inflammatory cytokine (Frisardi et al. 2010). Increased levels of amyloid precursor proteins have been correlated with levels of other proinflammatory cytokines, including interleukin (IL)-1β, IL-6 and 8 (Frisardi et al. 2010).

With respect to vascular dementia, this should not be thought of as an isolated phenomenon, but may be part of a larger complex of systemic vascular pathology that has been referred to as the 'metabolic syndrome'. Components of the metabolic syndrome include obesity, hypertension and diabetes (Frisardi et al. 2010; Obunai et al. 2007). A key component of this systemic vascular inflammatory syndrome is the adipocyte. Adipocytes should be regarded as generators of inflammation (through the creation of inflammatory cytokines and adipose-related hormones), rather than merely as repositories of excess energy in the form of fat. Through the production of these proinflammatory substances, adipocytes also can control the action of immunocytes, which may be important in systemic atherosclerosis.

Vitamin D is not actually a vitamin, but a steroid hormone that has an anti-inflammatory action on adipocytes. Vitamin D hinders adipocyte development and promotes adipocyte apoptosis. After vitamin D is hydroxylated, it promotes the entrance of calcium into the cell, causing the enzymes capsace-12 and calpase to initiate apoptosis (Sergeev 2009). Vitamin D also causes a loss of uncoupling protein 2, located in the inner mitochondrial membrane – inducing hydrogen ion efflux, which also promotes apoptosis (Zemel & Sun 2008).

Vitamin D may also have direct neuroprotective effects on the brain. Many cells that facilitate neuronal function (such as glial cells and macrophages) possess the hydroxylase that can convert vitamin D into its active form (Naveilhan et al. 1993; Neveu et al. 1994). Neuronal anti-inflammatory and antioxidant defences may be assisted by vitamin D (Buell & Dawson-Hughes 2008). Cells throughout the brain, especially dopaminergic neurons, contain vitamin D receptors, which (when activated) augment levels of neurotrophins, such as neurotrophin-E and glial cell line-derived neurotrophic factor (GDNF) (Cherniack et al. 2009; Eyles et al. 2009; Kalueff & Tuohimaa 2007). Vitamin D raises the population of suppressor T-cells, which suppresses inflammation (Ardizzone et al. 2009; Lisak et al. 2009; Smolders et al. 2009). Studies have also shown that animals injected with vitamin D or neurons grown in culture

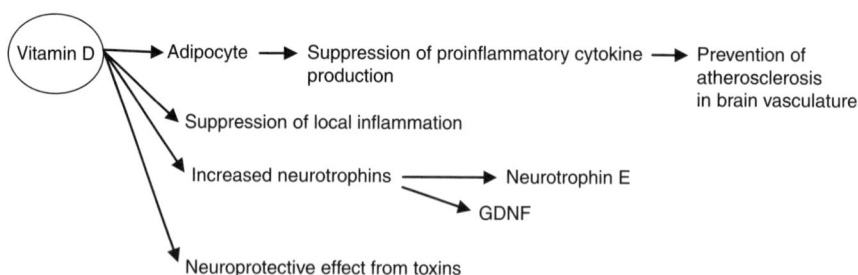

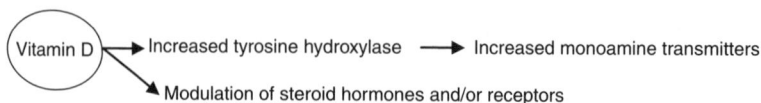

Figure 5.2 Summary of possible mechanisms of vitamin D action on the brain

with vitamin D are shielded from the effects of neurotoxins (Garcion et al. 2002; Ibi et al. 2001; Shinpo et al. 2000; Taniura et al. 2006; Wang et al. 2001).

Figure 5.2 presents a summary of possible mechanisms linking vitamin D deficiency to impaired neurocognitive functioning in dementia and depression.

In addition, polymorphisms of vitamin D receptor genes have been associated with cognitive illness (Gezen-Ak et al. 2007; Kuningas et al. 2009). Specifically, the haplotype 2(Bat), BsmI, Taq I has been associated with poorer performance on neuropsychological tests (Kuningas et al. 2009). ApaI has been correlated with an increased risk of Alzheimer's disease, but the combination of both Taq I and Apa was associated with a reduced risk of Alzheimer's (Gezen-Ak et al. 2007).

Depression

Vitamin D may be related to the pathogenesis of depression (Cherniack et al. 2009). No basic scientific investigative research has elucidated a mechanism by which inadequate vitamin D might cause depression, but hypotheses have been advanced (Bertone-Johnson 2009). One such hypothesis, which posits that depression is the result of a lack of sufficient monamine transmitters, suggests that vitamin D deficiency delays the expression of the tyrosine hydroxylase gene, which encodes the enzyme that converts monoamine precursors into norephinephrine and serotonin (Belmaker 2008; Bertone-Johnson 2009). Vitamin D does protect animal neurons from toxins that are structurally related to dopamine (Bertone-Johnson 2009).

A second hypothesis posits that vitamin D (which is itself a steroid hormone) acts upon neural steroid receptors to alter the adverse effects of stress on steroid regulatory hormones in the brain, precluding another postulated cause of depression (Belmaker 2008; Bertone-Johnson 2009). It is possible that vitamin D binding with neural vitamin D receptors also modulates steroid receptors (Bertone-Johnson 2009; see Figure 5.2 for a summary of possible mechanisms of vitamin D deficiency as a cause of depression).

Several epidemiological investigations have implied a relationship between low vitamin D levels and the presence of depression. In the Longitudinal Aging Study Amsterdam (LASA), comprising 1282 persons aged 65–95 years, the mean serum vitamin D level among subjects was 21 ng/ml (Hoogendijk et al. 2008). Those participants who had diagnosis of minor or major depression had a 14 per cent lower mean vitamin D concentration than others without the diagnosis. Symptoms were more severe in subjects with lower vitamin D levels ($p < 0.03$). In another previously mentioned survey, the InCHIANTI study, an increase in depressive symptoms over a six-year period was associated with lower vitamin D levels (Milaneschi et al. 2010). Those women whose levels were less than 20 ng/ml had twice the risk of depression as those with higher vitamin D levels, and men with deficient vitamin D levels had a 1.6 greater likelihood of becoming depressed compared with men whose vitamin D levels were more replete. In another survey, the 2005 Health Survey for England, among the 2070 participants with severely deficient vitamin D levels (< 10 ng/ml), those with lower vitamin D concentrations had a greater number of positive responses on the Geriatric Depression Scale ($r = -0.14$, $p < 0.001$) after adjustment for vitamin D intake, general health, chronic illness, socio-economic status, body mass index, season and age (Stewart & Hirani 2010). Among a cohort of 7358 individuals with cardiovascular disease aged over 50 years (mean age 73.1), those whose vitamin D levels were < 15 ng/ml were 2.7 times as likely to be depressed compared with those whose levels were > 50 ng/ml, while subjects whose vitamin D levels were between 16 and 30 ng/ml had a 2.15 times greater risk of depression than those whose levels were > 50 ng/ml (May et al. 2010).

Other studies from the Far East indicate a limited relationship between vitamin D and depression. A Chinese investigation noted an overall negative association between vitamin D levels and symptoms of depression in 3262 community-dwelling elderly people aged 50–70 years (Pan et al. 2009). However, the addition of covariates such as geographic location, marital status and smoking nullified this association. Another study was performed with Japanese municipal workers who were between 21 and 67 years old (Nanri et al. 2009). A negative association was observed between vitamin D levels and incidence of depression in November, but in July (during which time there was more sunlight) this association disappeared.

A few investigations have examined a possible relationship between depression and vitamin D levels in small groups of women. When depressive symptoms were measured in nine women with vitamin D levels of < 16 ng/ml for a year, a relationship was observed between depression symptom scores and seasonal variation in vitamin D levels (Shipowick et al. 2009).

There have also been small prospective trials in the use of vitamin D to treat seasonal affective disorder (SAD). In the winter, 250 women aged 43–72 years were given a dose of 400 IU of cholecalciferol (Harris & Dawson-Hughes 1993). However, depression symptom scores in this study were unchanged. In a controlled trial, 2117 English women with SAD were treated with 800 IU vitamin D a day or a placebo for six months (Dumville et al. 2006). Again, the supplementation did not affect symptoms of depression, although blood vitamin D levels were not measured in this or the prior study. An additional eight subjects with SAD were provided a single dose of 100,000 IU vitamin D and compared with seven subjects who were treated with phototherapy (Gloth et al. 1999). Vitamin D levels increased in those given vitamin D, from a mean of 11.0 ng/ml to 19.1 ng/ml ($p < 0.003$) and from 13.7 ng/ml to 18.6 ng/ml ($p < 0.007$) in the phototherapy-treated subjects. Both those who took vitamin D and those who used phototherapy had a significant decrease in depression symptom scores, but those who took vitamin D were more likely to demonstrate improvement (a 74 per cent improvement rate among subjects receiving vitamin D, p < 0.0005, vs. 36 per cent response, p < 0.01 improvement among those receiving phototherapy). In another study, a negative correlation between scores on the eight-question atypical Hamilton depression subscale and vitamin D concentrations ($r^2 = 0.26$, p = 0.05) was observed in 44 healthy Australians (mean age 22, age range 18–43) who received either two doses of cholecalciferol (400 IU or 800 IU) or placebo for five days during the winter (Lansdowne & Provost 1998). Participants who took vitamin D also showed a significantly greater improvement on the Positive and Negative Affect Schedule test than those who consumed a placebo.

Epidemiological studies comparing the incidence of depressive symptoms in certain geographical locations and average vitamin D intake (based on estimates of sunlight exposure at different latitudes) have yielded inconclusive results (Mersch et al. 1999). Several studies have ascertained that there is a lower incidence of SAD in North America and Europe at latitudes closer to the equator (Lingjarde et al. 1986; Mersch et al. 1999; Potkin et al. 1986; Rosen et al. 1990; Terman 1988), but other studies have failed to replicate this finding (Levitt & Boyle 2002; Magnusson & Axelsson 1993; Mersch et al. 1999; Partonen et al. 1993). Pooled results of studies in North America do indicate a relationship between the incidence of SAD and distance from the equator (r = 0.90, p = 0.003), but this has not been found on other continents (Mersch et al. 1999). Differences in daylight hours, heat, cloud cover and sampling methods may contribute to the difficulty in the interpretation of evidence from these investigations. Furthermore, light intensity (illuminance) may be a greater contributor to depressive symptoms than total daylight hours (Axelsson et al. 2004).

Therefore, there have been several epidemiological studies showing an association between depression and vitamin D, but a handful of prospective trials investigating the use of vitamin D in the treatment of SAD have yielded mixed results. However, these studies comprised short trials, most of which used low doses of supplementation.

Bipolar Illness and Schizophrenia

Some epidemiological studies suggest an association between vitamin D and bipolar illness and schizophrenia. The symptoms of bipolar disease more commonly manifest themselves during the winter, when less sunlight is available (Schaffer et al. 2003). However, the incidence of bipolar illness does not change significantly with latitude (Schaffer et al. 2003). In one investigation, in which 17 Australian patients with bipolar disease or depression were compared with another cohort of 861 healthy persons from the same part of the country, the mean vitamin D levels were significantly lower among the psychiatric patients (18.8 ng/ml) than in the healthy individuals (32.4 ng/ml) (Berk et al. 2007; Pasco et al. 2001).

Several more studies have reported an inverse association between vitamin D and schizophrenia. Prenatal vitamin D deficiency has been theorised to be related to schizophrenia (Altschuler 2001; Kinney et al. 2009; McGrath 1999). Geographical variation in the incidence of schizophrenia does occur. Specifically, children born in regions of the USA furthest from the equator in winter have the greatest probability of developing schizophrenia (Kinney et al. 2009; Torrey et al. 1977). Further, immigrants from Caribbean nations to nations at higher latitudes were more likely to give birth to schizophrenic children (Jarvis 1998). However, schizophrenia was not correlated with sunlight exposure in two studies of 30,000 children in Wales and Scotland – although the possible influence of breastfed milk on vitamin D consumption was not evaluated (Kendell & Adams 2002). The incidence of schizophrenia was reduced in young boys in Finland who had consumed a vitamin D supplement (McGrath et al. 2004). In an epidemiological investigation in Denmark, newborns with a serum vitamin D concentration in the lowest and highest quintiles both had a greater risk of developing schizophrenia (McGrath et al. 2010). In addition, foetal rats lacking vitamin D share anatomical and motor function similarities with individuals suffering from schizophrenia (Meyer & Feldon 2009). However, there have been no published reports of prospective trials using vitamin D in children to prevent schizophrenia.

Therefore, to date, only a small number of epidemiological studies have investigated schizophrenia and bipolar illness, all of which imply indirectly an association between illness and vitamin D status. More detailed epidemiological and prospective trials need to be performed.

Conclusions

While many studies suggest a role for vitamin D in the pathogenesis of many forms of cognitive and mental illness, causation has yet to be ascertained. In addition, there have been no prospective trials of the use of vitamin D levels to treat symptoms of cognitive or mental illness, except for the inconclusive results of trials which used low levels of vitamin D in relatively small samples to treat SAD.

Large prospective investigations are therefore clearly warranted. With regard to dementia and schizophrenia, an important question that still needs to be resolved is the optimal participant group. It is possible that the pathological changes that develop in dementia occur over a long period of time before symptoms appear, and so early intervention may be advantageous to delineate a meaningful change. Similarly, studies conducted thus far imply that maternal or early childhood vitamin D supplementation may be necessary to impact upon schizophrenia.

Even in the absence of mental illness, I have recommended that all older persons be tested and supplemented with 2000 IU vitamin D if their vitamin D levels were < 32 ng/ml, in order to prevent morbidity and potentially mortality (Cherniack et al. 2008; see Figure 5.3). Since my original proposal, other investigations have implied that more than 40 ng/ml may be desirable

Measure vitamin D and calcium levels   or   Supplement with 2000 IU
cholecalciferol daily (no further testing)

If ≥ 32 ng/ml, no
further intervention

Add 2000 IU daily,
reassess after three months,

If still < 32 ng/ml, check compliance
increase daily dose by 1000–2000 U
reassess every three months

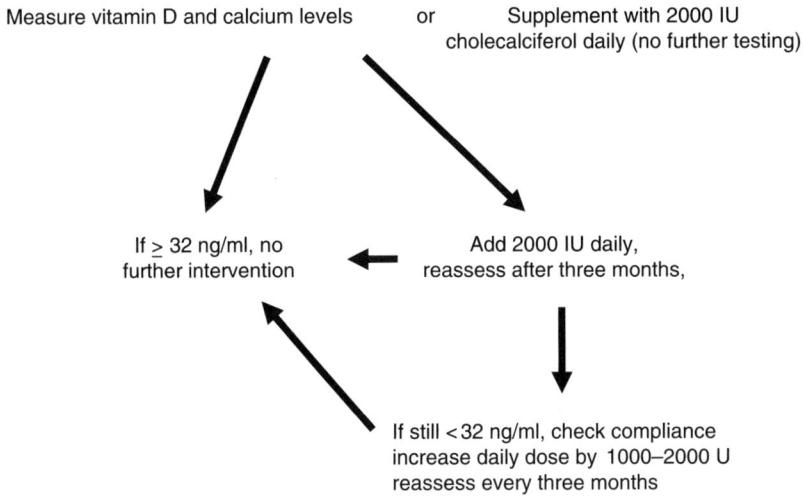

Figure 5.3 Vitamin D treatment algorithm
Source: Adapted from Cherniack et al. (2008).

for additional health benefits (Cannell & Hollis 2008; Cannell et al. 2008; Ginde et al. 2009; Grant 2010; Sabetta et al. 2010). In order to achieve this level, attention to diet alone is unlikely to ensure vitamin D sufficiency (given the difficulty of obtaining enough vitamin D through diet). Sunlight is an alternative, but access and the risk of skin cancer remain problems. Therefore, supplements will often be necessary.

A recent report by the Institute of Medicine has caused considerable controversy by claiming that a much lower level of serum vitamin D concentration is sufficient, thereby recommending lower vitamin D daily input (Institute of Medicine 2010). However, even this report (which established its conclusions focusing primarily on vitamin D requirements for bone health) recognised that the elderly are particularly vulnerable to insufficiency, and that extra supplementation might therefore be needed in this group. The possibility also exists that different organ systems have different vitamin D requirements for optimal health. It remains to be determined if the requirements of vitamin D for the adequate functioning of the brain will contribute to this debate, and how our growing understanding of the vitamin D needs of the brain will impact on our recognition of how much vitamin D is necessary to preserve human health through old age.

References

Altschuler, E. L. (2001). Low maternal vitamin D and schizophrenia in offspring. *Lancet, 358*(9291), 1464.

Annweiler, C., Allali, G., Allain, P., Bridenbaugh, S., Schott, A. M., Kressig, R. W., et al. (2009a). Vitamin D and cognitive performance in adults: a systematic review. *Eur J Neurol, 16*(10), 1083–1089.

Annweiler, C., Schott, A. M., Allali, G., Bridenbaugh, S. A., Kressig, R. W., Allain, P., et al. (2009b). Association of vitamin D deficiency with cognitive impairment in older women. Cross-sectional study. *Neurology, 74*(1): 27–32.

Ardizzone, S., Cassinotti, A., Trabattoni, D., Manzionna, G., Rainone, V., Bevilacqua, M., et al. (2009). Immunomodulatory effects of 1,25-dihydroxyvitamin D3 on TH1/TH2 cytokines in inflammatory bowel disease: an in vitro study. *Int J Immunopathol Pharmacol, 22*(1), 63–71.

Axelsson, J., Ragnarsdottir, S., Pind, J., & Sigbjornsson, R. (2004). Daylight availability: a poor predictor of depression in Iceland. *Int J Circumpolar Health, 63*(3), 267–276.

Batieha, A., Khader, Y., Jaddou, H., Hyassat, D., Batieha, Z., Khateeb, M., et al. (2011). Vitamin D status in Jordan: dress style and gender discrepancies. *Ann Nutr Metab, 58*(1), 10–18.

Belmaker, R. H. (2008). The future of depression psychopharmacology. *CNS Spectr, 13*(8), 682–687.

Berk, M., Sanders, K. M., Pasco, J. A., Jacka, F. N., Williams, L. J., Hayles, A. L., et al. (2007). Vitamin D deficiency may play a role in depression. *Med Hypotheses, 69*(6), 1316–1319.

Bertone-Johnson, E. R. (2009). Vitamin D and the occurrence of depression: causal association or circumstantial evidence? *Nutr Rev, 67*(8), 481–492.

Bjorkman, M. P., Sorva, A. J., & Tilvis, R. S. (2009). Does elevated parathyroid hormone concentration predict cognitive decline in older people? *Aging Clin Exp Res, 22*(2), 164–169.

Buell, J. S., & Dawson-Hughes, B. (2008). Vitamin D and neurocognitive dysfunction: preventing "D"ecline? *Mol Aspects Med, 29*(6), 415–422.

Buell, J. S., Dawson-Hughes, B., Scott, T. M., Weiner, D. E., Dallal, G. E., Qui, W. Q., et al. (2010). 25-Hydroxyvitamin D, dementia, and cerebrovascular pathology in elders receiving home services. *Neurology, 74*(1), 18–26.

Buell, J. S., Scott, T. M., Dawson-Hughes, B., Dallal, G. E., Rosenberg, I. H., Folstein, M. F., et al. (2009). Vitamin D is associated with cognitive function in elders receiving home health services. *J Gerontol A Biol Sci Med Sci, 64*(8), 888–895.

Cannell, J. J., & Hollis, B. W. (2008). Use of vitamin D in clinical practice. *Altern Med Rev, 13*(1), 6–20.

Cannell, J. J., Hollis, B. W., Zasloff, M., & Heaney, R. P. (2008). Diagnosis and treatment of vitamin D deficiency. *Expert Opin Pharmacother, 9*(1), 107–118.

Cherniack, E.P., Florez, H., Roos , B.A., Troen, B.R., Levis, S. (2008). Hypovitaminosis D in the elderly: from bone to brain. *J Nutr Health Aging, 12*(6), 366–373.

Cherniack, E. P., Levis, S., & Troen, B. R. (2008). Hypovitaminosis D: a widespread epidemic. *Geriatrics, 63*(4), 24–30.

Cherniack, E. P., & Troen, B. R. (2008). Calciotropic Hormones. In G. Duque & D. P. Kiel (Eds.), *Senile Osteoporosis: Advances in Pathophysiology and Therapeutic Approach* (pp. 34–46). London: Springer-Verlag.

Cherniack, E. P., Troen, B. R., Florez, H. J., Roos, B. A., & Levis, S. (2009). Some new food for thought: the role of vitamin D in the mental health of older adults. *Curr Psychiatry Rep, 11*(1), 12–19.

Dickstein, D. L., Walsh, J., Brautigam, H., Stockton, S. D., Jr., Gandy, S., & Hof, P. R. (2010). Role of vascular risk factors and vascular dysfunction in Alzheimer's disease. *Mt Sinai J Med, 77*(1), 82–102.

Dumville, J. C., Miles, J. N., Porthouse, J., Cockayne, S., Saxon, L., & King, C. (2006). Can vitamin D supplementation prevent winter-time blues? A randomised trial among older women. *J Nutr Health Aging, 10*(2), 151–153.

Eyles, D. W., Feron, F., Cui, X., Kesby, J. P., Harms, L. H., Ko, P., et al. (2009). Developmental vitamin D deficiency causes abnormal brain development. *Psychoneuroendocrinology,34*, S247–S257.

Frisardi, V., Solfrizzi, V., Seripa, D., Capurso, C., Santamato, A., Sancarlo, D., et al. (2010). Metabolic-cognitive syndrome: a cross-talk between metabolic syndrome and Alzheimer's disease. *Ageing Res Rev, 9*(4), 399–417.

Gannage-Yared, M. H., Chemali, R., Yaacoub, N., & Halaby, G. (2000). Hypovitaminosis D in a sunny country: relation to lifestyle and bone markers. *J Bone Miner Res, 15*(9), 1856–1862.

Garcion, E., Wion-Barbot, N., Montero-Menei, C. N., Berger, F., & Wion, D. (2002). New clues about vitamin D functions in the nervous system. *Trends Endocrinol Metab, 13*(3), 100–105.

Gezen-Ak, D., Dursun, E., Ertan, T., Hanagasi, H., Gurvit, H., Emre, M., et al. (2007). Association between vitamin D receptor gene polymorphism and Alzheimer's disease. *Tohoku J Exp Med, 212*(3), 275–282.

Ginde, A. A., Scragg, R., Schwartz, R. S., & Camargo, C. A., Jr. (2009). Prospective Study of Serum 25-Hydroxyvitamin D Level, Cardiovascular Disease Mortality, and All-Cause Mortality in Older U.S. Adults. *J Am Geriatr Soc 57*(9), 1595–1603.

Gloth, F. M., 3rd, Alam, W., & Hollis, B. (1999). Vitamin D vs broad spectrum phototherapy in the treatment of seasonal affective disorder. *J Nutr Health Aging, 3*(1), 5–7.

Grant, W. B. (2010). Relation between prediagnostic serum 25-hydroxyvitamin D level and incidence of breast, colorectal, and other cancers. *J Photochem Photobiol B, 101*(2), 130–136.

Harris, S., & Dawson-Hughes, B. (1993). Seasonal mood changes in 250 normal women. *Psychiatry Res, 49*(1), 77–87.

Hobbs, R. D., Habib, Z., Alromaihi, D., Idi, L., Parikh, N., Blocki, F., et al. (2009). Severe vitamin D deficiency in Arab-American women living in Dearborn, Michigan. *Endocr Pract, 15*(1), 35–40.

Holick, M. F. (2005). The vitamin D epidemic and its health consequences. *J Nutr, 135*(11), 2739S–2748S.

Hollis, B. W. (2005). Circulating 25-hydroxyvitamin D levels indicative of vitamin D sufficiency: implications for establishing a new effective dietary intake recommendation for vitamin D. *J Nutr, 135*(2), 317–322.

Hoogendijk, W. J., Lips, P., Dik, M. G., Deeg, D. J., Beekman, A. T., & Penninx, B. W. (2008). Depression is associated with decreased 25-hydroxyvitamin D and increased parathyroid hormone levels in older adults. *Arch Gen Psychiatry, 65*(5), 508–512.

Ibi, M., Sawada, H., Nakanishi, M., Kume, T., Katsuki, H., Kaneko, S., et al. (2001). Protective effects of 1 alpha,25-(OH)(2)D(3) against the neurotoxicity of glutamate and reactive oxygen species in mesencephalic culture. *Neuropharmacology, 40*(6), 761–771.

Institute of Medicine. (2010). *Dietary Reference Intakes for Calcium and Vitamin D.* November, Washington, DC: National Academy Press.

Jarvis, E. (1998). Schizophrenia in British immigrants: recent draws its bearings from some of the most robust findings, issues and implications. *Transcultural Psychiatry, 35*, 39–74.

Kalueff, A. V., & Tuohimaa, P. (2007). Neurosteroid hormone vitamin D and its utility in clinical nutrition. *Curr Opin Clin Nutr Metab Care, 10*(1), 12–19.

Kendell, R. E., & Adams, W. (2002). Exposure to sunlight, vitamin D and schizophrenia. *Schizophr Res, 54*(3), 193–198.

Kinney, D. K., Teixeira, P., Hsu, D., Napoleon, S. C., Crowley, D. J., Miller, A., et al. (2009). Relation of schizophrenia prevalence to latitude, climate, fish consumption, infant mortality, and skin color: a role for prenatal vitamin d deficiency and infections? *Schizophr Bull, 35*(3), 582–595.

Kuningas, M., Mooijaart, S. P., Jolles, J., Slagboom, P. E., Westendorp, R. G., & van Heemst, D. (2009). VDR gene variants associate with cognitive function and depressive symptoms in old age. *Neurobiol Aging, 30*(3), 466–473.

Lansdowne, A. T., & Provost, S. C. (1998). Vitamin D3 enhances mood in healthy subjects during winter. *Psychopharmacology (Berl), 135*(4), 319–323.

Lee, D. M., Tajar, A., Ulubaev, A., Pendleton, N., O'Neill, T. W., O'Connor, D. B., et al. (2009). Association between 25-hydroxyvitamin D levels and cognitive performance in middle-aged and older European men. *J Neurol Neurosurg Psychiatry, 80*(7), 722–729.

Levitt, A. J., & Boyle, M. H. (2002). The impact of latitude on the prevalence of seasonal depression. *Can J Psychiatry, 47*(4), 361–367.

Lingjarde O, B. T., Hansen, T., Gotestam, K.G. (1986). Seasonal affective disorder and midwinter insomnia in the far north: studies on two related chronobiological disorders in Norway. *Clin Neuropharmacol, 9*(Suppl. 4), 187–189.

Lisak, R. P., Benjamins, J. A., Bealmear, B., Nedelkoska, L., Studzinski, D., Retland, E., et al. (2009). Differential effects of Th1, monocyte/macrophage and Th2 cytokine mixtures on early gene expression for molecules associated with metabolism, signaling and regulation in central nervous system mixed glial cell cultures. *J Neuroinflammation, 6*, 4.

Llewellyn, D. J., Lang, I. A., Langa, K. M., Muniz-Terrera, G., Phillips, C. L., Cherubini, A., et al. (2010). Vitamin D and risk of cognitive decline in elderly persons. *Arch Intern Med, 170*(13), 1135–1141.

Llewellyn, D. J., Langa, K. M., & Lang, I. A. (2009). Serum 25-hydroxyvitamin D concentration and cognitive impairment. *J Geriatr Psychiatry Neurol, 22*(3), 188–195.

Magnusson, A., & Axelsson, J. (1993). The prevalence of seasonal affective disorder is low among descendants of Icelandic emigrants in Canada. *Arch Gen Psychiatry, 50*(12), 947–951.

May, H. T., Bair, T. L., Lappe, D. L., Anderson, J. L., Horne, B. D., Carlquist, J. F., et al. (2010). Association of vitamin D levels with incident depression among a general cardiovascular population. *Am Heart J, 159*(6), 1037–1043.

McGrath, J. (1999). Hypothesis: is low prenatal vitamin D a risk-modifying factor for schizophrenia? *Schizophr Res, 40*(3), 173–177.

McGrath, J., Saari, K., Hakko, H., Jokelainen, J., Jones, P., Jarvelin, M. R., et al. (2004). Vitamin D supplementation during the first year of life and risk of schizophrenia: a Finnish birth cohort study. *Schizophr Res, 67*(2–3), 237–245.

McGrath, J., Scragg, R., Chant, D., Eyles, D., Burne, T., & Obradovic, D. (2007). No association between serum 25-hydroxyvitamin D3 level and performance on psychometric tests in NHANES III. *Neuroepidemiology, 29*(1–2), 49–54.

McGrath, J. J., Eyles, D. W., Pedersen, C. B., Anderson, C., Ko, P., Burne, T. H., et al. (2010). Neonatal vitamin D status and risk of schizophrenia: a population-based case-control study. *Arch Gen Psychiatry, 67*(9), 889–894.

Mersch, P. P., Middendorp, H. M., Bouhuys, A. L., Beersma, D. G., & van den Hoofdakker, R. H. (1999). Seasonal affective disorder and latitude: a review of the literature. *J Affect Disord, 53*(1), 35–48.

Meyer, U., & Feldon, J. (2009). Epidemiology-driven neurodevelopmental animal models of schizophrenia. *Prog Neurobiol, 90*(3), 285–326.

Milaneschi, Y., Shardell, M., Corsi, A. M., Vazzana, R., Bandinelli, S., Guralnik, J. M., et al. (2010). Serum 25-hydroxyvitamin D and depressive symptoms in older women and men. *J Clin Endocrinol Metab, 95*(7), 3225–3233.

Mishal, A. A. (2001). Effects of different dress styles on vitamin D levels in healthy young Jordanian women. *Osteoporos Int, 12*(11), 931–935.

Nanri, A., Mizoue, T., Matsushita, Y., Poudel-Tandukar, K., Sato, M., Ohta, M., et al. (2009). Association between serum 25-hydroxyvitamin D and depressive symptoms in Japanese: analysis by survey season. *Eur J Clin Nutr, 63*(12), 1444–1447.

Naveilhan, P., Neveu, I., Baudet, C., Ohyama, K. Y., Brachet, P., & Wion, D. (1993). Expression of 25(OH) vitamin D3 24-hydroxylase gene in glial cells. *Neuroreport, 5*(3), 255–257.

Neveu, I., Naveilhan, P., Menaa, C., Wion, D., Brachet, P., & Garabedian, M. (1994). Synthesis of 1,25-dihydroxyvitamin D3 by rat brain macrophages in vitro. *J Neurosci Res, 38*(2), 214–220.

Obunai, K., Jani, S., & Dangas, G. D. (2007). Cardiovascular morbidity and mortality of the metabolic syndrome. *Med Clin North Am, 91*(6), 1169–1184.

Oudshoorn, C., Mattace-Raso, F. U., van der Velde, N., Colin, E. M., & van der Cammen, T. J. (2008). Higher serum vitamin D3 levels are associated with better cognitive test performance in patients with Alzheimer's disease. *Dement Geriatr Cogn Disord, 25*(6), 539–543.

Pan, A., Lu, L., Franco, O. H., Yu, Z., Li, H., & Lin, X. (2009). Association between depressive symptoms and 25-hydroxyvitamin D in middle-aged and elderly Chinese. *J Affect Disord, 118*(1–3), 240–243.

Partonen, T., Partinen, M., & Lonnqvist, J. (1993). Frequencies of seasonal major depressive symptoms at high latitudes. *Eur Arch Psychiatry Clin Neurosci, 243*(3–4), 189–192.

Pasco, J. A., Henry, M. J., Nicholson, G. C., Sanders, K. M., & Kotowicz, M. A. (2001). Vitamin D status of women in the Geelong Osteoporosis Study: association with diet and casual exposure to sunlight. *Med J Aust, 175*(8), 401–405.

Potkin, S. G., Zetin, M., Stamenkovic, V., Kripke, D., & Bunney, W. E., Jr. (1986). Seasonal affective disorder: prevalence varies with latitude and climate. *Clin Neuropharmacol, 9*(Suppl. 4), 181–183.

Przybelski, R. J., & Binkley, N. C. (2007). Is vitamin D important for preserving cognition? A positive correlation of serum 25-hydroxyvitamin D concentration with cognitive function. *Arch Biochem Biophys, 460*(2), 202–205.

Querfurth, H. W., & LaFerla, F. M. (2010). Alzheimer's disease. *N Engl J Med, 362*(4), 329–344.

Rondanelli, M., Trotti, R., Opizzi, A., & Solerte, S. B. (2007). Relationship among nutritional status, pro/antioxidant balance and cognitive performance in a group of free-living healthy elderly. *Minerva Med, 98*(6), 639–645.

Rosen, L. N., Targum, S. D., Terman, M., Bryant, M. J., Hoffman, H., Kasper, S. F., et al. (1990). Prevalence of seasonal affective disorder at four latitudes. *Psychiatry Res, 31*(2), 131–144.

Sabetta, J. R., DePetrillo, P., Cipriani, R. J., Smardin, J., Burns, L. A., & Landry, M. L. (2010). Serum 25-hydroxyvitamin d and the incidence of acute viral respiratory tract infections in healthy adults. *PLoS One, 5*(6), e11088.

Schaffer, A., Levitt, A. J., & Boyle, M. (2003). Influence of season and latitude in a community sample of subjects with bipolar disorder. *Can J Psychiatry, 48*(4), 277–280.

Sergeev, I. N. (2009). 1,25-Dihydroxyvitamin D3 induces Ca2+-mediated apoptosis in adipocytes via activation of calpain and caspase-12. *Biochem Biophys Res Commun, 384*(1), 18–21.

Shinpo, K., Kikuchi, S., Sasaki, H., Moriwaka, F., & Tashiro, K. (2000). Effect of 1,25-dihydroxyvitamin D(3) on cultured mesencephalic dopaminergic neurons to the combined toxicity caused by L-buthionine sulfoximine and 1-methyl-4-phenylpyridine. *J Neurosci Res, 62*(3), 374–382.

Shipowick, C. D., Moore, C. B., Corbett, C., & Bindler, R. (2009). Vitamin D and depressive symptoms in women during the winter: a pilot study. *Appl Nurs Res, 22*(3), 221–225.

Smolders, J., Thewissen, M., Peelen, E., Menheere, P., Tervaert, J. W., Damoiseaux, J., et al. (2009). Vitamin D status is positively correlated with regulatory T cell function in patients with multiple sclerosis. *PLoS One, 4*(8), e6635.

Stewart, R., & Hirani, V. (2010). Relationship between vitamin D levels and depressive symptoms in older residents from a national survey population. *Psychosom Med, 72*(7), 608–612.

Taniura, H., Ito, M., Sanada, N., Kuramoto, N., Ohno, Y., Nakamichi, N., et al. (2006). Chronic vitamin D3 treatment protects against neurotoxicity by glutamate in association with upregulation of vitamin D receptor mRNA expression in cultured rat cortical neurons. *J Neurosci Res, 83*(7), 1179–1189.

Terman, M. (1988). On the question of mechanism in phototherapy for seasonal affective disorder: considerations of clinical efficacy and epidemiology. *J Biol Rhythms, 3*(2), 155–172.

Torrey, E. F., Torrey, B. B., & Peterson, M. R. (1977). Seasonality of schizophrenic births in the United States. *Arch Gen Psychiatry, 34*(9), 1065–1070.

U.S. Department of Agriculture. (2009). *National Nutrient Database for Standard Reference, Release 22 (IU) Content of Selected Foods per Common Measure, Vitamin D Sorted by Nutrient Content.* Retrieved 31 December 2009, from http://www.ars.usda.gov/SP2UserFiles/Place/12354500/Data/SR22/nutrlist/sr22w324.pdf

Wang, J. Y., Wu, J. N., Cherng, T. L., Hoffer, B. J., Chen, H. H., Borlongan, C. V., et al. (2001). Vitamin D(3) attenuates 6-hydroxydopamine-induced neurotoxicity in rats. *Brain Res, 904*(1), 67–75.

Wilkins, C. H., Birge, S. J., Sheline, Y. I., & Morris, J. C. (2009). Vitamin D deficiency is associated with worse cognitive performance and lower bone density in older African Americans. *J Natl Med Assoc, 101*(4), 349–354.

Wilkins, C. H., Sheline, Y. I., Roe, C. M., Birge, S. J., & Morris, J. C. (2006). Vitamin D deficiency is associated with low mood and worse cognitive performance in older adults. *Am J Geriatr Psychiatry, 14*(12), 1032–1040.

Zemel, M. B., & Sun, X. (2008). Calcitriol and energy metabolism. *Nutr Rev, 66*(10 Suppl. 2), S139–S146.

CHAPTER 6

Exploring B Vitamins Beyond Early Adulthood

Jonathon Lee Reay

Chapter Overview

- B vitamins are essential for one-carbon metabolism.
- It has been hypothesised that one-carbon metabolism is a biological correlate of dementia.
- It has been suggested that homocysteine levels are a biological marker of one-carbon metabolism.
- Elevated homocysteine levels are often coupled with low vitamin B status.
- Epidemiological studies have identified elevated homocysteine levels as a risk factor for dementia.
- Vitamin intervention studies have provided little evidence of the effectiveness of vitamin therapy.

Introduction

It is clear from the evidence presented in Chapter 4 that the nine water-soluble B vitamins are integral to cell function and healthy development in early life and into early adulthood. This chapter will consider three specific B vitamins – folic acid (B9), pyridoxine (B6) and cobamides (B12) – and their role on psychological function beyond early adulthood. I will consider the evidence from epidemiological studies and randomised control trials (RCTs).

Over the last two decades, there has been an explosion of research exploring the therapeutic role of B vitamins for cognitive function in the elderly. This interest has developed as a result of numerous epidemiological studies demonstrating associations between low dietary vitamin B status, high homocysteine (Hcy) levels and impaired cognitive function in the elderly. Despite the initial excitement surrounding the discovery of these associations, RCTs have yet to provide compelling evidence for vitamin B therapy as a cognitive enhancer in the elderly. However, it can be concluded from extant findings that vitamin B supplementation lowers Hcy levels and increases vitamin B status in those who do not suffer absorption problems. The lack of evidence from RCTs may be a result of the lack of such research combined with disparity in (a) treatment dose, (b) study duration, and (c) outcome measures, together with heterogeneity between samples in those studies that have been conducted to date. More research is clearly warranted.

Mechanisms of Action

In 1992, two papers suggested that raised concentrations of Hcy may be a biological marker of abnormal one-carbon metabolism (see Figure 6.1) and that this may play a role in the aetiology of Alzheimer's disease (AD) (McCaddon & Kelly 1992; Regland & Gottfries 1992). Support for this hypothesis comes from epidemiological evidence in healthy and patient populations showing a negative association between Hcy levels and cognitive abilities. In addition, Hcy level has been reported to be an independent risk factor for AD (see Smith 2008). As shown in Figure 6.1, one-carbon metabolism refers to the generation of one-carbon units, normally from serine, through association with a folic acid-derivative, tetrahydrofolate (THF), to form 5, 10-methylenetetrahydrofolate and then 5-methyltetrahydrofolate. This is used to methylate Hcy, in a reaction catalysed by vitamin B12, and

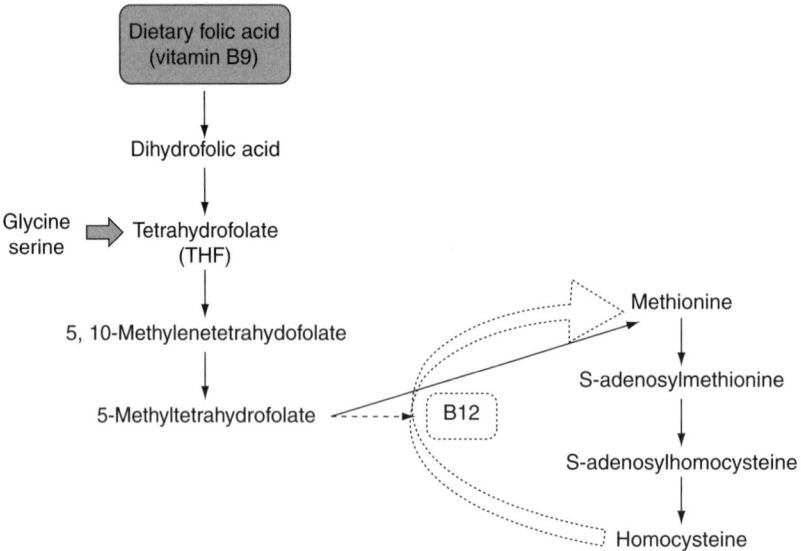

Figure 6.1   Illustration of two important roles of vitamin B9 and vitamin B12 in one-carbon metabolism

Vitamin B9 is the parent molecule of THF. Vitamin B12 catalyses the methylation reaction between 5-methyltetrahydrofolate and homocysteine to form methionine.

is used in the synthesis of methionine. Smith (2008) outlines 12 biologically plausible mechanisms to explain the association between B vitamins, Hcy and dementia. However, Selhub et al. (2010) state that the evidence for such mechanisms relies on animal and cell culture models and that there is 'no supportive evidence to suggest that the concentrations of Hcy in the cerebrospinal fluid (CSF) reach levels which could be considered neurotoxic, based on in vitro data' (Serot et al. 2005).

Epidemiological Evidence

The interest in vitamin B status and cognitive function gained momentum when reports started to document a relationship between the two in the elderly. Such evidence comes from a series of epidemiological studies, which will be discussed below.

Healthy Elderly

Goodwin et al. (1983) were the first to link nutritional status (using food record diaries and blood levels) and cognitive function in the healthy elderly population (mean age 71 years; n = 260). Goodwin and colleagues investigated simple correlations between nutrient blood levels (protein, ascorbate, thiamine, riboflavin, pyridoxine, B12, niacin and folate) and cognitive function (Russell Revision of the Wechsler Memory Test; Halstead-Reitan Categories Test). Using blood levels, the authors reported significant correlations between nutritional status and cognitive function for two nutrients (riboflavin and ascorbate) which could account for 2–3 per cent of the variance in cognitive function. Using the participants' nutritional status, the researchers then extracted the bottom 5 and 10 per cent of the sample and compared their blood nutrient levels to those in the rest of the sample (i.e. 90–95 per cent of the participants). Based on variance in levels of ascorbate and B12 vitamins, there were significant group differences on the Russell Revision of the Wechsler Memory Test. Following this first identification of a possible link between these nutritional factors and cognitive function, subsequent associations have been reported which suggest a link between vitamin B and cognitive function in older adults (for a review, see Selhub et al. 2000).

One issue that has hindered the evaluation of the link between vitamin B intake and cognitive function is the fact that low vitamin B levels and high Hcy levels are generally concomitant. Further research has attempted to investigate the relative importance of each of the B vitamins and Hcy for successful cognitive function. Riggs et al. (1996) investigated the relationship between Hcy, vitamins B6, B9 and B12 and cognitive function in 70 males (mean age 66 years) as part of a Normative Ageing Study. Results revealed that lower concentrations of vitamins B9 and B12 and a higher concentration of Hcy were associated with reduced spatial copying skills. Importantly, however, Hcy provided the strongest predictor of performance. In addition, results revealed higher concentrations of vitamin B6 to be related to better memory performance, leading the authors to suggest that the different B vitamins and Hcy may have differential effects on cognitive function. An interesting observation from this study was that these associations were observed despite few participants having low levels of vitamin B12 (< 200 ng/l) and vitamin B9 (< 3 µg/l).

In similar research, Tucker et al. (2005) used data from the Veterans' Affairs Normative Ageing Study (NAS) to investigate associations between Hcy and vitamins B6, B9 and B12 with cognitive decline in 321 men (mean age 67 years; mean Mini Mental State Exam (MMSE) 27.2) over a three-year period. In line with the results of Riggs et al. (1996), Tucker and colleagues reported that higher concentrations of vitamins B9 and B12 and lower concentrations of Hcy were associated with better spatial copying skills. They also suggested that vitamin B6 was associated with spatial copying skills and verbal fluency. However, when considering the relative importance of the different vitamins on different components of cognition, Tucker et al. reported vitamin B9 to be the only predictor of performance once the contribution of other vitamins was controlled for through covariance. In extending their research, Tucker and colleagues reported that only when vitamin B9 levels drop to < 20 nmol/l is there a significant loss of spatial copying ability.

The impact of the different B vitamins on different cognitive modalities has been further emphasised by Feng et al. (2006), who investigated the relative associations between cognitive function and vitamins B9 and B12 and Hcy levels in 451 high-functioning elderly (> 55 years old) Chinese volunteers from the Singapore Longitudinal Ageing Study (MMSE ≥ 24). In this study, higher levels of Hcy were associated with poorer performance in the Block Design task and the Symbol Digit Modality Test, and higher levels of vitamin B9 were associated with better performance in the Rey Auditory Verbal Learning test and verbal fluency. Further, levels of vitamin B12 were not associated with performance on any of the cognitive tasks. Therefore, it appears that levels of the different B vitamins may influence different domains of cognition.

It is worth mentioning one study that investigated these associations across a broader age group, rather than restricting the sample to older adults. Using a sample of 2871 participants from the Northern Manhattan Study, Wright et al. (2004) reported an association between higher Hcy levels and lower mean MMSE scores for adults older than 65 years, but not for adults between the ages of 40 and 64 years. This finding suggests that Hcy status may have a greater impact on cognitive function with increasing age. Further, adjusting for vitamin B12 deficiency and socio-demographic factors, the mean MMSE was 2.2 points lower for each unit increase in the log Hcy level.

Based on the studies summarised in this section, it is apparent that there are associations between levels of B vitamins, Hcy levels and cognitive function in older healthy adults. There are also suggestions of independent and differential effects of some B vitamins on sub-components of cognitive processing. Further research is needed to fully understand these associations. The next section will consider evidence of a relationship between vitamin B status and the risk of developing mild cognitive impairment (MCI) and dementia.

MCI and Dementia

The chance of developing MCI and/or dementia in later life is exacerbated by a number of risk factors. Although the biggest risk is increased age, there is substantial research investigating lifestyle risk factors, one of these being nutritional deficiencies. Droller and Dossett (1959) first reported a link between a low vitamin B12 level and dementia that was independent of pernicious anaemia. Subsequent to this, other studies attempted to establish the relative importance of each B vitamin and Hcy as risk factors for developing MCI and dementia. For example, Quadri et al. (2004) examined associations between plasma Hcy, vitamin B9, vitamin B12 and the probability of developing MCI, dementia of the Alzheimer's type (AD) or vascular dementia. Quadri and colleagues categorised 228 participants into three groups: an elderly control group (n = 55; mean age 76 years); those diagnosed with MCI (n = 81; mean age 76 years) and those diagnosed with dementia (n = 92; mean age 80 years). Results revealed that vitamin B9 concentrations were lower in the dementia group and MCI group compared with the control group. Hcy was also higher in the dementia group compared with the elderly control group, and the proportion of people in the dementia group with higher than normal Hcy level (< 14.6 μmol/l) was greater compared with the control group. The sample was further sub-categorised using tertiary split, and odds ratios (OR) were calculated. Results suggested that those with the highest Hcy levels (> 14.6 μmol/l) and the lowest vitamin B9 levels (<13.5 nmol/l) were more than three times likely to develop AD.

In further research, Ramos et al. (2005) used the Sacramento Area Latino Study on Ageing (SALSA) to investigate the relationship between vitamin B9 status, cognitive function and dementia diagnosis in 1789 community-dwelling individuals of Latino ethnic background over 60 years of age. Employing a four-stage regression analysis, results revealed that vitamin B9 status was positively associated with performance in all seven cognitive tasks (model 1), and was positively associated with six cognitive tasks after Hcy (model 2) and vitamin B12 and creatinine (model 3) were controlled for statistically. However, when demographic variables and depression scores were added (model 4), vitamin B9 only correlated with 3MSE (measure of global cognitive ability) and delayed recall. Secondary analyses demonstrated that the OR (controlling for Hcy, vitamin B12, creatinine, demographic variables and depression score) for low 3MSE score (< 78) and dementia diagnosis decreased with increasing vitamin B9 concentrations. The OR for Hcy did not remain significant after controlling for vitamin B9. In addition, Haan et al.

(2007) investigated the relationship between Hcy, vitamin B9, vitamin B12 and the risk of developing dementia or cognitive impairment without dementia (CIND) in 1405 participants aged 60–101 years old. In contrast to Ramos et al. (2005), Hann et al. reported that it is Hcy (not vitamin B9) that is associated with a greater risk of dementia or CIND, and that higher vitamin B12 concentrations may reduce this risk. To further complicate matters, Mooijaart et al. (2005) also investigated the relationship between Hcy, vitamins B12 and B9 and cognitive decline. A total of 599 elderly participants completed a test battery at 85 years of age and annually thereafter until they were 89 years of age. Hcy, vitamin B12 and vitamin B9 levels were measured at the first (i.e. 85 years) and last (i.e. 89 years) testing sessions. A cross-sectional analysis at 85 years showed a negative relationship between Hcy and MMSE, positive associations between vitamin B9 and cognitive function, but no significant relationship between vitamin B12 and cognitive function. A prospective analysis showed that concentrations of Hcy, vitamin B12 and vitamin B9 at 85 years were not predictive of cognitive decline over the study period. However, it should be noted that participants were not screened for health or demographic variables (shown to be important by Ramos et al. 2005). It is difficult to reconcile such opposing results. However, some of the differences may be due to different methods and the populations studied.

One potentially important methodological confound between studies in this area is the method used to measure vitamin B status. This potential conflict is highlighted by the results of Clarke et al. (2007), who investigated the associations between cognitive decline and levels of vitamins B12 and B9 and Hcy by measuring holoTC (a marker of reduced vitamin B12 status), total Hcy, methylmalonic acid (MMA), vitamin B12 and vitamin B9 levels. Cognitive function was assessed in 1648 participants at baseline (mean age 75 years; mean MMSE 26.2) and again at a minimum of three points in time over a ten-year period. Of the original participants, 691 survived until the ten-year follow-up (mean age 72 years; mean MMSE 27.3). Cross-sectional analyses at baseline and at the ten-year follow-up revealed significant associations between several factors; for example, between total homocysteine (tHcy) and cognitive function; between holoTC and cognitive function; and between MMA and cognitive function. Longitudinal analyses revealed that a doubling in holoTC concentrations (50–100 pmol/l) was associated with a 30 per cent slower rate of cognitive decline. Conversely, a doubling in tHcy (from 10 to 20 μmol/l) or MMA (0.25 μmol/l) was associated with > 50 per cent more rapid cognitive decline. Interestingly, in contrast to the previous studies, Clarke and colleagues found no association between cognitive function and total vitamin B12 or B9 levels. The authors suggest that these may be less accurate markers of vitamin B status compared with holoTC and MMA, high levels of which are indicative of vitamin B12 deficiency. An additional consideration is that evidence in the literature suggests that the relationship between vitamin B12 status and cognitive function in older individuals may be mediated by Apolipoprotein E (APOE). Feng et al. (2009) investigated this possibility.

A total of 539 adults (mean age 65 years) completed a number of cognitive function tests, including the MMSE. The MMSE was completed at baseline (all participants had had an MMSE > 21) and approximately 18 months and 38 months thereafter. Results suggested that the relationship between vitamin B12 and MMSE score was stronger in the carriers of the APOE $\varepsilon4$ than non-carriers, suggesting that APOE $\varepsilon4$ moderates this relationship. However, it should be noted that this moderating effect was not present when the relationship was analysed in a subsample of 416 'cognitively normal' (MMSE score of < 24) adults.

It is clear from extant findings that Hcy levels and vitamin B status are related to MCI and dementia in the elderly. However, to date, the nature of the relative contribution of each of these factors to cognitive function remains unclear, and further research is needed. As high levels of Hcy and low levels of B vitamins have been related to cognitive decline in the elderly, there has been a great deal of interest in whether vitamin B supplementation can slow age-related cognitive deficits.

Vitamin B Supplementation Studies

There has been significant research investigating whether vitamin B supplementation can aid cognitive function during the ageing processes. The majority of the evidence obtained using an objective, experimental methodology has provided no evidence of efficacy for vitamin B supplementation. However, it is clear that more research is needed to answer two fundamental questions: (a) the optimal dose required, and (b) the length of the treatment period. To date, these questions have not been sufficiently investigated or satisfactorily answered. Research needs to explore further the interaction between approved medical treatments for dementia and nutritional supplements while controlling for important extraneous variables (e.g. genetic and demographic variables). In addition to this, I propose that future research investigates the possibility of 'critical periods'.

Methodological issues are a problem for the interpretation of many previous vitamin B supplementation studies. For example, Yukawa et al. (2001) assessed the effects of 60 days' treatment with vitamin B9 (15 mg/day) in 36 (mean age 56 years old) vitamin B9-deficient volunteers (defined as < 4 ng/ml serum folate), all with neurological disease. Results suggested that B9 administration improved neurological symptoms in 24 of the 36 cases after two months (eight of these individuals having been diagnosed with dementia). However, Yukawa et al. did not include a placebo control group in their study. Similarly, Nilsson et al. (2001) investigated the effects of oral supplementation of vitamin B12 (1 mg/day) and B9 (5 mg/day) for two months. In total, 39 patients (mean age 78 years) were classified with mild, moderate or severe dementia (categorised according to DSM III-R [*Diagnostic and Statistical Manual of Mental Disorders 3* revised] criteria). The patients with severe dementia were too ill to participate and were excluded from the study. The remaining 28 patients were sub-categorised into two groups based on their Hcy levels (those having a plasma Hcy level < 19.9 μmol/l and those above). Results revealed that the patients with high Hcy (> 19.9 μmol/l) showed clinical improvements after oral supplementation with vitamin B12. However, once again, the study failed to implement a placebo control group.

In studies that have used more rigorous methods, the results are much clearer. The conclusions obtained from these scientific investigations indicate that vitamin B supplementation has no effect on cognitive function alone or in combination (for a review, see Malouf & Grimly-Evan 2008, 2009; Malouf & Areosa Sastre 2009). However, it should be noted that individual studies have varied substantially with regards to treatment, treatment dose, length of treatment regime, study population and method of assessment of cognitive function. In addition, some studies have used a very small sample size, which inevitably increases the probability of reporting findings that support a null hypothesis, due to decreased power. Some of the relevant problems will be highlighted in the remainder of this chapter, where the evidence for monotherapy and multivitamin therapy will be considered.

Monotherapy with Vitamins B6, B9 and B12

The majority of monotherapy studies have elected to study vitamin B9 and have kept the experimental design simple (something that I would advocate) by comparing one B vitamin with a placebo. However, one study has investigated the effect of vitamins B6, B9 and B12 against a placebo. Bryan et al. (2002) used a placebo-controlled design to investigate the effect of 35 days of vitamin B supplementation on cognitive function and mood in three age ranges: (a) young adults (20–30 years old; n = 56); (b) middle-aged adults (45–55 years old; n = 80) and (c) older adults (65–92 years old; n = 75). Participants were randomly allocated to one of four treatment conditions: (a) vitamin B9 (0.75 mg); (b) vitamin B12 (0.015 mg); (c) vitamin B6 (75 mg) or (d) placebo. Results revealed no significant evidence of a treatment effect on cognitive function, and no significant effect on subjective mood.

Vitamin B6

Few studies have investigated the therapeutic value of vitamin B6 as a monotherapy. Deijen et al. (1992) used a placebo-controlled trial to investigate the effects of 12 weeks' daily ingestion of vitamin B6 (20 mg). In total, 76 healthy volunteers (mean age 73 years) were assigned to either the treatment or the placebo group, using a matched pairs methodology. Of the 76 participants, 12 in the placebo group and 4 in the treatment group were defined as marginally B6 deficient at baseline (Pyridoxal 5' phosphate (PLP) < 20 nmol/l or α-aspartic aminotransferase (EAST) > 1.98). However, these participants were retained in the study, thereby raising some doubts about the match-pairing methods used (participants were reportedly matched on age, vitamin B6 status and IQ). Results provided minimal support for a treatment effect, using multiple bivariate correlations to examine changes in cognitive performance and PLP baseline levels. Although the authors did report some significant effects, they failed to control for multiple comparisons and did not report the associations observed in the placebo group.

Vitamin B9

Numerous studies have investigated the therapeutic value of vitamin B9 as a monotherapy. However, the disparity in methods makes it almost impossible to form a rational conclusion, other than one that requests more research to be carried out. Pathansali et al. (2006) used a placebo-controlled design to investigate the effect of daily ingestion of vitamin B9 (5 mg) over a four-week period on cognitive function in 24 healthy older adults (mean age 73 years; MMSE > 27) with normal baseline vitamin B9 levels (6.3 ± 2.4 µg/l). Results demonstrated no effect of vitamin B9 on psychomotor function. Vitamin B9 levels were raised by the treatment and Hcy levels were lowered. Over a longer treatment period and implementing a larger dose, Sommer et al. (2003) used a placebo-controlled design to investigate the effects of daily vitamin B9 treatment (10 mg) over a ten-week supplementation period in seven patients (mean age 77 years old) suffering from dementia (classified by *Diagnostic and Statistical Manual of Mental Disorders 3* revised (DSM-III-R)), who also presented low vitamin B9 levels (defined as serum B9 between 2 and 5 mcg/l or red blood cell (RBC) B9 between 127 and 452 mcg/l and B12 above 200 ng/l). Results showed no effect of B9 supplementation compared with the placebo. Connelly et al. (2008) investigated a longer treatment period but reduced the daily dose. In a placebo-controlled trial, they investigated the effect of vitamin B9 treatment on 41 (mean age 76 years; mean MMSE 23.49) participants with probable AD (meeting the National Institute of Neurological and Communicative Disorders and Stroke (NINCDS)-Alzheimer's Disease and Related Disorders Association (ADRDA) diagnostic criteria). Participants were allocated to one of two groups: daily treatment of 1 mg of vitamin B9 or a placebo for six months. At the same time, all participants started cholinesterase inhibition (ChI) treatment. However, the type of ChI and dose varied across participants, depending on their clinical response. Results showed that 16 out of 23 participants in the vitamin B9 group and 7 out of 18 participants in the placebo group were classified as NICE (National Institute for Health and Clinical Excellence) responders (defined as having good response after six months' treatment, according to NICE criteria (NICE 2001)). Within-group change from baseline showed improvements following vitamin B9 supplementation on activities of daily living and social behaviours, but no change in MMSE score. However, it is very difficult to delineate the effects of ChI in this study (and possible interaction with B9 supplementation), given the personalised medical treatment.

In a large study, Durga et al. (2007) implemented a placebo-controlled, between-subjects design to investigate the effects of daily vitamin B9 supplementation (0.8 mg) for three years on cognitive function in 818 individuals (mean age 60 years: MMSE at baseline > 24). A total of 404 participants were allocated to the intervention group. Within-group comparisons of change data revealed better cognitive performance following vitamin B9 treatment. However, as with Connelly et al., within-group comparisons were not supported by between-group differences, and the results should therefore be treated with some caution.

Vitamin B12

In a very small study, Seal et al. (2002) used a placebo-controlled design to investigate the effects of two doses of vitamin B12 (0.01 mg or 0.05 mg) ingested daily for one month by 31 older adults (mean age 81.4 years; mean MMSE 18.23) with subnormal vitamin B12 levels (serum B12 between 100 and 150 pmol/l). Results revealed that a daily intake of 0.05 mg of vitamin B12 increased serum B12 but did not affect cognitive function. In a larger study conducted over a longer period of time and using a larger dose, Hvas et al. (2004) used a placebo-controlled trial to investigate the effect of three months' treatment with vitamin B12 (1 mg intravenous). A total of 140 participants (mean age 74.5 years; mean MMSE = 26.5) were randomly assigned to either treatment or placebo groups. Participants had increased plasma methylmalonic acid (P-MMA) (0.4–2 μmol/l) and high tHcy (mean 13 μmol/l) at baseline. Results revealed no significant effect of treatment on cognitive performance.

Multi B-Vitamin Therapy

A number of studies have investigated the effect of vitamin B treatment which comprises two or more B vitamins. Stott et al. (2005) implemented a placebo-controlled trial to investigate the effect of a 12-week vitamin supplementation programme on Hcy levels (assessed at baseline and 3 months later) and cognitive function (assessed at baseline and 6 and 12 months after randomisation) in 185 elderly (> 65 years old) patients with vascular disease. Patients were randomly allocated to a placebo condition or one of seven different treatment conditions. Treatments were: (a) vitamin B9 (2.5 mg) plus vitamin B12 (0.5 mg); (b) vitamin B6 (25 mg); and (c) riboflavin (25 mg). Seven treatment comparisons were undertaken, comprising the three treatment conditions alone (i.e. a, b and c) and in combination with each other (i.e. a + b, a + c, b + c, a + b + c). The results suggested that vitamin B9 with vitamin B12 could decrease Hcy levels but had no effect on cognitive function.

Other studies have combined the intake of B vitamins with other vitamins that are not from this category. In a large study, over a long treatment period, Clarke et al. (2003) used a placebo-controlled trial to investigate the effects of aspirin (81 mg), vitamin B (B9 at 2 mg; B12 at 1 mg) and vitamins C and E (200 mg and 500 mg) ingested daily for 12 weeks. In total, 149 participants (median age = 75 years; median MMSE = 21; median Hcy = 12.4 μmol/l) were randomly allocated to treatment and placebo groups (6 different groups), with 142 of these individuals returning for a follow-up. At baseline (N = 149), data were split by age (< 75 and > 75 years old) and severity of cognitive impairment (interquartile split). Results revealed a negative correlation between cognitive function and Hcy levels, and a positive association between vitamin B9 and cognitive function, but no association between vitamin B12 and cognitive function (these associations remained after adjustment for age). After 12 weeks of supplementation with vitamin B (B9 at 2 mg; B12 at 1 mg), results revealed significant increases in B9 and B12 and significant decreases in Hcy. However, there was no effect on cognitive function. Lewerin et al. (2005) used a placebo-controlled trial to investigate the effect of four months of treatment (B12 at 0.5 mg; B9 acid at 0.8 mg and B6 at 3 mg) on nine cognitive tasks (digit span forward and backward, identical forms, visual reproduction, synonyms, block design, digit symbol, Thurstone's picture memory and Figure classification task) and five indices of movement (movement time, postural phase, locomotor phase, manual phase and simultaneity index) in 209 community-dwelling elderly volunteers (mean age 75.41 years; mean tHcy = 16.95 μmol/l; MMSE was not completed). At baseline, tHcy and MMA correlated with cognitive function and movement; however, vitamin B9 and B12 levels did not. Four months of treatment had no effect on cognitive function, but did decrease tHcy and MMA levels.

Eussen et al. (2006) implemented a placebo-controlled, between-subjects design to investigate the effect of a 24-week dietary supplementation period on cognitive function in elderly (> 70 years old) participants with mild vitamin B12 deficiency (defined as B12 concentration between 100 and 200 pmol/l, or between 200 and 300 pmol/l with concomitant high MMA levels and low creatinine concentrations). Treatments consisted of: (a) vitamin B12 (1 mg daily; n = 54); (b) vitamin B12 + B9 (1 mg + 0.4 mg; n = 51) and (c) placebo (n = 57). Results demonstrated no effect of either treatment on cognitive function.

Over a longer treatment period and using larger doses, Aisen et al. (2008) used a placebo-controlled between-subject design to investigate the effect of vitamin B supplementation (5 mg of B9, 25 mg of B6, 1 mg of B12) daily for 18 months on cognitive decline in those with mild to moderate AD (mean age 76 years; MMSE between 14 and 26). In total, 202 participants received the active treatment and 138 received the placebo. Results revealed no beneficial effect of treatment. Similar conclusions were made by McMahon et al. (2006), who investigated the effects of a two-year daily treatment regime (1 mg of B9, 0.5 mg of B12, 10 mg of B6) on Hcy and verbal and non-verbal cognitive function in participants aged 65 years and older with a baseline Hcy level of at least 13 μmol/l. Results revealed a reduction in Hcy following treatment at 6, 12, 18 and 24 months after randomisation; however, there was no effect on cognitive function. Smith et al. (2010) implemented another study to investigate whether 24-month supplementation with high doses of B vitamins (B9 at 0.8 mg/d, B12 at 0.5 mg/d, B6 at 20 mg/d) resulted in slowing the rate of atrophy in elderly (> 70 years old) participants with MCI in comparison to the placebo. Results revealed that treatment preceded reduction in tHcy and a 30 per cent reduction in the rate of atrophy, which increased to 53 per cent reduction in those who had shown the greatest baseline levels of tHcy (>13 μmol/l). There was no effect of treatment on those with the lowest baseline levels of tHcy (< 9.5 μmol/l). The authors do not report results pertaining to performance in cognitive tests. In a study which has involved one of the longest supplementation periods, Kang et al. (2008) used a placebo-controlled design to investigate the effects of vitamin B supplementation (comprising 2.5 mg of B9, 1 mg of B12 and 50 mg of B6) in 2009 elderly (mean age 72 years) women with cardiovascular disease (CVD) and CVD risk factors (1002 allocated to the treatment group) over a five-and-a-half year period using a telephone cognitive battery measuring: (a) general cognition (Telephone Interview of Cognitive Status – TICS); (b) verbal memory (delayed recall of the TICS ten-word list and the immediate and delayed recalls of the East Boston Memory Test) and (c) category fluency (asked to name as many animals as possible in one minute). Results revealed no effect of treatment.

Conclusions and Future Directions

As summarised in this chapter, there appears to be some robust epidemiological evidence linking vitamin B status with cognitive decline. However, I propose that there is a lack of evidence from the existing intervention studies to support the notion that vitamin B therapy impacts positively on age-related cognitive decline. This last point notwithstanding, subsequent to treatment with B vitamins, a number of studies have reported a rise in vitamin B levels and a fall in Hcy level, both of which can be considered risk factors for dementia. In addition to this, as noted earlier, Smith et al. (2010) reported that 24 months of treatment with B vitamins (B9, B12 and B6) resulted in a slowing in brain atrophy in a group of older adults (>70 years old) suffering MCI. This is an important finding. It is clear from these and other findings reported in this chapter (together with inconsistencies that have been highlighted) that further research is warranted in this domain.

References

Aisen, P.S., Schneider, L.S., Sano, M., Diaz-Arrastia, R., van Dyck, C.H., Weiner, M.F., Bottiglieri, T., Jin, S., Stokes, K.T., Thomas, R.G. & Thal, L.J. (2008). High-dose B vitamin supplementation and cognitive decline in Alzheimer disease: a randomized controlled trial. *JAMA, 300*(15), 1774–83.

Bryan, J., Calvaresi, E. & Hughes, D. (2002). Short-term folate, vitamin B-12 or vitamin B-6 supplementation slightly affects memory performance but not mood in women of various ages. *Journal of Nutrition, 132*(6), 1345–56.

Clarke, R., Birks, J., Nexo, E., Ueland, P.M., Schneede, J., Scott, J., Molloy, A. & Evans, J.G. (2007). Low vitamin B-12 status and risk of cognitive decline in older adults. *American Journal of Clinical Nutrition, 86*(5), 1384–91.

Clarke, R., Harrison, G. & Richards, S. (2003). Effect of vitamins and aspirin on markers of platelet activation, oxidative stress and homocysteine in people at high risk of dementia. *Journal of International Medicine, 254*(1), 67–75.

Connelly, P.J., Prentice, N.P., Cousland, G. & Bonham, J. (2008). A randomised double-blind placebo-controlled trial of folic acid supplementation of cholinesterase inhibitors in Alzheimer's disease. *Inernationalt Journal of Geriatric Psychiatry, 23*(2), 155–60.

Deijen, J.B., van der Beek, E.J., Orlebeke, J.F. & van den Berg, H. (1992). Vitamin B-6 supplementation in elderly men: effects on mood, memory, performance and mental effort. *Psychopharmacology (Berlin), 109*(4), 489–96.

Droller, H. & Dossett, J.A. (1959). Vitamin B12 levels in senile dementia and confusional states. *Geriatrics, 14*(6), 367–73.

Durga, J., van Boxte,l M.P., Schouten, E.G., Kok, F.J., Jolles, J., Katan, M.B. & Verhoef, P. (2007). Effect of 3-year folic acid supplementation on cognitive function in older adults in the FACIT trial: a randomised, double blind, controlled trial. *Lancet, 369*, 208–16.

Eussen, S.J., de Groot, L.C., Joosten, L.W., Bloo, R.J., Clarke, R., Ueland, P.M., Schneede, J., Blom, H.J., Hoefnagels, W.H., & van Staveren, W.A. (2006). Effect of oral vitamin B-12 with or without folic acid on cognitive function in older people with mild vitamin B-12 deficiency: a randomized, placebo-controlled trial. *American Journal of Clinical Nutrition, 84*(2), 361–70.

Feng, L., Li, J., Yap, K.B., Kua, E.H. & Ng, T.P. (2009). Vitamin B-12, apolipoprotein E genotype, and cognitive performance in community-living older adults: evidence of a gene-micronutrient interaction. *American Journal of Clinical Nutrition, 89*(4), 1263–8. Epub 2009 February 25.

Feng, L., Ng, T.P., Chuah, L., Niti, M. & Kua, E.H. (2006). Homocysteine, folate, and vitamin B-12 and cognitive performance in older Chinese adults: findings from the Singapore Longitudinal Ageing Study. *American Journal of Clinical Nutrition, 84*(6), 1506–12.

Goodwin, J.S, Goodwin, J.M. & Garry, P.J. (1983). Association between nutritional status and cognitive functioning in a healthy elderly population. *JAMA, 249*(21), 2917–21.

Haan, M.N., Miller, J.W., Aiello, A.E., Whitmer, R.A., Jagust, W.J., Mungas, D.M., Allen, L.H. & Green, R. (2007). Homocysteine, B vitamins, and the incidence of dementia and cognitive impairment: results from the Sacramento Area Latino Study on Aging. *American Journal of Clinical Nutrition 85*(2), 511–7.

Hvas, A.M., Juul, S., Lauritzen, L., NexÃ, E. & Ellegaard, J. (2004). No effect of vitamin B-12 treatment on cognitive function and depression: a randomized placebo controlled study. *Journal of Affective Disorders 81*(3), 269–73.

Kang, J.H., Cook, N., Manson, J., Buring, J.E., Albert, C.M. & Grodstein, F. (2008). A trial of B vitamins and cognitive function among women at high risk of cardiovascular disease. *American Journal of Clinical Nutrition, 88*(6), 1602–10.

Lewerin, C., Matousek, M., Steen, G., Johansson, B., Steen, B. & Nilsson-Ehle, H. (2005). Significant correlations of serum homocysteine and serum methylmalonic acid with movement and cognitive performance in elderly subjects but no improvement from short-term vitamin therapy: a placebo-controlled randomized study. *American Journal of Clinical Nutrition, 81*, 1155–62.

Malouf, R. & Areosa Sastre, A. (2009). Vitamin B12 for cognition. *Cochrane Database of Systematic Reviews, 2003(3)*, CD004326. Review. Update in: *Cochrane Database of Systematic Reviews*, (1): CD004326.

Malouf, R. & Grimley Evans, J. (2008). Folic acid with or without vitamin B12 for the prevention and treatment of healthy elderly and demented people. *Cochrane Database of Systematic Reviews, October 8*(4), CD004514. Review.

Malouf, R. & Grimley Evans, J. (2009). The effect of vitamin B6 on cognition. *Cochrune Database of Systems Review*, (4), CD004393. Review.

McCaddon, A. & Kelly, C.L. Alzheimer's disease (1992). a "cobalaminergic" hypothesis. *Medical Hypotheses, 37*(3), 161–5.

McMahon, J.A., Green, T.J., Skeaff, C.M., Knight, R.G., Mann, J.I. & Williams, S.M. (2006). A controlled trial of homocysteine lowering and cognitive performance. *New England Journal of Medicine, June 29, 354*(26), 2764–72.

Mooijaart, S.P., Gussekloo, J., Frölich, M., Jolles, J., Stott, D.J., Westendorp, R.G. & de Craen, A.J. (2005). Homocysteine, vitamin B-12, and folic acid and the risk of cognitive decline in old age: the Leiden 85-Plus study. *American Journal of Clinical Nutrition, 82*(4), 866–71.

Nilsson, K., Gustafson, L. & Hultberg, B. (2001). Improvement of cognitive functions after cobalamin/folate supplementation in elderly patients with dementia and elevated plasma homocysteine. *International Journal of Geriatric Psychiatry, 16*(6), 609–14.

Pathansali, R., Mangoni, A.A., Creagh-Brown, B., Lan, Z.C., Ngow, G.L., Yuan, X.F., Ouldred, E.L., Sherwood, R.A., Swift, C.G. & Jackson, S.H. (2006). Effects of folic acid supplementation on psychomotor performance and hemorheology in healthy elderly subjects. *Archives of Gerontology and Geriatrics, 43*(1): 127–37. Epub 15 December 2005.

Quadri, P., Fragiacomo, C., Pezzati, R., Zanda, E., Forloni, G., Tettamanti, M. & Lucca, U. (2004).Homocysteine, folate, and vitamin B-12 in mild cognitive impairment, Alzheimer disease, and vascular dementia. *American Journal of Clinical Nutrition, 80*(1), 114–22.

Ramos, M.I., Allen, L.H., Mungas, D.M., Jagust, W.J., Haan, M.N., Green, R. & Miller, J.W. (2005). Low folate status is associated with impaired cognitive function and dementia in the Sacramento Area Latino Study on Aging, *American Journal of Clinical Nutrition, 82*(6), 1346–52.

Regland, B. & Gottfries, C.G. (1992). Slowed synthesis of DNA and methionine is a pathogenetic mechanism common to dementia in Down's syndrome, AIDS and Alzheimer's disease? *Medical Hypotheses, 38*(1): 11–9.

Riggs, K.M., Spiro, A., 3rd., Tucker, K. & Rush, D. (1996). Relations of vitamin B-12, vitamin B-6, folate, and homocysteine to cognitive performance in the Normative Aging Study. *American Journal of Clinical Nutrition, 63*(3), 306–14.

Seal, E.C., Metz, J., Flicker, L. & Melny, J. (2002). A randomized, double-blind, placebo-controlled study of oral vitamin B12 supplementation in older patients with subnormal or borderline serum vitamin B12 concentrations. *Journal of the American Geriatric Society, 50*(1), 146–51.

Selhub, J., Bagley, L.C., Miller, J. & Rosenberg, I.H. (2000). B vitamins, homocysteine, and neurocognitive function in the elderly. *American Journal of Clinical Nutrition, 71*(2), 614S–620S.

Selhub, J., Troen, A., Rosenberg, I.H. (2010). B vitamins and the aging brain. *Nutrition Reviews, 68* (Suppl. 2).

Serot, J.M., BarbÃ, F., Arning, E., Bottiglieri, T., Franck, P., Montagne, P. & Nicolas, J.P. (2005). Homocysteine and methylmalonic acid concentrations in cerebrospinal fluid: relation with age and Alzheimer's disease. *Journal of Neurology, Neurosurgery and Psychiatry, 76*(11):1585–7.

Smith, A.D. (2008). The worldwide challenge of the dementias: a role for B vitamins and homocysteine? *Food and Nutrition Bulletin, June 29*(Suppl. 2), S143–72.

Smith, A.D., Smith, S.M., de Jager, C.A., Whitbread, P., Johnston, C., Agacinski, G., Oulhaj, A., Bradley, K.M., Jacoby, R. & Refsum, H. (2010). Homocysteine-lowering by B vitamins slows the rate of accelerated brain atrophy in mild cognitive impairment: a randomized controlled trial. *PLoS One, 5*(9), e12244.

Sommer, B.R., Hoff, A.L. & Costa, M.J. (2003). Folic acid supplementation in dementia: a preliminary report. *Journal of Geriatric Psychiatry and Neurology, 16*(3), 156–9.

Stott, D.J., MacIntosh, G., Lowe, G.D, Rumley, A., McMahon, A.D., Langhorne, P., Tait, R.C., O'Reilly, D.S., Spilg, E.G., MacDonald, J.B., MacFarlane, P.W. & Westendorp, R.G. (2005). Randomized controlled trial of homocysteine-lowering vitamin treatment in elderly patients with vascular disease. *American Journal of Clinical Nutrition, 82*(6), 1320–6.

Tucker, K.L., Qiao, N., Scott, T., Rosenberg, I. & Spiro, A., 3rd. (2005). High homocysteine and low B vitamins predict cognitive decline in aging men: the Veterans Affairs Normative Aging Study. *American Journal of Clinical Nutrition, 82*(3), 627–35.

Wright, C.B., Lee, H.S., Paik, M.C., Stabler, S.P., Allen, R.H. & Sacco, R.L. (2004). Total homocysteine and cognition in a tri-ethnic cohort: the Northern Manhattan Study. *Neurology, 63*(2): 254–60.

Yukawa, M., Naka, H., Murata, Y., Katayama, S., Kohriyama, T., Mimori, Y. & Nakamura, S. (2001). Folic acid-responsive neurological diseases in Japan. *J Nutr Sci Vitaminol (Tokyo), 47*(3): 181–7.

PART II

Macronutrients

CHAPTER 7

Protein Deficiency During Development: Implications for Cognitive Function

Emma Jones

Chapter Overview

- This chapter will define the key terms relevant to protein deficiency and identify some of the difficulties encountered in specifying the role of protein in the developmental problems experienced by malnourished and undernourished children.
- The chapter will then discuss the biological role of dietary proteins, the importance of these proteins during development and the potential mechanisms by which protein could influence cognitive function.
- Finally, this chapter will explore research that examines the effects of protein malnutrition in humans and animals. Both animal and human research have contributed to our understanding of the effects of protein deficiency, however each also has its limitations. These issues are examined.

Introduction

Proteins are composed of one or more polypeptide chains, which are long chains of amino acids. Small chains, consisting of two or more amino acids, are called peptides. The structure and function of a protein is determined by the sequence of amino acids and how the chains are folded. Dietary proteins are metabolised into their constituent amino acids in the alimentary canal (Moughan 2005). These amino acids are either re-synthesised into new proteins or oxidised (Moughan 2005). Humans can synthesise some amino acids but a number of *essential amino acids* cannot be synthesised and must be obtained from the diet (e.g. leucine, tryptophan and tyrosine).

There are a number of terms used in the literature to refer to different types of nutritional deficit, for example, undernutrition, malnutrition, intrauterine growth restriction and protein energy malnutrition (PEM). Each term has a very precise definition. Moreover, different markers have been used to infer/identify children suffering from such nutritional deficits, for example, height, weight and head size.

Protein energy malnutrition is a term used in early literature to refer to deficiencies in dietary protein. This definition has since been revised. It was recognised that what was termed PEM actually included chronic energy malnutrition in combination with micronutrient deficiencies and infections (Grantham-McGregor et al. 1999). In children, PEM is defined as being two standard deviation (SD) under normal weight for age (underweight), normal height for age (stunting) or weight for height (wasting) (Grantham-McGregor et al. 1999; Müller & Krawinkel 2005). There is strong evidence that stunting in particular is related to protein deficiency (for a discussion, see Grantham-McGregor et al. 1999). In children, there are two main conditions specifically associated with protein malnutrition: kwashiorkor and marasmus. Kwashiorkor is characterised by malnutrition with oedema (swelling due to excess water); marasmus is characterised by severe wasting (Müller & Krawinkel 2005). Both conditions are forms of protein malnutrition which may be related to a lack of dietary nutrients in addition to reduced protein caused by infections and diarrhoea. Grantham-McGregor et al. (1999) report that early malnutrition is associated with significant cognitive and behavioural deficiencies. These can be somewhat ameliorated with considerable improvements to the child's environment, for example, in the case of adoption.

Undernutrition refers to a general lack of nutrients, malnutrition refers to deficiencies in specific nutrients, and intrauterine growth restriction (Walker et al. 2007) refers to deficiencies in nutrition during critical periods of brain development. Undernutrition has been associated with reduced brain weight and reduced synapses in the visual cortex (Peeling & Smart 1994a, 1994b) and, in humans, early post-natal undernutrition has been shown to influence the development of pyramidal cells in the motor cortex (Cordero et al. 1993).

Methods to infer in utero nutritional deficits include low birthweight (LBW), small for gestational age (SGA) and microcephaly (a disorder characterised by small head size). In developing countries, 11 per cent of births are LBW (Walker et al. 2007).

A great deal of research has examined how LBW and other measures of malnutrition influence the development of an infant/child. For example, underweight and stunting have been associated with cognitive impairment (Walker et al. 2007), and neonatal head size is strongly associated with intelligence quotient (IQ) at the age of seven years (Grantham-McGregor et al. 2007). Kuklina et al. (2006) demonstrated that height-for-age, weight-for-age and head circumference were associated with psychomotor and mental development. In addition, malnutrition is associated with impaired cognition at 7–10 years (Miranda et al. 2007). It is important to note that Miranda et al. found stunting was not associated with cognitive deficits in this age group. For a discussion, see Victoria et al. (2008).

In animal research, a technique to induce hippocampal kindling has been employed to examine the effects of malnutrition and protein deprivation on brain function. Kindling is the repeated administration of electrical stimulation to brain regions, particularly the limbic system, which causes the regions to be more susceptible to seizures (Austin-Lafrance et al. 1991; Bronzino et al. 1986). Kindling has been used as an animal model to study a diverse range of brain functions, including epilepsy, learning and memory, and brain plasticity (Bronzino et al. 1986). Some of this research is discussed in the 'Animal Studies' section of this chapter.

Functions of Protein During Development

Amino acids have a number of important functions in the body. For example, they supply energy and are involved in cell regulatory processes, such as contributing to the synthesis of hormones and neurotransmitters, and the formation of catalysts (e.g. Lai 1988). Some essential amino acids are neurotransmitters (e.g. aspartic acid; Lai 1988), while others are precursors to neurotransmitters; for example, tryptophan is a precursor to serotonin. Amino acid availability can directly influence brain neurotransmitter levels; for instance, serotonin synthesis in the brain relies on circulating tryptophan (Growdon & Wurtman 1979; Lajtha et al. 2007; Schaechter & Wurtman 1989, 1990), with serotonin synthesis being directly related to circulating tryptophan levels (Lajtha et al. 2007). To a lesser extent, this occurs with dopamine synthesis in response to tyrosine levels (Lajtha et al. 2007).

Dietary proteins play an essential role in development, particularly during prenatal and early neonatal periods. The development and structure of the brain and nervous system is reliant on amino acids which are used:

1. To synthesise structural proteins for cellular growth and development
2. For molecular and cellular differentiation
3. For synapse formation and myelination
4. For synthesis of enzymes, peptide hormones and neurotransmitters. Some amino acids are neurotransmitters (Tonkiss et al. 1993).

There are well-defined critical periods during gestation when nutrient availability must be tightly regulated. Excess or deficiency during these times can have significant effects on development, with some authors suggesting that these early experiences may contribute to disorders in later life, such as schizophrenia (see Brown & Susser 2008 for a review; Dauncey & Bicknell 1999). Georgieff (2006) suggests that weeks 24–42 of gestation are particularly susceptible to nutrient deficiency, especially protein. Specific regions of the brain under development at this time include the hippocampus (which mediates learning and memory processes), visual and auditory cortices, and the striatum (responsible for motor behaviour, addiction and learning). For instance, in a series of studies examining a rat model of prenatal protein deficiency, Tonkiss and colleagues (1993) reported impairments in the performance of tasks related to visual acuity and impaired learning ability. The authors reported that prenatally malnourished animals performed worse on the Morris Water Maze and subsequent studies suggested that the reason may be impaired visual acuity, because adults who had been malnourished struggled to discriminate vertical from horizontal stripes.

In terms of neurophysiology, protein malnutrition is not associated with gross changes in brain structure, but effects occur at the molecular and cellular level. Undernutrition and intrauterine malnutrition, for example, have been shown to influence myelination and affect the size and number of brain cells (Lai & Lewis 1980; Lai et al. 1980; Meberg 1981). Moreover, a body of work by Bronzino and colleagues (Bronzino et al. 1983, 1986, 1989, 1990, 1996, 1999) has demonstrated the effects of protein malnutrition on hippocampal cell development and function (discussed in the 'Animal Studies' section of this chapter).

The specific mechanisms by which early nutrition generally influences later outcomes may be related to neonatal programming, whereby lifetime structures and functions are laid down during critical developmental periods (e.g. Lucas 1998), with mechanisms including adaptive changes in gene expression, specific cloning of adaptive cells in certain tissues and 'differential proliferation of specific cell types'; that is, cells being dispersed differently in different tissues depending on early adaptations (Dauncey & Bicknell 1999; Lucas 1998). Early nutritional programming and its impact on later life function and health has been well documented in animals and there is some evidence of this in human studies (see Lucas 1998 for a review).

Although the long-term effects of undernutrition and malnutrition have been well documented, the specific influence of protein deficiency during these critical periods is unclear. Rodent models have been used to examine the effects of protein deficiency during gestation and early infancy. These findings are presented in Table 7.1 and discussed in the 'Animal Studies' section of this chapter. In terms of the effects in human populations, large epidemiological studies have demonstrated negative effects of PEM on developmental outcomes (discussed further in the 'Functions of Protein during Development' and 'Human Studies' sections of this chapter). Drawing conclusions from human research has been problematic. Human studies often suffer from methodological issues, and findings are complicated by considerable confounders. For example, much of the research in human populations has taken place in developing countries. Samples are malnourished and tend to live in poverty. Along with nutritional deficiencies, they also experience high disease and infection rates. All of these factors have considerable effects on development. These studies have been summarised in Table 7.1 and are discussed in the 'Human Studies' section.

Recovery from early insults appears to be possible, though conditional. Although many effects of pre- and early post-natal protein malnutrition appear

Table 7.1 Studies examining the influence of pre- and post-natal protein intake in behaviour, brain function and performance

Reference	Design	Sample	Methods	Results	Results
			Protein manipulation	Brain and/or behaviour outcome measures	
Almeida et al. (1996)	Animal	Female Sprague-Dawley rats and offspring	Five weeks prior to mating dams randomised to either: adequate protein diet = 25% casein, or LP diet = 6% casein. Pups from LP dams fostered by dams on 25% protein diet. Pups either prenatally malnourished and then adequately nourished or control (well nourished pre- and post-natally)	Body weight, elevated T-maze	At birth and age 70-days malnourished rats weighed less than controls (gap narrowed later). Prenatal malnutrition did not increase avoidance latency over trails whereas controls did. Female controls reduced avoidance latency on retention test whereas other groups did not.
Almeida et al. (2001)	Animal	61 male and 50 female Wistar Rats	Diets administered during lactation to dams and pups: 6% protein or 16% protein diet. After weaning pups randomised to one of three groups: 1. Control (W) = continued to receive 16% protein 2. Malnourished (M) = continued to receive 6% protein and 3. Previously malnourished (PM) – had been malnourished prior to weaning and then given 16% protein diet. Until 38-days of age.	Body weight, social behaviours: pin, walk-over, Allogroom, side-mount, anogenital sniff, and non-social behaviours: rear	Body weight, age 35-days: W > M and PM. Malnutrition associated with mixed results in social behaviour. PM associated with reversal of changes in non-playful and non-social behaviours (rear) but increased playful social behaviour (pin). Gender effects were observed only on side-mount (higher incidence in males) and walk-over (higher incidence in females).

Table 7.1 (Continued)

Reference	Design	Sample	Methods	Results	Results
			Protein manipulation	Brain and/or behaviour outcome measures	
Bennis-Taleb et al. (1999)	Animal	Female Wistar rats and offspring	Gestation until parturition: control group (C) = 20% protein diet; low protein group (LP) = 8% protein diet. After birth third group: Recuperation (R) = some LP animals transferred to 20% protein diet. Following weaning – maintained on respective diets. Assessed age 110-days.	Body and brain weight, DNA, protein and lipid concentrations and vascularisation of the cerebral cortex	Lower blood vessel density in cerebral cortex in LP than C. Maintained into adulthood regardless of whether 8% or 20% protein diet was give postnatally. Body and brain weights lower in LP animals (birth and adulthood). Lower DNA in forebrain and higher in cerebellum of LP pups. LP adults had lower brain levels of DNA, protein, cholesterol and phospholipids which were restored by a normal postnatal diet.
Bronzino et al. (1996)	Animal	Female Sprague-Dawley rats and offspring	Diet provided 5-weeks prior to mating and throughout gestation and lactation. Male sires provided with same diet for 1-week prior to mating. C = 25% casein, test diet = 6% casein supplemented with methionine. Pups were cross fostered so that pups from 6% dams were fostered by control dams and pups born to control dams were fostered by other control dams.	Measures of LTP: EPSP and PSA following a tetanisation paradigm which involved stimulation of medial perforant pathway/ denate granule cell synapse at age 15, 30 and 90 days	Age 15 days: malnourished significantly reduced LTP than controls. Age 30 and 90 days: half of the malnourished animals exhibited reduced LTP and the other half were not different from controls.

Bronzino et al. (1997)	Animal	Female Sprague-Dawley rats and offspring	Diet provided 5-weeks prior to mating and throughout gestation and lactation. Male sires provided with same diet for 1-week prior to mating. C = 25% casein, test diet = 6% casein supplemented with methionine. Pups were cross fostered so that pups from 6% dams were fostered by control dams and pups born to control dams were fostered by other control dams.	Measures of LTP: EPSP and PSA following a tetanisation paradigm which involved stimulation of medial perforant pathway. Up to 24 hours following tetanisation.	No differences between animals prior to tetanisation paradigm. Reduces PSA enhancement in malnourished animals indicating a problem converting EPSPs into PSAs
Camargo et al. (2005)	Animal	276 male Wistar rats	Lactation phase: Well nourished (16% protein), malnourished (6% protein). Weaning onwards: well-nourished (W) – maintained on 16% protein; malnourished (M) – maintained on 6% protein and previously malnourished (PM) – malnourished during lactation then maintained on 16% protein following weaning	Social behaviour: pinning, wrestling, walk-over and rear	Reduced body weight in M and PM into adulthood. Wrestling and walk-over higher in M. W better at demonstrating pinning indicating increased dominance.
Chen et al. (1997)	Animal	Female Sprague-Dawley rats and offspring	Diet provided 5-weeks prior to mating and throughout gestation and lactation. Male sires provided with same diet for 1-week prior to mating. C = 25% casein, test diet = 6% casein supplemented with methionine. Pups were cross fostered so that pups from 6% dams were fostered by control dams and pups born to control dams were fostered by other control dams.	Concentration of monoamine neurotransmitters and their metabolites and precursors in the hippocampal formation, striatum, brain stem and cerebral cortex in rats aged: 1-, 15-, 30-, 45-, 90- and 220-days	Significant variation of neurotransmitters, metabolites and precursors throughout development. Prenatal protein malnutrition was associated with increased tryptophan levels in hippocampal formation of newborns and reduced tyrosine levels in the striatum of 30-day old rats

Table 7.1 (Continued)

Reference	Design	Sample	Methods Protein manipulation	Results Brain and/or behaviour outcome measures	Results
Renade et al. (2008)	Animal	Swiss albino female mice and their offspring	From 6 weeks prior to conception, and then throughout gestation and lactation, mice were randomized to one of four groups: C (including 34.2% casein), CED, PD (including 7.1% casein) or ID	Body weight, hippocampal volume, RAM: reference memory, working memory	Body weight lower and Hippocampal volumes significantly reduced in three experimental groups compared to C. Body weight lower 10-days prior to trial but over the course of the trial means were similar. CED – sparing of reference memory. PD – most impaired working memory.
Tonkiss et al. (1993)	Animal	Female rats and offspring	5-weeks before mating, throughout gestation. LP = 6% protein, C = 25% casein	Homing, Morris water maze, CTA, reversal learning in T-maze, differential reinforcement of low rates (DFL) operant task. LTP and kindling in hippocampus.	By 9 and 11-days old, C were better at finding home compared to those who have been malnourished i.e. either pre- or post-natally or both. Prenatally malnourished performed worse on Morris water maze. CTA, T-maze and DRL suggest malnourished impaired at changing learned responses. Differences in LTP in malnourished animals and they were more susceptible to seizures but require more kindle stimulations before reaching kindled state.
Adair (1985)	Bacon Chow Study (Taiwan)	225 women	Nutrient-rich supplement or placebo given after birth of first child through lactation of second child	Physical, motor, mental and dental development	Supplementation with Bacon Chow enhanced motor development but there were no effects on mental development or IQ

Engle et al. (1993)	Epidemiological: INCAP longitudinal study – 1969–1977	1614 children <7 years in Guatemala	Atole- protein and calorie supplement or fresco – calories only. Villages were randomized to a supplement (rather than participants being randomized)	Infant battery – mental and motor development (Bayley, Gesell, Psyche, Cattell, and Merrill-Palmer scales). Preschool battery (ages 3, 4, 5, 6 and 7 years) – 22 tests including embedded figures, verbal inferences, vocabulary naming and recognition, memory for sentences, memory for digits, draw a line slowly, reversal discrimination learning, impossible puzzle, memory for objects, knox cubes; and for 5–7 year olds: memory for designs, incidental and intentional learning, hepatic-visual matching, matching familiar figures, block design, animal houses, elimination of odd figure, face-hands touching, incomplete figures and conservation of: matter, area and quantity.	First analysis combined groups. Calorie intake – high, medium, low. Mental development and motor skills associated with calorie intake at a number of time points. Atole villages had higher calorie and protein intake and performed better on many of the tasks. Aged 3–7 years language, memory, perception and composite scores were positively related to calorie intake. Effects more likely in children who were pre- and post-natally supplemented. Another analysis examined total amount of supplement consumed over lifetime as a predictor in 3–6 year olds. Only found effects in boys. Second analysis: Atole boys had significantly higher mental scores at 15 months and atole girls higher motor scores at 24 months. At 3–5 years, atole children better on combined, generalised verbal and perceptual-motor performance at 48 and 60 months. Effects of supplement more pronounced in low-SES compared to higher-SES.
Heys et al. (2010)	Epidemiological	13,513 adults aged 50+ in China	Childhood meat-eating assessed by questionnaire	10-word list-learning task	Adjusted for age, sex, education, childhood and adulthood socioeconomic position and physical activity. Daily meat-eating in childhood significantly related to DR score and eating meat monthly, weekly or daily associated with IR score.

Table 7.1 (Continued)

Reference	Design	Sample	Methods	Results	Results	Results
			Protein manipulation	Brain and/or behaviour outcome measures		
Huiman et al. (2003)	Cross sectional, longitudinal, follow-up, INCAP study. 1996–1999	143 females from INCAP study. Age 22-29 years	See Engle et al. (1993)	EA		Atole had increased general knowledge, numeracy, comprehension and vocabulary. Completing primary school was a stronger predictor of EA
Hoddinott et al. (2008)	INCAP follow-up. 2002–2004	1424 male and females from INCAP study. Age 25-42 years	See Engle et al. (1993)	Economic productivity: estimates of annual income, hours worked and average hourly wage		Atole before age 3 associated with higher hourly rates of pay for males.
Joos et al. (1982)	RCT: The Bacon Chow Study	294 pregnant women in Taiwan	High calorie and protein supplement (800kcal, 40g protein per day) or placebo (until 1970 contained 6g protein per day. After it contained 80kcal with no protein) administered three weeks following delivery of first infant, through lactation and then gestation and lactation of second infant.	Mental and motor development (Bayley scales of infant development)		No significant effects of supplement on mental scores. Experimental group had higher motor scores than placebo – further analysis indicates that most differences due to ability to raise to sitting, pull itself up until standing and bring 2 objects together at midline

Martorell (1995)	INCAP follow-up	REVIEW OF SUPPLEMENT			
Pollitt et al. (1995)	INCAP follow-up. 1988–1999	636 individuals from INCAP study. Age 13–19 years	See Engle et al. (1993)	Literacy, numeracy, general knowledge, 2 standardized educational achievement tests, Raven's Progressive Matrices and tests of information processing: SRT, CRT, memory RT and paired associates	Exposure to Atole during pre- and early post-natal period associated with increased numeracy, general knowledge, reading, vocabulary and better information processing.
Stein et al. (2008)	INCAP follow-up. 2002–2004	1448 individuals from INCAP study.	See Engle et al. (1993)	Schooling, reading comprehension, Inter American Series, Raven Progressive Matrices.	Children who had atole from birth to 24-months demonstrated increased performance on InterAmerican Series and Raven's progressive Matrices.
Waber (1985)	Supplementation study	433 families	Received supplement: 6mo–3 yrs, 3rd trimester-6mo, throughout study, education program only, 3rd trimester- 3 yrs and control.	The Griffiths test of infant development.	All supplementation improved performance especially motor. Supplementation reduced gap in cognition caused by SES.

C = control, CED = chronic energy deficiency, CRT = choice reaction time, CTA= conditioned taste aversion, DNA = deoxyribonucleic acid, DR= delayed recall, EA = educational achievement, EPSP = Excitatory Postsynaptic Potential; ID = iron deficiency, IR= immediate recall, LTP = long-term potentiation LP = low protein, PD = protein deficiency, PSA = population spike amplitude, RAM = radial arm maze, RCT = randomised controlled trial, RT = reaction time, SES = socioeconomic status, SRT = simple reaction time.

to be permanent, stimulating and enriched environments can reduce some of the negative effects (see Dauncey & Bicknell 1999).

The remainder of this chapter will concentrate on the importance of protein during the critical periods of pre- and neo-natal development, as these periods have been shown to be particularly sensitive to nutritional insults due to the rapid development occurring. This is not a comprehensive review of the literature. Where necessary, the reader is directed towards relevant reviews.

Prevalence

Grantham-McGregor et al. (2007) reported that '200 million children ... in developing countries do not reach their developmental potential and do poorly in school' (p. 60). There are a number of reasons for this, including poor nutrition and associated physiological deficits, such as stunting and wasting. It was estimated that in 2000–2, 852 million people worldwide were undernourished, with the majority (815 million) in developing countries (Müller & Krawinkel 2005). Some changes have been reported, with the number of cases reducing in China but increasing in other developing nations, with the overall numbers remaining the same.

Animal Studies

Selection Criteria

Research included in this section was conducted between 1996 and 2011. It focuses on protein administration/deficiency in the pre- and early post-natal period. For a review of the general effects of nutrition (including protein), see Dauncey & Bicknell (1999). For a discussion of how protein malnutrition influences development and function of the hippocampal formation, see Morgane et al. (2002). See Table 7.1. for a summary of relevant studies.

Selected Research

In early studies, malnutrition was elicited in pups by removing them from their mothers (e.g. Krigman & Hogan 1976). However, maternal separation per se influences development, so this method was abandoned in later research. The standard procedure employed in the most recent animal studies involves the random allocation of dams (female rat mothers/mothers to be) to a high or low protein diet during gestation and lactation. Sometimes, the dams will begin the diet before mating. At birth, the pups are either maintained on the same diet as their mother or cross-fostered to a mother on a different diet and maintained during the post-natal period on this diet for the duration of the experimental period. If the protein is casein, then low protein diets are usually supplemented with methionine (e.g. Galler & Tonkiss 1991).

Early research (see Dauncey & Bicknell 1999 for a review) demonstrated the effects of pre- and early post-natal protein malnutrition on electrically evoked response latencies, with greater latencies in malnourished animals. Brain and body weights were reduced in malnourished animals compared with controls. However, in one study, brain weight was maintained and this may have been due to the small amount of protein in the dams' diet.

Recent research is consistent with some of these early findings (see Table 7.1). Rats born to well or adequately nourished (i.e. adequate protein intake) dams weighed more at birth than low-protein counterparts (Almeida et al. 1996; Bennis-Taleb et al. 1999; Camargo & de Sousa Almeida 2005). Overall, well-nourished rats, compared with low-protein rats, exhibited: increased dominance and reduced wrestling and walk-over behaviours (Camargo & de Sousa Almeida 2005); increased avoidance latency (which the authors interpreted as representing an increase in anxiety) in the elevated T-maze (Almeida et al. 1996); increased playful social behaviours but decreased non-playful social behaviours (Almeida & De Araújo 2001); improved home finding (Tonkiss et al. 2003); higher cerebral cortex blood vessel density; higher DNA in the forebrain and lower DNA in the cerebellum of pups. Adults exhibited increased brain levels of protein, DNA phospholipids and cholesterol (Bennis-Taleb et al. 1999). Low-protein rats were significantly impaired on the Morris Water Maze (discussed in more detail below) with an inability to change learned responses (Tonkiss et al. 1993). Reduced body weight and hippocampal volumes have been observed in chronic energy-deficient, protein-deprived and iron-deficient mice (Renade et al. 2008), however, in this study, behavioural effects were mixed, with chronic energy deficiency being associated with protection of memory function, and protein deficiency associated with impairments in working memory (note, this is a larger study and only the findings related to protein have been included in Table 7.1). In addition, some sex differences have been reported; for example, well-nourished females exhibit reduced avoidance latencies in retention tests (Almeida et al. 1996).

Recovery from malnourishment has been examined by cross-fostering half of the offspring of low-protein dams to well- or adequately nourished dams and then maintaining them on this diet throughout lactation and the post-natal period. No evidence of recovery was reported (Camargo & de Sousa Almeida 2005; Tonkiss et al. 2003). Findings suggest some potential recovery in some social behaviours (Almeida & De Araújo 2001). In addition, lower brain levels of DNA, protein, cholesterol and phospholipids in previously malnourished adult rats were restored by a normal post-natal diet (Bennis-Taleb et al. 1999) and findings from Almeida et al. (1996) suggest that previously malnourished rats have lower levels of anxiety compared with controls.

In a rat model of in uterine protein deficiency, Tonkiss et al. (1993) reported that rats which were protein deprived in utero but then suckled from a healthy mother, demonstrated impaired visual discrimination and inflexibility in extinguishing learned responses when no longer appropriate. Furthermore, these animals also had altered neurophysiology, specifically in terms of hippocampal processing.

Using the hippocampal kindling paradigm, Bronzino and colleagues have demonstrated that prenatal protein deprivation influences hippocampal plasticity (Bronzino et al. 1991). Other studies have shown that it also impairs long-term potentiation (Austin et al. 1986; Bronzino 1997; Bronzino et al. 1996). In addition, Chen et al. (1997) have investigated the effects of prenatal protein malnutrition monoamine neurotransmitter content in the hippocampus, striatum, brain stem and cortex. Rats that were protein malnourished exhibited increased tryptophan levels in the hippocampus at birth and reduced tyrosine in the striatum at age 30 days. In work by Bronzino and colleagues (e.g. Bronzino 1997; Bronzino et al. 1996) and Chen et al. (1997), the animals were cross-fostered so that all animals received a normal postnatal diet (25 per cent protein). Consequently, these findings suggest that neurological changes occur during the prenatal period that have lasting effects on postnatal brain function regardless of postnatal diet.

Human Studies

Selection Criteria

With the notable exception of two large influential studies, the Bacon Chow Study and the Institute of Nutrition of Central America and Panama (INCAP) study, the literature discussed here has been selected on the basis of year (between 1996 and 2011); its focus, being primarily on protein deficiency and/or supplementation; and the timing of interventions, being the prenatal or the early neonatal period. For epidemiological studies, PEM occurs during the prenatal or early postnatal period. Where appropriate, the reader is directed to existing reviews for further details.

General Research

Epidemiological studies have often examined the effects of malnutrition in developing nations, where it is prevalent. Often, children born in malnourished communities have LBW, are SGA or may have microcephaly. In many epidemiological studies, these measures have been employed to infer the existence of nutritional deficits (see Grantham-McGregor et al. 1999 for a review). Here, we concentrate on studies specifically focused on PEM.

Evidence is complicated and consists mostly of epidemiological studies (see Grantham-McGregor et al. 1999 for a review). Some randomised controlled trials (RCTs) have examined the influence of protein administration in existing malnourished populations, for example the Bacon Chow Study (reported by Joos et al. 1983). See Table 7.1. for details of human studies.

In general, protein from meat has been associated with earlier walking (Kuklina et al. 2004). Breast milk fortified with protein leads to increased weight and head circumference (Arslanoglu et al. 2006), and a study by de Rooij and colleagues (2010) indicates that prenatal undernutrition associated with famine impairs adult cognitive function.

Meat is an important source of protein. Recently, Heys et al. (2010) examined the influence of meat intake during childhood on cognitive function. They asked 13,513 adults aged over 50 years about their meat intake during childhood. The study was conducted in China where, during participants' childhood, eating meat was less common than in the USA during the same period. The authors asked the participants to report how frequently they had eaten meat as a child and tested their immediate and delayed recall (five minutes post encoding) for words. Individuals who reported eating meat frequently during their childhood exhibited enhanced immediate and delayed recall. This study relies on self-reporting meat intake and it is unclear whether the protein content of meat is responsible for the later-life cognitive superiority given the other benefits of eating meat (e.g. iron). In addition, it is debateable as to whether five minutes is a sufficient delay to assess delayed recall. However, this is a large study and it certainly supports previous research indicating the benefits of meat consumption for cognition and specifically protein.

The Bacon Chow Study, reported by Joos et al. (1983) was an RCT conducted in Taiwan. A total of 294 pregnant women were given a protein supplement or placebo during their first pregnancy, throughout lactation and then during pregnancy and lactation of their second infant. The mental and motor development of the infants was subsequently examined, when the infants were eight months of age (see Table 7.1 for details of the assessments employed). Protein supplementation was found to enhance motor performance compared with the placebo. However, no effects on mental function were reported. Specifically, the aspects of motor performance enhanced were raising to sitting, pull up until standing and bringing two objects together.

Institute of Nutrition of Central America and Panama (INCAP): 1949–99

By far the longest-running study of prenatal and childhood protein supplementation is the INCAP study, conducted between 1969 and 1977. The initial phases of the study are reviewed in Engle et al. (1993) and Ramirez-Zea et al. (2010). Discussion of the first phase of the study (see below) is based on that reported in these two papers. The original sample has been subjected to numerous follow-up cross-sectional studies over the intervening years. Some of these are described in Table 7.1 and will be discussed below. For reviews of the 50-year programme and its impact, see Martorell (2010) and Ramirez-Zea et al. (2010).

Severe PEM during the pre- and early post-natal stage of development has been shown to be associated with impaired cognitive and motor development (see, for example, Grantham-McGregor 1995 for a review). When INCAP first began, there were two theories of the potential mechanisms by which this occurred:

1. Neural mechanism – neurons fail to develop properly due to reduced protein during important developmental stages
2. Developmental mechanism – previously it has been difficult to separate the neural effects from many coexisting environmental variables such as poverty and environmental stimulation.

The INCAP study aimed to identify the precise contribution of protein malnutrition to neural development by administering protein/energy to similar groups of mothers and children in Guatemala. It was recognised that many Guatemalan communities suffer malnutrition and that this is reflected in the diminutive stature of Mayan Guatemalan people who are among the shortest people in the world (Martorell 2010). The INCAP study was of an independent groups design and it examined the effects of a protein-plus-calories supplement (atole) with a calories-only supplement (fresco) (see Table 7.1 in Ramirez-Zea et al. 2010 for nutritional information of the supplements). The researchers implemented controls for social stimulation during supplement administration (placebo administration) and self-selection. They controlled self-selection by randomly assigning whole villages to a treatment group rather than assigning individuals. They selected four villages that were similar and matched on village size, with a large village and a small village receiving atole and a large and a small village receiving fresco. Potential subjects included all children who were under seven years old at the start of the study or who were born during the study. Once recruited into the study, if applicable, children were assessed at birth, then at 6 and 15 months of age and then annually from age 24 months until they were 7 years old or the cessation of the study. In total, 2392 children were recruited into the study (Ramirez-Zea et al. 2010).

The initial phase of the study included measures of diet and supplement intake, anthropometry, mental development, physical fitness, morbidity and maturation (Ramirez-Zea et al. 2010). Cognitive assessments included memory (words, objects and designs), inhibition and conservation. Early reports revealed that at 6, 15 and 24 months of age, mental development was significantly associated with nutrient intake. In addition, at 15 and 24 months of age, there were significant differences in motor skills between groups with better and worse calorie intake. In both cases, higher calorie intake was associated with enhanced performance (Engle et al. 1993). Results showed that individuals from the atole villages exhibited higher calorie and protein intakes than those from the fresco villages. Moreover, they performed better on many of the tasks. In 3–7 year olds, where calorie intake was included as a continuous variable, it was shown that higher intake was associated with benefits on a number of tasks, including language, memory (as above) and perception, with effects being more likely in children who were supplemented pre- and early post-natally. However, Engle et al. note that early reports suffer from statistical issues. For example, one report of the effects of supplementation in infancy combined the atole and the fresco groups and stratified participants according to adequacy of calorie intake (high, medium or low). Consequently, comparisons did not differentiate between the treatment groups. Some effects reported in the small villages were only observed in girls, and Engle et al. (1993) point out the difficulty selecting appropriate alpha levels for multiple comparisons. Some observed effects could conceivably, therefore, reflect type 1 error.

The second period of analysis occurred later and compared the treatment groups directly. This research revealed that, controlling for socio-economic status (SES), atole boys had higher mental functioning at 15 months of age and atole girls demonstrated enhanced motor function at 24 months of age. Overall, the atole children exhibited enhanced motor function compared with their fresco counterparts at 24 months. Older atole children demonstrated enhanced ability on combined, generalised verbal and perceptual motor performance. In addition, the supplement-mediated enhancement was more pronounced in low SES individuals compared with those with a higher SES (reported in Engle et al. 1993).

A special edition of the *Journal of Nutrition* in 1995 included a number of follow-ups to date of the INCAP participants. See Martorell (1995) for a review of these. Some are discussed below.

Atole was shown in this group of studies to reduce the physiological effects of low protein, such as stunting (Martorell 1995; Schroeder et al.1995).

Of the original INCAP participants, 636 were followed up in 1988–9, when they were aged 13–19 years (e.g. Pollitt et al. 1995). For details of tasks administered, see Table 7.1. When adjusted for SES and formal education, findings indicated that atole supplementation during the pre- and early post-natal period was associated with increased numeracy, general knowledge, reading, vocabulary and information processing. Note, some of these participants were still at school when followed up.

In 1996–9, Li et al. (2003) followed up 143 of the original, female participants who were exposed to supplements during the prenatal period until two years of age. The age range was 22–29 years. This study only looked at educational achievement (EA). When adjusted for SES and schooling, atole increased general knowledge, numeracy, comprehension and vocabulary. Both supplementation with atole and completing primary school were strong independent predictors of EA, with completing primary school being the strongest predictor.

A further follow-up was conducted between 2002 and 2004 (Hoddinott et al. 2008), involving 1424 of the original participants, aged 25–42 years.

This study examined the economic productivity of the participants in terms of estimated annual income, hours worked and average hourly wage. Supplementation with atole before the age of three years was associated with higher hourly pay for males.

Stein et al. (2008) reported that the atole group had increased intelligence in adulthood when adjusted for schooling and SES, in addition to other factors that are thought to influence childhood intelligence, such as parental school attainment. The overall message from the INCAP studies is that protein supplementation can have profound benefits on the development and lives of children and that the benefits are most pronounced when children are exposed to protein at a younger age.

However, the studies have been criticised (not least by their own authors). In addition to the issues that have been raised already in this section, the authors have pointed out further limitations with the INCAP research. First, the research has largely focused on mental development while ignoring motor development (e.g. Pollitt 2000). Second, Ramirez-Zea et al. (2010) point out that the individuals in the fresco group consumed more intervention than the individuals in the atole group because they treated it like a drink. The effect was increased vitamin and mineral intake as both interventions were fortified with vitamins and minerals.

Conclusions

Impaired nutrition can have lasting consequences for infants and young children. Protein is particularly important, and adequate protein must be available during critical periods of development. Studies demonstrate that many years following protein deprivation, individuals may still experience disadvantages and impairments in many aspects of their performance and suffer reduced quality of life. For example, increased protein in early life is associated with enhanced cognitive function in childhood and adulthood, in addition to higher rates of pay!

Rather than specific domains being affected by protein, deprivation research appears to indicate a generalised impairment following protein malnutrition (see, for example, Peeling & Smart 1994a), and generalised enhancement in supplemented individuals. However, it is conceivable that the timing, duration and extent of the deprivation may have domain-specific effects. In addition to global cognitive deficits, research demonstrates the importance of protein for motor development. This effect is consistently observed. Although rodent research has been concentrated on hippocampal effects (with other regions largely ignored) hippocampally mediated tasks are not the only tasks represented in human studies of protein supplementation.

The mechanisms are still poorly understood. In terms of neurochemistry, sufficient protein is required during gestation and infancy to lay down the foundations of a healthy brain structure. During these critical periods, protein is tightly regulated with deviations resulting in cognitive and behavioural problems. The timing, duration and amount of protein for optimal development have yet to be determined. Deficits resulting from deprivation during the critical periods may be maintained throughout life and may, for example, contribute to mental disorders such as schizophrenia (Dauncey & Bicknell 1999). It must be remembered that protein deficiency rarely occurs in isolation and that children deficient in protein are subject to multiple nutritional deficits and illnesses.

Compounding these nutritional deficits are the environmental insults that such children often also endure. The influence of environmental factors can be highlighted by the fact that, although infant brains are plastic, allowing for the possibility of recovery from protein malnutrition, full reversal of symptoms only occurs when the diet and environment are both changed (Grantham-McGregor et al. 1999). Supplementation may somewhat ameliorate symptoms.

It may be interesting for future research in this area to begin to examine the specific window of opportunity for different effects and determine whether deprivation at different times during brain development differentially influence cognitive development.

References

Almeida, S. S., & De Araújo, M. (2001). Postnatal protein malnutrition affects play behavior and other social interactions in juvenile rats. *Physiology and Behavior*, *74*(1–2), 45–51.

Almeida, S. S., Tonkiss, J., & Galler, J. R. (1996). Prenatal protein malnutrition affects avoidance but not escape behavior in the elevated T-maze test. *Physiology and Behavior*, *60*(1), 191–195.

Arslanoglu, S., Moro, G. E., & Ziegler, E. E. (2006). Adjustable fortification of human milk fed to preterm infants: does it make a difference? *Joural of Perinatology*, *26*(10), 614–621.

Austin, K. B., Bronzino, J., & Morgane, P. J. (1986). Prenatal protein malnutrition affects synaptic potentiation in the dentate gyrus of rats in adulthood. *Developmental Brain Research*, *29*(2), 267–273.

Austin-Lafrance, R. J., Morgane, P. J., & Bronzing, J. D. (1991). Prenatal protein malnutrition and hippocampal function: Rapid kindling. *Brain Research Bulletin*, *27*(6), 815–818.

Bennis-Taleb, N., Remacle, C., Hoet, J. J., & Reusens, B. (1999). A low-protein isocaloric diet during gestation affects brain development and alters permanently cerebral cortex blood vessels in rat offspring. *The Journal of Nutrition*, *129*(8), 1613–1619.

Bronzino, J. (1997). Effects of prenatal protein malnutrition on hippocampal long-term potentiation in freely moving rats. *Experimental Neurology*, *148*, 317–323.

Bronzino, J. D., Austin, K., Siok, C. J., Cordova, C., & Morgane, P. J. (1983). Spectral analysis of neocortical and hippocampal EEG in the protein malnourished rat. *Electroencephalography and Clinical Neurophysiology*, *55*(6), 699–709.

Bronzino, J., Austin La France, R. J., Franceschini, R. J., & Morgane, P. J. (1989). Altered neual circuit activity: prenatal protein malnutrition and the paired pulse response.

Bronzino, J. D., Austin-LaFrance, R. J., & Morgane, P. J. (1990). Effects of prenatal protein malnutrition on perforant path kindling in the rat. *Brain Research, 515*(1–2), 45–50.

Bronzino, J. D., Austin-LaFrance, R. J., Morgane, P. J., & Galler, J. R. (1991). Effects of prenatal protein malnutrition on kindling-induced alterations in dentate granule cell excitability: I. Synaptic transmission measures. *Experimental Neurology, 112*(2), 206–215.

Bronzino, J. D., Austin La France, R. J., Morgane, P. J., & Galler, J. R. (1996). Diet-induced alterations in the ontogeny of long-term potentiation. *Hippocampus, 6*(2), 109–117.

Bronzino, J. D., Austin-Lafrance, R. J., Siok, C. J., & Morgane, P. J. (1986). Effect of protein malnutrition on hippocampal kindling: Electrographic and behavioral measures. *Brain Research, 384*(2), 348–354.

Bronzino, J. D., Blaise, J. H., Mokler, D. J., Galler, J. R., & Morgane, P. J. (1999). Modulation of paired-pulse responses in the dentate gyrus: effects of prenatal protein malnutrition. *Brain Research, 849*(1–2), 45–57.

Brown, A. S., & Susser, E. S. (2008). Prenatal nutritional deficiency and risk of adult schizophrenia. *Schizophrenia Bulletin, 34*(6), 1054–1063.

Camargo, L. M. M., & de Sousa Almeida, S. (2005). Early postnatal protein malnutrition changes the development of social play in rats. *Physiology and Behavior, 85*(3), 246–251.

Chen, J.-C., Turiak, G., Galler, J., & Volicer, L. (1997). Postnatal changes of brain monoamine levels in prenatally malnourished and control rats. *International Journal of Developmental Neuroscience, 15*(2), 257–263.

Cordero, M. E., D'Acuña, E., Benveniste, S., Prado, R., Nuñez, J. A., & Colombo, M. (1993). Dendritic development in neocortex of infants with early postnatal life undernutrition. *Pediatric Neurology, 9*(6), 457–464.

Dauncey, M. J., & Bicknell, R. J. (1999). Nutrition and neurodevelopment: Mechanisms of developmental dysfunction and disease in later life. *Nutrition Research Reviews, 12*(02), 231–253.

de Rooij, S. R., Wouters, H., Yonker, J. E., Painter, R. C., & Roseboom, T. J. (2010). Prenatal undernutrition and cognitive function in late adulthood. *Proceedings of the National Academy of Sciences, 107*(39), 16881–16886.

Engle, P. L., Gorman, K., Martorell, R., & Pollitt, E. (1993). Infant and preschool psychological development. *Food and Nutrition Bulletin, 14*(3), 201–214.

Galler, J. R., & Tonkiss, J. (1991). Prenatal protein malnutrition and maternal behavior in Sprague-Dawley rats. *The Journal of Nutrition, 121*(5), 762–769.

Georgieff, M. K. (2006). Early brain growth: Macronutrients for the developing brain. *Neoreviews, 7*(7), e334–e343.

Grantham-McGregor, S., Cheung, Y. B., Cueto, S., Glewwe, P., Richter, L., & Strupp, B. (2007). Developmental potential in the first 5 years for children in developing countries. *The Lancet, 369*(9555), 60–70.

Grantham-McGregor, S. M. (1995). A review of studies of the effect of severe malnutrition on mental development. *Journal of Nutrition, 125*(8 Suppl.), 2233S–2238S.

Grantham-McGregor, S. M., Fernald, L. C., & Sethuraman, K. (1999). Effects of health and nutrition on cognitive and behavioural development in children in the first three years of life. Part 1: Low birth weight, breastfeeding, and protein-energy malnutrition. *Food and Nutrition Bulletin., 20*, 53–75.

Growdon, J. H., & Wurtman, R. J. (1979). Dietary influences on the synthesis of neurotransmitters in the Brain. *Nutrition Reviews, 37*(5), 129–136.

Heys, M., Jiang, C., Schooling, C., Zhang, W., Cheng, K., Lam, T., et al. (2010). Is childhood meat eating associated with better later adulthood cognition in a developing population? *European Journal of Epidemiology, 25*(7), 507–516.

Hoddinott, J., Maluccio, J. A., Behrman, J. R., Flores, R., & Martorell, R. (2008). Effect of a nutrition intervention during early childhood on economic productivity in Guatemalan adults. *The Lancet, 371*(9610), 411–416.

Joos, S. K., Pollitt, E., Mueller, W. H., & Albright, D. L. (1983). The Bacon Chow study: Maternal nutritional supplementation and infant behavioral development. *Child Development, 54*(3), 669–676.

Krigman, M. R., & Hogan, E. L. (1976). Undernutrition in the developing rat: Effect upon myelination. *Brain Research, 107*(2), 239–255.

Kuklina, E. V., Ramakrishnan, U., Stein, A. D., Barnhart, H. H., & Martorell, R. (2004). Growth and diet quality are associated with the attainment of walking in rural guatemalan infants. *The Journal of Nutrition, 134*(12), 3296–3300.

Kuklina, E. V., Ramakrishnan, U., Stein, A. D., Barnhart, H. H., & Martorell, R. (2006). Early childhood growth and development in rural Guatemala. *Early Human Development, 82*(7), 425–433.

Lai, K. S. (1988). Biochemical and clinical aspects of amino acids – a brief review. *Journal of the Hong Kong Medical Association, 40*(3), 226–235.

Lai, M., & Lewis, P. D. (1980). Effects of undernutrition on myelination in rat corpus callosum. *The Journal of Comparative Neurology, 193*(4), 973–982.

Lai, M., Lewis, P. D., & Patel, A. J. (1980). Effects of undernutrition on gliogenesis and glial maturation in rat corpus callosum. *The Journal of Comparative Neurology, 193*(4), 965–972.

Lajtha, A., Oja, S. S., Schousboe, A., Saransaari, P., Cansev, M., & Wurtman, R. J. (2007). *Handbook of Neurochemistry and Molecular Neurobiology* (pp. 59–97). Berlin Heidelberg: Springer.

Li, H., Barnhart, H. X., Stein, A. D., & Martorell, R. (2003). Effects of early childhood supplementation on the educational achievement of women. *Pediatrics, 112*(5), 1156–1162.

Lucas, A. (1998). Programming by early nutrition: an experimental approach. *The Journal of Nutrition, 128*(2), 401S–406S.

Martorell, R. (1995). Results and implications of the INCAP follow-up study. *The Journal of Nutrition, 125*(4 Suppl.), 1127S–1138S.

Martorell, R. (2010). Physical growth and development of the malnourished child: contributions from 50 years of research at INCAP. *Food and Nutrition Bulletin, 31*, 68–82.

Meberg, A. (1981). Somatic growth and brain development. *Neonatology, 39*(5–6), 272–284.

Miranda, M. C., Nóbrega, F. J., Sato, K., Pompéia, S., Sinnes, E. G., & Bueno, O. F. A. (2007). Neuropsychology and malnutrition: a study with 7 to 10 years-old children in a poor community. *Revista Brasileira de Saúde Materno Infantil, 7*, 45–54.

Morgane, P. J., Mokler, D., & Galler, J. R. (2002). Effects of prenatal protein malnutrition on the hippocampal formation. *Neuroscience and Biobehavioural Reviews, 26*, 471–483.

Moughan, P. J. (2005). Dietary protein quality in humans – An overview. *Journal of AOAC international, 88*(2), 874–876.

Müller, O., & Krawinkel, M. (2005). Malnutrition and health in developing countries. *CMAJ, 173*(3), 279–286.

Peeling, A. N., & Smart, J. L. (1994a). Review of literature showing that undernutrition affects the growth rate of all processes in the brain to the same extent. *Metabolic Brain Disease, 9*(1), 33–42.

Peeling, A. N., & Smart, J. L. (1994b). Successful prediction of immediate effects of undernutrition throughout the brain growth spurt on capillarity and synapse-to-neuron ratio of cerebral cortex in rats. *Metabolic Brain Disease, 9*(1), 81–95.

Pollitt, E. (2000). Developmental sequel from early nutritional deficiencies: conclusive and probability judgements. *The Journal of Nutrition, 130*(2), 350.

Pollitt, E., Gorman, K., Engle, P. L., Rivera, J. A., & Martorell, R. (1995). Nutrition in early life and the fulfillment of intellectual potential. *The Journal of Nutrition, 125*(4 Suppl.), 1111S–1118S.

Ramirez-Zea, M., Melgar, P., & Rivera, J. A. (2010). INCAP oriente longitudinal study: 40 years of history and legacy. *The Journal of Nutrition, 140*(2), 397–401.

Ranade, S. C., Rose, A., Rao, M., Gallego, J., Gressens, P., & Mani, S. (2008). Different types of nutritional deficiencies affect different domains of spatial memory function checked in a radial arm maze. *Neuroscience, 152*(4), 859–866.

Schaechter, J. D., & Wurtman, R. J. (1989). Tryptophan availability modulates serotonin release from rat hypothalamic slices. *Journal of Neurochemistry, 53*(6), 1925–1933.

Schaechter, J. D., & Wurtman, R. J. (1990). Serotonin release varies with brain tryptophan levels. *Brain Research, 532*(1–2), 203–210.

Schroeder, D., Martorell, R., Rivera, J. A., Ruel, M., & Habicht, J. (1995). Age differences in the impact of nutritional supplementation on growth. *Journal of Nutrition, 125*(4 Suppl.), 1051S–1059S.

Stein, A. D., Wang, M., DiGirolamo, A., Grajeda, R., Ramakrishnan, U., Ramirez-Zea, M., et al. (2008). Nutritional supplementation in early childhood, schooling, and intellectual functioning in adulthood: a prospective study in guatemala. *Archives of Pediatrics and Adolescent Medicine, 162*(7), 612–618.

Tonkiss, J., Bonnie, K. E., Hudson, J. L., Shultz, P. L., Duran, P., & Galler, J. R. (2003). Ultrasonic call characteristics of rat pups are altered following prenatal malnutrition. *Developmental Psychobiology, 43*(2), 90–101.

Tonkiss, J., Galler, J., Morgane, P. J., Bronzino, J. D., & Austin-Lafrance, R. J. (1993). Prenatal protein malnutrition and postnatal brain functiona. *Annals of the New York Academy of Sciences, 678*(1), 215–227.

Victora, C. G., Adair, L., Fall, C., Hallal, P. C., Martorell, R., Richter, L., et al. (2008). Maternal and child undernutrition: consequences for adult health and human capital. *The Lancet, 371*(9609), 340–357.

Walker, S. P., Wachs, T. D., Meeks Gardner, J., Lozoff, B., Wasserman, G. A., Pollitt, E., et al. (2007). Child development: Risk factors for adverse outcomes in developing countries. *The Lancet, 369*(9556), 145–157.

CHAPTER 8

Carbohydrates, Glucose and Cognitive Performance

Michael A. Smith, Jonathan K. Foster and Leigh M. Riby

Chapter Overview

- Glucose is the brain's primary fuel.
- Carbohydrates have been observed to modulate cognitive performance.
- Glucose has been associated with enhanced memory performance, most notably in the elderly, but also in younger individuals when memory materials are encoded under conditions of divided attention.
- Ingestion of breakfast foods with a lower glycaemic load (GL) has been demonstrated to improve attention, most notably in children.
- Several mechanisms of action have been proposed, but more work is needed to determine the precise neurocognitive mechanism(s) by which carbohydrates influence cognitive performance.

Introduction

Glucose is a key energy source for the body and serves as the brain's primary fuel. Glucose is metabolised in the liver from ingested monosaccharide carbohydrates, such as glucose, fructose and galactose. Over the past three decades, many researchers have begun to investigate the role of glucose and carbohydrate ingestion in the modulation of cognitive performance across the lifespan. This chapter will begin by providing some background on carbohydrates and carbohydrate metabolism. It will then discuss studies which have investigated the influence of glucose administration on acute cognitive performance in (a) children, (b) young adults, and (c) elderly individuals, and will also discuss some studies which have investigated the role of longer-term dietary carbohydrate intake and its effects on cognitive functioning. Finally, this chapter will consider how dietary intake of carbohydrates other than glucose may influence cognitive performance across the lifespan.

Carbohydrates and Carbohydrate Metabolism

Carbohydrates are present in numerous different processed and unprocessed foods. Large quantities of carbohydrates are found in grain-based foods (e.g. bread, cereals, rice and pasta), potatoes, other vegetables, fruits and dairy products, as well as highly processed foods, such as soft drinks and sweets. A distinction is often made between simple and complex carbohydrates. Complex carbohydrates have a relatively more complex chemical structure (hence the name) and tend to be found in natural foods, such as grains. Simple carbohydrates are typically found in processed foods. Simple carbohydrates are metabolised more easily than complex carbohydrates; therefore, they enter the bloodstream much more quickly, resulting in a rapid increase in blood glucose concentration. Conversely, complex carbohydrates are gradually broken down by the liver, resulting in a more prolonged release of glucose into the bloodstream, with a typically lower peak in blood glucose concentration.

In order to maintain homeostasis, it is important for a healthy individual to maintain a blood glucose concentration within a reasonably narrow range. Upon ingestion, carbohydrates are transported to the liver, where they are (a) mobilised via the bloodstream to cells around the body for immediate use as fuel, (b) broken down into glycogen for storage in the liver and muscle cells or (c) converted to fat. These mechanisms enable homeostasis by ensuring that an adequate supply of stored glucose is available for use as fuel when required. Release of glucose into the bloodstream from the liver is mediated by the hormone insulin. It has been observed that during periods of increased brain activity (e.g. heightened cognitive demand) circulating blood glucose decreases markedly (Donohoe & Benton 1999a; Fairclough & Houston 2004; Scholey et al. 2001, 2006). The authors of these studies have suggested that the ingestion of glucose prior to undertaking a cognitively demanding task may provide the rapid boost in 'brain fuel' that is required to cope with the cognitively demanding conditions. A number of comprehensive reviews have been published on this phenomenon, which has been termed the 'glucose memory facilitation effect' (e.g. Smith et al. 2011b).

Glucose Enhancement of Cognitive Performance

The aforementioned 'glucose memory facilitation effect' has been well established by means of several empirical studies conducted over the past 20 years. The typical methodology employed in this area is to measure cognitive performance via one or more cognitive tasks during the ten-minute to two-hour time window following oral administration of either (a) a glucose drink (most often comprising 25 g or 50 g glucose) or (b) an aspartame or saccharine placebo drink (Smith et al. 2011b). Studies which have been conducted in (a) children and adolescents, (b) young adults, and (c) older adults will be discussed herein.

Children and Adolescents

Children tend to exhibit a higher basal cerebral metabolic rate than older individuals (Chiron et al. 1992), due to the larger brain size of children, relative to their body weight, in comparison to adults (Benton & Stevens 2008). On this basis, children may be particularly sensitive to the potential cognitive benefits arising from acute glucose ingestion. In concordance with this argument, one study has reported glucose enhancement of memory in infants as young as 2–4 days old (Horne et al. 2006). In this study, infants given a dose of 2 g/kg glucose were seen to turn their head less frequently to the source of a repeatedly spoken word, suggesting that glucose ingestion facilitated habituation to the stimulus in these babies. While, to the best of our knowledge, this is the only study which has considered the influence of glucose ingestion on cognitive performance in infants, many studies have considered the role of glucose as a cognitive enhancer in older children.

Benton and colleagues observed glucose enhancement of attention following a 25 g GL in children aged between six and seven years on a reaction time task (Benton et al. 1987). A subsequent study from the same laboratory failed to replicate these findings in a group of children aged between nine and ten years of age (Benton & Stevens 2008). However, most of the previous research investigating glucose enhancement of memory in children within the pre-adolescent age range has investigated cognitive performance following breakfast meals, which differ with regard to the speed by which glucose is released into the bloodstream. This phenomenon is referred to as 'glycaemic index' (GI) and is discussed in detail elsewhere in this volume (see Chapters 9 and 10). The majority of previous studies conducted in this area have found that breakfast meals which are associated with slower glucose release have conferred cognitive benefits in children, most notably in the domain of attention (Benton et al. 2007; Ingwersen et al. 2007; Mahoney et al. 2005). These studies are reviewed in depth by Ingwersen in her chapter on breakfast (see Chapter 9). Taken together, these findings suggest that glucose reliably enhances cognitive performance in pre-adolescent children, and may be useful for fuelling the brain in order to enable maintenance of attention throughout a prolonged period of cognitive activity, such as that which children would typically experience during a morning in the classroom.

In addition to the body of research which has been conducted in younger children, a substantial body of literature has emerged which relates to glucose enhancement of memory in adolescents. In fact, one of the first studies to report a positive effect of glucose on cognitive performance used an adolescent sample. In this previous study, ingestion of an oral glucose tolerance test (OGTT) preparatory breakfast and 150 g glucose enhanced recall of low- and high-imagery paired associates, relative to a fasting control condition (Lapp 1981). More recently, research from our own laboratory has thoroughly investigated glucose influences on verbal episodic memory in healthy adolescents. Generally, we have found that a 25 g glucose dose is effective as an enhancer of verbal episodic memory performance in healthy adolescents, when items are encoded under conditions of divided attention (Smith & Foster 2008a; Smith et al. 2009, 2011a). Figure 8.1 depicts typical findings from these studies. These findings support those from work in adults which suggest that glucose only reliably facilitates memory under dual-task conditions (Sünram-Lea et al. 2002b). This may be related to the notion that healthy young adults are operating at their 'cognitive peak'; therefore, glucose would only be effective in improving performance when such individuals face increased cognitive demands that allow 'room for improvement' (Foster et al. 1998). Previous research which has been conducted in adults will be discussed in further detail below.

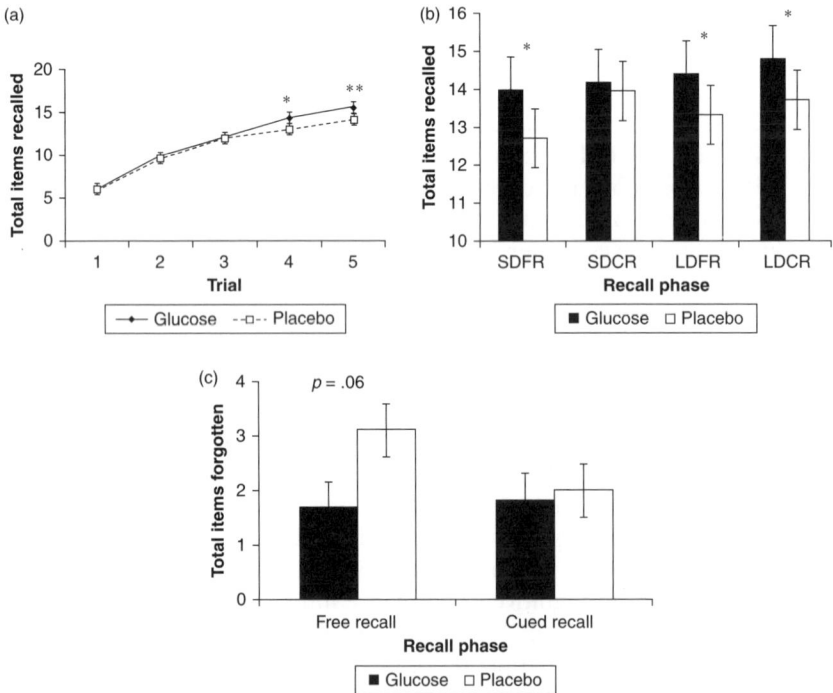

Figure 8.1   Glucose influences on verbal episodic memory performance, as measured by the California Verbal Learning Test (CVLT), in adolescents

Figure 8.1a shows the total number of items recalled on each immediate free recall trial of the CVLT, subsequent to ingestion of (a) glucose, or (b) placebo. Glucose ingestion significantly improved performance on immediate free recall trials 4 and 5, relative to placebo. Figure 8.1b shows that relative to placebo, ingestion of oral glucose enhanced short delay free recall (SDFR), long delay free recall (LDFR) and long delay cued recall (LDCR) performance on the CVLT, but not short delay cued recall (SDCR) performance. Figure 8.1c shows a trend towards enhanced CVLT free recall, one-week post encoding, subsequent to pre-encoding ingestion of glucose, relative to placebo. There were no long-term effects of glucose ingestion on cued recall performance. Data from Smith (2009) and Smith et al. (2011a), collected in adolescent males aged 14–17 years. Figures adapted and reproduced with permission from Smith et al. (2011a). $^{*}p < 0.05$, $^{**}p < 0.01$.

Our previous work in adolescents has uncovered some novel findings relating to the glucose memory facilitation effect. For example, in one study, we found that glucose enhanced only verbal episodic memory performance in adolescents with relatively higher trait anxiety (Smith et al. 2011a). This finding is in concordance with work in adults which suggests that glucose most reliably enhances verbal episodic memory in individuals with some degree of cognitive impairment (due to ageing or cognitive disorders), given that a relationship between negative affective states and cognitive impairment has been established previously (McEwen & Sapolsky 1995; Sala et al. 2004). Moreover, a further novel finding in our studies with adolescents is that glucose enhances event-related potential (ERP) components of both recollection and familiarity (Smith et al. 2009; see Figure 8.2). Recollection refers to recognition memory accompanied by spatio-temporal contextual details, and is thought to be mediated by the hippocampus. Familiarity does not involve this degree of episodic richness, and is subserved by non-hippocampal brain regions, including the perirhinal cortex (Aggleton & Brown 2006). Therefore, our ERP findings in adolescents have implications for the 'hippocampus hypothesis', which suggests that glucose specifically targets the hippocampus in modulating cognitive performance (Riby & Riby 2006; Sünram-Lea et al. 2008). It is also noteworthy that in our study, the familiarity ERP component was small in the placebo condition. This is unsurprising, given previous reports that familiarity does not develop until very late in childhood (Czernochowski et al. 2005). However, the observation of a significantly larger ERP familiarity effect in the glucose condition suggests that glucose may actually elicit cognitive processes at stages of development prior to those at which specific cognitive processes would be expected to emerge.

Further work which we have conducted in adolescents has involved the administration of breakfast cereals, which have differed with regard to GI, prior to tasks of verbal episodic memory completed under conditions of divided attention (Smith & Foster 2008b). Contrary to previous work which has typically found that low GI meals have a beneficial effect on attention and episodic memory in children and adults (see Chapter 9 for further details), we found that ingestion of a high GI breakfast prior to encoding resulted in fewer words being forgotten following a 100-minute post-encoding delay (in fact, glucose was associated with recall of items at this long delay phase that hadn't been remembered at a previous recall phase, see Figure 8.3). We suggest that only the relatively larger acute increase in blood glucose concentration provided by the high GI breakfast was sufficient to enhance cognitive processing, given that encoding took place in this study under dual-task conditions (to-be-remembered words were presented concurrently with a motor task). This relates to the argument presented earlier that glucose may only be effective in improving performance under increased cognitive demands that allow 'room for improvement'.

In summary, it appears that glucose is effective in facilitating memory performance in children and adolescents. Our work in adolescents suggests

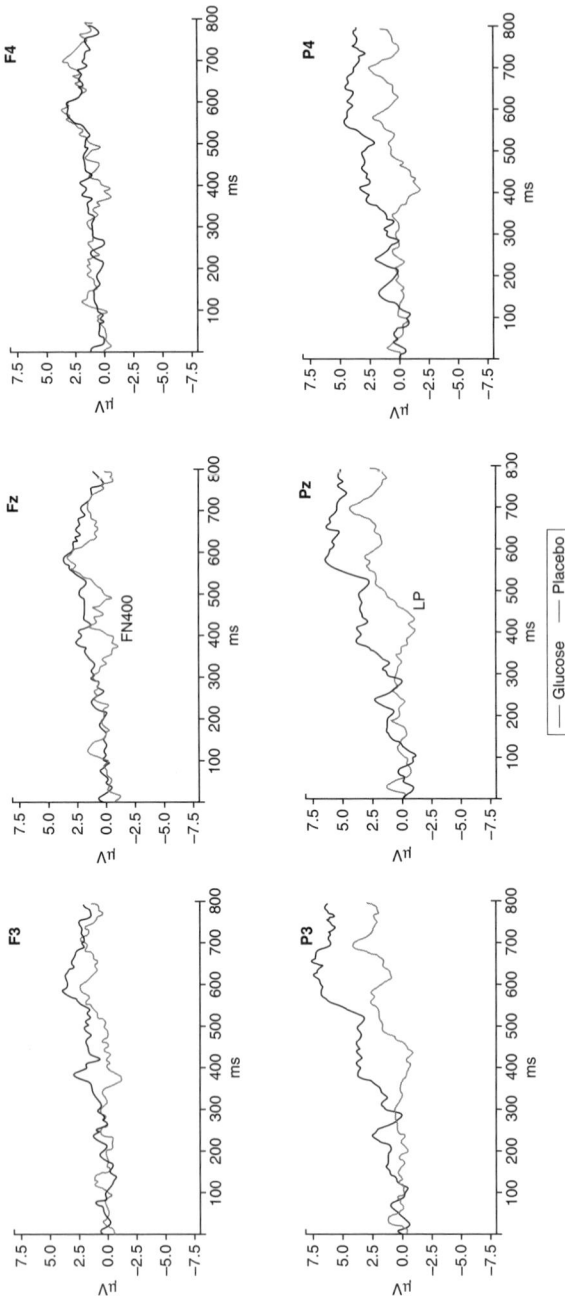

Figure 8.2 ERP old–new difference waveforms at three frontal electrodes (F3, Fz, F4) and three parietal electrode sites (P3, Pz, P4), subsequent to the ingestion of (a) glucose or (b) placebo

The difference waveforms were significantly larger in amplitude over the frontal sites between the 300–500 ms latency range (labelled FN400) for the glucose condition, relative to the placebo condition. This suggests that glucose significantly enhanced familiarity based recognition memory processing. The difference waveforms were also significantly larger in amplitude over the left parietal sites between the 400–800 ms latency range (labelled LP) for the glucose condition, suggesting that recollection-based recognition memory processing was also enhanced by glucose.

Source: Reproduced with permission from Smith et al. (2009).

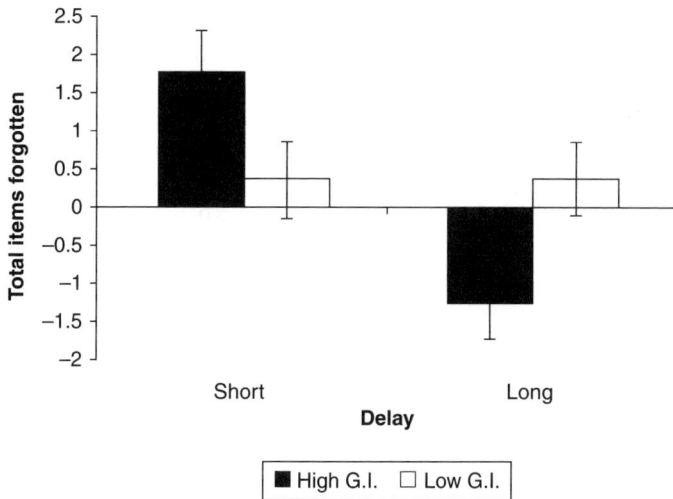

Figure 8.3 Mean number of items remembered/forgotten for a group who were administered (a) a low GI breakfast cereal meal or (b) a high GI breakfast cereal meal (± SE)

There was a significant time x treatment interaction effect, in that those participants who consumed the high GI breakfast cereal meal recalled significantly more items following a long delay (100 minutes) than following a short delay (60 minutes). Further, at the long delay, significantly more items were remembered by the high GI group, relative to the low GI group.
Source: Reproduced with permission from Smith & Foster (2008b).

that glucose may only be reliable in enhancing verbal episodic memory performance when memory materials are encoded under conditions of divided attention. The following section will elaborate on this point, via discussion of a series of studies in young adults which support the notion that dual-task conditions may be essential for reliably observing the glucose memory facilitation effect.

Young Adults

A number of studies over the past 20 years have reported that glucose enhances verbal episodic memory performance in healthy young adults (Benton et al. 1994; Foster et al. 1998; Meikle et al. 2004, 2005; Messier et al. 1998; Morris 2008; Owen et al. 2010; Parker & Benton 1995; Riby et al. 2006, 2008a; Sünram-Lea et al. 2001, 2002a, 2002b, 2004, in press). In addition, studies in healthy young adults have investigated whether glucose can facilitate performance across a wider range of cognitive domains than those explored in children, including attention (Benton 1990; Meikle et al. 2004; Reay et al. 2006), face recognition (Metzger 2000), semantic memory (Riby et al. 2006), prospective memory (Riby et al. 2011), verbal fluency (Donohoe & Benton

1999a), visuospatial functioning (Scholey & Fowles 2002), visuospatial long-term memory (Sünram-Lea et al. 2001, 2002a, 2002b) and working memory (Hall et al. 1989; Kennedy & Scholey 2000; Meikle et al. 2004; Reay et al. 2006; Scholey et al. 2001; Sünram-Lea et al. 2002b, 2004, in press).

With regard to the notion that glucose only reliably enhances memory performance when memory materials are encoded under conditions of divided attention, a number of studies have investigated verbal episodic memory performance under dual-task conditions following administration of a glucose drink. Typically, participants are required to perform a motor task involving sequences of hand movements during presentation of verbal stimuli (Foster et al. 1998; Riby et al. 2006; Sünram-Lea et al. 2001, 2002a, 2002b, 2004), analogous to that of our aforementioned studies in adolescent participants. While all of these studies have reported that oral glucose ingestion improves memory, some studies which have not incorporated a divided attention paradigm have failed to observe a significant effect of glucose on verbal episodic memory performance (Azari 1991; Benton & Owens 1993; Hall et al. 1989; Manning et al. 1997; Scholey & Kennedy 2004; Scholey et al. 2001; Winder & Borrill 1998). In one study which involved administration of a verbal episodic memory task concurrently with a tracking task, glucose ingestion was associated with improved performance on the tracking task, but not the memory task (Scholey et al. 2009). Taken together, these findings suggest that glucose enhancement of cognitive performance, particularly in the domain of verbal episodic memory, may be dependent upon whether the cognitive tests are administered under dual-task conditions. Alternatively, other studies in healthy young adults have suggested that it may not be divided attention per se, but rather cognitive demand more generally, which is important in terms of reliably observing a glucose enhancement effect. For example, task difficulty also seems to be an important determining factor in whether a facilitatory effect of glucose is observed (Kennedy & Scholey 2000; Meikle et al. 2004, 2005; Scholey et al. 2001). It has been suggested that more demanding cognitive tasks are associated with a greater depletion in blood glucose concentration (Donohoe & Benton 1999b; Fairclough & Houston 2004; Scholey et al. 2001, 2006). Indeed, the glucose memory facilitation effect may be attributable to the supply of glucose 'topping up' circulating glucose resources under conditions of high cognitive demand (Scholey et al. 2006).

An important series of findings to arise from the young adult work in this area are from neuroimaging studies which have attempted to determine the specific brain regions involved in the glucose memory facilitation effect. A functional magnetic resonance imaging (fMRI) study in patients with schizophrenia, conducted by Stone and colleagues (2005), found that parahippocampal activation was significantly enhanced during verbal encoding, relative to placebo, suggesting a role for the medial temporal lobe (MTL) in the mediation of the glucose memory facilitation effect. Further, in a more recent fMRI study by Parent and colleagues (in press), the increased activation of several MTL brain regions was also observed during a visual encoding task, demonstrating further support for the role of the MTL in mediating the glucose enhancement effect. Moreover, these previous fMRI findings are supported by ERP findings in healthy young adults. During an oddball task, Riby and colleagues (2008b) observed significant amplitude reduction of an ERP component that is thought to be modulated by the hippocampus (P3b). This finding was interpreted as demonstrating that glucose enhances memory by decreasing the cognitive resources required for memory updating (Riby et al. 2008b). A further fMRI study adopted a similar argument after decreased prefrontal activation was observed during a monitoring task subsequent to a drink comprising a combination of glucose and caffeine (Serra-Grabulosa et al. 2010).

Taken together, the most notable findings in this area from young adults are that (a) glucose appears to enhance performance across a range of cognitive domains, most notably episodic memory, (b) the glucose memory facilitation effect is most reliably demonstrated when two concurrent tasks are administered, or possibly, when the task demands are high more generally, and (c) neuroimaging findings support a role for the MTL in the mediation of the glucose memory facilitation effect.

Older Adults

Ageing is typically associated with some degree of cognitive decline, including memory loss (Craik 1994; Grady & Craik 2000; Salthouse 2003; Winocur 1988; Riby et al. 2004). In recent years, a plethora of research has been conducted to investigate potential lifestyle and dietary changes which may reduce the detrimental effects of ageing on cognitive performance. Such research is of vital importance in terms of decreasing the social and economic costs of cognitive ageing.

Many of the early studies investigating the influence of glucose ingestion on cognitive performance were conducted with elderly individuals. A key rationale for this seminal early work was that the ageing process involves dysregulation of a number of key hormones involved in both memory and glucoregulation, such as adrenaline (Gold 2005; Korol & Gold 1998). Further, impaired glucoregulation is a typical feature of ageing (Awad et al. 2002; Messier 2004; Parsons & Gold 1992). Similar to the studies in young adults, modulation of verbal episodic memory subsequent to glucose ingestion has been investigated in older adults relatively more frequently than other cognitive domains. This may be due to the fact that memory problems are a notable concern for many elderly individuals, as suggested above. The majority of the studies conducted have reported an improvement in verbal episodic memory (Hall et al. 1989; Manning et al. 1990, 1992, 1997, 1998; Parsons and Gold 1992; Riby et al. 2004, 2006). However, few studies have found evidence for glucose facilitation of other cognitive domains. One study by Allen and colleagues (1996) found that a 50 g glucose dose enhanced design fluency, verbal fluency and visual memory in a sample ranging between 61 and 87 years of age. Addition-ally, Messier and colleagues (1997) found that glucose ingestion improved attention in older adults. Beyond these findings, there is limited evidence to suggest that glucose enhances non-episodic memory cognitive domains (for a review, see Smith et al. 2011b).

Given the prevalence of cognitive dysfunction among elderly individuals, it would be of interest to investigate whether modulation of long-term car-bohydrate intake could improve chronic cognitive performance in elderly individuals. For example, manipulation of dietary carbohydrate and fibre intake could lower the GL of meals, which may improve chronic glucoregulation in elderly individuals and possibly result in improved cognitive performance. A recent Cochrane review reported that no randomised controlled trials have been conducted to date which could extend current knowledge on whether carbohydrate intake could influence chronic cognitive performance in older individuals (Ooi et al. 2011). This Cochrane review focused specifi-cally on mild cognitive impairment (MCI), a transient state of cognitive decline between healthy cognition and dementia. However, our understanding of how carbohydrate intake influences cognitive performance in older adults more generally is limited. This should be a priority for future research in this area.

It is worthwhile noting that glucose facilitation in the elderly can also be extended to pathological ageing. Work from our own lab on MCI has found story recall (a measure of episodic memory and known to decline in ageing) to be facilitated after the intake of a drink containing 25 g glucose. Whether 'deficit' groups such as MCI can particularly benefit from glucose ingestion remains debatable, since the glucose facilitation effect in this study was not exaggerated in MCI compared with a control group of older adults. Elsewhere, work on dementia has yielded similar results. For instance, Manning et al. (1993) found enhanced performance in both moderate and severe Alzheimer's disease (AD). In that study, cognitive facilitation was found on tests including recall and recognition of words, narrative prose and face recognition. For a review of the glucose facilitation effect in 'deficit' populations, see Smith et al. (2011b).

Long-Term Studies

As mentioned above, our understanding of the ways in which long-term dietary carbohydrate intake influences cognitive performance is limited. How-ever, in recent years, researchers have begun to consider the role of weight loss diets on cognitive performance. Such diets tend to involve restricted carbohy-drate intake and meals and snacks with a low GL. Chapter 10 in this volume provides a comprehensive review of this body of work. While results of these studies are somewhat mixed, it seems that a severely carbohydrate-restricted diet can be detrimental for long-term cognitive performance. Research by D'Anci and colleagues (2009) has found support for the notion that low car-bohydrate diets are associated with a reduction in memory and processing speed. Further, Cheatham and colleagues (2009) failed to find a cognitive benefit from low GL weight-loss diets. This is despite an abundance of previous reports from acute studies that ingestion of a low GI meal is associated with improvements in cognitive performance, most notably attention, during the period following ingestion of the meal. Carbohydrate-restricted dieting is obviously a special case, and as suggested above, more work is clearly needed to ascertain the relationship between dietary carbohydrate intake, GL and cognitive performance across the lifespan.

Mechanisms

A number of theories pertaining to the neurocognitive mechanisms underlying the glucose memory facilitation effect have been proposed. Three of these mechanisms will be discussed herein: (a) insulin receptors, (b) hippocampal acetylcholine (Ach) synthesis, and (c) potassium adenosine triphosphate (K_{ATP}) channel function.

Insulin

Insulin receptors are densely populated in the hippocampus relative to other brain regions (Unger et al. 1989). Increases in circulating blood glucose subsequent to carbohydrate ingestion trigger insulin release from the pancreas, and therefore increase circulating insulin. Insulin has been observed previously to improve acute memory performance when administered directly (Martins et al. 2006; Reger et al. 2006, 2008a, 2008b; Watson and Craft 2004), confirming that insulin can act as a cognitive enhancer in its own right. This has led to suggestions that glucose enhances memory via modulation of insulin release and subsequent insulin effects on the brain. This hypothesis currently remains speculative, due to the difficulty associated with increasing blood glucose in the absence of a subsequent increase in insulin release.

ACh Synthesis

ACh is a major neurotransmitter which serves many important functions in the autonomic and central nervous systems. Given that glucose metabolism is centrally involved in the synthesis of ACh, it has been investigated as a candidate mechanism underlying the glucose memory facilitation effect (Messier 2004). Studies in rodents have found that administration of a glucose dose dependently increases both cognitive performance and ACh output (Ragozzino et al. 1996, 1998). Additionally, low glucose doses which would not typically enhance performance when administered independently demonstrate a cognitive facilitation effect when administered with choline, a further ACh precursor metabolite in mice (Kopf et al. 2001). Taken together, these findings suggest a role for ACh in the modulation of cognitive performance subsequent to glucose administration. However, these findings do not provide conclusive evidence that glucose directly exerts its effects on cognition via ACh, and the pathway by which such a mechanism operates is likely to be complex.

K_{ATP} *Channel Function*

Related to the previous suggestion that glucose enhances memory via its effects on the synthesis of ACh, it has also been suggested that glucose modulates cognitive performance via its role in the regulation of K_{ATP} channel function. Administration of glucose, in addition to other substances which are associated with K_{ATP} channel blockade in rats, results in an increase in hippocampal ACh output and enhanced cognitive performance (Stefani and Gold 2001; Stefani et al. 1999). While these observations support the hypothesis that glucose exerts an effect on cognitive performance by regulating K_{ATP} channel function, and thus increasing hippocampal ACh output, these findings should again be treated with caution. This is because these studies do not investigate the direct effects of glucose administration on K_{ATP} channel function, but only observe a similar effect on ACh output between glucose and other K_{ATP} channel regulators.

Other Carbohydrates

Thus far, this chapter has focused primarily on glucose effects on cognitive performance across the lifespan. While studies investigating chronic glucose consumption and cognitive performance are lacking, evidence from several studies suggests that ingestion of glucose can enhance cognitive performance acutely. In addition to this body of work, some studies have also considered the influence of other dietary carbohydrates on cognitive performance at various stages of the lifespan, including fructose and sucrose. Some of these studies are discussed herein.

Fructose

Fructose is a monosaccharide which is naturally found in high levels in fruit, vegetables, honey and sugar cane. Evidence from both acute (Messier & White 1987; Rodriguez et al. 1994, 1999) and long-term (Messier et al. 2007) animal studies has supported a role for fructose enhancement of cognitive performance. In this latter study, a high fructose diet administered for three months was associated with faster learning on an operant conditioning task (Messier et al. 2007). By contrast, it has been suggested that increased fructose intake across the lifespan is a risk factor for dementia (Siervo et al. 2011; Stephan et al. 2010). The evidence for such an association is based largely upon epidemiological observations of dietary fructose intake and dementia prevalence, typically at a national, rather than an individual, level (Stephan et al. 2010). However, there are obvious methodological shortcomings associated with such a gross observational research method, and an investigation of this question using a large-scale longitudinal cohort study would be beneficial. While no direct mechanism has yet been determined for the relationship between dementia and fructose intake, fructose, via a number of indirect pathways, increases insulin release, which can lead to insulin resistance and other cardiovascular risk factors for dementia (Stephan et al. 2010).

Sucrose

Sucrose is commonly known as table sugar and is synthesised from fructose and glucose. In addition to the vast literature which now exists relating to glucose effects on cognitive performance, some evidence is beginning to emerge from the animal literature which suggests an association between a high sucrose diet and cognitive impairment. For example, rats fed a high sucrose diet for six weeks demonstrated relatively poor spatial memory (Jurdak et al. 2008), and in a further study by this same group, rats fed a high sucrose diet for eight weeks showed poor recognition memory performance (Jurdak & Kanarek 2009). In both of these studies, the animals in the high sucrose condition exhibited relatively poorer glucoregulatory efficiency subsequent to the dietary intervention period, relative to those fed a standard chow (Jurdak & Kanarek 2009; Jurdak et al. 2008). On this basis, it may well be that insulin resistance is the mechanism underlying the association between sucrose intake and cognitive performance. Work from our own lab has found that middle-aged adults who report high-sugar sweet and drink intake as part of their normal diet have poorer glucose regulation ability, which impacts on their cognitive functioning (Riby et al. 2008a). However, epidemiological studies should investigate this association as a first step to better understanding the relationship underlying sucrose-induced cognitive impairment.

Summary and Conclusions

In summary, evidence from numerous human studies now exists which supports the notion that glucose ingestion enhances cognitive performance across the lifespan. In younger adults, glucose is observed to enhance cognitive performance across a number of cognitive domains, but only reliably under conditions of divided attention. In older adults, findings to date suggest that cognitive enhancement of memory subsequent to glucose ingestion is limited to the domain of verbal episodic memory. In adolescents, research has primarily focused on the enhancement of attention subsequent to ingestion of a low GI meal, relative to a high GI meal, which has implications for the nutritional composition of breakfast and lunchtime meals prior to sustained cognitive activity in the classroom. Future research should focus on the long-term influence of diets differing with regard to carbohydrate composition. Some animal work has incorporated dietary manipulation of sucrose and fructose composition and investigated the impact of such manipulations on cognitive performance. However, more work needs to be done to further elucidate these relationships, and the mechanisms underlying the association between long-term carbohydrate intake and cognitive performance across the lifespan.

In conclusion, the literature reviewed in this chapter suggests a relationship between acute carbohydrate ingestion and cognitive performance, as well as between long-term carbohydrate intake and cognition. However, these relationships are complex, and more work is needed, particularly relating to longer-term studies of dietary intake across the lifespan, to further enhance our understanding of the associations which have been established.

References

Aggleton, J. P. & Brown, M. W. (2006). Interleaving brain systems for episodic and recognition memory. *Trends in Cognitive Sciences, 10*, 455–463.

Allen, J. B., Gross, A. M., Aloia, M. S. & Billingsley, C. (1996). The effects of glucose on nonmemory cognitive functioning in the elderly. *Neuropsychologia, 34*, 459–465.

Awad, N., Gagnon, M., Desrochers, A., Tsiakas, M. & Messier, C. (2002). Impact of peripheral glucoregulation on memory. *Behavioral Neuroscience, 116*, 691–702.

Azari, N. P. (1991). Effects of glucose on memory processes in young adults. *Psychopharmacology, 105*, 521–524.

Benton, D. (1990). The impact of increasing blood glucose on psychological functioning. *Biological Psychology, 30*, 13–19.

Benton, D., Brett, V. & Brain, P. F. (1987). Glucose improves attention and reaction to frustration in children. *Biological Psychology, 24*, 95–100.

Benton, D., Maconie, A. & Williams, C. (2007). The influence of the glycaemic load of breakfast on the behaviour of children in school. *Physiology and Behavior, 92*, 717–724.

Benton, D. & Owens, D. S. (1993). Blood glucose and human memory. *Psychopharmacology, 113*, 83–88.

Benton, D., Owens, D. S. & Parker, P. Y. (1994). Blood glucose influences memory and attention in young adults. *Neuropsychologia, 32*, 595–607.

Benton, D. & Stevens, M. K. (2008). The influence of a glucose containing drink on the behavior of children in school. *Biological Psychology, 78*, 242–245.

Cheatham, R. A., Roberts, S. B., Das, S. K., Gilhooly, C. H., Golden, J. K., Hyatt, R., et al. (2009). Long-term effects of provided low and high glycemic load low energy diets on mood and cognition. *Physiology and Behavior, 98*, 374–379.

Chiron, C., Raynaud, C., Maziere, B., Zilbovicius, M., Laflamme, L., Masure, M.-C., et al. (1992). Changes in regional cerebral blood flow during brain maturation in children and adolescents. *The Journal of Nuclear Medicine, 33*, 696–703.

Craik, F. I. M. (1994). Memory changes in normal aging. *Current Directions in Psychological Science, 3*, 155–158.

Czernochowski, D., Mecklinger, A., Johansson, M. & Brinkmann, M. (2005). Age-related differences in familiarity and recollection: ERP evidence from a recognition memory study in children and young adults. *Cognitive, Affective, and Behavioral Neuroscience, 5*, 417–433.

D'Anci, K. E., Watts, K. L., Kanarek, R. B. & Taylor, H. A. (2009). Low-carbohydrate weight-loss diets. Effects on cognition and mood. *Appetite, 52*, 96–103.

Donohoe, R. T. & Benton, D. (1999a). Cognitive functioning is susceptible to the level of blood glucose. *Psychopharmacology, 145*, 378–385.

Donohoe, R. T. & Benton, D. (1999b). Declining blood glucose levels after a cognitively demanding task predict subsequent memory. *Nutritional Neuroscience, 2*, 413–424.

Fairclough, S. H. & Houston, K. (2004). A metabolic measure of mental effort. *Biological Psychology, 66*, 177–190.

Foster, J. K., Lidder, P. G. & Sünram, S. I. (1998). Glucose and memory: fractionation of enhancement effects. *Psychopharmacology, 137*, 259–270.

Gold, P. E. (2005). Glucose and age-related changes in memory. *Neurobiology of Aging, 26*, S60–S64.

Grady, C. L. & Craik, F. I. M. (2000). Changes in memory processing with age. *Current Opinion in Neurobiology, 10*, 224–231.

Hall, J. L., Gonder-Frederick, L. A., Chewning, W. W., Silvera, J. & Gold, P. E. (1989). Glucose enhancement of performance on memory tests in young and aged humans. *Neuropsychologia, 27*, 1129–1138.

Horne, P., Barr, R. G., Valiante, G., Zelazo, P. R. & Young, S. N. (2006). Glucose enhances newborn memory for spoken words. *Developmental Psychobiology, 48*, 574–582.

Ingwersen, J., Defeyter, M. A., Kennedy, D. O., Wesnes, K. A. & Scholey, A. B. (2007). A low glycaemic index breakfast cereal preferentially prevents children's cognitive performance from declining throughout the morning. *Appetite, 49*, 240–244.

Jurdak, N. & Kanarek, R. B. (2009). Sucrose-induced obesity impairs novel object recognition learning in young rats. *Physiology and Behavior, 96*, 1–5.

Jurdak, N., Lichtenstein, A. H. & Kanarek, R. B. (2008). Diet-induced obesity and spatial cognition in young male rats. *Nutritional Neuroscience, 11*, 48–54.

Kennedy, D. O. & Scholey, A. B. (2000). Glucose administration, heart rate and cognitive performance: effects of increasing mental effort. *Psychopharmacology, 149*, 63–71.

Kopf, S. R., Buchholzer, M. L., Hilgert, M., Löffelholz, K. & Klein, J. (2001). Glucose plus choline improve passive avoidance behaviour and increase hippocampal acetylcholine release in mice. *Neuroscience, 103*, 365–371.

Korol, D. L. & Gold, P. E. (1998). Glucose, memory, and aging. *American Journal of Clinical Nutrition, 67*, 764S–771S.

Lapp, J. E. (1981). Effects of glycemic alterations and noun imagery on the learning of paired associates. *Journal of Learning Disabilities, 14*, 35–38.

Mahoney, C. R., Taylor, H. A., Kanarek, R. B. & Samuel, P., 2005. Effect of breakfast composition on cognitive processes in elementary school children. *Physiology and Behavior, 85*, 635–645.

Manning, C. A., Hall, J. L. & Gold, P. E. (1990). Glucose effects on memory and other neuropsychological tests in elderly humans. *Psychological Science, 1,* 307–311.

Manning, C. A., Parsons, M. W., Cotter, E. M. & Gold, P. E., 1997. Glucose effects on declarative and nondeclarative memory in healthy elderly and young adults. *Psychobiology, 25,* 103–108.

Manning, C. A., Parsons, M. W. & Gold, P. E. (1992). Anterograde and retrograde enhancement of 24-h memory by glucose in elderly humans. *Behavioral and Neural Biology, 58,* 125–130.

Manning, C., Ragozzino, M. & Gold, P. (1993). Glucose enhancement of memory in patients with probable senile dementia of the Alzheimer's type. *Neurobiology of Ageing, 14,* 523–528.

Manning, C. A., Stone, W. S., Korol, D. L. & Gold, P. E. (1998). Glucose enhancement of 24-h memory retrieval in healthy elderly humans. *Behavioural Brain Research, 93,* 71–76.

Martins, I. J., Hone, E., Foster, J. K., Sünram-Lea, S. I., Gnjec, A., Fuller, S. J., et al. (2006). Apolipoprotein E, cholesterol metabolism, diabetes, and the convergence of risk factors for Alzheimer's disease and cardiovascular disease. *Molecular Psychiatry, 11,* 721–736.

McEwen, B. S. & Sapolsky, R. M. (1995). Stress and cognitive function. *Current Opinion in Neurobiology, 5,* 205–216.

Meikle, A., Riby, L. M. & Stollery, B. (2004). The impact of glucose ingestion and gluco-regulatory control on cognitive performance: a comparison of younger and middle aged adults. *Human Psychopharmacology, 19,* 523–535.

Meikle, A., Riby, L. M. & Stollery, B. (2005). Memory processing and the glucose facilitation effect: the effects of stimulus difficulty and memory load. *Nutritional Neuroscience, 8,* 227–232.

Messier, C. (2004). Glucose improvement of memory: a review. *European Journal of Pharmacology, 490,* 33–57.

Messier, C., Gagnon, M. & Knott, V. (1997). Effect of glucose and peripheral glucose regulation on memory in the elderly. *Neurobiology of Aging, 18,* 297–304.

Messier, C., Pierre, J., Desrochers, A. & Gravel, M. (1998). Dose-dependent action of glucose on memory processes in women: effect on serial position and recall priority. *Cognitive Brain Research, 7,* 221–233.

Messier, C., Whately, K., Liang, J., Du, L. & Puissant, D. (2007). The effects of a high-fat, high-fructose, and combination diet on learning, weight, and glucose regulation in C57BL/6 mice. *Behavioural Brain Research, 178,* 139–145.

Messier, C. & White, N. M. (1987). Memory improvement by glucose, fructose, and two glucose analogs: a possible effect on peripheral glucose transport. *Behavioral and Neural Biology, 48,* 104–127.

Metzger, M. M. (2000). Glucose enhancement of a facial recognition task in young adults. *Physiology and Behavior, 68,* 549–553.

Morris, N. (2008). Elevating blood glucose level increases the retention of information from a public safety video. *Biological Psychology, 78,* 188–190.

Ooi, C. P., Loke, S. C., Yassin, Z. & Hamid, T. A. (2011). Carbohydrates for improving the cognitive performance of independent-living older adults with normal cognition or mild cognitive impairment. *Cochrane Database of Systematic Reviews,* CD007220.

Owen, L., Finnegan, Y., Hu, H., Scholey, A. B. & Sunram-Lea, S. I. (2010). Glucose effects on long-term memory performance: duration and domain specificity. *Psychopharmacology, 211,* 131–140.

Parent, M. B., Krebs-Kraft, D. L., Ryan, J. P., Wilson, J. S., Harenski, C. & Hamann, S. (in press). Glucose administration enhances fMRI brain activation and connectivity related to episodic memory encoding for neutral and emotional stimuli. Neuropsychologia.

Parker, P. Y. & Benton, D. (1995). Blood glucose levels selectively influence memory for word lists dichotically presented to the right ear. *Neuropsychologia, 33,* 843–854.

Parsons, M. W. & Gold, P. E. (1992). Glucose enhancement of memory in elderly humans: an inverted-U dose-response curve. *Neurobiology of Aging, 13,* 401–404.

Ragozzino, M. E., Pal, S. N., Unick, K., Stefani, M. R. & Gold, P. E. (1998). Modulation of hippocampal acetylcholine release and spontaneous alternation scores by intrahippocampal glucose injections. *Journal of Neuroscience, 18,* 1595–1601.

Ragozzino, M. E., Unick, K. E. & Gold, P. E. (1996). Hippocampal acetylcholine release during memory testing in rats: augmentation by glucose. *Proceedings of the National Academy of Sciences of the United States of America, 93,* 4693–4698.

Reay, J. L., Kennedy, D. O. & Scholey, A. B. (2006). Effects of Panax ginseng, consumed with and without glucose, on blood glucose levels and cognitive performance during sustained "mentally demanding" tasks. *Journal of Psychopharmacology, 20,* 771–781.

Reger, M. A., Watson, G. S., Frey, W. H., Baker, L. D., Cholerton, B., Keeling, M. L., et al. (2006). Effects of intranasal insulin on cognition in memory-impaired older adults: modulation by APOE genotype. *Neurobiology of Aging, 27,* 451–458.

Reger, M. A., Watson, G. S., Green, P. S., Baker, L. D., Cholerton, B., Fishel, M. A., et al. (2008a). Intranasal insulin administration dose-dependently modulates verbal memory and plasma amyloid-beta in memory-impaired older adults. *Journal of Alzheimers Disease, 13,* 323–331.

Reger, M. A., Watson, G. S., Green, P. S., Wilkinson, C. W., Baker, L. D., Cholerton, B., et al. (2008b). Intranasal insulin improves cognition and modulates beta-amyloid in early AD. *Neurology, 70,* 440–448.

Riby, L.M., Perfect, T.J. & Stollery, B. (2004). Evidence for disproportionate costs in older adults for episodic but not semantic retrieval. The Quarterly Journal of Experimental Psychology: Section A 47, 241–267.

Riby, L. M., Law, A. S., McLaughlin, J. & Murray, J. (2011). Preliminary evidence that glucose ingestion facilitates prospective memory performance. *Nutrition Research, 31,* 370–377.

Riby, L. M., McLaughlin, J., Riby, D. M. & Graham, C. (2008a). Lifestyle, glucose regulation and the cognitive effects of glucose load in middle-aged adults. *British Journal of Nutrition, 100,* 1128–1134.

Riby, L. M., McMurtrie, H., Smallwood, J., Ballantyne, C., Meikle, A. & Smith, E. (2006). The facilitative effects of glucose ingestion on memory retrieval in younger and older adults: is task difficulty or task domain critical? *British Journal of Nutrition, 95,* 414–420.

Riby, L. M., Meikle, A. & Glover, C. (2004). The effects of age, glucose ingestion and gluco-regulatory control on episodic memory. *Age and Ageing, 33,* 483–487.

Riby, L. M. & Riby, D. M. (2006). Glucose, ageing and cognition: The hippocampus hypothesis, In Ballesteros, S. (Ed.), *Age, Cognition and Neuroscience/Envejecimiento, Cognición y Neurociencia.* Madrid: UNED Varia.

Riby, L. M., Sünram-Lea, S. I., Graham, C., Foster, J. K., Cooper, T., Moodie, C., et al. (2008b). The P3b versus the P3a: an event-related potential investigation of the glucose facilitation effect. *Journal of Psychopharmacology, 22*, 486–492.

Riby, L. M., Marriott, A., Bullock, R., Hancock, J., Smallwood, J. & McLaughlin, J. (2009). The effects of glucose ingestion and glucose regulation on memory performance in older adults with mild cognitive impairment. *European Journal of Clinical Nutrition, 63*, 566–571.

Rodriguez, W. A., Horne, C. A., Mondragon, A. N. & Phelps, D. D. (1994). Comparable dose-response functions for the effects of glucose and fructose on memory. *Behavioral and Neural Biology, 61*, 162–169.

Rodriguez, W. A., Horne, C. A. & Padilla, J. L. (1999). Effects of glucose and fructose on recently reactivated and recently acquired memories. *Progress in Neuro-Psychopharmacology and Biological Psychiatry, 23*, 1285–1317.

Sala, M., Perez, J., Soloff, P., Ucelli di Nemi, S., Caverzasi, E., Soares, J. C., et al. (2004). Stress and hippocampal abnormalities in psychiatric disorders. *European Neuropsychopharmacology, 14*, 393–405.

Salthouse, T. A. (2003). Memory aging from 18 to 80. *Alzheimer Disease & Associated Disorders, 17*, 162–167.

Scholey, A. B. & Fowles, K. A. (2002). Retrograde enhancement of kinesthetic memory by alcohol and by glucose. *Neurobiology of Learning and Memory, 78*, 477–483.

Scholey, A. B., Harper, S. & Kennedy, D. O. (2001). Cognitive demand and blood glucose. *Physiology and Behavior, 73*, 585–592.

Scholey, A. B. & Kennedy, D. O. (2004). Cognitive and physiological effects of an "energy drink": an evaluation of the whole drink and of glucose, caffeine and herbal flavouring fractions. *Psychopharmacology, 176*, 320–330.

Scholey, A. B., Laing, S. & Kennedy, D. O. (2006). Blood glucose changes and memory: effects of manipulating emotionality and mental effort. *Biological Psychology, 71*, 12–19.

Scholey, A. B., Sünram-Lea, S. I., Greer, J., Elliott, J. & Kennedy, D. O. (2009). Glucose administration prior to a divided attention task improves tracking performance but not word recognition: evidence against differential memory enhancement? *Psychopharmacology, 202*, 549–558.

Serra-Grabulosa, J. M., Adan, A., Falcon, C. & Bargallo, N. (2010). Glucose and caffeine effects on sustained attention: An exploratory fMRI study. *Human Psychopharmacology, 25*, 543–552.

Siervo, M., Wells, J. C., Brayne, C. & Stephan, B. C. (2011). Reemphasizing the role of fructose intake as a risk factor for dementia. *Journals of Gerontology. Series A, Biological Sciences and Medical Sciences, 66*, 534–536.

Smith, M. A. (2009). *Glucose modulation of verbal episodic memory in adolescents.* Unpublished Doctor of Philosophy, University of Western Australia.

Smith, M. A. & Foster, J. K. (2008a). Glucoregulatory and order effects on verbal episodic memory in healthy adolescents after oral glucose administration. *Biological Psychology, 79*, 209–215.

Smith, M. A. & Foster, J. K. (2008b). The impact of a high versus a low glycaemic index breakfast cereal meal on verbal episodic memory in healthy adolescents. *Nutritional Neuroscience, 11*, 219–227.

Smith, M. A., Hii, H. L., Foster, J. K. & van Eekelen, J. A. M. (2011a). Glucose enhancement of memory is modulated by trait anxiety in healthy adolescent males. *Journal of Psychopharmacology, 25*, 60–70.

Smith, M. A., Riby, L. M., Sunram-Lea, S. I., van Eekelen, J. A. & Foster, J. K. (2009). Glucose modulates event-related potential components of recollection and familiarity in healthy adolescents. *Psychopharmacology (Berlin)*, *205*, 11–20.

Smith, M. A., Riby, L. M., van Eekelen, J. A. & Foster, J. K. (2011b). Glucose enhancement of human memory: a comprehensive research review of the glucose memory facilitation effect. *Neuroscience and Biobehavioral Reviews*, *35*, 770–783.

Stefani, M. R. & Gold, P. E. (2001). Intrahippocampal infusions of K-ATP channel modulators influence spontaneous alternation performance: relationships to acetylcholine release in the hippocampus. *The Journal of Neuroscience*, *21*, 609–614.

Stefani, M. R., Nicholson, G. M. & Gold, P. E. (1999). ATP-sensitive potassium channel blockade enhances spontaneous alternation performance in the rat: a potential mechanism for glucose-mediated memory enhancement. *Neuroscience*, *93*, 557–563.

Stephan, B. C., Wells, J. C., Brayne, C., Albanese, E. & Siervo, M. (2010). Increased fructose intake as a risk factor for dementia. *Journals of Gerontology. Series A, Biological Sciences and Medical Sciences*, *65*, 809–814.

Stone, W. S., Thermenos, H. W., Tarbox, S. I., Poldrack, R. A. & Seidman, L. J. (2005). Medial temporal and prefrontal lobe activation during verbal encoding following glucose ingestion in schizophrenia: a pilot fMRI study. *Neurobiology of Learning and Memory*, *83*, 54–64.

Sünram-Lea, S. I., Dewhurst, S. A. & Foster, J. K. (2008). The effect of glucose administration on the recollection and familiarity components of recognition memory. *Biological Psychology*, *77*, 69–75.

Sünram-Lea, S. I., Foster, J. K., Durlach, P. & Perez, C. (2001). Glucose facilitation of cognitive performance in healthy young adults: examination of the influence of fast-duration, time of day and pre-consumption plasma glucose levels. *Psychopharmacology*, *157*, 46–54.

Sünram-Lea, S. I., Foster, J. K., Durlach, P. & Perez, C. (2002a). The effect of retrograde and anterograde glucose administration on memory performance in healthy young adults. *Behavioural Brain Research*, *134*, 505–516.

Sünram-Lea, S. I., Foster, J. K., Durlach, P. & Perez, C. (2002b). Investigation into the significance of task difficulty and divided allocation of resources on the glucose memory facilitation effect. *Psychopharmacology*, *160*, 387–397.

Sünram-Lea, S. I., Foster, J. K., Durlach, P. & Perez, C. (2004). The influence of fat co-administration on the glucose memory facilitation effect. *Nutritional Neuroscience*, *7*, 21–32.

Sünram-Lea, S. I., Owen, L., Finnegan, Y. & Hu, H. (2011). Dose-response investigation into glucose facilitation of memory performance and mood in healthy young adults. *Journal of Psychopharmacology*, *25*, 1076–187.

Unger, J., McNeill, T. H., Moxley, R. T., White, M., Mosi, A. & Livingston, J. N. (1989). Distribution of insulin receptor-like immunnoreactivity in the rat forebrain. *Neuroscience*, *31*, 143–157.

Watson, G. S. & Craft, S. (2004). Modulation of memory by insulin and glucose: neuropsychological observations in Alzheimer's disease. *European Journal of Pharmacology*, *490*, 97–113.

Winder, R. & Borrill, J. (1998). Fuels for memory: the role of oxygen and glucose in memory enhancement. *Psychopharmacology (Berlin)*, *136*, 349–356.

Winocur, G. (1988). A neuropsychological analysis of memory loss with age. *Neurobiology of Aging*, *9*, 487–494.

CHAPTER 9

The Impact of Breakfast on Cognitive Performance in Children and Adults

Jeanet Ingwersen

Chapter Overview

- Research investigating the effects of breakfast on cognitive performance suggests a beneficial effect of breakfast consumption compared with breakfast omission.
- Recently, there has been an increased interest in whether breakfasts with different nutritional composition affect cognitive performance differently. Such differences in composition have particularly focused on differences in glycaemic index (GI) or glycaemic load (GL).
- Although some studies point towards positive effects of a low GI/GL, it is difficult to make definite conclusions due to methodological differences between the studies.
- This chapter examines previous breakfast research in both adults and children.

Introduction

Scientific investigation of the relationship between food consumption and cognitive function is a relatively new area of research. Research has investigated the effects of both macronutrients and micronutrients on cognitive function and it has been suggested that such nutritional manipulations can have beneficial effects on cognitive performance in both adults and children (Blom-Hoffman et al. 2004; Dye & Blundell 2002; Dye et al. 2000; Hoyland et al. 2008, 2009). Research has, for example, shown a positive effect of iodine supplementation on cognitive performance in iodine-deficient children (Van den Briel et al. 2000) and a positive effect of vitamin/mineral supplementation on attention in children (Haskell et al. 2008). However, contrary to these findings, other research has not found any effects of nutritional manipulation or has only found effects on a few out of a number of cognitive measures. Kennedy et al. (2009), for example, investigated the effects of the essential

158

fatty acid, omega-3, on mood and cognitive performance in children aged 10–12 years of age. Out of 15 cognitive outcome measures, the authors only found significant effects of the omega-3 supplement on one measure, speed of word recognition. Kennedy et al. concluded that the results were a chance effect and that the omega-3 supplement did not have an effect on cognitive performance in children.

One aspect of food consumption that has received increased attention is the consumption of breakfast and the effects on cognitive performance (Benton & Jarvis 2007; Benton & Sargent 1992; Benton & Stevens 2008; Busch et al. 2002; Mahoney et al. 2007, 2005; Muthayya et al. 2007; Smith et al. 1994; Vaisman et al. 1996; Wesnes et al. 2003; Widenhorn-Müller et al. 2008). Research has found benefits of breakfast consumption in a number of areas such as mood (Foster et al. 2007; Lloyd et al. 1996) and body weight. De la Hunty & Ashwell (2007) reviewed the literature on the relationship between the consumption of breakfast cereal and body weight. They found that for both adults and children, those who ate breakfast cereal on a regular basis were slimmer than people who regularly skipped breakfast. The UK Government is committed to promoting healthier lifestyles and is in particular committed to promoting healthy schools through good quality school meals (Ells et al. 2008). The Government is furthermore committed to improving learning and raising standards in schools, which makes breakfast research in children particularly pertinent due to the potential effect of breakfast in enhancing children's cognitive performance and consequently their achievements in school (Gathercole et al. 2004).

Research investigating the effects of breakfast on cognitive function in both adults and children has suggested that breakfast omission can have adverse effects on cognitive performance, whereas breakfast consumption can have beneficial effects (e.g. Benton & Sargent 1992; Smith et al. 1994; Wesnes et al. 2003). Recently, there has also been an increased interest in whether cognitive performance is differentially affected by breakfast composition (e.g. Benton et al. 2003; Ingwersen et al. 2007). Research has investigated the effects of a number of different components of breakfast, such as carbohydrate, protein and fat (e.g. Fischer et al. 2001, 2002). More recently, however, there has been increased interest in comparing the effects of carbohydrate-rich breakfasts as they can have differential effects on blood glucose levels depending on the quality of the carbohydrates (glycaemic response) which, in turn, can alter cognitive performance. This chapter examines research investigating the effects of breakfast on cognitive performance in both adults and children with particular focus on the effects of carbohydrate in terms of glucose and GI.

Glycaemic Response

The brain depends almost exclusively on glucose for energy. The cells in the brain are constantly active and, because the brain does not store glucose, the cells continually draw glucose from the supply in the fluids that surround them. Glucose is the product of the breakdown of food in the digestion process. Following food intake, glucose gets absorbed into the blood and distributed to the brain cells by the bloodstream to provide energy to the brain (Raven & Johnson 1992).

Because brain functioning is dependent on a constant supply of glucose, it is important that the body's blood glucose level is maintained within a reasonably narrow range (see Chapter 8). The body's ability to maintain constant blood glucose levels is primarily regulated by two pancreatic endocrine hormones that have opposite actions: insulin and glucagon. Insulin is produced and secreted by the beta cells of the pancreatic islets (small islands of endocrine cells in the pancreas) and glucagon is produced and secreted by the alpha cells of the pancreatic islets. Glucagon acts on the same cells as insulin, but has the opposite effects (Carlson 1999; Kalat 2001).

After food intake, glucose is absorbed from the food into the blood, and blood glucose levels rise. High blood glucose levels stimulate the pancreas to release insulin into the blood. Although there is always a low level of insulin being secreted by the pancreas, the amount being secreted increases as blood glucose levels rise, and fall as blood glucose levels decrease. The increased levels of insulin in the blood stimulate cells to absorb the glucose out of the blood to use as energy and to store as glycogen in the liver and muscles (Raven & Johnson 1992). Hence, insulin prevents large increases in blood glucose concentration and plays a major role in maintaining a steady blood glucose concentration.

In contrast, when blood glucose levels are low, such as in between meals, the pancreas secretes glucagon into the blood. As blood glucagon levels rise, the liver is stimulated to break down its stored glycogen reserves and release glucose into the bloodstream, with the net effect of increasing blood glucose levels (Carlson 1999). Thus, when blood glucose levels are high, the liver stores excess glucose as glycogen, which is released again as glucose into the bloodstream when blood glucose levels fall between meals. The interaction between insulin and glucagon secretions and the stimulation of liver function help to keep blood glucose concentrations constant.

How quickly carbohydrates in food are absorbed, how high blood glucose rises, and how quickly it returns to normal levels again is referred to as glycaemic response. A high glycaemic response is characterised by fast absorption, a rush of glucose into the blood and, in response to the high blood glucose levels, there is a sharp drop in blood glucose to below normal levels. A low glycaemic response, on the other hand, is characterised by slow absorption, a moderate rise in blood glucose, and a steady return to normal levels. Different foods have different effects on glycaemic response. A high GI food results in a high glycaemic response, and a low GI food results in a low glycaemic response (Jenkins et al. 2002; Sheard et al. 2004). Hence, there is a smaller and longer lasting rise in blood glucose following the consumption of low GI food compared with high GI food. In addition to the rate of

Table 9.1 Examples of high, medium and low GI food

Category	GI range	Examples (GI)
High GI	70 or above	Glucose (100), Potato – boiled (93), White bread (70), Watermelon (80), Coco Pops cereal (77)
Medium GI	56–69	Whole-meal pasta (58), Apricot (57), Sweet corn – boiled (60), Peach (56), Cous cous – boiled (69)
Low GI	55 or less	Milk- full fat (34), Lentils – green (37), Soya beans (15), Apple (40), All Bran cereal (42)

The GI values are taken from an international table of glycaemic index (Foster Powell et al. 2002) and an online GI database (The University of Sydney).

digestion and absorption of carbohydrates, the glycaemic response following food consumption also depends on the rate of gastric emptying. A number of factors, such as the starch, fat and protein content, can alter the GI of the food. A food's GI can, for example, be lowered by the fat content or presence of dietary fibre, whereas the presence of water increases the GI. Food is rated on a GI scale of 1 to 100, depending on its effect on blood glucose levels, and is categorised into high, medium and low GI food. Table 9.1 shows some examples of foods and their GI rating.

There is a wealth of research demonstrating that aspects of cognitive performance can be enhanced by an increase in blood glucose levels via the ingestion of a drink containing glucose (Riby 2004; Scholey & Kennedy 2004). Cognitive tasks that can be affected include reaction times (Owens & Benton 1994), memory (Sünram-Lea et al. 2002) and rapid visual information processing (Benton et al. 1994) (for a more detailed account of the effects of glucose on cognitive performance, see Chapter 8).

Based on the previous literature, one may suggest that different GI foods may have different effects on cognitive performance. Shortly following consumption of both high and low GI foods there is an immediate increase in blood glucose. Hence, under such conditions, one would predict that cognitive performance should be facilitated. However, after approximately 60–90 minutes (Vitapole 2001) one would expect that cognitive performance, following the intake of a low GI food, to be superior relative to a high GI food. This is due to the fact that the blood glucose levels, following a high GI food, have now plummeted back down to below normal levels; whereas, blood glucose levels following a low GI food are sustained and, hence, still supplying energy to the brain (Figure 9.1).

The main source of energy for the brain is glucose, and the main source of glucose is carbohydrate (Benton & Parker 1998) because it is broken down into simple sugars, such as glucose, during digestion. The interest in the effects of breakfast on cognitive function stems from the observation that the brain

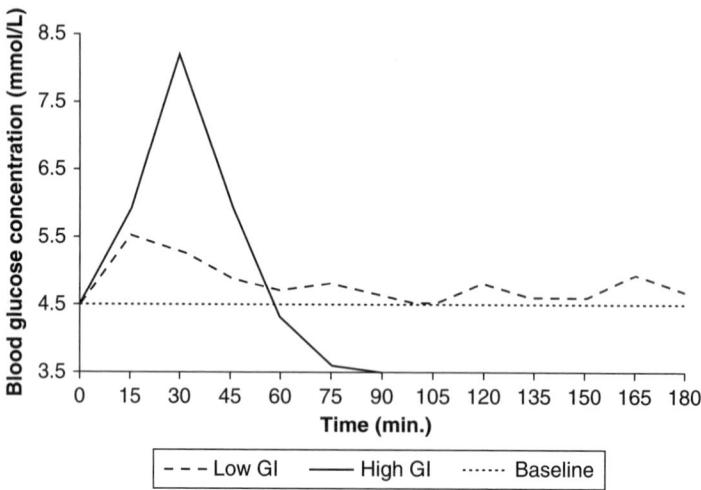

Figure 9.1   Blood glucose response after intake of high and low GI food
Source: Adapted from Jenkins et al. (2002)

is particularly starved of energy after an overnight fast and that breakfast can provide glucose to the brain, particularly since the main constituent of most breakfasts is carbohydrates (Dye et al. 2000).

The notion that food consumption can influence cognitive function is receiving increased support from research (Dye et al. 2000; Hoyland et al. 2008). The findings from such studies are of interest not only to researchers and scientists, but also to the food industry and food policy governing bodies in order to inform the general public of what might affect their own and their children's cognition. This has stimulated the growth in what could be labelled 'brain food', and importantly such findings have also encouraged changes in food policies and a subsequent improvement of food in schools.

Breakfast Consumption and Cognitive Performance in Adults

Some of the earliest studies addressing the importance of the role of breakfast consumption on cognitive performance were the Iowa Breakfast Studies (Tuttle et al. 1949, 1950, 1952, 1954) which were carried out in the USA in the late 1940s and early 1950s. From a series of studies investigating the effects of breakfast on cognitive performance in adults, children and the elderly, Tuttle and colleagues found that breakfast can enhance cognitive performance. Although their studies have been criticised for a number of reasons, such as inconsistent findings due to methodological weaknesses (e.g. small sample sizes), the results warranted further investigation and generated an increased interest in the effects of breakfast on cognitive performance.

Studies Comparing Breakfast Consumption Versus Breakfast Omission in Adults

Since Tuttle et al.'s studies in the late 1940s and early 1950s, a number of studies have investigated the effects of breakfast on cognitive function in adults. Much of this research has found improved cognitive performance following the consumption of breakfast compared with the omission of breakfast, although some results have found no effects. Benton & Sargent (1992) investigated the effects of breakfast versus no breakfast on memory (spatial memory and immediate word recall) in male and female participants aged 19–28 years. After an overnight fast, participants were tested under one of two conditions: they were either given no breakfast or a breakfast consisting of a milk-based nutritional drink ('Build Up', Nestlé) containing 327 kcal of energy, 18.5 g of protein, 37.7 g of carbohydrate and 12.2 g of fat. Tests of spatial memory and immediate word recall were given two hours after breakfast. The results showed that participants in the breakfast condition took less time to finish both the spatial memory task and the immediate recall task than participants in the no breakfast condition. However, the consumption of breakfast had no effect on the number of errors produced in either task. Benton & Sargent's (1992) results suggest that the consumption of breakfast can affect performance on memory tasks. Smith et al. (1994) carried out two experiments investigating the effects of breakfast and caffeine on cognitive performance, mood and cardiovascular functioning (only the effects of breakfast on cognitive function will be discussed here). A total of 48 male and female university students were assigned to a no-breakfast condition, a cooked breakfast condition or a cereal/toast breakfast condition. The cooked breakfast consisted of two scrambled eggs, two rashers of bacon, one slice of wholemeal toast with butter (451 kcal). The cereal/toast breakfast consisted of 25 g cornflakes, 150 ml semi-skimmed milk, two teaspoons of sugar and one slice of wholemeal toast with butter or marmalade (451 kcal). In Experiment 1, three cognitive tests (a simple reaction time task, a five-choice serial response task and a repeated digit vigilance task) measured sustained attention. Subjects were tested at baseline at approximately 8 a.m., and then they consumed breakfast and were tested again at 60 minutes and 120 minutes after baseline. Smith et al. (1994) found no significant effects of either breakfast on any of the sustained attention tasks in Experiment 1. Smith et al. (1994) argued that although the results showed no effect of breakfast on sustained attention, this did not mean that breakfast could not affect other cognitive domains, such as memory. Based on Benton & Sargent's (1992) findings that breakfast may improve memory, Smith et al. (1994) carried out Experiment 2 to examine whether breakfast might have an influence on memory rather than attention. Smith et al.

(1994) removed the cereal/toast condition in Experiment 2, leaving a comparison of the no-breakfast condition and the cooked breakfast condition (these two conditions were identical to Experiment 1). Baseline was measured at approximately 8 a.m. followed by breakfast and testing again at approximately 60 minutes and 120 minutes. The tests consisted of four memory tasks: free word recall (measure of episodic memory), delayed word recognition (measure of episodic memory), logical reasoning (part of the central executive which is a component of working memory) and semantic processing (measure of long-term memory). The results from Experiment 2 revealed a significant effect of breakfast on free word recall, with participants in the breakfast condition recalling significantly more words than participants in the no-breakfast condition one hour after breakfast consumption. At two hours post-breakfast this effect was still evident although not significant. For delayed word recognition they found a significant effect of breakfast on the number of false alarms, with fewer false alarms in the breakfast condition than the no-breakfast condition at both one and two hours post-breakfast. However, on the logical reasoning task, the results showed an opposite effect, with better performance (higher accuracy) for the no-breakfast condition compared with the breakfast condition. The results for the semantic processing tasks showed no effect of breakfast. Smith et al. (1994) concluded that breakfast had no effect on sustained attention and that the effects were limited to alterations in memory.

Research has suggested that the effects of nutritional manipulations are only observable for certain cognitive domains. In line with Smith et al.'s results, Hoyland et al. (2008) reported that effects of macronutrient manipulations are most often observed for memory. However, it is not easy to draw any definite conclusions about the preferential effects on different cognitive domains as it is difficult to compare the findings of previous research due to differences in cognitive measures and nutritional manipulations. An alternative explanation for Smith et al.'s finding that breakfast had an effect on memory and not attention is cognitive demand. It has been suggested that tasks with a higher cognitive demand (i.e. more difficult) are more sensitive to nutritional manipulations (e.g. Scholey et al. 2001). It is possible that the memory tasks employed in Smith et al.'s study were more difficult to perform than the attention tasks and, hence, breakfast only had an effect on memory. However, it is again difficult to draw any definite conclusions with regards to task demand as it is hard to compare the difficulty of tasks across domains.

Studies Comparing Different Breakfasts in Adults

Recently, there has been increased interest in the cognitive consequences of different breakfast types. Research has suggested that breakfast composition can affect cognitive performance. However, as with studies comparing breakfast versus no breakfast, research has also found no effects of breakfast composition or only effects on some outcome measures, making the evidence for the effects of breakfast equivocal. Lloyd et al. (1996) investigated the effect of breakfasts differing in fat and carbohydrate content on mood and cognitive performance in 16 male and female participants with a mean age of 26 years. Following a repeated measures design, participants received four breakfasts: a low fat/high carbohydrate breakfast, a medium fat/medium carbohydrate breakfast, a high fat/low carbohydrate breakfast and no breakfast. The series of cognitive tests consisted of the Bakan Task (version of rapid visual information processing with high working memory load), a Two-finger Tapping Task (test of motor speed), Free Word Recall Task (memory test) and a Simple Reaction Time Task (attention task). Although the authors don't offer much detail on the results of the cognitive assessments, it seems Lloyd et al. didn't find any effects of breakfast or of breakfast composition. This result supports previous research that has found no effects of breakfast on cognition. Consistent with Smith et al. (1994), Lloyd et al. found no effects of breakfast on attention. Contrary to Lloyd et al. however, Smith et al. did find an effect on memory.

GI is a concept that has received increased interest in scientific research as well as in the popular media. As mentioned earlier in this chapter, GI is a measure of the effect ingested food has on blood glucose levels: following the consumption of a high GI food, there is a sharp increase followed by a sharp decrease in blood sugar levels, whereas following the consumption of a low GI food, there is a smaller but longer-lasting increase in blood glucose (Ingwersen et al. 2007). Due to the plethora of research that has found that increased blood glucose levels can enhance cognitive performance (Scholey & Kennedy 2004), it is reasonable to suppose that the GI of food can affect cognitive performance through alterations in blood glucose. Some studies have specifically focused on comparing the GI of breakfasts. Benton et al. (2003) investigated the effect of the GI of breakfast on cognitive performance in female undergraduate students. Following an overnight fast, participants consumed either a low GI breakfast consisting of 50 g of plain biscuits or a high GI breakfast consisting of 50 g of a cereal bar. Immediate and delayed (10 minutes) memory was assessed by a word recall task at 30, 90, 150 and 210 minutes following breakfast. In the ten minutes between the immediate and delayed word recall, participants performed other tasks which were not reported in the paper. Measures of blood glucose were also taken before breakfast and at 20, 50, 80, 140, 200 and 230 minutes after breakfast. Benton et al. found that the high GI breakfast (referred to as high RAG by Benton et al. – rapidly available glucose) produced a distinct increase in blood glucose at 30 minutes after breakfast and then returned to baseline over the next 240 minutes. The low GI breakfast (referred to as high SAG – slowly available glucose), on the other hand, produced a smaller increase at 30 minutes and returned to baseline slightly faster than the high GI breakfast. On initial analysis, Benton et al. did not find any interaction between immediate and delayed word recall and combined the two into a global score for further analysis. When the authors analysed memory as a global score, they found an effect of meal composition at 150 and 210 minutes after breakfast but not at 30 and 90 minutes after breakfast. Participants recalled more words following the low GI breakfast than after the high GI breakfast at both 150 and 210 minutes.

Benton et al. concluded that their study supports the notion that a low GI breakfast can enhance memory from 150 to 210 minutes after consumption but not before. However, at 150 and 210 minutes after breakfast (when performance was better for the low GI than the high GI) the difference in blood glucose between the two breakfasts was negligible. So, although it appears that memory is affected by the GI of the breakfasts, it is difficult to say whether the underlying mechanism of these results are changes in blood glucose levels induced by the composition of the breakfasts because the blood glucose levels are more or less the same at 150 and 210 minutes after breakfast.

Interestingly, Benton et al. (2003) investigated the results further by splitting the word recall into recall of concrete words and abstract words. Their reasoning for this was that concrete words are easy to recall, whereas abstract words are more difficult to remember. Although a trend emerged, there was no effect of breakfast composition on the recall of concrete words. There was, however, a significant effect of breakfast composition on the recall of abstract words, with a better recall following the low GI breakfast compared with the high GI breakfast at 210 minutes. This result is in line with other research which argues that the effect of glucose and other macronutrient manipulations is more likely to manifest itself on tasks that are more cognitively demanding (Scholey et al. 2001). However, again, with blood glucose levels close to baseline after both breakfasts at 210 minutes, it is difficult to conclude that the results are due to blood glucose changes induced by breakfast consumption.

Further research has investigated the role of breakfast composition on performance. Nabb and Benton (2006) examined the effects of glucose tolerance and a variety of breakfasts designed to vary the delivery rate of glucose into the bloodstream. In short, they found support for the notion that a breakfast which releases glucose into the bloodstream at a slow rate (e.g. low GI) can improve memory. They also found that better glucose tolerance was associated with better memory. However, they did not find corresponding support for performance on attention tasks. On the contrary, they found better performance on reaction times, and vigilance was associated with higher levels of blood glucose.

Overall, research investigating the effects of breakfasts on cognitive performance has found some significant effects, whereas other results have not found any significant findings. Such differences in the findings could be due to methodological variations in the studies. Studies often vary in timing, both in terms of what time breakfast is consumed and what time cognitive tests are performed. Vaisman et al. (1996) carried out a study to examine the effects of timing of breakfast in children aged 11–13 years. They found that breakfast consumption enhanced cognitive performance if it was consumed 30 minutes prior to testing but not if consumed two hours before. Contrary to this result, Benton et al. (2003) found effects of breakfast at 150 and 210 minutes after breakfast and no effects at earlier test times. Another methodological criticism of studies investigating the effect of breakfast composition on cognitive performance is that breakfast manipulations are often not matched on a number of factors, such as fat or protein content, energy, quantity or taste. Furthermore, different studies tend to test different breakfasts. Such differences make it hard to determine which type of breakfast is more beneficial to cognitive performance. There are also inconsistencies in the findings of previous research in terms of which cognitive domains are affected by breakfast. As mentioned earlier in this chapter, most studies have found an effect on memory compared with other cognitive domains, such as attention (Hoyland et al. 2008). However, given that some studies have found effects on attention, it makes it difficult to draw definite conclusions as to whether breakfast has a preferential effect on specific cognitive domains.

Breakfast Consumption and Cognitive Performance in Children

Much of the research examining the relationship between breakfast consumption and cognitive performance has focused on children. The possibility that breakfast can enhance cognitive performance is particularly important in this population as it can have an impact on school performance.

Glucose utilisation changes throughout the lifespan and is 1.8 times larger in children aged 3–11 years than in adults. The brain glucose utilisation rate in children is 10.8 mg/min/100 g compared with 5.5 mg/min/100 g in adults. Similarly, the blood flow in a child's brain is 102 ml/min/100 g compared with 57 ml/min/100 g in adults (Kennedy & Sokoloff 1957). In support of these observations, Chugani (1994) found that in children aged 4–10 years, cerebral glucose utilisation was twice as high as in adults. He reported that there is a gradual decrease in this utilisation from 9–10 years until 16–18 years, when it reaches adult levels of utilisation. Furthermore, Chugani (1998) reported that these changes in glucose utilisation in children seem to coincide with the time at which various behaviours and cognitive skills emerge in children.

Due to such age-related changes in glucose utilisation (Chugani 1998), combined with developing cognitive skills, it is possible that breakfast consumption has a different effect in children than in adults. Hence, research investigating the relationship between breakfast and cognitive function in children is on the increase.

Studies Comparing Breakfast Consumption Versus Breakfast Omission in Children

A number of studies have investigated the effects of breakfast on children's cognitive performance, and the majority of these studies have found some positive effect of breakfast consumption (for reviews, see Hoyland et al. 2009;

Rampersaud, et al. 2005). Under controlled conditions in clinical research centres, Pollitt et al. (1982/83) investigated the effect of breakfast consumption versus breakfast omission in children aged 9–11 years. Measures of a series of cognitive tasks were taken at approximately 165 minutes after breakfast. Although the authors did find some non-significant results, they concluded that, overall, the results showed that breakfast omission has adverse effects on children's performance. In a similar study, Conners & Blouin (1983) compared breakfast consumption with breakfast omission in a group of adolescents. They found that performance on attention span and vigilance improved following breakfast intake. Furthermore, a recent study by Cooper et al. (2011) found positive effects of breakfast consumption compared with breakfast omission on cognitive performance (Visual search, Stroop task and Sterberg paradigm) in adolescents. In addition, Cooper et al. found that the effect of breakfast was particularly noticeable in the more difficult tasks, hence supporting the suggestion that tasks with a higher cognitive demand are more sensitive to nutritional manipulations (e.g. Scholey et al. 2001).

More recently, Wesnes et al. (2003) investigated the effects of breakfast on cognitive performance in children aged 9–16 years. In a repeated measures design, the children received Cheerios, Shreddies, a glucose drink or no breakfast on four consecutive days. The Cheerios provided 28.7 g of carbohydrate, of which 16.0 g was complex carbohydrate; the Shreddies provided 38.3 g of carbohydrate, of which 25.2 g was complex carbohydrate, and the glucose drink provided 38.3 g of carbohydrate. Cognitive performance was assessed through the Cognitive Drug Research (CDR) Computerised Assessment Battery, which consists of a series of computerised tests of attention and memory. The test battery was completed once before breakfast and again at 30, 90, 150 and 210 minutes following breakfast. Wesnes et al. found that the children's performance declined throughout the morning but that this decline was significantly reduced on measures of attention and episodic memory after the children had consumed either of the two cereals compared with the glucose drink and breakfast omission. Wesnes et al. concluded that breakfast, in the form of cereal, can have a positive effect on cognitive performance in school children in that it reduces the decline observed in cognitive performance across the morning. It is, however, noteworthy that they only found significant results in two out of five outcomes measures. Furthermore, Wesnes et al. tested children aged 9–16 years. This is an age range where the youngest children are likely to still have very high rates of cerebral glucose utilisation and cognitive functioning is still developing. However, Wesnes et al. did not include age as a factor or as a covariate in the analysis, and it is likely that the effects of breakfast could have asserted themselves differently depending on the age of the children.

Some research has found no effect of breakfast consumption on cognitive performance in children. Over two studies examining the effects of breakfast intake on school children, Dickie & Bender (1982) found no evidence that breakfast consumption was beneficial to performance over breakfast omission.

Further studies have, however, suggested that the benefits of breakfast on cognitive performance are only measurable in undernourished children (e.g. Jacoby et al. 1996; Pollitt et al. 1998; Simeon & Grantham-McGregor 1989). Chandler et al. (1995) examined breakfast versus no breakfast in 8–11-year-old undernourished and sufficiently nourished children from rural Jamaica. They found no effect of breakfast in the sufficiently nourished children. However, they did find an improvement in verbal fluency in the undernourished children after breakfast compared with no breakfast. Although, the significant effect was found only in one out of a number of cognitive measures, the results suggest that undernourished children are possibly more susceptible to the beneficial effects of breakfast than are adequately nourished children. On the contrary, Lopez et al. (1993) did not find any benefits of breakfast on cognitive performance in either adequately nourished or undernourished children in Chile.

Studies Comparing Different Breakfasts in Children

Research on the effects of different types of breakfasts on children's cognitive performance has increased over the last few years (for a review, see Hoyland et al. 2009). There has been a particular rise in the interest of breakfasts that have different effects on blood glucose levels, such as GI or GL (Benton et al. 2007; Ingwersen et al. 2007; Mahoney et al. 2005; Micha et al. 2010). As described earlier in this chapter, GI is a measure of the rise in blood glucose following the consumption of a given food. GL is a measure of the effect on blood glucose levels resulting from a given amount of that food (calculated by multiplying the GI of the food by the amount of available carbohydrate in the food and dividing the results by 100) (Benton et al. 2007). The reason for this increased interest in GI/GL, as outlined in the section on adult studies, is that a low GI or GL breakfast provides a more steady increase in blood glucose levels that lasts for longer than a high GI or GL breakfast. Hence, consumption of breakfasts with a low GI or GL should benefit cognitive performance, particularly in the late morning (Benton et al. 2007; Ingwersen et al. 2007).

Mahoney et al. (2005) investigated the effects of breakfast composition on cognitive performance in children aged 6–11 years. In a repeated measures design, the children were provided with oatmeal breakfast (38 g carbohydrate, 19 g sugar, 3 g fibre, 8 g protein, 2 g fat, 200 kcal), a ready-to-eat cereal (36 g carbohydrate, 22 g sugar, 1 g fibre, 5 g protein, 1.5 g fat, 200 kcal) or no breakfast. Their reasoning was that the oatmeal breakfast would provide more sustained levels of blood glucose, similar to a low GI breakfast. In their first experiment, the children were 9–11 years old, and in their second experiment, the children were 6–8 years old. Mahoney et al. found significant effects of breakfast compared with no breakfast in both age groups. They also found effects of breakfast composition in both age groups (better performance after low GI breakfast) although this effect was observed on more variables for the younger age group.

To investigate the effects of GI of breakfasts further, Ingwersen et al. (2007) compared two breakfast cereals with differing GI on cognitive performance in children aged 6–11 years. The children were assessed on a version of the CDR battery employed by Wesnes et al. (2003). The CDR battery consisted of a comprehensive battery of tasks in the following order: word presentation; immediate word recall; picture presentation; simple reaction time; digit vigilance; choice reaction time; spatial working memory; numeric working memory; delayed word recall; delayed word recognition and delayed picture recognition. The outcome measures from these tasks were combined into five factors which were derived by Wesnes et al. (2000). These five factors were Speed of Attention, Speed of Memory, Accuracy of Attention, Secondary Memory and Working Memory (speed of attention, accuracy of attention and secondary memory are referred to as power of attention, continuity of attention and quality of episodic secondary memory, respectively, in Wesnes et al. (2000, 2003)). In a counter-balanced, crossover design, children received the high GI cereal Coco Pops and the low GI cereal All Bran on two consecutive mornings (after an overnight fast). Performance was assessed prior to breakfast (9 a.m.), immediately after breakfast (9.40 a.m.) and at 60 (10.40 a.m.) and 120 (11.40 a.m.) minutes after breakfast. Analysis on change from baseline scores revealed a decline in the children's performance throughout the morning and that this decline was significantly reduced following the low GI breakfast on accuracy of attention and secondary memory. For accuracy of attention, there was a significant decline in performance following the high GI breakfast compared with the low GI breakfast at 120 minutes (11.40 a.m.) after breakfast consumption (Figure 9.2). On secondary memory, this benefit of the low GI cereal was evident throughout the morning, with a significantly smaller decline following the low GI breakfast immediately (9.40 a.m.) and at 120 minutes (11.40 a.m.) after breakfast (Figure 9.2).

These results support Mahoney et al.'s (2005) findings, suggesting that a low GI breakfast is beneficial to children's cognitive performance compared with a high GI breakfast. The study did not, on the other hand, find any significant differences between boys and girls, which Mahoney et al.'s study did. Wesnes et al. (2003) employed the CDR battery to a sample of children and although they compared breakfast consumption with a glucose drink and no breakfast, it is worth noting that, like Ingwersen et al. (2007), they found beneficial effects of breakfast on secondary memory and also on one of the attention factors. Furthermore, in line with previous research, Ingwersen et al. found only significant effects on two out of five measures.

Furthering the research investigating the effects of different breakfasts on cognitive performance, Benton et al. (2007) tested children's classroom behaviour, memory, reaction to frustration and ability to sustain attention following the intake of breakfasts that were designed to provide similar energy but differed in GL (high, medium and low GL). Benton et al. found that certain

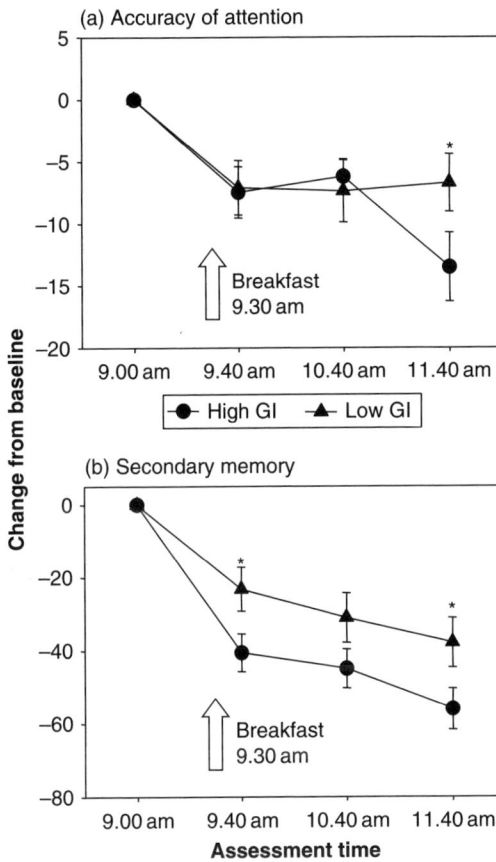

Figure 9.2 Mean (±SEM) change from baseline score on (a) accuracy of attention and (b) secondary memory

* Denotes a significant difference at $p < 0.05$.

classroom behaviours (time spent on task) were better in the first ten minutes of the observation following the low GL breakfast compared with the medium or high GL breakfasts. There were, however, no effects on other children's behaviours, such as time spent talking, fidgeting and out of their seat. Benton et al. furthermore found some mixed findings on the reaction to frustration, memory and attention. In contrast to these studies, Smith & Foster (2008) found a beneficial effect on memory following a high GI breakfast compared with a low GI breakfast in adolescents aged 14–17 years. Smith & Foster also measured blood glucose during this study and found that at the time when memory was found to be significantly better after the high GI than low GI, blood glucose was higher for the high GI breakfast. Smith et al.'s measures were taken at regular intervals following breakfast and the last blood glucose measurement was taken at 90 minutes and the last cognitive measures were taken at 100 minutes following breakfast. It is, however, possible that if blood glucose measures as well as cognitive measures had been taken for a longer period of time (e.g. at 120 and 150 minutes after breakfast), the blood glucose levels following the low GI breakfast would be higher than following the high GI breakfast as the latter would possibly drop to below baseline at a faster rate. Hence, it is possible that at these later time points memory may be better following the low GI breakfast rather than the high GI breakfast.

Breakfast Clubs

As noted earlier, the Government is committed to promoting healthier lifestyles and healthy schools with good quality school meals (Ells et al. 2008) as well as improving learning and raising standards in schools (Gathercole et al. 2004). Despite research suggesting that breakfast can be beneficial to children's cognitive performance and the common belief that breakfast is the most important meal of the day, many children skip breakfast. Defeyter et al. (2010) reported that a Taylor Nelson Sofres Family Panel found that, in the UK, 13 per cent of 0–10-year-old children skipped breakfast and among 11–16-year-old children as many as 19 per cent skipped breakfast. Furthermore, it has been reported that the trend for skipping breakfast is more common among children from a low socio-economic background and the main reason given for skipping breakfast is lack of time in the morning (Bidgood & Cameron 1992). School breakfast clubs or similar school breakfast programmes therefore provide the opportunity for children to get breakfast and hence to provide energy for the brain, which in turn might affect their cognitive performance and school achievements.

There is ample anecdotal evidence from teachers, pupils and parents that breakfast clubs have a positive impact on children, in terms of behaviour and school performance, as well as school attendance (Defeyter et al. 2010). As with the research on the effects of breakfast consumption on cognitive performance, the results from research on the effects of breakfast club attendance are mixed. Murphy et al. (1998) examined the relationship between school breakfast club attendance, behaviour and academic performance. Murphy et al. introduced a free breakfast programme in schools that already had school breakfast available but not for free. With regards to academic performance, they found that prior to the introduction of the free programme, children who had school breakfast sometimes or often had significantly higher maths scores than children who rarely or never had school breakfast. Four months after the introduction of the free breakfast programme, the number of children eating a school breakfast had doubled. At this point, Murphy et al. found that the children who had increased their school breakfast intake by participating in the free school breakfast programme had a significant increase in their maths scores. This suggests that attending a breakfast programme (or club) can affect children's cognitive function and hence academic performance. Such findings are further supported by research by Edwards & Evers (2001), who found that teachers report lower hyperactivity and increased attention in children who attend breakfast clubs.

On the other hand, there is research suggesting that there is no evidence that breakfast club attendance has a positive effect on cognitive functioning in children (e.g. Shemilt et al. 2004). Murphy et al. (2010) examined the effect of a free breakfast programme in Wales, UK. In relation to cognitive function, Murphy et al. measured the effects of breakfast club participation on episodic memory. Although Murphy et al. did find some positive effects, such as an improvement in nutritional intake, they did not find any effects on episodic memory. These results suggest that there are no cognitive benefits of attending a school breakfast club. It is, however, worth noting that the only cognitive assessment employed in this study was episodic memory and it could be that breakfast club attendance has an effect on other cognitive domains.

Conclusions and Directions for Future Research

This chapter has examined whether cognitive performance can be enhanced through breakfast consumption, whether different types of breakfast affect performance differently and whether attending a breakfast club can affect performance. The findings from previous research have been mixed, although there appears to be a consensus that breakfast consumption, breakfast clubs and breakfasts with a low GI/GL can have beneficial effects on performance. A number of methodological issues have, however, been identified in previous research. One of the major criticisms, particularly for studies looking at different breakfast compositions, is that breakfast manipulations are often not matched on aspects such as macronutrient content, energy or volume (Hoyland et al. 2009). In Smith et al.'s (1994) study, for example, the two breakfast conditions were matched on calorie content. However, the authors didn't report on any other component of the breakfasts and it is more than likely that there were differences in macronutrients such as fat and carbohydrate, which have been shown to have an effect on cognitive performance. In the study by Ingwersen et al. (2007) which investigated breakfasts with differing GI, the two breakfasts also had other compositional differences, such as protein, which could have an effect on cognitive performance (Dye et al. 2000). Because breakfast manipulations are not matched, it makes it difficult to draw definite conclusions as to whether the results are due to the experimental manipulation.

Another methodological issue with the previous research is that cognitive testing mainly focuses on attention and memory and overlooks other areas such as problem-solving abilities. Furthermore, the attention and memory tasks that have been investigated are so diverse that it makes comparisons between studies difficult. Other issues include study design (independent versus repeated measures), habitual breakfast habits, second meal effects and timing (Hoyland et al. 2009).

The findings from previous literature warrant further investigation into the effects of breakfast on cognitive performance in both adults and children. In light of the methodological issues identified, future research should attempt to compare carefully matched breakfasts in studies with controlled crossover designs. With respect to timing, further research is needed to investigate immediate effects compared with effects later in the morning (e.g. 30 minutes compared with 210 minutes after breakfast). As the number of breakfast clubs in the UK is growing (Defeyter et al. 2010) it is also essential that further research investigates the potential benefits of breakfast club attendance. Research evidencing the positive effects of breakfast clubs will help to secure funding for such clubs from sponsors and charities to keep the breakfast clubs running. Keeping breakfast clubs running is important not only because they might have an impact on children's school performance but also because they provide a safe place for children before school starts, they provide children with good nutrition and also provide children with a social, supportive environment where they can take part in a variety of activities (Defeyter et al. 2010).

Although there have been some mixed findings, it appears that breakfast consumption compared with breakfast omission has an effect on cognitive performance, particularly in undernourished children. It also appears that breakfast composition can affect performance and that, in particular, a breakfast consisting of slow-releasing glucose or having a low GI is beneficial compared with breakfasts consisting of rapidly available glucose or high GI. Furthermore, research has suggested that there are potential cognitive benefits for children who attend a breakfast club.

References

Benton, D., & Jarvis, M. (2007). The role of breakfast and a mid-morning snack on the ability of children to concentrate at school. *Physiology and Behavior 90*, 382–385.

Benton, D., Maconie, A., & Williams, C. (2007). The influence of the glycaemic load of breakfast on the behavior of children in school. *Physiology and Bheavior, 92I*, 717–724.

Benton, D. Owens, D. S., & Parker, P. Y. (1994). Blood glucose influences memory and attention in young adults. *Neuropsychologia, 32*, 95–607.

Benton, D., & Parker, P. Y. (1998). Breakfast, blood glucose, and cognition. *American Journal of Clinical Nutrition, 67*(Suppl.), 772S–778S.

Benton, D., Ruffin, M.-P., Lassel, T., Nabb, S., Messaoudi, M., Vinoy, S., Desor, D., & Lang, V. (2003). The delivery rate of dietary carbohydrates affects cognitive performance in both rats and humans. *Psychopharmacology, 166*, 86–90.

Benton, D., & Sargent, J. (1992). Breakfast, blood glucose and memory. *Biological Psychology, 33*, 207–210.

Benton, D., & Stevens, M. K. (2008). The influence of a glucose containing drink on the behaviour of children in school. *Biological Psychology, 78*, 242–245.

Bidgood, B. A., & Cameron, G. (1992). Meal/snack missing and dietary adequacy of primary school children. *Journal of the Canadian Dietetic Association, 53*(2), 164–8.

Blom-Hoffman, J., Kelleher, C., Power, T. J., & Leff, S. S. (2004). Promoting healthy food consumption among young children: Evaluation of a multi-component nutrition education program. *Journal of School Psychology, 42*, 45–60.

Busch, C. R., Taylor, H. A., Kanarek, R. B., & Holcomb, P. J. (2002). The effects of a confectionery snack on attention in young boys. *Physiology and Behavior, 77*, 333–340.

Carlson, N. R. (1999). *Foundations of physiological psychology*, 4th ed. Amherst, MA: Allyn and Bacon.

Chandler, A. K., Walker, S. P., Connolly, K. & Grantham-McGregor, S. M. (1995). School breakfast improves verbal fluency in undernourished Jamaican children. *Journal of Nutrition, 125*, 894–900.

Conners, C. K., & Blouin, A. G. (1983). Nutritional effects of behaviour of children. *Journal of Psychiatric Research, 17*, 193–201.

Cooper, S. B., Bandelow, S., & Nevill, M. E. (2011). Breakfast consumption and cognitive function in adolescent schoolchildren. *Physiology and Behavior, 103*, 431–439.

Chugani, H. T. (1994). Development of regional brain glucose metabolism in relation to behavior and plasticity. In: Dawson, G. and Fischer, K. W. (ed.) *Human Behavior and the Developing Brain*. New York: Guildford, pp. 153–75.

Chugani, H. T. (1998). A critical period of brain development: Studies of cerebral glucose utilization with PET. *Prevevntive Medicine, 27*, 184–188.

de la Hunty, A. & Ashwell, M. (2007). Are people who regularly eat breakfast cereals slimmer than those who don't? A systematic review of the evidence. *Nutrition Bulletin, 32*, 118–128.

Defeyter, M. A, Graham, P. L, Walton, J. & Apicella, T. (2010). Breakfast clubs: Availability for British schoolchildren and the nutritional, social and academic benefits. *Nutrition Bulletin, 35*, 245–253.

Dickie, N. H., & Bender, A. E. (1982). Breakfast and performance in schoolchildren. *British Journal of Nutrition, 48*, 483.

Dye, L., & Blundell, J. E. (2002). Functional foods: Psychological and behavioural functions. *British Journal of Nutrition, 88*(2), S187–S211.

Dye, L., Lluch, A., & Blundell, J. E. (2000). Macronutrients and mental performance. *Nutrition, 16*, 1021–1034.

Edwards, H. G., & Evers, S. (2001). Benefits and barriers associated with participation in food programs in three, low-income Ontario communities. *Canadian Journal of Dietetic Practice & Research, 62*, 76–81.

Ells, L. J., Hillier, F. C., Shucksmith, J., Crawley, H., Harbige, L., Shield, J., Wiggins, A., & Summerbell, C. D. (2008). A systematic review of the effect of dietary exposure that could be achieved through normal dietary intake on learning and performance of school aged children of relevance to UK schools. *British Journal of Nutrition, 100*, 927–936.

Fischer, K., Colombani, P. C., Langhans, W. & Wenk, C. (2001). Cognitive performance and its relationship with postprandial metabolic changes after ingestion of different macronutrients in the morning. *British Journal Nutrition, 85*, 393–405.

Fischer, K., Colombani, P. C., Langhans, W. & Wenk, C. (2002). Carbohydrate to protein ratio in food and cognitive performance in the morning. *Physiology and Behaviour, 75,* 411–423.

Foster, J. K., Smith, M. A., Woodman, M., Zombor, R., & Ashton, J. (2007). Impact of a wholegrain breakfast cereal meal on blood glucose level, mood and affect. *Food Australia, 59,* 593–596.

Gathercole, S. E., Pickering, S. J., Knight, C., & Stegmann, Z. (2004). Working memory skills and educational attainment: Evidence from national curriculum assessments at 7 and 14 years of age. *Applied Cognitive Psychology, 18,* 1–16.

Haskell, C. F., Scholey, AB., Jackson, P. A., Elliott, J. M., Defeyter, M. A., Greer, J., Robertson, B. C., Buchanan, T., Tiplady, B., & Kennedy, D. O. (2008). Cognitive and mood effects in healthy children during 12 weeks supplementation with multivitamin/minerals. *British Journal of Nutrition, 100,* 1086–1096.

Hoyland, A., Dye, L., & Lawton, C. L. (2009). A systematic review of the effect of breakfast on the cognitive performance of children and adolescents. *Nutrition Research Reviews, 22,* 220–243.

Hoyland, A., Lawton, C. L., & Dye, L. (2008). Acute effects of macronutrient manipulations on cognitive test performance in healthy young adults: A systematic research review. *Neuroscience and Biobehavioral Reviews, 32,* 72–85.

Ingwersen, J., Defeyter, M. A., Kennedy, D. O., Wesnes, K. A., & Scholey, A. B. (2007). A low glycaemic index breakfast cereal preferentially prevents children's cognitive performance from declining throughout the morning. *Appetite, 49,* 240–244.

Jacoby, E., Cueto, S. & Pollitt, E. (1996). Benefits of a school breakfast program among Andean children in Huaraz, Peru. *Food Nutrition Bulletin, 17,* 54–64.

Jenkins, D. J. A., Kendall, C. W. C., Augustin, L. S. A., Franceschi, S., Hamidi, M., Marchie, A., Jenkins, A. L., & Axelsen, M. (2002). Glycemic index: Overview of implications in health and disease. *American Journal of Clinical Nutrition, 76*(Suppl.), 266S–273S.

Kalat, J. (2001). *Biological psychology* (7th ed.). Belmont, CA: Wadsworth.

Kennedy, D. O., Jackson, P. A., Elliott, J. M., Scholey, A. B., Robertson, B. C., Greer, J., Tiplady, B., Buchanan, T., & Haskell, CF. (2009). Cognitive and mood effects of 8 weeks' supplementation with 400 mg or 1000 mg of the omega-3 essential fatty acid docosahexaenoic acid (DHA) in healthy children aged 10–12 years. *Nutritional Neuroscience, 12,* 48–56.

Kennedy, D., & Sokoloff, L. (1957). An adaption of the nitrous oxide method to the study of the cerebral circulation in children; normal values for cerebral blood flow and cerebral metabolic rate in childhood. *Journal of Clinical Investigation, 36,* 1130–1137.

Lloyd, H. M., Rogers, P. J., & Hedderley, D. I. (1996). Acute effects on mood and cognitive performance of breakfasts differing in fat and carbohydrate content. *Appetite, 27,* 151–164.

Lopez, I., de Andraca, I., Perales, C. G., Heresi, E., Castillo, M., & Colombo, M. (1993). Breakfast omission and cognitive performance of normal, wasted and stunted schoolchildren. *European Journal of Clinical Nutrition, 47*(8), 533–542.

Mahoney, C. R., Taylor, H. A., & Kanarek, R. B. (2007). Effect of an afternoon confectionery snack on cognitive processes critical to learning. *Physiology and Behavior, 90,* 344–352.

Mahoney, C. R., Taylor, H. A. Kanarek, R. B., & Samuel, P. (2005). Effect of breakfast composition on cognitive processes in elementary school children. *Physiology and Behaviour, 85,* 635–645.

Micha, R., Rogers, P. J., & Nelson, M. (2010). The glycaemic potency of breakfast and cognitive function in school children. *European Journal of Clinical Nutrition, 64,* 948–957.

Murphy, J. M., Pagano, M. E., Nachmani, J., Sperling, P., Kane, S., & Kleinman, R. E. (1998). The relationship of school breakfast to psychosocial and academic functioning: Cross-sectional and longitudinal observations in an inner-city school sample. *Archives of Paediatrics & Adolescent Medicine, 152*(9), 899–907.

Murphy, S., Moore, G. F., Tapper, K., Lynch, R., Clark, R., Rasianen, L., Desousa, C. & Moore, L. (2010). Free healthy breakfasts in primary schools: A cluster randomised controlled trial of a policy intervention in Wales, UK. *Public Health Nutrition, 14*(2), 219–226.

Muthayya, S., Thomas, T., Srinivasan, K., Rao, K., Kurpad, A. V., van Klinken, J-W., Owen, G., & de Bruin, E. A. (2007). Consumption of a mid-morning snack improves memory but not attention in school children. *Physiology and Behavior, 90,* 142–150.

Nabb, S., & Benton, D. (2006). The influence on cognition of the interaction between the macro-nutrient content of breakfast and glucose tolerance. *Physiology and Behavior, 87,* 16–23.

Owens, D. S., & Benton, D. (1994). The impact of raising blood glucose on reaction time. *Neuropsychobiology, 30,* 106–113.

Pollitt, E., Cueto, S., & Jacoby, E. R. (1998). Fasting and cognition in well- and undernourished schoolchildren: A review of three experimental studies. *American Journal of Clinical Nutrition, 67*(Suppl.), 779S–784S.

Pollitt, E., Lewis, N. L., Garza, C., & Shulman, R. J. (1982/83). Fasting and cognitive function. *Journal of Psychiatric Research, 17*(2), 169–174.

Rampersaud, G. C., Pereira, M. A., Girard, B. L., Adams, J., & Metzl, J. D. (2005). Breakfast habits, nutritional status, body weight, and academic performance in children and adolescents. *Journal of the American Dietetic Association, 105,* 743–760.

Raven, P. H., & Johnson, G. B. (1992). *Biology* (3rd ed.). St. Louis, MO: Mosby Year Book.

Riby, L. M. (2004). The impact of age and task domain on cognitive performance: A meta-analytic review of the glucose facilitation effect. *Brain Impairment, 5*(2), 145–165.

Scholey, A. B., Harper, s., & Kennedy, D. O. (2001). Cognitive demand and blood glucose. *Physiology and Behavior, 73,* 585–592.

Scholey, A. B., & Kennedy, D. O. (2004). Cognitive and physiological effects of an "energy drink": An evaluation of the whole drink and of glucose, caffeine and herbal flavouring fractions. *Psychopharmacology, 176,* 320–330.

Sheard, N. F., Clark, N. G., Brand-Miller, J. C., Franz, M. J., Pi-Sunyer, F. X., Mayer-Davis, E., Kulkarni, K., & Geil, P. (2004). Dietary carbohydrate (amount and type) in the prevention and management of diabetes. *Diabetes Care, 27*(9), 2266–2271.

Shemilt, I., Harvey, I., Shepstone, L., Swift, L., Reading, R., Mugford, M., et al. (2004). A national evaluation of school breakfast clubs: Evidence from a cluster randomized controlled trial and an observational analysis. *Child: Care, Health and Development, 30,* 413–427.

Simeon, D. T., & Grantham-McGregor, S. (1989). Effects of missing breakfast on the cognitive functions of school children of differing nutritional status. *American Journal of Clinical Nutrition, April, 49*(4), 646–653.

Smith, M. A., & Foster, J. K. (2008). The impact of a high versus a low glycaemic index breakfast cereal meal on verbal episodic memory in healthy adolescents. *Nutritional Neuroscience, 11*(5), 219.

Smith, A., Kendrick, A., & Maben, A. (1994). Effects of breakfast and caffeine on cognitive performance, mood and cardiovascular functioning. *Appetite, 22*, 39–55.

Sünram-Lea, S. I., Foster, J. K., Durlach,.P., & Perez, C. (2002). The effect of retrograde and anterograde glucose administration on memory performance in healthy young adults. *Behavioural Brain Research, 134*, 505–516.

Tuttle, W. W., Wilson, M., & Daum, K. (1949). Effect of altered breakfast habits on physiologic response. *Journal of Applied Physiology, 1*, 545.

Tuttle, W. W., Daum, K., Myers, L., & Martin, C. (1950). Effect of omitting breakfast on the physiologic response of men. *Journal of the American Dietetic Association, 26*, 332–335.

Tuttle, W. W., Daum, K., Imig, C. J. Randall, B., & Scumacher, M. T. (1952). Effect of omitting breakfast on the physiologic response of the aged. *Journal of the American Dietetic Association, 28*, 117.

Tuttle, W. W., Daum, K., Larsen, R., Salzano, J., & Roloff, L. (1954). Effect on schoolboys of omitting breakfast. Physiological responses, attitudes and scholastic attainments. *Journal of American Dietetic Association, 30*, 674–677.

Vaisman, N., Voet, H., Akivis, A., & Vakil, E. (1996). Effect of breakfast timing on the cognitive functions of elementary school students. *Archives of Pediatrics & Adolescent Medicine, October, 150*(10), 1089–1092.

van den Briel, T., West, C. E., Bleichrodt, N., van de Vijver, F. J. R., Ategbo, E. A., & Hautvast, J. G. A. J. (2000). Improved iodine status is associated with improved mental performance of schoolchildren in Benin. *American Journal of Clinical Nutrition, 72*, 1179–85.

Vitapole, D. (2001). *Glycaemic Index & Health: The Quality of the Evidence.* London: Libbey.

Wesnes, K. A., Ward, T., McGinty, A. & Petrini, O. (2000). The memory enhancing effects of a Ginkgo biloba/Panax ginseng combination in healthy middle aged volunteers. *Psychopharmacology, 152*, 353–361.

Wesnes, K. A., Pincock, C., Richardson, D., Helm, G., & Hails, S. (2003). Breakfast reduces declines in attention and memory over the morning in schoolchildren. *Appetite, 41*, 329–331.

Widenhorn-Müller, K., Hille, K., Klenk, J., & Weiland, U. (2008). Influence of having breakfast on cognitive performance and mood in 13- to 20-year-old high school students: Results of a crossover trial. *Pediatrics, 122*(2), 279–284.

CHAPTER 10

Reduced-Calorie Diets and Mental Performance in Adults

Kristen E. D'Anci

Chapter Overview

- Although changes in mood and cognition have been reported with weight-loss diets, the underlying mechanism responsible for these changes remains to be elucidated.
- It is proposed that reduced-calorie diets negatively affect mood and mental performance via intrusions of food-related thoughts and via unpleasant physical sensations of hunger.
- Recent work has examined the role of macronutrient composition of weight-loss diets.
- Diets that are high in any one macronutrient have variable effects on performance; meals high in carbohydrates increase subjective ratings of sleepiness, meals high in protein reduce fatigue, and meals high in fat can improve accuracy.
- This chapter considers current research and future directions in this area, with particular emphasis on energy intake and the most appropriate balanced diet for maintaining mental performance and mood.

Introduction

Weight loss continues to be a popular yet elusive goal for many individuals. People choose to follow weight-loss diets for a variety of reasons, including improved health, improved self-esteem and to improve appearance. The promotion of weight loss is a multi-billion dollar industry in the USA alone, and many reducing diets support the notion that eliminating certain foods, such as sugar, fat or refined carbohydrate, or that increasing foods, such as protein, are the 'magic bullet' for health and slimness. However, the consequences of following weight-loss diets can include both transient and long-term changes in mood and cognitive performance. It is likely that the process of following a weight-loss regime affects both psychological processes, such as attention, and physiological function, such as glucose regulation, thus negatively affecting mood and mental performance.

The study of weight-reduction diets and brain function is not as simple as looking at energy restriction or macronutrient composition. The short-term, or local, effects of recently ingested food can influence mood and thinking. In the very short term, in the order of minutes, intake of a caloric food supplies energy to the brain, principally in the form of blood glucose. The direct supply of energy to the brain is crucial for brain function; hypoglycemia produces such immediate effects as anxiety, dizziness and confusion. Blood flow to the brain brings not only glucose, but also oxygen, amino acids and nutrients that function directly or indirectly at neurons. Over time, nutrients alter neural cell membranes, neurotransmission and oxidative reactions, and provide the building blocks of neural tissue and its repair. In the long term, diets deficient in one or more nutrients may lead to mood disturbances, including depression and anxiety. Furthermore, when looking at the long-term effects of weight-loss diets, issues such as dietary preferences, palatability, social dynamics and personal perceptions of self-image and self-efficacy all contribute to an individual's cognitions and behaviours. For example, when comparing highly restrictive or demanding diets, such as those which require close attention to nutrient proportions or the elimination of foods, with more balanced diets, the restrictive diets can lead to social isolation and, potentially, to negative mood. Alternatively, inability to adhere to a weight-loss regime can lead to feelings of inadequacy or poor self-esteem, which can also impact on cognitive performance. Further complicating the interpretation of studies examining diet and cognitive performance is the vast repertoire of cognitive tasks and measurements employed in different experiments. When looking at a number of different experiments, it is clear that while similar cognitive domains may be studied, such as attention, spatial memory, verbal memory, psychomotor function and executive function, the types of tests used can vary from study to study. Finally, in evaluating the evidence linking weight-loss diets with cognitive performance, it is important to consider that many of the studies examining *purely cognitive processes* compare dieters versus non-dieters. However, the majority of studies examining the effects of a *specific* diet on cognitive performance use *only* individuals engaged in dieting behaviour.

Cognitive Processes

From a psychological perspective, it is thought that reduced-calorie diets negatively affect mood and mental performance via intrusions of food-related thoughts and via unpleasant physical sensations of hunger (Green et al. 1994, 2003; Jones & Rogers 2003; Kemps et al. 2005; Vreugdenburg et al. 2003). These models are primarily based on Baddeley's model of working memory (Baddeley & Hitch 1974). In this model, a theoretical central executive controls and integrates information handled by two parallel systems: the visuospatial sketchpad, which processes incoming visual and spatial information, and the phonological loop, which processes auditory, verbal and written information. Within this model, cognitive processing becomes less efficient when two tasks are processed within the same domain. With respect to dieting behaviour, internal articulation, or 'self-talk', related to food would tax the phonological loop. Additionally, inhibition of food-related ideation would potentially tax the central executive and would potentially lead to poorer overall cognitive performance.

It is hypothesised that the demands of dieting and the preoccupation with distracting food-related thoughts or physical sensations of hunger increase cognitive load, resulting in impaired processing of other, competing information. In support of this hypothesis, women who reported being on a weight-loss diet showed impairments in tasks assessing working memory, such as the Tower of London task, mental arithmetic and letter string recall, relative to women not on a diet. These decrements in performance may not be related to physiological changes from dieting, and impairments may be seen even in the absence of significant weight loss (Green et al.1994, 2003; Jones & Rogers 2003; Kemps & Tiggemann,2005; Vreugdenburg et al. 2003). The negative effects of dieting behaviour appear to be limited to processing in the phonological loop and at the level of the central executive, but visuospatial processing is relatively spared. For example, women dieters performed as well as non-dieters on a mental rotation task, but dieters' performance was impaired on the Tower of London task (Green et al. 2003). As described above, these observations are in line with the proposal that mental processing related to 'dietary restraint', such as diet-related internal articulation, and attention to and suppression of food-related thoughts specifically tax the central executive and the phonological loop slave system. In further support of this proposal, some research shows that experimental caloric restriction does not impact on cognitive performance to the same degree as 'naturalistic dieting' (Kemps et al. 2005; Lieberman et al. 2008). In a double-blind, crossover trial, when participants were unaware of how many calories they received over a two-day period, cognitive performance, mood and fatigue were not affected even after consuming a minimal amount of calories relative to when they consumed a normal amount of calories (Lieberman et al. 2008). These findings suggest that hunger alone is not the causative factor in disrupting cognitive function in individuals undergoing energy restriction. Additionally, although some of the symptoms of food deprivation are typically reversed by ingestion of food, consumption of 'diet-threatening' foods, such as chocolate bars or doughnuts, can negatively impact on verbal memory, body image and mood in dieters and restrained eaters (Hayes et al. 2011; Jones & Rogers 2003). More specifically, within 30 minutes of consuming a doughnut, a food perceived to be unhealthy, women had more negative scores for state body image relative to women who consumed a banana, or who consumed no food. Moreover, women who consumed a banana showed reductions in tension within 30 minutes of consuming the food, but women who consumed a doughnut did not show reductions in tension. These changes in state body image and mood scores were more pronounced in women scoring high for dietary restraint (Hayes et al. 2011). These findings indicate that mental processing of dietary restraint and self-image is ongoing, and imply that consumption of foods perceived to be unhealthy has immediate effects on mood and self-image. The impact of food restriction on food-related cravings and behaviours appears to be driven, to some degree, by the type of food restricted. More specifically, individuals who restrict proteins tend to report cravings for high-protein foods, and individuals who restrict carbohydrates tend to crave high-carbohydrate foods (Coelho et al. 2006).

The above evidence in dieters versus non-dieters outlines a clear role for additional processing loads to alter cognitive performance in individuals actively engaged in dieting. However, one question related to these observations is: will different types of weight-loss diets differentially alter food-related distracting thoughts? When comparing participants following a low carbohydrate diet and participants following a balanced 'exchange'-type diet, there were no differences between food and non-food items in either a food-related Stroop task assessing selective attention or on a food paired-associates task, which assessed cued memory and retrieval, suggesting that food preoccupations were not significantly altered by diet type (D'Anci, et al. 2009). These observations are limited with respect to previous studies (Green et al. 2003; Kemps et al. 2005), however, as there was no non-dieting condition in this study to serve as a comparison.

Glucose, Carbohydrates and Brain Function

Glucose is the brain's primary fuel, but it is not stored in the brain (Sieber & Traystman 1992; Wenk 1989). All digestible carbohydrates are ultimately broken down into simple sugars, primarily glucose. After absorption from the gastrointestinal tract, glucose is carried in the bloodstream to the liver, brain and other tissues where it is further utilised. Although the brain has a high demand for glucose, it lacks enzymes that are present in the liver for converting amino acids and fats into glucose. As such, the brain is dependent on circulating blood glucose for fuel and experiences consequences related to fluctuating blood glucose levels (Cox et al. 2002). To meet the glucose requirements of the brain, and based on the average minimum amount utilised by the brain, the US Recommended Daily Allowance (RDA) for carbohydrate is 130 g/day (Food and Nutrition Board 2002). Glucose is stored in limited quantities as glycogen in muscle and liver, and glycogen is converted back into glucose and released into the bloodstream as needed. However, as glycogen stores can be depleted within 1–2 days, a continual dietary source of carbohydrates is required to replenish these stores. If carbohydrates are not consumed, and glycogen stores are depleted, body fat will be metabolised into ketone bodies, which can then be used by the brain as fuel. Research in young animals shows that a ketogenic diet reduces spontaneous seizure activity following administration of lithium/pilocarpine, but the diet also results in reduced brain growth and impairment in visuospatial tasks (Zhao et al. 2004). Clinically, ketogenic diets are used to manage epilepsy and seizures, and the negative consequences of a high ketogenic diet are in many cases preferable to high seizure activity (Cantello et al. 2007; Freitas et al. 2007; Hartman et al. 2007). In older individuals at risk of developing cognitive decline, ketogenic diets are of benefit in mental performance, perhaps, in part, by correcting hyperinsulinemia (Krikorian et al. 2010).

It is generally well accepted that alterations in glucose intake and blood glucose levels translate into altered mood and mental performance (see Chapter 8 in this volume for a comprehensive account of glucose and cognitive performance). Hunger leads to cognitive decrements not only because of short-term changes in nutrient status, such as hypoglycaemia, but also because attention shifts from tasks at hand to the physical signs of hunger and the drive to eat. For most individuals, skipping meals increases hunger, irritability and distractibility, and decreases alertness and motivation (Neely et al. 2004). In general, eating reduces hunger, and attention then shifts back to the tasks at hand.

Of particular relevance to energy-restricted diets, an acute reduction in blood glucose levels, as might result from food restriction, can lead to feelings of anxiety, dizziness and confusion, and decrements in cognitive performance (D'Anci & Kanarek 2006; Hoyland et al. 2008). These negative effects of low blood glucose are typically reversed soon after eating. While reductions in the availability of glucose lead to decrements in behaviour, acute intakes of sugars and foods containing significant amounts of carbohydrates increase blood glucose levels and the availability of glucose to the brain, which are in turn associated with improvements in mood and cognitive behaviour. In particular, intake of high carbohydrate foods enhances subjective reports of energy and facilitates performance in tests requiring sustained attention (Benton & Nabb 2003; Lieberman et al. 2002; Mahoney et al. 2007). Different macronutrient composition of the diet produces differing effects on alertness and mental function (Wells & Read 1996). For example, in the short-term, eating meals high in carbohydrates and relatively low in protein can result in sleepiness or fatigue (Fischer et al. 2002; Nabb & Benton 2006), whereas the addition of some protein improves alertness and cognitive performance (Fischer et al. 2002). These considerations are important when reducing calories and deciding on meal-to-meal food choices for dietetic meals.

The degree to which and the rate at which consumption of food causes blood glucose to rise and remain elevated has become an important topic in research on the behavioural effects of the diet. Generally speaking, foods with a lower glycaemic index (GI) or with greater proportions of complex carbohydrates or protein may be more beneficial in the short term in improving cognitive performance and alertness than foods with a greater proportion of simple sugars (Benton & Nabb 2003; Fischer et al. 2002; Mahoney et al.

2005). In the long-term, dieters who follow a low GI diet report less hunger or less depression than those following a higher GI diet (Bellisle et al. 2007; Cheatham et al. 2009). However, over the long term, low glycaemic diets do not appear to be more beneficial for cognitive performance than high glycaemic diets (Cheatham et al. 2009), although reductions in hunger may confer some benefit in the ability to follow a diet (Bellisle et al. 2007).

Low Carbohydrate Diets and Brain Function

Low carbohydrate diets have surged in popularity over the past decade or so. Typically, these diets promote high intake of protein while severely limiting carbohydrate to less than 60 g per day (Last & Wilson 2006). Although carbohydrates are later reintroduced, most low carbohydrate diets require some ongoing degree of carbohydrate restriction. Proponents of low carbohydrate diets claim increased fat metabolism and superior weight loss with minimal hunger or cravings relative to low fat, high carbohydrate diets. Indeed, intake of protein is rated as more satiating than fat or carbohydrate in the short term (Bertenshaw et al. 2008; Poppitt et al. 1998). In the long term, individuals following a low carbohydrate diet report less hunger than individuals following a low fat diet (McClernon et al. 2007), but do not report significantly different levels of hunger than individuals following a balanced, reduced-energy diet (D'Anci et al. 2009). In contrast to popular conceptions about low carbohydrate diets, such diets have not been shown to be superior to other energy-restricted diets in terms of weight loss (Last & Wilson 2006), and the preponderance of individuals who maintain long-term weight loss do not follow a low carbohydrate diet (Wing & Phelan 2005).

When restricting carbohydrate, individuals typically report the majority of negative symptoms, such as fatigue, irritability and cravings, within the first 48–72 hours of initiating the diet. This timeline corresponds with the depletion of the body's glycogen stores. During this time, blood glucose decreases and fat is metabolised into ketones, which can be used for fuel in the absence of glucose. In one study, short-term memory, processing speed and spatial memory were impaired within one week of beginning a low carbohydrate weight-loss diet relative to a more balanced reducing diet (D'Anci et al. 2009). Memory decrements were reversed when the low carbohydrate dieters reintroduced carbohydrate into the diet (D'Anci et al. 2009). Specifically, individuals following a low carbohydrate diet remembered significantly fewer items on a fictitious map during carbohydrate restriction relative to participants following a balanced diet (Figure 10.1). Memory was impaired both for immediate recall and for long-term recall. When completing the memory task after reintroduction of carbohydrates, performance for the two groups was similar for short-term recall. However, participants in the low carbohydrate condition performed worse when tested for memory of a map learned during the carbohydrate restriction phase. This observation is not surprising, if something is

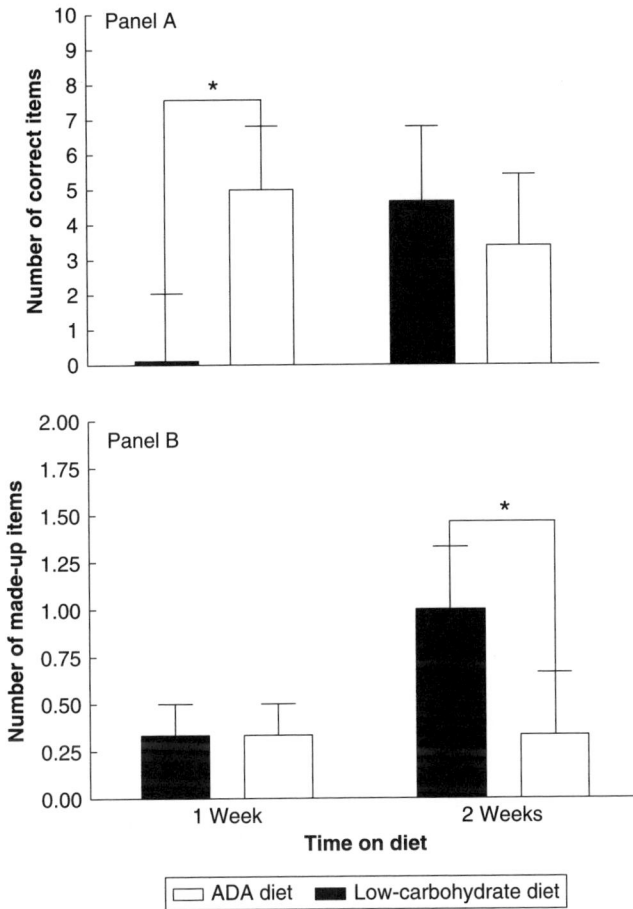

Figure 10.1   Impact of carbohydrate restriction on short- and long-term spatial memory

Participants were asked to study a map with fictitious continents and learn the names of items located on the maps. Short-term recall was assessed immediately after presentation of the map, and long-term recall was assessed one week later. Panel A: Correct placement of items in the short-term map recall task (Mean + SEM). Participants following a low-carbohydrate condition placed significantly fewer items in the correct location on a fictitious map within one week of beginning a diet relative to participants following an American Dietetic Association diet pattern (ADA). Panel B: Number of made-up locations in the long-term map recall task (Mean + SEM). When asked to complete blank maps from the previous week, low-carbohydrate participants included more made-up items than ADA participants.
N.B. Data are normalized to baseline performance. Adapted from D'Anci et al., 2009.

poorly learned, subsequent recall will be affected. In other work, after consuming a very low energy ketogenic diet, participants had significantly impaired executive function relative to those participants following a very low energy non-ketogenic diet, and the most severe decrements were seen within the first week of following the diet (Wing et al. 1995). In other work, after two weeks of following low carbohydrate diets, individuals reported greater fatigue and increased negative mood relative to those following a non-ketogenic diet (White et al. 2007). In a large-scale, long-term trial, improvements in mood were observed in participants following a low carbohydrate diet and in those following a low fat diet for eight weeks. After one year of following the diets, working memory improved relative to baseline in both dietary conditions, but mood continued to be improved only in the low fat condition (Brinkworth et al. 2009). Compared with low carbohydrate diets, long-term adherence to the Dietary Approaches to Stop Hypertenstion (DASH) diet, in combination with weight loss and aerobic exercise, is associated with improvements in learning, memory and psychomotor speed relative to a control diet (Smith et al. 2010).

Dietary Fat and Brain Function

While vilified by some weight-loss programmes and lauded by others, intake of at least some dietary fat, or lipids, is crucial for normal brain function. Approximately two-thirds of the dry weight of the brain is comprised of lipids. Fatty acids are important in neuronal membrane structure, neurotransmission and ion channel function (Worobey 2006). High fat diets, such as ketogenic diets used in the treatment of epilepsy, are clinically useful, but are not commonly seen in weight reduction. This may relate, in part, to the relative difficulty in adhering to a high fat but reduced energy diet. Low carbohydrate diets, as described above, typically incorporate high levels of protein as well, which lends variety to the diet. While low carbohydrate diets can also be high in fat, the focus is more directly on limiting carbohydrates and increasing protein.

Not all fats are equal with respect to their role in brain functioning. Diets rich in monounsaturated and polyunsaturated fats are associated with improved brain functioning. As an example of the positive aspects of fat intake, docosahexaeonic acid (DHA) and eicosapentaenoic acid (EPA) omega-3 fatty acids, found in fatty cold-water fish, are especially important for brain function (Worobey 2006). Low intake or deficiencies of these compounds have been associated with decrements in mental function (Freemantle et al. 2006; see Chapter 12 for an account of the consumption of fish and omega-3 on cognitive performance in older adults). High dietary intake of saturated and/or *trans*-fatty acids is associated with vascular disease, which increases risk of stroke, and some studies show an increased risk of mild cognitive impairment (Eskelinen et al. 2008; Kalmijn et al. 2004), but this relationship is not consistent (Naqvi et al. 2011). General recommendations from the US dietary guidelines are to reduce or minimise intake of saturated and *trans*-fatty acids, and to choose dietary sources of polyunsaturated and monounsaturated fatty acids. When reducing energy intake, incorporating a greater proportion of omega-3 fatty acids to other fats in the diet may be beneficial to brain function.

In the absence of energy restriction, short-term intake of very high fat, low carbohydrate diets result in impairments in attention, mental processing speed and increased negative mood relative to diets containing less fat (Holloway et al. 2011). In trials examining the effect of low carbohydrate, high fat diets, the effects of long-term adherence to a high fat ketogenic diet are less clear. In overweight adults, weight loss over eight weeks was greater in individuals following a low carbohydrate, high fat diet relative to those following a high carbohydrate, low fat diet (Halyburton et al. 2007). Mood improved in both conditions, possibly due to weight loss, but mental processing speed improved to a greater degree in the high carbohydrate, low fat condition relative to the low carbohydrate, high fat condition. In overweight adolescents, over 12 weeks, weight loss was similar in low carbohydrate/high fat, low carbohydrate/low fat, and high carbohydrate/low fat conditions. Improvements were seen in emotional, academic and psychosocial functioning in both low fat conditions but not in the high fat condition (Yackobovitch-Gavan et al. 2008).

Weight Loss, Bariatric Surgery, Weight Maintenance and Brain Function

A number of trials have observed improved cognitive performance and mood with weight loss. Weight loss following energy restricted diets is associated with improvements in word recall (Kretsch et al. 1997), decision making (Witbracht et al. 2011), composite scores of neurocognitive function (Smith et al. 2010) and working memory (Brinkworth et al. 2009). Some work, however, shows impairments in reaction time following long-term energy reduction (Kretsch et al. 1997). Although research in the area of bariatric surgery and cognitive outcome is limited, some work indicates that individuals who undergo bariatric surgery show improvements in verbal memory relative to baseline (Gunstad et al. 2010). Bariatric surgery is also associated with improvements in psychosocial functioning and social interactions, as well as improvements in depression scores and quality of life (Karlsson et al. 1998, 2007). Studies that examine the relationship between long-term maintenance of weight loss and psychological function are limited. Weight loss is associated with improvements in mood and psychosocial function (Karlsson et al. 2003), and maintenance of at least 10 per cent body weight loss for up to ten years improves psychosocial function and quality of life (Karlsson et al. 2007; Kaukua et al. 2003).

Conclusion

While the act of following a weight reduction diet may temporarily disrupt mental function and mood relative to non-dieting, weight loss itself can confer some benefits on cognition and mood. In the short term, restriction of carbohydrates or adherence to a ketogenic diet results in impairments in a number of cognitive domains, including spatial learning and memory, working memory and executive function, relative to balanced diets. In the long term, however, the decision to follow a traditional balanced diet, a low fat diet, a low glycaemic diet or a carbohydrate-controlled diet may have little consequence on mental performance. More important considerations for weight reduction diets are likelihood of successful adherence to the diet and metabolic consequences of the diet. Current research indicates that weight loss improves insulin resistance and glucose handling, and improves cardiovascular risk profiles. Replacing fat intake with either carbohydrate or protein produces health benefits (Kratz et al. 2010), and over the long term, health outcomes with ketogenic diets, low carbohydrate diets, low fat diets and more balanced diets appears to be similar (Dansinger et al. 2005) and may lie in the eating preference of the dieter.

Overweight and obesity are risk factors for type 2 diabetes, cardiovascular disease and hypertension – all of which can negatively impact on brain function. Additionally, obesity itself could directly influence brain function through inflammation, oxidative stress and alterations in hippocampal function. Overweight and obesity also carry a significant stigma in modern times (Puhl & Brownell 2003; Puhl & Heuer 2010), and the psychosocial impact of experiencing negative stereotypes and attitudes directly affects mood and self-esteem. Overweight and obese individuals, particularly women, commonly experience weight-related ridicule and teasing, which can lead to anxiety, stress and depression (Goldfield et al. 2010; Keddie 2011). Furthermore, overweight and obese individuals suffer from a negative 'halo effect', and individuals who are overweight are perceived to be less capable, likeable and intelligent than lean counterparts (Latner et al. 2007; Puhl & Heuer 2010). Weight loss is commonly rewarded with compliments from others and reduction in stigmatising attitudes. It seems likely, therefore, that changes in mood and cognitive performance following weight loss are related not only to physiological changes but also to changes in psychosocial dynamics.

References

Baddeley, A. D., & Hitch, G. J. (1974). Working memory. In G. H. Bower (Ed.), *The psychology of learning and motivation: Advances in research and theory* (Vol. 8, pp. 47–89). London: Academic Press.

Bellisle, F., Dalix, A. M., De Assis, M. A., Kupek, E., Gerwig, U., Slama, G., et al. (2007). Motivational effects of 12-week moderately restrictive diets with or without special attention to the Glycaemic Index of foods. *British Journal of Nutrition, 97*(4), 790–798.

Benton, D., & Nabb, S. (2003). Carbohydrate, memory, and mood. *Nutrition Reviews, 61*(5 Pt 2), S61–S67.

Bertenshaw, E. J., Lluch, A., & Yeomans, M. R. (2008). Satiating effects of protein but not carbohydrate consumed in a between-meal beverage context. *Physiology and Behavior, 93*(3), 427–436.

Brinkworth, G. D., Buckley, J. D., Noakes, M., Clifton, P. M., & Wilson, C. J. (2009). Long-term effects of a very low-carbohydrate diet and a low-fat diet on mood and cognitive function. *Archives of Internal Medicine, 169*(20), 1873–1880.

Cantello, R., Varrasi, C., Tarletti, R., Cecchin, M., D'Andrea, F., Veggiotti, P., et al. (2007). Ketogenic diet: Electrophysiological effects on the normal human cortex. *Epilepsia, 48*(9), 1756–1763.

Cheatham, R. A., Roberts, S. B., Das, S. K., Gilhooly, C. H., Golden, J. K., Hyatt, R., et al. (2009). Long-term effects of provided low and high glycemic load low energy diets on mood and cognition. *Physiology and Behavior, 98*(3), 374–379.

Coelho, J. S., Polivy, J., & Herman, C. P. (2006). Selective carbohydrate or protein restriction: Effects on subsequent food intake and cravings. *Appetite, 47*(3), 352–360.

Cox, D., Gonder-Frederick, L., McCall, A., Kovatchev, B., & Clarke, W. (2002). The effects of glucose fluctuation on cognitive function and QOL: The functional costs of hypoglycaemia and hyperglycaemia among adults with type 1 or type 2 diabetes. *International Journal of Clinical Practice Supplement, 129*, 20–26.

D'Anci, K. E., Watts, K. L., Kanarek, R. B., & Taylor, H. A. (2009). Low-carbohydrate weight-loss diets. Effects on cognition and mood. *Appetite, 52*(1), 96–103.

D'Anci, K. E., & Kanarek, R. B. (2006). Dietary sugar and behavior. In John Worobey, Beverly J. Tepper, Robin B. Kanarek (Eds.), *Nutrition and behavior, a multidisciplinary approach* (pp. 162–178). Oxfordshire: CABI Publishing.

Dansinger, M. L., Gleason, J. A., Griffith, J. L., Selker, H. P., & Schaefer, E. J. (2005). Comparison of the Atkins, Ornish, Weight Watchers, and Zone diets for weight loss and heart disease risk reduction: A randomized trial. *JAMA, 293*(1), 43–53.

Eskelinen, M. H., Ngandu, T., Helkala, E. L., Tuomilehto, J., Nissinen, A., Soininen, H., & Kivipelto, M. (2008). Fat intake at midlife and cognitive impairment later in life: A population-based CAIDE study. *Interntional Journal of Geriatric Psychiatry, 23*(7), 741–747.

Fischer, K., Colombani, P. C., Langhans, W., & Wenk, C. (2002). Carbohydrate to protein ratio in food and cognitive performance in the morning. *Physiology and Behavior, 75*(3), 411–423.

Food and Nutrition Board. (2002). *Dietary reference intakes for energy, carbohydrate, fiber, fat, fatty acids, cholesterol, protein, and amino acids (macronutrients).* Washington, DC: National Academy of Sciences Press.

Freemantle, E., Vandal, M., Tremblay-Mercier, J., Tremblay, S., Blachere, J. C., Begin, M. E., et al. (2006). Omega-3 fatty acids, energy substrates, and brain function during aging. *Prostaglandins, Leukotrienes, and Essential Fatty Acids, 75*(3), 213–220.

Freitas, A., Paz, J. A., Casella, E. B., & Marques-Dias, M. J. (2007). Ketogenic diet for the treatment of refractory epilepsy: A 10 year experience in children. *Arquivos de Neuro-Psiquiatria, 65*(2B), 381–384.

Goldfield, G., Moore, C., Henderson, K., Buchholz, A., Obeid, N., & Flament, M. (2010). The relation between weight-based teasing and psychological adjustment in adolescents. *Paediatrics and Child Health, 15*(5), 283–288.

Green, M. W., Jones, A. D., Smith, I. D., Cobain, M. R., Williams, J. M., Healy, H., et al. (2003). Impairments in working memory associated with naturalistic dieting in women: No relationship between task performance and urinary 5-HIAA levels. *Appetite, 40*(2), 145–153.

Green, M. W., Rogers, P. J., Elliman, N. A., & Gatenby, S. J. (1994). Impairment of cognitive performance associated with dieting and high levels of dietary restraint. *Physiology and Behavior, 55*(3), 447–452.

Gunstad, J., Strain, G., Devlin, M. J., Wing, R., Cohen, R. A., Paul, R. H., et al. (2010). Improved memory function 12 weeks after bariatric surgery. *Surgery for Obesity and Related Diseases*, doi:10.1016/j.soard.2010.09.015

Halyburton, A. K., Brinkworth, G. D., Wilson, C. J., Noakes, M., Buckley, J. D., Keogh, J. B., et al. (2007). Low- and high-carbohydrate weight-loss diets have similar effects on mood but not cognitive performance. *American Journal of Clinical Nutrition, 86*(3), 580–587.

Hartman, A. L., Gasior, M., Vining, E. P., & Rogawski, M. A. (2007). The neuropharmacology of the ketogenic diet. *Pediatric Neurology, 36*(5), 281–292.

Hayes, J. F., D'Anci, K. E., & Kanarek, R. B. (2011). Foods that are perceived as healthy or unhealthy differentially alter young women's state body image. *Appetite*, doi:10.1016/j.appet.2011.05.323

Holloway, C. J., Cochlin, L. E., Emmanuel, Y., Murray, A., Codreanu, I., Edwards, L. M., et al. (2011). A high-fat diet impairs cardiac high-energy phosphate metabolism and cognitive function in healthy human subjects. *American Journal of Clinical Nutrition, 93*(4), 748–755.

Hoyland, A., Lawton, C. L., & Dye, L. (2008). Acute effects of macronutrient manipulations on cognitive test performance in healthy young adults: A systematic research review. *Neuroscience and Biobehavioral Reviews, 32*(1), 72–85.

Jones, N., & Rogers, P. J. (2003). Preoccupation, food, and failure: An investigation of cognitive performance deficits in dieters. *International Journal of Eating Disorders, 33*(2), 185–192.

Kalmijn, S., van Boxtel, M. P., Ocké, M., Verschuren, W. M., Kromhout, D., & Launer, L. J. (2004). Dietary intake of fatty acids and fish in relation to cognitive performance at middle age. *Neurology, 62*(2), 275–280.

Karlsson, J., Sjostrom, L., & Sullivan, M. (1998). Swedish obese subjects (SOS) – an intervention study of obesity. Two-year follow-up of health-related quality of life (HRQL) and eating behavior after gastric surgery for severe obesity. *International Journal of Obesity and Related Metabolic Disorders, 22*(2), 113–126.

Karlsson, J., Taft, C., Ryden, A., Sjostrom, L., & Sullivan, M. (2007). Ten-year trends in health-related quality of life after surgical and conventional treatment for severe obesity: The SOS intervention study. *International Journal of Obesity (London), 31*(8), 1248–1261.

Karlsson, J., Taft, C., Sjostrom, L., Torgerson, J. S., & Sullivan, M. (2003). Psychosocial functioning in the obese before and after weight reduction: Construct validity and responsiveness of the obesity-related problems scale. *International Journal of Obesity and Related Metabolic Disorders, 27*(5), 617–630.

Kaukua, J., Pekkarinen, T., Sane, T., & Mustajoki, P. (2003). Health-related quality of life in obese outpatients losing weight with very-low-energy diet and behaviour modification: A 2-y follow-up study. *International Journal of Obesity and Related Metabolic Disorders, 27*(9), 1072–1080.

Keddie, A. M. (2011). Associations between severe obesity and depression: Results from the National Health and Nutrition Examination Survey, 2005–2006. *Prevention of Chronic Disease, 8*(3), A57.

Kemps, E., & Tiggemann, M. (2005). Working memory performance and preoccupying thoughts in female dieters: Evidence for a selective central executive impairment. *British Journal of Clinical Psychology, 44*(Pt 3), 357–366.

Kemps, E., Tiggemann, M., & Marshall, K. (2005). Relationship between dieting to lose weight and the functioning of the central executive. *Appetite, 45*(3), 287–294.

Kratz, M., Weigle, D. S., Breen, P. A., Meeuws, K. E., Burden, V. R., Callahan, H. S., et al. (2010). Exchanging carbohydrate or protein for fat improves lipid-related cardiovascular risk profile in overweight men and women when consumed ad libitum. *Journal of Investigative Medicine, 58*(5), 711–719.

Kretsch, M. J., Green, M. W., Fong, A. K., Elliman, N. A., & Johnson, H. L. (1997). Cognitive effects of a long-term weight reducing diet. *International Journal of Obesity and Related Metabolic Disorders, 21*(1), 14–21.

Krikorian, R., Shidler, M. D., Dangelo, K., Couch, S. C., Benoit, S. C., & Clegg, D. J. (2010). Dietary ketosis enhances memory in mild cognitive impairment. *Neurobiology of Aging*. Retrieved from http://www.ncbi.nlm.nih.gov/pubmed/21130529

Last, A. R., & Wilson, S. A. (2006). Low-carbohydrate diets. *American Family Physician, 73*(11), 1942–1948.

Latner, J. D., Simmonds, M., Rosewall, J. K., & Stunkard, A. J. (2007). Assessment of obesity stigmatization in children and adolescents: Modernizing a standard measure. *Obesity (Silver Spring), 15*(12), 3078–3085.

Lieberman, H. R., Caruso, C. M., Niro, P. J., Adam, G. E., Kellogg, M. D., Nindl, B. C., et al. (2008). A double-blind, placebo-controlled test of 2 d of calorie deprivation: Effects on cognition, activity, sleep, and interstitial glucose concentrations. *American Journal of Clinical Nutrition, 88*(3), 667–676.

Lieberman, H. R., Falco, C. M., & Slade, S. S. (2002). Carbohydrate administration during a day of sustained aerobic activity improves vigilance, as assessed by a novel ambulatory monitoring device, and mood. *American Journal of Clinical Nutrition, 76*(1), 120–127.

Mahoney, C. R., Taylor, H. A., & Kanarek, R. B. (2007). Effect of an afternoon confectionery snack on cognitive processes critical to learning. *Physiology and Behavior, 90*(2–3), 344–352.

Mahoney, C. R., Taylor, H. A., Kanarek, R. B., & Samuel, P. (2005). Effect of breakfast composition on cognitive processes in elementary school children. *Physiology and Behavior, 85*(5), 635–645.

McClernon, F. J., Yancy, W. S., Jr., Eberstein, J. A., Atkins, R. C., & Westman, E. C. (2007). The effects of a low-carbohydrate ketogenic diet and a low-fat diet on mood, hunger, and other self-reported symptoms. *Obesity (Silver Spring), 15*(1), 182–187.

Nabb, S. L., & Benton, D. (2006). The effect of the interaction between glucose tolerance and breakfasts varying in carbohydrate and fibre on mood and cognition. *Nutritional Neuroscience, 9*(3–4), 161–168.

Naqvi, A. Z., Harty, B., Mukamal, K. J., Stoddard, A. M., Vitolins, M., & Dunn, J. E. (2011). Monounsaturated, trans, and saturated fatty acids and cognitive decline in women. *Journal of the American Geriatric Society, 59*(5), 837–843.

Neely, G., Landstrom, U., Bystrom, M., & Junberger, M. L. (2004). Missing a meal: Effects on alertness during sedentary work. *Nutrition and Health, 18*(1), 37–47.

Poppitt, S. D., McCormack, D., & Buffenstein, R. (1998). Short-term effects of macronutrient preloads on appetite and energy intake in lean women. *Physiology and Behavior, 64*(3), 279–285.

Puhl, R. M., & Brownell, K. D. (2003). Psychosocial origins of obesity stigma: Toward changing a powerful and pervasive bias. *Obesity Reviews, 4*(4), 213–227.

Puhl, R. M., & Heuer, C. A. (2010). Obesity stigma: Important considerations for public health. *American Journal of Public Health, 100*(6), 1019–1028.

Sieber, F. E., & Traystman, R. J. (1992). Special issues: Glucose and the brain. *Critical Care Medicine, 20*(1), 104–114.

Smith, P. J., Blumenthal, J. A., Babyak, M. A., Craighead, L., Welsh-Bohmer, K. A., Browndyke, J. N., et al. (2010). Effects of the dietary approaches to stop hypertension diet, exercise, and caloric restriction on neurocognition in overweight adults with high blood pressure. *Hypertension, 55*(6), 1331–1338.

Vreugdenburg, L., Bryan, J., & Kemps, E. (2003). The effect of self-initiated weight-loss dieting on working memory: The role of preoccupying cognitions. *Appetite, 41*(3), 291–300.

Wells, A. S., & Read, N. W. (1996). Influences of fat, energy, and time of day on mood and performance. *Physiology and Behavior, 59*(6), 1069–1076.

Wenk, G. L. (1989). An hypothesis on the role of glucose in the mechanism of action of cognitive enhancers. *Psychopharmacology (Berlin), 99*(4), 431–438.

White, A. M., Johnston, C. S., Swan, P. D., Tjonn, S. L., & Sears, B. (2007). Blood ketones are directly related to fatigue and perceived effort during exercise in overweight adults adhering to low-carbohydrate diets for weight loss: A pilot study. *Journal of the American Dietetic Association, 107*(10), 1792–1796.

Wing, R. R., & Phelan, S. (2005). Long-term weight loss maintenance. *American Journal of Clinical Nutrition, 82*(1 Suppl.), 222S–225S.

Wing, R. R., Vazquez, J. A., & Ryan, C. M. (1995). Cognitive effects of ketogenic weight-reducing diets. *International Journal of Obesity and Related Metabolic Disorders, 19*(11), 811–816.

Witbracht, M. G., Laugero, K. D., Van Loan, M. D., Adams, S. H., & Keim, N. L. (2011). Performance on the Iowa Gambling Task is related to magnitude of weight loss and salivary cortisol in a diet-induced weight loss intervention in overweight women. *Physiology and Behavior*, doi:10.1016/j.physbeh.2011.04.035

Worobey, J. (2006). Direct effects of nutrition on behavior: Brain-behavior connections. In John Worobey, Beverly J. Tepper, Robin B. Kanarek (Eds.), *Nutrition and behavior: A multidisciplinary approach* (pp. 25–42). Oxfordshire: CABI Publishing.

Yackobovitch-Gavan, M., Nagelberg, N., Demol, S., Phillip, M., & Shalitin, S. (2008). Influence of weight-loss diets with different macronutrient compositions on health-related quality of life in obese youth. *Appetite, 51*(3), 697–703.

Zhao, Q., Stafstrom, C. E., Fu, D. D., Hu, Y., & Holmes, G. L. (2004). Detrimental effects of the ketogenic diet on cognitive function in rats. *Pediatric Research, 55*(3), 498–506.

CHAPTER 11

Water, Hydration Status and Cognitive Performance

Caroline J. Edmonds

Chapter Overview

- Water is crucial for life and it is important to maintain optimal hydration in order to allow the body to work effectively.
- Dehydration to 2 per cent or more loss of body weight has negative effects on cognition.
- Recent research suggests that drinking water has a positive effect on cognitive performance.
- The mechanisms by which dehydration and water consumption affect cognition have yet to be established.

Introduction

Water is crucial for life; without water, humans can survive for a few days at most (Lunn & Foxen 2008). A large proportion of our body weight is formed of water, approximately 60 per cent on average (Jéquier & Constant 2010) and the brain is formed of just under 80 per cent water (McIlwain & Bachelard 1985). Thus, one might expect that hydration is related to good cognitive performance and that dehydration is related to poor cognitive performance. Indeed, we often hear that water is 'good for us' and that it might help us to concentrate or improve our attention (BBC 2000; Clarke 2007; Macrae 2009). But does research support that assertion? Should fluids be consumed when we need to ensure optimal cognitive performance? This chapter will discuss human water requirements, the effect of dehydration and water consumption on cognitive performance in adults and in children and potential mechanisms for these effects.

193

Human Water Requirements, Thirst Mechanism and Measurement of Hydration Status

Human Water Requirements

It is important to maintain optimal hydration in order that the body can work effectively (Benelam & Wyness 2010; Jéquier & Constant 2010; Popkin et al. 2010). Optimal hydration, also called euhydration, occurs when there is a balance between water consumed and water lost or excreted (Grandjean & Campbell 2004). Water plays a role in many bodily functions, including metabolism, excretion of waste products and regulating body temperature (Grandjean & Campbell 2004). Humans consume water via both food and drink, and the body also makes a small amount of metabolic water. Water is lost via urine, faeces, respiration and evaporation. A loss of body water that is not replenished with fluid ingested via food or drink will lead to dehydration.

Water requirements are affected by a number of factors, including the amount of physical exertion our daily lives entail, the environment in which we live, age, sex and diet. Greater levels of physical activity and living in a warmer climate will lead to greater water loss and therefore higher water requirements. The water requirements of children and adults differ, as discussed later in this chapter. The data presented in Table 11.1 show how different foods contain different amounts of water. The proportion of fluid obtained from drink and especially food varies according to an individual's diet because foods differ in the amount of fluid they contain, but food generally provides about 20 per cent of our water intake (Benelam & Wyness 2010; although see Sichert-Hellert et al. 2010).

Table 11.1 Water intake from often consumed food and drink

Proportion of water (%)	Food or drink
90–100	Water, tea, coffee, diet soft drinks, vegetable juice
90–95	Beer, wine
85–90	Milk, regular soft drinks, fruit juice
80–85	Strawberries, melon, grapefruit, peaches, pears, oranges, apples, grapes, cucumbers, lettuce, celery, tomatoes, cabbage, broccoli, onions, carrots
70–80	Yoghurt, fish, bananas, potatoes, sweetcorn
65–80	Rice, pasta, eggs
50–60	Ice cream, pizza
40–50	Cheese
30–45	Bread, bagels, biscuits
1–10	Crisps, crackers, dried fruit, popcorn
1–5	Seeds, nuts

Data from Grandjean & Campbell (2004).

Table 11.2 Human water-requirements in ml/day according to European Food Safety Authority (EFSA), Institute of Medicine (IOM) and World Health Organisation (WHO)[1]

Age	Sex	EFSA (2008)[2]	IOM (2005)[3]	WHO (2005)[4]
6–12 months	M & F	800–1,000	800	
1–2 years	M & F	1,100–1,200	1,300	
2–3 years	M & F	1,300	1,300	
4–8 years	M & F	1,600	1,700	
9–13 years	F	1,900	2,100	
	M	2,100	2,400	
Adult	F	2,000	2,700	2,200
	M	2,500	3,700	2,900

[1] EFSA and IOM data sourced from Jequier & Constant (2010). WHO data sourced from WHO (2005).
[2] Dietary reference intake values, indicating adequate intake.
[3] Adequate intake from drinking water, other beverages and water in food.
[4] Volume of water recommended for hydration in a sedentary, temperate environment.

The majority of our fluid intake comes from drinks. Guidelines for water consumption have been developed by the World Health Organization (2005), the American Institute of Medicine and the European Food Safety Authority (the latter two both cited in Jéquier & Constant 2010). The data presented in Table 11.2 show how these guidelines indicate suggested consumption for adults and children. In childhood, the amounts are the same for males and females, but these diverge with increasing age, with males having higher water demands than females. Although it is widely believed that adults should drink 6–8 glasses of water a day (approximately 1.5–2 litres), we do not need to drink this amount in addition to the fluids that we take in from food and other drinks (Lunn & Foxen 2008; Valtin 2002). Guidelines from the British Nutrition Foundation suggest that water should be the most frequent choice of drink for both children and adults because it provides calorie-free hydration and does not contain sugar or acids that can cause dental cavities (Benelam & Wyness 2010).

Thirst Mechanism

The sensation of thirst identifies the need to consume fluids. Total body water is the term used to refer to the amount of water in the body, which is distributed between intracellular fluid (ICF; the fluid inside cells) and extracellular fluid (ECF; the fluid outside cells). Thirst is triggered as a result of changes observed by physiological mechanisms, with osmoreceptors responding to deficits in ICF and ECF volume. When water in either compartment is low, thirst is triggered (for a thorough review of these mechanisms, see McKinley & Johnson 2004). As well as being a response to thirst, we consume fluids for psychological and cultural reasons. The sensation of drinking is pleasurable, with palatability influenced by factors such as colour, flavour, temperature, odour and texture (Grandjean & Campbell 2004). The types of fluids consumed are, to an extent, culturally determined. For example, the British are reknowned as a nation of tea drinkers. Furthermore, drinking fluids is a learned behaviour (Jéquier & Constant 2010), which children must acquire.

Dehydration and Measurement of Hydration Status

While dehydration occurs when body water output is greater than body water intake (Whitney & Rolfes 1996), it is extremely difficult to assess all water intakes and outputs outside of a clinical setting. How, then, should we ascertain when an individual is dehydrated and the extent of any dehydration? Dehydration, in both children and adults, can be indicated by a number of symptoms, including poor mood, fatigue and headache (Kleiner 1999). More severe dehydration can be indicated by shrivelled skin, clumsiness and delerium (Kleiner 1999). However, these indicators are not wholly reliable because they can also be indicative of factors other than dehydration (e.g. headache could indicate illness). Furthermore, these symptoms do not indicate the *extent* of dehydration; for this, measurement is necessary. Some methods, such as examining the colour of urine (Mentes et al. 2006) and bioelectrical impedance (Pialoux et al. 2004), do not provide a reliable measure of small changes in hydration status (Jéquier & Constant 2010). While small changes in hydration status are difficult to measure (Shirrefs 2000; Speedy et al. 2001), it is desirable to do so, because subtle changes in cognition might occur with smaller levels of dehydration.

There are a number of more accurate methods for assessing the extent of dehydration. A change in body weight due to water loss has been argued to be one of the most practical measures of hydration status (D'Anci et al. 2009; Grandjean & Campbell 2004), particularly in field studies when it converges with subjective assessment of thirst (D'Anci et al. 2009). This is one of the most commonly adopted measures. Change in body weight is only appropriate for assessing static dehydration; a one-off measurement (Cheuvront et al. 2010). Other methods that involve serial measurement may be more precise and appropriate to measure dynamic dehydration. These include examining plasma, serum, urine or saliva osmolality (Cheuvront et al. 2010; Oppliger et al. 2005; for a review, see Jéquier & Constant 2010). Osmolality provides a measure of solute concentration in a fluid. It is defined in terms of the number of osmoles (osm) of solute per litre of solution (osm/l). Urine osmolality is termed Uosm. A higher concentration of solutes (a larger value) indicates greater dehydration. There is no accepted method for assessing hydration status at a population level (Popkin et al. 2010) and the choice of measurement may vary according to the outcome measures of interest, with multiple measures being used to increase validity (Grandjean & Campbell 2004).

Hydration Status and Cognitive Performance and Mood in Adults

Dehydration, Cognitive Performance and Mood in Adults

We consume fluid from many different drinks, but when assessing the effects of fluid consumption on cognition, this chapter will focus solely on the effects of water, because including other drinks would confound the effect of hydration with that of factors such as caffeine and glucose (see Chapter 8 for glucose; Chapter 14 for caffeine; Chapter 13 for energy drinks). Studies in which dehydration of 2 per cent or greater loss of body weight was induced by exercise under hot conditions report that cognitive performance is negatively affected by dehydration. For example, Sharma et al. (1986) required a group of eight 20-year-old males to exercise in hot conditions in order to dehydrate to 1 per cent, 2 per cent and 3 per cent loss of body weight. Performance on a memory test and psychomotor stylus test, which assessed perceptual-motor coordination, decreased in a manner that was proportional to the level of dehydration, with poorer performance detectable at 2 per cent and 3 per cent loss of body weight, but not at a loss of 1 per cent. Performance on a letter-symbol coding task, which assessed processing speed, also decreased, but not significantly so. Gopinathan et al. (1988) also examined exercise-induced dehydration and similarly found that performance decreased proportionally to the degree of dehydration. Participants exercised in hot conditions to dehydrate to 1 per cent, 2 per cent, 3 per cent and 4 per cent loss of body weight. Word recognition, used to assess memory; serial addition, used to assess mathematical calculation; and performance on the trail-making tests, which assess executive functions, were all significantly impaired compared with baseline at 2 per cent loss of body weight, but not at lower levels, and serial addition and trail-making performance decreased further at 4 per cent loss of body weight.

D'Anci et al. (2009) also found that dehydration of less than 2 per cent loss of body weight did not affect cognition, as measured by a range of tests, but did affect mood. Cognitive performance was not sensitive to exercise-induced dehydration of between 1 per cent and 2 per cent loss of body weight, with performance decrements not observed on short-term memory, assessed using digit span; simple reaction time, which requires a speeded response to a stimulus; map planning, which assesses executive functions; mathematical computation or attention. The only hydration effect was observed when males, but not females, made fewer errors when dehydrated on a choice reaction time task which required a speeded response to one of a range of stimuli. There were, however, consistent negative effects of dehydration on subjective assessment of mood, with dehydrated participants reporting higher levels of anger, fatigue, depression, tension and confusion, and lower levels of vigour, when compared with their ratings when hydrated. These results might suggest that mood, or perhaps tests of mood, may be more sensitive to dehydration than cognitive performance (or perhaps tests of cognitive performance). Alternatively, it could be that mood measurements are more susceptible to demand characteristics. Further, mood, particularly the level of confusion, may affect cognitive performance. One possibility is that mood may moderate the relationship between dehydration and poor performance on cognitive tasks, perhaps by detracting attention, and this should be addressed by future research.

One interpretative issue in these studies is that it is difficult to tease apart the effects on cognition of dehydration and the means by which dehydration was induced, be it by exercise or heat. Tomporowski (2003) reported an extensive review of the effects of exercise on cognition and concluded that the literature is contradictory, with some studies reporting no effects of exercise on cognition (Bard & Fleury 1978; Fleury et al. 1981), others reporting a negative effect (Isaacs & Polhlman 1991; Wrisberg & Herbert 1976) and some reporting that intense exercise enhances cognitive performance (Aks1998; Allard et al. 1989; Arcelin et al. 1997; Hancock & McNaughton 1986; Hogervorst et al. 1996). A recent meta-analysis found that exercise has a general negative effect on cognitive performance for the first 20 minutes of exercise, and thereafter has a positive effect, but this was dependent on the type and duration of exercise and the type of cognitive task (Lambourne & Tomporowski 2010). Environmental heat also appears to affect cognition. For example, performance on tasks requiring concentration is negatively affected by an increase in temperature of just a few degrees (Wyon et al. 1979), while performance on memory tasks has been shown to improve with a small increase in temperature and decline with a larger increase in temperature (Wyon et al. 1979). Thus, the evidence suggests that, within certain parameters, heat exposure and exercise may have confounding effects on cognitive performance and mood.

Cian and colleagues systematically evaluated the effects on cognition of dehydration induced by both heat and exercise (Cian et al. 2000, 2001). Cian et al. (2000) tested participants in each of four conditions: dehydration via heat, dehydration via exercise, hyperhydration and a condition in which euhydration was maintained. Both dehydration conditions were to a 2.8 per cent loss of body weight, a level at which previous research has found cognitive performance to be impaired. Performance on a number of cognitive tasks was negatively affected by dehydration, including judgement of line length (visuospatial performance), digit span (short-term memory) and joystick tracking (motor task). Importantly, these performance impairments occurred in both dehydration conditions, with a similar decrement compared with baseline. Two tests, long-term picture memory and choice reaction time, were not sensitive to dehydration. Ratings of subjective mood were similarly affected; participants in both dehydration conditions rated themselves as significantly more tired and having a lower mood than those in the hydration conditions. The finding that there were no differences between dehydration conditions on performance on either the cognitive tests or the mood scales might be taken to imply that these were due to dehydration and not the manner by which it was induced. Cian et al. (2001) confirmed that there were no performance differences between dehydration induced by exercise and that induced by heat stress; dehydration negatively affected performance on digit span, judgement of line length and ratings of fatigue, but not joystick tracking. The two modes of inducing dehydration may have resulted in similarly sized effects on cognition by coincidence, thus, this correlational evidence is not conclusive. However, if supported by more conclusive evidence, these findings could suggest that the method of inducing dehydration did not lead to different effects on the areas of cognition and mood tested, thus implying that dehydration per se was likely to be responsible for the performance impairments.

Fluid Restriction

Dehydration can also be induced by fluid restriction. Studies using this method have shown negative effects on mood. Shirrefs et al. (2004) tested young adults at baseline, after which, fluid restriction began. At 24 hours and 37 hours, they showed approximately 2 per cent and 2.7 per cent loss of body weight, respectively. Subjective reporting of physiological symptoms increased with fluid restriction, including thirst, reporting of a dry and unpleasant mouth, and headache. The ability to concentrate, alertness and fatigue were all negatively affected. Similar results were reported by Szinnai et al. (2005), who had young adults restrict fluid for 24 hours and dehydrate to a body weight loss of 1–2 per cent, a level at which increases in plasma and urine osmolality were also observed. Ratings of subjective thirst, tiredness, and effort and concentration increased significantly, while alertness ratings decreased compared with those in a control condition. These studies suggest that perceived mood is sensitive to low levels of dehydration that have been induced by fluid restriction.

Cognitive performance, however, has not yet been shown to be sensitive to dehydration induced by fluid restriction. While Szinnai et al. (2005) found that dehydration negatively affected mood, performance on the following cognitive tests did not change under conditions of fluid restriction: a choice reaction time task; an auditory serial addition task, which assesses sustained attention; Stroop task, which assesses inhibitory control; and a manual tracking task, assessing motor control. There was some evidence of sex-specific findings; auditory evoked potentials showed no overall dehydration effect, but females were slower when dehydrated, and males faster. This lack of sensitivity could simply be because the extent of dehydration was not sufficient; Szinnai et al.'s (2005) sample were dehydrated to 1–2 per cent loss of body weight, a level at which cognitive performance does not appear sensitive to dehydration. It also takes longer to achieve dehydration by water restriction than exercise (28 hours compared with 2 hours), giving the young participants longer to adapt to a water deficit that is slowly induced, which may mean that they do not show cognitive impairments (Szinnai et al. 2005).

There may be confounding factors that affect performance in these studies. For example, participants may increase cognitive effort when feeling the negative effects of dehydration, and this may have different effects on the range of cognitive functions assessed. Szinnai et al. suggest that the negative effects on subjective mood (tiredness and alertness), but lack of sensitivity for cognitive tests, could indicate that participants increase effort when feeling the negative effects of water deprivation. Szinnai et al.'s findings of higher ratings of effort and concentration when dehydrated support this interpretation. This concurs with Wyon et al.'s (1979) argument that the negative effect of heat stress on arousal is countered by an increase in mental effort. There may also be individual differences in effort, and different physiological mechanisms may be involved in dehydration induced in different ways. It is possible that fatigue may moderate the relationship between dehydration and cognitive impairment in a manner similar to that proposed for mood. The effect of fatigue, effort and concentration should be addressed by future research.

To summarise, it appears that cognitive effects of dehydration are observed when the degree of dehydration is at or above a 2 per cent loss of body weight, while negative effects of mood are observed at dehydration of 1 per cent loss of body weight. Because loss of percentage body weight may not be the most sensitive measure of dehydration, future studies should assess effects on cognitive performance and measure dehydration using more sensitive measures. Certain aspects of cognition seem to be affected, including memory, mathematical calculation, visuospatial performance, perceptual-motor skills, executive function and mood. However, it is too early to make firm conclusions from what is currently a small number of studies and further work is required to elucidate the range of mental functions that are affected by dehydration.

Water Consumption and Cognitive Performance and Mood in Adults

Research into the effects of water consumption on cognition is in its infancy, with very few intervention studies. Two studies have examined the effect of drinking water on cognitive performance in adults who have not been purposely dehydrated. Rogers et al. (2001) found evidence for the effect of water consumption on cognition after using water as a control substance in studies designed to investigate the effects of caffeine-containing drinks. In their study, 60 adult participants aged in their twenties and thirties were given nothing, 120 ml or 330 ml water and initial thirst and performance was measured on a rapid visual information processing (RVIP) task, which assesses sustained attention. Cognitive testing took place again at 25 and 50 minutes after water consumption, with similar results at both time points. Participants who rated themselves as having a high initial thirst showed a dose-related improvement in performance. However, participants with low initial thirst showed a dose-related impairment in performance. High and low thirst participants in the no-drink group did not differ on RVIP performance initially, but the high thirst group performed more poorly than the low thirst group in the subsequent tests, suggesting that their higher level of dehydration negatively affected sustained attention. However, Neave et al. (2001) conducted a similar study in a smaller group of young adults ($N = 24$) and found that performance on the computerised battery of tests that they employed, including measures of attention, memory and reaction time, was not sensitive to water consumption. This may be due to methodological differences between the two studies, including the sample size, age group, tasks used and the timing of the administration of water.

In conclusion, it appears that cognitive performance is affected by both dehydration and water consumption, with effects observed on tasks assessing verbal memory, sustained attention, motor skills, executive function and mood. Lieberman (2007) described the effects of hydration status on cognition as 'contradictory and inconsistent' and this could occur because different cognitive skills are differently affected by hydration status (this notion is explored in the mechanisms section, below). It is important to remember that there are still few studies in this area and future research should not try to narrow down the cognitive domains tested too soon. While widening the range of cognitive functions assessed, it would be helpful if future research used similar tasks in order that cross-study comparisons can be made. This could be achieved both through the use of standardised assessments and through sharing of non-standardised, experimental tasks.

Hydration Status and Cognitive Performance in Children

While studies examining dehydration and cognition in adults employ interventions to induce dehydration, studies in children do so via voluntary dehydration, which is defined as a condition in which one simply does not consume enough fluid for the body's needs (Adolph 1947; Greenleaf & Sargent 1965). Voluntary dehydration is examined in children both because it could be viewed as ethically difficult to purposely dehydrate children and because children may be more likely than adults to be in a state of dehydration voluntarily because they simply do not consume sufficient fluids. Studies of children in Israel have found that approximately 70 per cent of school-aged children are chronically dehydrated (defined as Uosm exceeding 800 mosm/kg H_2O; Bar-David et al. 2009; Phillip et al. 1993). Using the same criteria, Bar-David et al. (2005) found that 55 per cent of their sample of children in Israel were dehydrated. In more temperate climates, similarly high estimates have been reported, again using the same criteria as above; Fadda et al. (2008) reported that 83 per cent of a sample of Sicilian schoolchildren were dehydrated.

Dehydration and Cognitive Performance in Children

Children and adults have different water demands, and children are particularly susceptible to dehydration for both physiological and social reasons (for a review, see D'Anci et al. 2009). Children have proportionally larger total body water content; adult males have around 60 per cent total body water compared with total body mass (50 per cent in the case of females), while infants born at term comprise approximately 75 per cent body water at birth (Benelam & Wyness 2010). This decreases during the first year of life and then remains stable until puberty. Children also have a higher surface area to volume ratio than adults, rendering them more susceptible to skin temperature changes in response to environmental temperature increases (Bar-Or 1989). Furthermore, they have a higher respiratory and metabolic rate compared with adults, which results in a higher proportion of water lost through the lungs. Children's renal function is immature at birth and this affects the ability to maintain homeostasis with regards to body hydration. In addition, children do not display the same sensitivity to thirst as adults; a 1–2 per cent increase in plasma osmolality stimulates thirst in adults, but children lack this sensitivity, which needs to be learned, and consequently will go long periods without drinking (Box & Landman 1994). They may also fail to drink enough to replenish body water after exercise (Bar-Or et al. 1980). Furthermore, the body mechanisms for cooling and tolerating heat are immature in childhood; children show poor acclimatisation to heat (Falk & Dotan 2008; Bytomski & Squire 2003) and are at higher risk of dehydrating while exercising (D'Anci et al. 2006). Finally, children are reliant on caregivers for access to drinks, but not all caregivers are sensitive to the signs of dehydration (Gittelman et al. 2004).

Evidence suggests that children are not consuming sufficient fluids. Kaushik et al. (2007) observed 145 Year 2 (6–7-year-old) and Year 5 (9–10-year-old) children from six UK schools and found that only 29 per cent of children were consuming the minimum expected fluid intake. The school's policy on access to drinking water affected consumption, with children in schools allowing 'free access' (water allowed on desks) drinking more than those whose schools restricted access. While access to water in schools is required by law in the UK (The Education (Nutritional Standards and Requirements for School Food) (England) Regulations 2007), the form and frequency of access are not specified. Guidelines on the most appropriate access would be likely to increase the fluid consumption of children. The prevalence of dehydration has not yet been established in the UK or the USA, but given the higher risks compared with adults, and insufficient fluid consumption, it is likely to be a similarly large proportion to that previously reported.

Two studies have found that voluntary dehydration is associated with poorer performance on digit span tasks in children; these assess short-term memory. Bar-David et al. (2005) studied 51 children who are 10–12-year-old in a warm climate (Israel). Urine samples were collected on arrival at school and at lunchtime, to test hydration status, and cognitive tests were conducted immediately after these collections. Hydration status assessed at lunchtime affected cognitive performance, with scores on the digit span memory test greater in the more hydrated group. The better hydrated group's mean scores were higher than the dehydrated group's mean scores on making groups and verbal analogies, both of which assess semantic flexibility, and number addition, which assesses mathematical calculation, but not significantly so. Performance on the hidden figures task, which assesses visual attention, seemed to be insensitive to dehydration. Fadda et al. (2008) in their study of 168 Italian schoolchildren also found that hydration status, measured by change in urine osmolality from morning to lunchtime, was negatively correlated with changes in performance on digit span; children with a higher level of dehydration showed poorer performance. These combined studies suggest that dehydration is associated with poorer working memory performance in children.

Water Consumption and Cognitive Performance in Children

Given that dehydration seems to have similar negative impacts on cognition in children to those observed in adults, studies have begun to explore whether water consumption has similar positive effects on performance in children to those observed in adults. Some anecdotal evidence suggests that regular and improved access to drinking water may aid the cognitive performance of schoolchildren (BBC 2000). Kaushik et al.'s (2007) study suggested that teachers' believe that water consumption is related to cognitive performance in children. They found that the majority of teachers, when questioned about the children in their class, agreed with the statement, 'I believe children are more focused when they are not thirsty'. There is a growing body of experimental evidence that suggests that children's cognitive performance improves after water consumption.

Benton & Davies (unpublished findings) found that nine-year-old children spent a greater proportion of time focused 'on task' when given 200 ml water to drink (78.8 per cent) compared with an occasion when they had not been given a drink (53.0 per cent), suggesting that attention or concentration are positively affected by water consumption. Edmonds & Burford (2009) found that performance on a letter cancellation task, which assesses visual attention in a speeded task, and a consecutive spot the difference task that assesses visual memory, were positively affected by water consumption. Of 58 children aged seven or eight, half were offered 250 ml water, while the remainder were not offered a drink. Children were tested 20 minutes after water consumption, at which point performance benefits were shown for those who had consumed water. Those children also rated their thirst as lower. Two further tasks were not sensitive to water consumption; a test of story memory and a visuomotor tracking task. A post hoc analysis suggested that there may be a dose-response effect: when the performance of children who drank all 250 ml was compared with those who drank less, it was found that children who drank the larger amount of water performed better than children who drank the smaller amount, but these initial findings require replication.

A further study found that children who had a drink of water showed a greater improvement in performance from baseline to test than those who were not offered additional water. Edmonds & Jeffes (2009) tested 23 children aged 6 and 7 years, offering 500 ml water and testing both at baseline and 40 minutes after water consumption. Similarly to Edmonds & Burford, they also found that performance on a letter cancellation task was positively affected by water consumption with higher scores at test than baseline for children in the water group. Furthermore, performance on a simultaneous spot the difference task that assessed visual attention (not from memory in this case) was also positively affected. Performance on visuomotor tracking and a visual memory task were not sensitive to water consumption. In the case of visual memory, this was likely to be a result of the task being generally too easy with scores at ceiling. There was some limited evidence that the water consumption may have positively affected happiness ratings.

Benton & Burgess (2009) also found positive effects of water consumption on verbal memory; their tasks required verbal recall of previously viewed pictures of objects. Forty 8-year-old children were studied on two separate days; on day one they were given a drink of 300 ml water, while they received no drink on the other day. Testing started between 20 and 34 minutes after water consumption. Performance on a test of object memory was better on the day on which water was consumed, and a larger water consumption effect was observed for delayed recall (5-minute delay) compared with immediate recall. A sustained attention task was not sensitive to water consumption.

There are some interpretive issues with these studies that should be the focus of future research. One centres on the demand characteristics of the studies; do children given extra water do better because they try harder? Children in both the Edmonds & Burford and Edmonds & Jeffes studies were unaware that they were being treated differently to their classmates, which might reduce any demand characteristic effect. Although measuring effort might at first glance appear to be a way of assessing demand characteristics, this is unlikely to be straightforward. In some of the dehydration studies reviewed above, the explanation for a lack of effect on cognitive task with an associated increase in fatigue is that participants who are dehydrated try harder and thus also feel more tired (Szinnai et al. 2005). Therefore, perceived effort may not be an uncomplicated test of demand characteristics. A second interpretative issue centres around the relationship between hydration and diet. Well-hydrated children may eat a more balanced diet, which might lead them to perform better on cognitive tests (Bellisle 2004). A German survey found that children who were better hydrated not only drank more water, but also consumed less energy from fat, than their less well-hydrated peers (Stahl et al. 2007). This is also likely to be an issue in studies of adults.

In conclusion, initial research findings suggest that both dehydration and water consumption affect cognitive performance in children, in a manner similar to that observed in adults. In children, the specific cognitive processes affected include visual attention, visual memory and verbal memory. However, these studies are few in number and further research is necessary, both to confirm the findings and to rule out alternative explanations.

Assuming that hydration is responsible, potential mechanisms should be evaluated.

Mechanisms by which Hydration Status Affects Cognition

Theoretical accounts for the effects of hydration status on cognitive performance have not yet been established, but some mechanisms have been proposed to explain the effects of dehydration on cognition (for reviews, see D'Anci et al. 2006; Wilson and Morley 2003). These explanations are speculative, with research in this area still in its infancy. Explanations for the effect of water consumption on cognitive performance have not yet been proposed.

A simple explanation for the effects of dehydration on cognition was suggested by Cohen (1983), who proposed that dehydration may divert attention from task performance and result in negative effects on cognition. This proposal assumes a capacity model of information processing, such as Kahneman's (1973) model of divided attention. This model suggests that attention is a finite resource and that if processing capacity is used by one process, less will be available for others. Thus, if the physical sensation of thirst were attentionally demanding, there would be fewer attentional resources available for cognition, and this might explain the negative effects of dehydration on cognition.

Alternatively, it has been suggested that dehydration may affect cognition via the effect of hormones (see D'Anci et al. 2006; Wilson & Morley 2003). For example, the stress hormone cortisol is elevated as a result of increased dehydration (Sawka et al. 1984), and elevated cortisol has been associated with impaired cognitive function (Greendale et al. 2000; Kirschbaum et al. 1996); also see Vedhara et al. (2000). Animal studies have shown that administration of stress hormones (glucocortocoids) can result in dendritic atrophy in the hippocampus, with associated cognitive decrements (Raber 1998). While this is a chronic effect, rather than a transient one, it does suggest avenues for future research on potential neural mechanisms by which dehydration may affect cognition, via the effects of cortisol.

D'Anci et al. (2006) also suggest an explanation for some of the contradictory effects of hydration on cognition, where in the case of some aspects of cognition, dehydration has a negative effect (Bar-David et al. 2005; Cian et al. 2000, 2001; Gopinathan et al. 1988; Sharma et al. 1986) and in other cases has a statistically non-significant effect (Bar-David et al. 2005; Cian et al. 2000, 2001; Gopinathan et al. 1988; Sharma et al. 1986; Szinnai et al. 2005). Physiological processes can have either inhibitory or excitatory effects, which might lead to positive or negative effects on cognition. For example, arginine vasopressin, which is released by the hypothalamus, activates the thirst response, and there is some evidence that vasopressin may be associated with better cognitive performance by increasing attention and arousal (Van Londen et al. 1998). Thus, increases in vasopressin could explain some of the neutral, statistically non-significant or positive effects of dehydration on cognition.

In a similar vein, animal studies have shown that chronic dehydration increases the release of gamma-Aminobutyric acid (GABA) (inhibitory) and glutamate (excitatory) neurotransmitters (Di & Tasker 2004). These potential mechanisms, with a focus on their different outcomes, should be the subject of future research.

Some recent studies have used magnetic resonance imaging (MRI) to explore the effect of hydration status on the brain. Protective, regulating mechanisms protect the brain against dramatic dehydration, but it can result in a decrease in brain volume, which can be observed both through computerised analysis techniques and even by visual inspection of scans (Duning et al. 2005). Duning et al. (2005) found that adult participants who did not consume fluid for 16 hours showed a reduction in brain volume, and that rehydration increased brain volume to pre-test levels. Dehydration seems to be associated with changes in ventricular volume (Dickson et al. 2005; Kempton et al. 2009, 2010). These effects on ventricular volume may be moderated by the degree of dehydration, with low levels (up to 2.2 per cent loss of body mass) being associated with decreases in volume, and more severe dehydration being associated with increases in ventricular volume (Dickson et al. 2005). The volume of ventricular fluid may increase as a result of an increase in osmolality of extracellular fluid (Kempton et al. 2009). These structural effects are interesting, but is there a link between hydration status, the brain and task performance?

While there are many studies examining the neural localisation of function using MRI (Gazzaniga 2009), to date, there is only one imaging study that explores the relationship between brain function and cognitive function that occurs under conditions of dehydration. Kempton et al. (2010) found an association between changes in cerebral blood flow after dehydration measured by functional MRI and performance on a measure of executive functions (the Tower of London). There was a stronger increase in the blood-oxygen-level-dependent (BOLD) signal in the fronto-parietal region in a dehydration condition than in a control condition, although task performance per se was not affected. Kempton et al. argue that the increase in cerebral blood flow under conditions of dehydration might be due to increased neuronal effort being necessary to achieve the same level of performance while dehydrated. This is likely to be a fruitful area for future research.

Conclusions

Hydration status, as shown by studies examining both dehydration and water consumption, is related to cognitive performance in both adults and children. Dehydration in adults is associated with poor performance on tasks assessing perceptual-motor coordination, verbal memory, mathematical calculation, visuospatial skills and mood. In children, dehydration is related to poorer verbal short-term memory. Fewer studies have examined the effects of water consumption, but in thirsty adults, having a drink of water improves sustained attention. In children, water consumption improves performance on visual attention, visual and verbal short-term memory tasks. It is hard to compare the effects of hydration status on adults and children because there are few studies, particularly in children. However, memory seems to be affected by hydration status generally and further research should examine whether this is specific to short- or long-term memory and whether it is modality specific. It should also be noted that the list of cognitive processes affected by hydration refers to the primary function of the tasks; the tasks employed require multiple aspects of cognition, thus the effects are not clear-cut because many of the tasks used are not process pure.

Further research should also describe and quantify the water consumption effect; this should explore the cognitive processes affected, and the amount of water necessary, and the interval between drinking and optimal performance should be considered, alongside research into the mechanisms involved, both for dehydration and water consumption. The long-term effects of dehydration and water consumption on cognition should also be examined with regard to outcomes, such as workplace and academic performance. In particular, if water consumption can be shown to have a reliable, positive effect on cognitive performance in children, it could be an extremely straightforward and cost-effective way to improve the school performance of children in developed countries.

Acknowledgements

I would like to thank Paula Booth, Rosanna Crombie, and especially Dr Mark Gardner, for their helpful comments.

References

Adolph, E. F. (1947). *Physiology of man in the desert*. New York: Interscience.

Aks, D. J. (1998). Influence of exercise on visual search: Implications for mediating cognitive mechanisms. *Perceptual and Motor Skills, 87,* 771–783.

Allard, F., Brawley, L., Deakin, J., & Elliot, F. (1989). The effect of exercise on visual attention performance. *Human Performance, 2,* 131–145.

Arcelin, R., Brisswalter, J., & Delignierres, D. (1997). Effects of physical exercise duration on decision-making performance. *Journal of Human Movement Studies, 32,* 123–140.

Bar-David, Y., Urkin, J., & Kozminsky, E. (2005). The effect of voluntary dehydration on cognitive functions of elementary school children. *Acta Paediatrica, 94,* 1667–1673.

Bar-David, Y., Urkin, J., Landau, D., Bar-David, Z., & Pilpel, D. (2009). Voluntary dehydration among elementary school children residing in a hot arid environment. *Journal of Human Nutrition and Dietetics, 22,* 455–460.

Bar-Or, O. (1989). Temperature regulation during exercise in children and adolescents. In C. V. Gisolfi & D. R. Lamb (Eds.), *Youth, exercise and sport* (pp. 335–367). Indianapolis: Benchmark.

Bar-Or, O., Dotan, R., Inbar, O., Rotshtein, A., & Zonder, H. (1980). Voluntary hypohydration in 10 to 12 year old boys. *Journal of Applied Physiology, 48,* 104–108.

Bard, C., & Fleury, M. (1978). Influence of imposed metabolic fatigue on visual capacity components. *Perceptual and Motor Skills, 47,* 1283–1287.

BBC, n. o. (2000, Retrieved 28 August 2008). *Water improves school test results,* from http://news.bbc.co.uk/1/hi/education/728017.stm

Bellisle, F. (2004). Effects of diet on behaviour and cognition in children. *British Journal of Nutrition, 92*(Suppl. 2), S227–S232.

Benelam, B., & Wyness, L. (2010). Hydration and health: A review. *Nutrition Bulletin, 35,* 3–25.

Benton, D., & Burgess, N. (2009). The effect of the consumption of water on the memory and attention of children. *Appetite, 53*(1), 143–146.

Benton, D., & Davies, J. (unpublished findings). The hydration of children and their behavior in school. *Submitted for publication.*

Box, V., & Landman, J. (1994). Children who have no breakfast. *Health Education, 4,* 10–13.

Bytomski, J. R. & Squire, D. L. (2003). Heat illness in children. *Current Sports Medicine Reports, 2,* 320–324.

Cheuvront, S. N., Ely, B. R., Kenefick, R. W., & Sawka, M. N. (2010). Biological variation and diagnostic accuracy of dehydration assessment markers. *American Journal of Clinical Nutrition, 92*(3), 565–573.

Cian, C., Barraud, P. A., Melin, B., & Raphel, C. (2001). Effects of fluid ingestion on cognitive function after heat stress or exercise-induced dehydration. *International Journal of Psychophysiology, 42*(3), 243–251.

Cian, C., Koulmann, N., Barraud, P. A., Raphel, C., Jimenez, C., & Melin, B. (2000). Influence of variations in body hydration on cognitive function: Effect of hyperhydration, heat stress, and exercise-induced dehydration. *Journal of Psychophysiology, 14*(1), 29–36.

Clarke, J. (2007). Will water make you top of the class? *Daily Mail.* London.

Cohen, S. (1983). After effects of stress on human performance during a heat acclimatization regimen. *Aviation, Space, and Environmental Medicine, 54,* 709–713.

D'Anci, K. E., Constant, F., & Rosenberg, I. H. (2006). Hydration and cognitive function in children. *Nutrition Reviews, 64,* 457–464.

D'Anci, K. E., Mahoney, C. R., Vibhakar, A., Kanter, J. H., & Taylore, H. A. (2009). Voluntary dehydration and cognitive performance in trained college athletes. *Perceptual and Motor Skills, 109,* 251–269.

Di, S., & Tasker, J. G. (2004). Dehydration-induced synaptic plasticity in magnocellular neurons of the hypothalamic supraoptic nucleus. *Endocrinology, 145,* 5141–5149.

Dickson, J. M., Weavers, H. M., Mitchell, N., Winter, E. M., Wilkinson, I. D., Van Beek, E. J. R., et al. (2005). The effects of dehydration on brain volume – preliminary results. *International Journal of Sports Medicine, 26,* 481–485.

Duning, T., Kloska, S., Steinstrater, O., Kugel, H., Heindel, W., & Knecht, S. (2005). Dehydration confounds the assessment of brain atrophy. *Neurology, 64,* 548–550.

Edmonds, C. J., & Burford, D. (2009). Should children drink more water? The effects of drinking water on cognition in children. *Appetite, 52,* 776–779.

Edmonds, C. J., & Jeffes, B. (2009). Does having a drink help you think? 6–7 year old children show improvements in cognitive performance from baseline to test after having a drink of water. *Appetite, 53*, 469–472.

The Education (Nutritional Standards and Requirements for School Food) (England) Regulations. (2007).

Fadda, R., Rappinett, G., Grathwohl, D., Parisi, M., Fanari, R., & Schmitt, J. A. J. (2008). The benefits of drinking supplementary water at school on cognitive performance in children, *41st Annual Meeting of the International Society for Developmental Psychobiology*. Washington, DC.

Falk, B., & Dotan, R. (2008) Children's thermoregulation during exercise in the heat – a revisit. *Applied Physiology Nutrition and Metabolism, 33*, 420–427.

Fleury, M., Bard, C., Jobin, J., & Carriere, L. (1981). Influence of different types of physical fatigue on a visual detection task. *Perceptual and Motor Skills, 53*, 723–730.

Gazzaniga, M. S. (Ed.). (2009). *The cognitive neurosciences* (4th ed.). Cambridge, MA: MIT Press.

Gittelman, M. A., Mahabee-Gittens, M., & Gonzalez-del-Rey, J. (2004). Common medical terms defined by parents: Are we speaking the same language? *Pediatric Emergency Care, 20*, 754–758.

Gopinathan, P., Pichan, G., & Sharma, V. (1988). Role of dehydration in heat stress-induced variations in mental performance. *Archives of Environmental Health, 43*(1), 15–17.

Grandjean, A. C., & Campbell, A. J. (2004). *Hydration: Fluids for life. A monograph by the North American Branch of the International Life Science Institute*. Washington, DC: ILSI North America.

Greendale, G. A., Kritz-Silverstein, D., Seeman, T., & Barrett-Connor, E. (2000). Higher basal cortisol predicts verbal memory loss in postmenopausal women: Rancho Bernardo Study: Brief Reports. *Journal of the American Geriatrics Society, 48*, 1655–1658.

Greenleaf, J. E., & Sargent, F. (1965). Voluntary dehydration in man. *Journal of Applied Physiology, 20*, 719–724.

Hancock, S., & McNaughton, L. (1986). Effects of fatigue on ability to process visual information by experienced orienters. *Perceptual and Motor Skills, 62*, 491–498.

Hogervorst, E., Riedel, W. J., Jeukendrup, A., & Jolles, J. (1996). Cognitive performance after strenuous physical exercise. *Perceptual and Motor Skills, 83*, 479–488.

Isaacs, L. D., & Polhlman, E. L. (1991). Effects of exercise intensity on an accompanying timing task. *Journal of Human Movement Studies, 20*, 123–131.

Jéquier, E., & Constant, F. (2010). Water as an essential nutrient: The physiological basis of hydration. *European Journal of Clinical Nutrition, 64*(2), 115–123.

Kahneman, D. (1973). *Attention and effort*. New York: Springer-Verlag.

Kaushik, A., Mullee, M. A., Bryant, T. N., & Hill, C. M. (2007). A study of the association between children's access to drinking water in primary schools and their fluid intake: Can water be "cool" in school? *Child: Care, Health and Development, 33*(4), 409–415.

Kempton, M. J., Ettinger, U., Foster, R., Williams, S. C. R., Calvert, G. A., Hampshire, A., et al. (2011). Dehydration affects brain structure and function in healthy adolescents. *Human Brain Mapping, 32*, 71–79.

Kempton, M. J., Ettinger, U., Schmechtig, A., Winter, E. M., Smith, L., McMorris, T., et al. (2009). Effects of acute dehydration on brain morphology in healthy humans. *Human Brain Mapping, 30*, 291–298.

Kirschbaum, C., Wolk, O. T., May, M., Wippich, W., & Hellhammer, D. H. (1996). Stress- and treatment-induced elevations of cortisol levels associated with impaired declarative memory in healthy adults. *Life Sciences, 58,* 1475–1483.

Kleiner, S. M. (1999). Water: An essential but overlooked nutrient. *Journal of the American Dietetic Association, 99*(2), 200–206.

Lambourne, K., & Tomporowski, P. D. (2010). The effect of exercise-induced arousal on cognitive task performance: A meta-regression analysis. *Brain Research, 1341,* 12–24.

Lieberman, H. R. (2007). Hydration and cognition: A critical review and recommendations for future research. *Journal of the American College of Nutrition, 26*(Suppl. 5), 555S–561S.

Lunn, J., & Foxen, R. (2008). How much water do we really need? *Nutrition Bulletin, 33,* 336–342.

Macrae, F. (2009). For better exam results simply have a drink of water, *Daily Mail.* London.

McIlwain, H., & Bachelard, H. S. (1985). *Biochemistry and the central nervous system.* Edinburgh: Churchill Livingstone.

McKinley, M. J., & Johnson, A. K. (2004). The physiological regulation of thirst and fluid intake. *Physiology and Behavior, 19,* 1–6.

Mentes, J. C., Wakefield, B., & Culp, K. (2006). Use of a urine color chart to monitor hydration status in nursing home residents. *Biological Research for Nursing, 7*(3), 197–203.

Neave, N., Scholey, A. B., Emmett, J. R., Moss, M., Kennedy, D. O., & Wesnes, K. A. (2001). Water ingestion improves subjective alertness, but has no effect on cognitive performance in dehydrated healthy young volunteers. *Appetite, 37*(3), 255–256.

Oppliger, R. A., Magnes, S. A., Popowski, L. A., & Gisolfi, C. V. (2005). Accuracy of urine specific gravity and osmolality as indicators of hydration status. *International Journal of Sport Nutrition and Exercise Metabolism, 15,* 236–251.

World Health Organization. (2005). Nutrients in drinking water. (Vol. ISBN 92 4 159398 9). Geneva, Switzerland: WHO Press. Retrieved from http://www.who.int/water_sanitation_health/dwq/nutrientsindw.pdf

Phillip, M., Chaimovitz, C., Singer, A., & Golinsky, D. (1993). Urine osmolality in nursery school children in hot climate. *Israel Journal of Medical Sciences, 29,* 104–106.

Pialoux, V., Mischler, I., Mounier, R., Gachon, P., Ritz, P., Coudert, J., et al. (2004). Effect of equilibriated hydration changes on total body water estimates by bioelectrical impedance analysis. *British Journal of Nutrition, 91,* 153–159.

Popkin, B. M., D'Anci, K. E., & Rosenberg, I. H. (2010). Water, hydration, and health. *Nutrition Reviews, 68*(8), 439–458.

Raber, J. (1998). Detrimental effects of chronic hypothalamic-pituitary-adrenal axis activation. *Molecular Neurobiology, 18,* 1–22.

Rogers, P. J., Kainth, A., & Smit, H. J. (2001). A drink of water can improve or impair mental performance depending on small differences in thirst. *Appetite, 36*(1), 57–58.

Sawka, M. N., Francesconi, R. P., Pimental, N. A., & Pandolf, K. B. (1984). Hydration and vascular fluid shifts during exercise in the heat. *Journal of Applied Physiology, 56*(9), 1–96.

Sharma, V. M., Sridharan, K., Pichan, G., & Panwar, M. R. (1986). Influence of heat-stress induced dehydration on mental functions. *Ergonomics, 29*(6), 791–799.

Shirrefs, S. M. (2000). Markers of hydration status. *Journal of Sports Medicine and Physical Fitness, 40*, 80–84.

Shirrefs, S. M., Merson, S. J., Fraser, S. M., & Archer, D. T. (2004). The effects of fluid restriction on hydration status and subjective feelings in man. *British Journal of Nutrition, 91*, 951–958.

Sichert-Hellert, W., Kersting, M., & Manz, F. (2010). Fifteen year trends in water intake in German children and adolescents: Results of the DONALD Study. Dortmund Nutritional and Anthropometric Longitudinally Designed Study. *Acta Paediatrica, 90*(7), 732–737.

Speedy, D. B., Noakes, T. D., & Schneider, C. (2001). Exercise-associated hyponatremia: A review. *Emergency Medicine, 13*, 17–21.

Stahl, A., Kroke, A., Bolzenius, K., & Manz, F. (2007). Relation between hydration status in children and their dietary profile – results from the DONALD study. *European Journal of Clinical Nutrition, 61*, 1386–1392.

Szinnai, G., Schachinger, H., Arnaud, M. J., Linder, L., & Keller, U. (2005). Effect of water deprivation on cognitive-motor performance in healthy men and women. *American Journal of Physiology: Regulatory, Integrative & Comparative Physiology, 58*, R275–R280.

Tomporowski, P. D. (2003). Effects of acute bouts of exercise on cognition. *Acta Psychologica, 112*, 297–324.

Valtin, H. (2002). "Drink at least eight glasses of water a day." Really? Is there scientific evidence for "8 × 8"? *American Journal of Physiology – Regulatory, Integrative and Comparative Physiology, 283*, R993–R1004.

Van Londen, L., Goekoop, J. G., Zwinderman, A. H., Lanser, J. B. K., Wiegant, V. M., & De Wied, D. (1998). Neurospcyhological performance and plasma cortisol, arginine vasopressin, and oxytocin in patients with major depression. *Psychological Medicine, 28*, 275–284.

Vedhara, K., Hyde, J., Gilchrist, I. D., Tytherleigh, M., & Plummer, S. (2000). Acute stress, memory, attention and cortisol. *Psychoneuroendocrinology, 25*, 535–549.

Whitney, E., & Rolfes, S. R. (1996). *Understanding nutrition*. New York: West Publishing Company.

Wilson, M. M. G., & Morley, J. E. (2003). Impaired cognitive function and mental performance in mild dehydration. *European Journal of Clinical Nutrition, 57*, S24–S29.

Wrisberg, C. A., & Herbert, W. G. (1976). Fatigue effects on the timing performance of well practiced subjects. *Research Quarterly, 47*, 839–844.

Wyon, D. P., Anderson, I., & Lundqvist, G. R. (1979). The effects of moderate heat stress on mental performanace. *Scandinavian Journal of Work, Environment and Health, 5*, 352–361.

CHAPTER 12

Consumption of Fish and Omega-3 Polyunsaturated Fatty Acids: Impact on Cognitive Function in Older Age and Dementia

Louise A. Brown

Chapter Overview

- Research assessing the potential benefits of omega-3 polyunsaturated fatty acid (PUFA) intake on cognitive ability in older adults has produced inconsistent findings.
- Some promising evidence does exist; in some cases, specific domains of cognitive function have exhibited a benefit, for example, executive function or processing speed.
- Some evidence suggests a protective role of omega-3 PUFA in cognitive impairment and dementia, at least in the short term; however, complex interactions may exist, for example, with Apolipoprotein E (ApoE) genotype.
- A limited number of randomised controlled trials (RCTs) have largely produced disappointing results; more RCTs of very long duration may be necessary.

Introduction

There is growing interest in the possibility that nutrients from the everyday diet may help to protect against age-related cognitive decline and dementia. Nutritional substances, such as omega-3 PUFAs, antioxidant vitamins, and fruit and vegetable polyphenols are being investigated for their potential to help preserve brain structure and function in older age (see Brown et al. 2010 for a review). It is therefore possible that better cognitive ageing may result from simple adjustments to the diet. However, it is important that we consider the empirical evidence regarding the particular nutritional substances that may enhance cognition, as well as the possible mechanisms by which they operate.

Fish consumption and omega-3 (ω-3, also referred to as n-3) PUFAs, which are most abundant within oily fish, have been implicated in cognitive well-being in later life, resulting in a recent increase in research on this topic. This chapter reviews the evidence regarding the possible associations between fish and omega-3 PUFA intake and cognitive function in older age, and will focus on research involving human participants. Studies have adopted different approaches in terms of the way in which cognitive functions are assessed, not only with respect to the particular assessments used, but also with respect to whether performance is assessed between individuals at a particular time (i.e. a cross-sectional approach) or within the same group of individuals over time (i.e. a longitudinal approach). Furthermore, some studies have investigated variation in function among cognitively healthy individuals, while others have focused specifically on the relative risk of developing cognitive impairment and dementia. Evidence resulting from each of these approaches will be addressed in turn. The first section provides a brief overview of the mechanisms through which omega-3 PUFAs may exert cognitive benefits.

Mechanisms of Omega-3 Action

While the typical European diet is believed to have decreased in omega-3 PUFA intake over millennia, the intake of omega-6 fatty acids has increased in recent decades, with a shift towards increased use of vegetable oil in the diet (Sanders 2000). Omega-3 PUFAs are essential nutrients for the human body. They are key structural and functional components in cell membranes (Connor 2000), particularly benefitting cardiovascular and visual function, as well as being a requirement for development and maintenance of the central nervous system (Connor 2000; Lukiw & Bazan 2008; Uauy & Dangour 2006).While the precise mechanisms of the putative omega-3 PUFA action on brain ageing and cognitive function are not yet known, a number of hypotheses have been proposed. Cardiovascular disease is a major risk factor for unhealthy brain ageing, and omega-3 PUFAs may modulate this risk in a number of ways; for example, by lowering elevated blood plasma triglycerides (Cunnane et al. 2009), by preventing arrhythmias and through antithrombotic action (Connor 2000). Furthermore, through modulation of oxidative and inflammatory processes, PUFAs may promote neuronal repair and survival, and help to prevent neurodegeneration (Lukiw & Bazan 2008; Mazza et al. 2007). Not only are omega-3 PUFAs – in particular, docosahexaenoic acid (DHA) – present in abundance in the healthy brain, but decreased neuronal DHA is associated with age-related cognitive decline and Alzheimer's disease (AD). In AD, decreased DHA levels have been observed in the hippocampus, which is a crucial brain structure for the function of memory (Lukiw et al.

2005). Specifically related to AD, in a recent review, Boudrault et al. (2009) highlighted a number of possible mechanisms of omega-3 PUFA action, including anti-inflammatory effects, neuroprotection and altered cellular signalling. DHA is believed to counteract the inflammatory effects of arachidonic acid (AA, an omega-6 PUFA) by limiting the expression of enzymes involved in the AA inflammatory pathway. Neuroprotectin D1 (NPD1) is derived from DHA and is associated with increased cell survival (see Lukiw et al. 2005). Also, DHA may affect the processing of amyloid precursor protein (APP). AD is associated with the pathological processing of APP, which causes the overproduction and accumulation of β-amyloid peptide, and the formation of senile plaques in the brain. Therefore, with a variety of candidate mechanisms for beneficial effects, researchers have been prompted to question whether cognitive performance may be better maintained into older age if increased supplies of omega-3 PUFAs are available to the brain.

Omega-3 PUFAs are most abundant in oily fish, such as salmon, mackerel and kippers (Scientific Advisory Committee on Nutrition SACN – 2004; see Table 12.1). Because oily fish is a particularly rich source of omega-3 PUFAs, in particular DHA and eicosapentaenoic acid (EPA), in the UK the Food Standards Agency has advised that at least two portions of fish are consumed per week, one of which should be oily. While women past childbearing age and men may consume up to four portions of oily fish per week, girls and women of childbearing age are advised to limit their intake to one to two portions per week. This is because oily fish may contain contaminants that can build up in the body over time (SACN 2004; see also Mozaffarian & Rimm 2006). The recommendations are, of course, based upon the wider range of health benefits

Table 12.1 Estimated long-chain omega-3 PUFA content (g/100 g) of a selection of popular fish types

	Long-chain omega-3 PUFAs
Oily fish	
Herring	1.3
Kipper	2.5
Mackerel – fresh	1.9
Salmon – fresh	2.7
Salmon – canned and smoked	1.5
Sardines – canned	1.6
Trout – fresh	1.2
Tuna – fresh	1.5
Lean/white fish	
Cod – fresh	0.3
Haddock – fresh	0.2
Plaice and whiting – fresh	0.3
Tuna – canned	0.4
Sole – fresh	0.1

SACN (2004).

that result from oily fish consumption. However, this chapter focuses on the evidence pertaining specifically to the cognitive benefits of fish and omega-3 PUFA intake that may be observed in older age.

Cognitive Performance

Research has investigated the associations between fish and omega-3 PUFA intake and cognitive performance. These studies and the methodologies employed are summarised in Table 12.2. Typically, in this type of study, data on fish, or specifically omega-3 PUFA intake, are collected at the same time as the cognitive data, although intake data may pre-exist the cognitive testing. These data then allow the investigation of whether or not a statistical association (correlation, relationship) exists between the extent of fish or omega-3 PUFA intake and cognitive performance. Often, there are a number of measures of cognitive performance, too, and this may allow specification of the cognitive functions that benefit. However, if an association is found to exist between fish or omega-3 PUFA intake and cognitive function, it is quite possible that other variables also associated with intake are affecting cognitive function. For example, an indirect relationship between fish consumption and cognitive function may in fact be due to an association that exists between fish consumption and life-long intelligence, with more intelligent individuals making healthier decisions regarding diet and lifestyle. Therefore, potentially confounding variables (such as age, sex, education level, energy intake, physical activity, smoking) and medical variables (such as diabetes, stroke, heart disease) are typically taken into account statistically, so that we may be more confident that we are observing a true relationship between the key variables under investigation.

The Mini-Mental State Examination (MMSE; Folstein et al. 1975) is a commonly used assessment of overall cognitive health that is quick and simple to administer and offers direct comparison across studies. It is often used in research to screen for the presence of dementia. The total possible score is 30. A score of 24 or less is typically taken to indicate dementia, although the exact cut-off that is used may vary from study to study. Research that involves cognitive decline within the 'normal' or healthy range would typically use screening tools such as the MMSE to confirm that only healthy participants are included in the study and that any variation in performance is within the range of interest.

Dangour et al. (2009) designed the OPAL (Older People and n-3 Long-chain polyunsaturated fatty acids) study to investigate omega-3 PUFA supplementation in healthy older adults (MMSE ≥ 24). At the baseline (starting point) of the study, it was possible to investigate the relationship between reported fish consumption (based on both frequency and type of fish consumed) and cognitive function (measured by immediate and delayed memory, processing speed and executive (general attentional) function) in

Table 12.2 Summary of research findings in relation to cognitive performance and cognitive change (in order of discussion in text)

Study	Design	Main findings
Cognitive performance		
Dangour et al. (2009)	– 867 participants aged 70–79 years – fish intake and cognitive function (immediate and delayed memory, processing speed, executive function) assessed	Associations between fish intake and (i) executive function and (ii) global cognitive function
Kalmijn et al. (2004)	– over 1600 participants aged 45–70 years – fish intake and cognitive function (global function, memory, processing speed, cognitive flexibility) assessed	(i) Oily fish and (ii) EPA and DHA intake both associated with lower risk of impairment in processing speed and global function. No associations between cognition and total PUFAs or total fish intake
Nurk et al. (2007)	– over 2000 participants aged 70–74 years – fish intake and cognitive function (episodic memory, semantic memory, executive function, processing speed, visuospatial executive function, and global function [MMSE]) assessed	Lean and oily fish intake both exhibited dose-dependent relationships with MMSE score, episodic memory, processing speed, and visuo-spatial ability
van de Rest et al. (2009)	– 1025 males with a mean age of 68 years* – oily fish and omega-3 PUFA intake, and cognitive function (derived from a variety of measures) assessed	No associations observed

Cognitive change

Study	Description	Findings
Beydoun et al. (2007)	– 2251 participants aged 50–64 years at baseline provided blood plasma omega-3 PUFA concentration data – cognitive decline assessed over approximately nine years in delayed verbal memory, executive function, and psychomotor speed	Lower plasma omega-3 concentration associated with higher risk of cognitive decline in executive function, but not in processing speed or delayed memory
Heude et al. (2003)	– 246 participants aged 63–74 years at baseline provided erythrocyte membrane fatty acid composition data – global cognitive decline (MMSE) assessed over four years	Higher omega-3 PUFA and DHA concentrations associated with lower risk of cognitive decline, while higher omega-6 PUFAs was a risk factor
van Gelder et al. (2007)	– 210 participants aged 70–89 years at baseline provided fish, EPA, and DHA intake data – cognitive decline (MMSE) assessed over five years	Fish intake associated with less cognitive decline. A dose-response relationship between EPA + DHA intake and protection from decline
Vercambre et al. (2009)	– over 4800 females with a mean age of 66 years were administered dietary assessment at baseline – cognitive decline measured 13 years later via a questionnaire completed by a close friend or relative assessing memory, attention, language, and visuospatial skills	Those exhibiting recent cognitive decline had in the past consumed less fish. Greater odds of cognitive decline were associated with (i) lower consumption of omega-3 PUFAs, and (ii) higher omega-6 to omega-3 ratio

218

Table 12.2 (Continued)

Study	Design	Main findings
van de Rest et al. (2009)	– 313 of the participants outlined above* were followed up over 6 years	No longitudinal relationships identified between the range of cognitive functions and either oily fish or omega-3 PUFA intake
Morris et al. (2005)	– more than 3700 participants aged 65 years and above were followed up over 6 years – dietary intake data collected – global cognitive function (derived from various measures) assessed	Dose-response effect of fish intake on rate of cognitive decline. Cognitive decline not associated with omega-3 PUFA intake
Kesse-Guyot et al. (2011)	– nearly 3300 participants aged 45 years and above provided dietary intake data at baseline – cognitive performance assessed 13 years later by the MMSE, immediate and delayed verbal memory, and self-reporting	No associations between cognitive function and (i) fish or (ii) omega-3 intake. Lower intakes of (i) omega-3, (ii) EPA, and (iii) DHA all related to self-report of cognitive difficulties
Dullemeijer et al. (2007)	– 404 participants aged 50–70 years provided blood plasma omega-3 PUFA concentration data at baseline – cognitive performance (memory, processing speed, executive function) assessed over three years	Lower blood plasma omega-3 concentrations related to greater cognitive decline in two measures of processing speed but not in memory or executive function

867 participants aged 70–79. Dangour et al. (2009) reported significant positive associations between fish consumption and memory performance, but these were attenuated beyond significance after statistical adjustment for a number of variables, including psychological health. Only the associations with executive function and global cognitive function remained significant. Such associations are promising, though, particularly when taking into account that central executive function tends to be particularly sensitive to adult ageing (e.g. see Hedden & Gabrieli 2004).

Another positive finding in relation to variation in cognitive performance came from Kalmijn et al. (2004), who showed that in a large sample (over 1600 participants) of adults aged 45–70 years, intake of oily fish and, more specifically, fish omega-3 PUFAs (EPA and DHA) was related to lower risk of impairment in overall cognitive function and processing speed. This was the case even in this sample for whom the observed impairment was mild and the participants were still relatively young. Interestingly, total PUFAs, and consumption of fish in general, were not related to cognition, suggesting a specific benefit of the omega-3 PUFAs from oily fish.

On the other hand, a specific role for fish oils is not supported by Nurk et al. (2007). In this study, cognitive performance was assessed via a number of tests in over 2000 Norwegian participants aged 70–74 years, who generally had high intakes of fish products, as measured by a food frequency questionnaire. A dose-dependent relationship was observed between cognitive performance and fish consumption for four of the six tests (episodic memory, information-processing speed, visuospatial executive function and MMSE). When this was broken down by the type of fish consumed, the associations were stronger for non-processed lean fish and oily fish, as compared with processed fish (e.g. fish cakes, fish fingers), fish as a sandwich spread and fish oil consumption. Indeed, all tests apart from the MMSE were associated with oily fish consumption. However, the same pattern of findings was present for the consumption of lean (white) fish, which questions whether fish oil is responsible for the observed associations.

In fact, null findings recently came from van de Rest et al. (2009), who investigated possible relationships between a number of cognitive measures and fatty fish and omega-3 PUFA intake levels in a sample of over 1000 males with a mean age of 68 years. No associations were observed between either fatty fish or omega-3 PUFAs and performance across three cognitive factors (memory and language, processing speed and visuospatial attention).

Overall, research assessing cognitive performance in older adults in relation to fish consumption has been inconsistent. Specifically, although associations have tended to be observed, what is not clear is the extent to which the associations are specifically due to oily fish and omega-3 PUFA intake. To illustrate, because Nurk et al. (2007) demonstrated the same pattern of associations for oily fish as for white fish, the possibility of residual confounding, or the involvement of variables other than specifically fish oil, is raised. However,

Dangour et al. (2009) were able to demonstrate a role for specifically oily fish and omega-3 PUFAs on the age-sensitive cognitive ability of executive functioning.

Cognitive Change

Longitudinal research is useful because the same individuals are observed on more than one occasion, allowing assessment of cognitive change from baseline to follow-up, with a number of follow-ups perhaps taking place. Also, rather than simply relying on reported intake, another (possibly more accurate) approach to investigating the relationship between omega-3 PUFAs and cognitive function is to measure concentrations present in the blood. Using this method, Beydoun et al. (2007) showed that lower baseline blood plasma omega-3 PUFA concentration was related to a higher risk of cognitive decline in the domain of executive function (measured by verbal fluency), but no associations were observed in psychomotor speed (digit-symbol substitution) or verbal learning. While this finding is comparable to that of Dangour et al. (2009) discussed above, the participants in the study by Beydoun et al. (2007) were middle-aged and had exhibited little cognitive decline, so it is possible that stronger effects would have been observed later on in the lifespan. The authors did note, however, that the observed association was particularly strong for those individuals who were seen to be under increased oxidative stress; that is, hypertensive and dyslipidemic (high cholesterol and triglycerides) individuals.

Heude et al. (2003) assessed the association between blood cell (erythrocyte) membrane fatty acid composition and cognitive decline, as measured by the MMSE, over a 4-year period in 246 adults who were aged 63–74 years at baseline. While higher proportions of saturated fatty acid and omega-6 PUFAs were related to increased risk of cognitive decline, a higher concentration of omega-3 (and specifically DHA) was associated with a lower risk of cognitive decline. Further positive findings came from van Gelder et al. (2007), who investigated the associations between fish consumption, EPA intake, and DHA intake, and 5-year cognitive decline in 210 Dutch participants aged 70–89 at baseline. Notably, the authors took into account the relatively small amounts of omega-3 fatty acids found in foods other than fish (Meyer et al. 2003). The authors found that, while cognitive function did not differ at baseline between those who did and did not consume fish, greater cognitive decline was observed five years later in those with no fish consumption. Furthermore, no difference existed in cognitive function among the tertiles representing level of EPA + DHA intake at baseline. However, a dose-response relationship was observed between the tertiles of EPA + DHA intake and cognitive decline; that is, the greater the intake of EPA and DHA, the greater the cognitive benefit. The authors therefore concluded that a moderate intake of EPA + DHA may be protective against age-related cognitive decline.

A recently published longitudinal study benefited not only from a long follow-up period of 13 years, but also from a very large sample of over 4800 older females with a mean age of approximately 66 years at the time of dietary assessment (i.e. in 1993; Vercambre et al. 2009). The outcome measures, taken in 2006, were recent cognitive decline (alterations in the performance of tasks specifically related to memory and attention, and visuospatial and language skills) and the functional impact of cognitive decline (impacts on everyday tasks such as telephone use, shopping, ability to self-administer medications), as reported by a close friend or relative. The authors found that the participants with recent cognitive decline had in the past consumed significantly lower amounts of fish. Furthermore, the odds of recent cognitive decline became greater with decreasing omega-3 PUFA intake, and those exhibiting cognitive decline had a greater omega-6 to omega-3 ratio.

Not all longitudinal studies investigating cognitive change have supported the role of omega-3 PUFAs, though. Van de Rest et al. (2009) assessed the associations between fatty fish intake (as well as specifically omega-3 PUFA intake) and cognitive status over a 6-year period in a sample of older Boston males with an average age of 68 years at baseline. At baseline, 1025 participants were included in the study, but this reduced to 313 by the 6-year follow-up. Cognitive function was measured via a battery of tests assessing the domains of memory, language, speed of processing, executive function, and visuospatial ability, and global cognitive function was measured using the MMSE. Dietary intake was assessed using a food frequency questionnaire. The initial cross-sectional analyses showed no association between fatty fish or omega-3 consumption and cognitive performance, and the analysis of cognitive change also revealed no associations. Fish consumption was, however, relatively high in this sample; at baseline, mean fish consumption was 2.4 servings of total fish a week and 1.3 servings of fatty fish a week. It is therefore possible that associations will only be observed when sufficient variation in consumption exists within the sample being studied, such that 'floor' and 'ceiling' levels of fish consumption are avoided.

Other research has provided more equivocal evidence for the role of omega-3 PUFAs. For example, Morris et al. (2005) investigated the associations between fish consumption and omega-3 PUFA intake and cognitive change over 6 years in more than 3700 Chicago adults aged 65 years and over. Approximately one year from baseline, participants completed a food frequency questionnaire. Cognitive performance was measured via the MMSE, memory (immediate and delayed), and information-processing speed; the four measures were standardised and averaged into a global measure of cognitive performance. Notably, in this sample, fish consumption was low to moderate, with 43 per cent of participants consuming less than one fish meal a week and only 21 per cent consuming two or more. Compared with those consuming less than one fish meal a week, cognitive decline was 10 per cent slower in those consuming fish once a week, and 13 per cent slower in those consuming fish at least twice a week. The authors investigated whether the pattern of associations with fish was related to more general healthy eating habits, but the results were not changed when fruit and vegetable consumption was taken into account. Omega-3 PUFAs were not significantly associated with cognitive decline in this study, but more precise measurement of the particular fish consumed (i.e. white and oily) would have been beneficial.

Recently, Kesse-Guyot et al. (2011) assessed fish and omega-3 consumption in nearly 3300 French participants aged 45 years or over at baseline, and assessed the relationship with cognitive performance (as measured by the MMSE and an immediate and delayed verbal memory test) and self-reported memory problems 13 years later. No associations were found between fish or omega-3 consumption and the measures of cognitive performance. However, omega-3 intake was negatively related to the self-reporting of cognitive difficulties (i.e. higher intake was related to fewer complaints about cognitive performance); this was also the case for individual intakes of EPA and DHA. Again, if only very low intakes of omega-3 are related to cognitive decline, then this study may have been relatively insensitive to any associations, because there were relatively high intakes of omega-3 in the sample overall (Kesse-Guyot et al. 2011).

As suggested by some of the findings in the previous section, it is possible that only specific cognitive domains are susceptible or sensitive to the action of omega-3 PUFAs. Rather than measuring global cognitive decline, Dullemeijer et al. (2007) argued that individual cognitive domains must be assessed instead. Another benefit of this study is that blood plasma omega-3 PUFA concentrations were measured directly and related to cognition. The authors assessed cognitive performance at baseline (at which participants had a mean age of 60 years) and cognitive change over 3 years in the domains of memory, sensorimotor speed, complex speed, information-processing speed, and executive function. In 807 participants at baseline, no significant associations were observed for any of the five cognitive domains. However, in 404 participants followed up over three years, higher blood plasma omega-3 concentrations at baseline were associated with less decline in two of the speed-related measures (complex speed and sensorimotor speed). While omega-3 PUFAs were not predictive of change in any of the other domains, the complex speed measure was derived from tasks that would appear to draw upon executive abilities (i.e. the Concept Shifting Test and the Stroop task). Notably, the sample exhibited high plasma homocysteine and low serum vitamin B12 concentrations because the participants were targeted for an randomised controlled trial (RCT) on folate supplementation. Although the cognitive change data were performed only on the placebo group of the trial (i.e. those who did not receive supplements), the participants may have had relatively unhealthy dietary habits (Dullemeijer et al. 2007).

Longitudinal research assessing cognitive change has therefore produced mixed findings. However, most studies are suggestive of a beneficial role of fish and/or omega-3 PUFAs. Notably, in the one study providing no evidence of any associations (van de Rest et al. 2009), baseline MMSE scores were fairly high (i.e. approximately 28) and mean age was relatively young (i.e. 68 years). The potential for observing cognitive change within the population sample under investigation is therefore likely to impact on study outcomes.

Cognitive Impairment and Dementia

Pathological processes in cognitive ageing are characterised by marked deficits in memory that may be difficult to differentiate from healthy ageing at the earliest stages of the onset of disease. MCI (Mild Cognitive Impairment) is a transitional state in which cognitive deficits first become apparent, after which further progression of cognitive decline would eventually lead to a diagnosis of dementia. The most common form of the dementias is AD, which, in its initial stages, is associated with damage to the medial temporal lobe areas of the brain (more specifically, the hippocampus; see Hedden & Gabrieli 2004 for a review).

Some earlier research showed promise in demonstrating that fish consumption is positively related to cognitive function and reduced risk of dementia in older age (see Table 12.3 for a summary of research studies and methodologies employed). Barberger-Gateau et al. (2002) followed up 1416 French participants, aged 68 years and over at baseline, for whom data existed on the frequency of the consumption of fish or seafood. Over the 7 years of follow-up, 170 cases of dementia had been identified. Increasing consumption of fish or seafood was found to be related to decreased risk of dementia, including AD (see also Kalmijn et al. 1997). More recently, this inverse association between fish intake and dementia was also observed in nearly 15,000 participants sampled from a number of low- and middle-income countries (Latin America, China, India; Albanese et al. 2009). Morris et al. (2003) also observed that, over a follow-up of approximately 4 years, consumption of fish at least once a week was associated with 60 per cent less risk of AD compared with those who rarely or never consumed fish, in a sample aged 65–94 years at baseline. Furthermore, total intake of omega-3 PUFAs and DHA were both associated with less risk (see also Lopez et al. 2011).

Cherubini et al. (2007) assessed blood plasma omega-3 PUFA levels in 935 Italian adults with a mean age of 76 years. Individuals were divided into three groups: (1) cognitively healthy, (2) with cognitive impairment but no diagnosis of dementia, or (3) with dementia. Dementia sufferers were found to possess lower plasma omega-3 PUFA levels than those with normal cognitive function. This finding supports earlier work that noted lower concentrations of EPA, DHA, total omega-3 PUFAs, and lower omega-3 to omega-6 ratio in AD, 'other' dementia, and cognitive impairment groups compared with individuals with normal cognitive function (Conquer et al. 2000). Conquer et al. (2000) stated that such disturbances in omega-3 PUFA concentrations may signal oxidative damage occurring early on in the pathological process. These studies are limited by the lack of dietary assessment in the participants and

Table 12.3 Summary of research findings in relation to cognitive impairment and dementia and supplementation (in order of discussion in text)

Study	Design	Main findings
Cognitive impairment and dementia		
Barberger-Gateau et al. (2002)	– 1416 participants aged 68 years and above provided fish and seafood intake data at baseline – participants followed up for cognitive assessment over seven years	Fish intake at least once a week associated with reduced risk of dementia
Kalmijn, Launer et al. (1997)	– nearly 5400 participants aged 55 years or above (the Rotterdam cohort)^ administered dietary assessment – dementia screening took place an average of 2.1 years later	Higher fish intake associated with reduced risk of dementia
Albanese et al. (2009)	– nearly 15,000 participants aged 65 and above from low- and middle-income countries (China, India, Cuba, Dominican Republic, Venezuela, Mexico, and Peru) – dietary and cognitive assessments administered	Higher fish intake associated with reduced prevalence of dementia
Morris et al. (2003)	– 815 participants aged 65–94 provided dietary intake data – clinical evaluation took place an average of 2.3 years later	(i) Fish intake at least once a week, (ii) total intake of omega-3 PUFAs, and (iii) DHA intake all associated with less risk of AD. No associations involving EPA

Study	Method	Findings
Lopez et al. (2011)	– 266 participants aged 67–100 years provided blood plasma DHA data and cognitive assessment took place – at a previous wave (approximately two years earlier) dietary data had been collected	High (i) blood plasma DHA concentrations, and (ii) dietary DHA intake both associated with reduced odds of dementia as well as specifically AD
Cherubini et al. (2007)	– 935 participants with a mean age of 76 years provided data on blood plasma omega-3 PUFA concentration – participants categorised as (1) cognitively healthy, (2) with cognitive impairment but no dementia diagnosis, or (3) with dementia	Participants with dementia possessed lower plasma omega-3 PUFA levels than those with normal cognitive function
Conquer et al. (2000)	– 84 patients with diagnosis of AD, 'other' dementia, cognitive impairment without dementia, or normal cognitive function – data collected on blood plasma omega-3 PUFA	Lower concentrations of EPA, DHA, total omega-3 PUFAs, and lower omega-3 to omega-6 ratio in individuals with dementia, AD and cognitive impairment compared with normals
Devore et al. (2009)	– the Rotterdam cohort (see above)ˆ were followed up over a mean period of ten years	Higher fish and omega-3 PUFA intakes associated with lower dementia risk over a shorter follow-up period (0–8 years) but not at the longer follow-up (9–14 years)
Engelhart et al. (2002)	– the Rotterdam cohortˆ were followed up over six years	No associations observed for dementia risk

Table 12.3 (Continued)

Study	Design	Main findings
Roberts et al. (2010)	– over 1200 participants aged 70 years or above were administered cognitive assessments and categorised as having normal cognitive function, MCI, or dementia – dietary intake data also collected	Dose-response relationship between omega-3 PUFA intake and reduced odds of MCI
Kalmijn, Feskens et al. (1997)	– 342 male participants aged 69–89 years at baseline when dietary intake data collected – global cognitive function assessed via MMSE at baseline and after a three-year follow-up	No significant associations between fish or omega-3 PUFA intake and (i) cognitive impairment at baseline or (ii) cognitive decline
Barberger-Gateau et al. (2007)	– over 8000 participants aged 65 years or above at baseline when food frequency questionnaire administered – dementia screening conducted over the next four years	Association between fish intake at least once a week and reduced dementia risk more pronounced in ApoE ε4 non-carriers. In ApoE ε4 non-carriers, omega-6 intake associated with increased dementia risk, when not counterbalanced by omega-3 intake
Huang et al. (2005)	– over 2200 participants with a mean age of approximately 72 years at baseline – food frequency questionnaire administered at baseline – dementia screening over a mean follow-up of 5.4 years	Associations between oily fish intake and decreased dementia and, specifically AD risk, in ApoE ε4 non-carriers attenuated below significance after confounding variables taken into account

227

Study	Sample/Method	Findings
Kröger et al. (2009)	– 663 participants with a mean age of approximately 81 years at study entry, when blood samples provided – dementia screening over a median follow-up of 4.9 years	No associations between dementia or specifically AD risk and erythrocyte membrane concentrations of (i) total omega-3 PUFAs,(ii) DHA, and (iii) EPA. Results unchanged when ApoE ε4 status taken into account
Laurin et al. (2003)	– 174 participants aged 65 years and above at baseline, when blood sampling provided – clinical evaluation conducted at baseline and at follow-up approximately five years later	Baseline analyses taking ApoE ε4 into account showed no protective effects of blood plasma omega-3 PUFA concentrations. In prospective analyses, higher EPA concentrations in the cognitively impaired, and higher total omega-3 PUFAs in dementia cases (n.b., insufficient data to take ApoE ε4 into account)
Samieri et al. (2008)	– over 1200 participants aged 65 or above at baseline – blood samples and dietary data provided at baseline – dementia screening conducted over four years of follow-up	After adjustments for variables, including ApoE ε4, increased blood plasma concentrations of EPA, but not DHA or total omega-3 PUFAs, significantly associated with decreased incidence of dementia. Increased AA to DHA concentration related to increased dementia risk

Table 12.3 (Continued)

Study	Design	Main findings
Supplementation and RCTs		
Whalley et al. (2004)	– 350 participants aged 64 years provided food supplement use data – cognitive function assessed in domains of non-verbal reasoning, verbal memory, executive function, processing speed, and visuospatial executive function – nested case-control study compared erythrocyte membrane fatty acid concentrations in 60 fish oil users with 60 matched non-users	After controlling variables including childhood IQ, use of food supplements associated with better performance only in processing speed. In the case-controlled study, fish oil users performed better than non-users in visuospatial executive function and, when childhood IQ was controlled for, higher DHA to AA ratio associated with visuospatial executive function and IQ in older adulthood
Whalley et al. (2008)	– 113 participants aged 64 years provided erythrocyte membrane fatty acid concentration and ApoE genotype data – cognitive assessment as above, followed up at 66 and 68 years	After controlling variables, including childhood IQ, total omega-3 PUFAs benefitted cognition at age 64 and from 64 to 68 years, but only in ApoE ε4 non-carriers
van de Rest et al. (2008)	RCT: 1800 mg/d or 400 mg/d of EPA + DHA, or a placebo over 26 weeks in 302 participants aged 65 years or above – assessed cognitive domains of attention, sensorimotor speed, memory, and executive function	No differences observed. A possibly spurious finding was that only in ApoE ε4 carriers, attention performance was improved by both supplement doses
Dangour et al. (2010)	RCT: 700 mg/d DHA + EPA or placebo over 2 years in 867 participants aged 70–79 years – assessed cognitive domains of memory, processing speed, and executive function	No observed cognitive benefits

Study	Details	Findings
Yurko-Mauro et al. (2010)	RCT: 900 mg/d DHA or placebo over 24 weeks in 485 participants, aged 55 years and above – participants had complained of memory problems and exhibited age-related cognitive decline (impaired performance on logical memory subtest of Wechsler Memory Scale III)	Improvements observed for measures of learning and memory but not working memory or executive function
Fruend-Levi et al. (2006)	RCT: 2320 mg/d DHA + EPA or placebo over 6 months in 174 patients with mild to moderate AD (mean age 74 years) – primary measures were MMSE and cognitive portion of Alzheimer's disease Assessment Scale (ADAS-Cog)	No differences in cognitive decline overall, but a slower rate of decline in MMSE scores was observed in the patients with very mild impairment (MMSE > 27, n = 32)
Quinn et al. (2010)	RCT: 2 g/d of DHA or placebo over 18 months in 295 patients with mild to moderate AD (mean age 76 years) – primary measures were ADAS-cog and Clinical Dementia Rating scale	No differences overall, but possible cognitive benefits for ApoE ε4 non-carriers

the difficulty in establishing whether the lower omega-3 PUFAs is specifically due to disease processes and/or altered dietary intake either before or after disease onset. Nevertheless, the evidence is useful when used in conjunction with other observational findings adopting different approaches, and indicates that altered omega-3 PUFA plasma concentration is a risk factor for cognitive impairment and dementia.

One recent study provided additional insight over a longer follow-up period. Devore et al. (2009) investigated fish and omega-3 PUFA intake and their relationships with long-term dementia risk. Nearly 5400 Dutch participants, aged 55 years or over and free of dementia at baseline, provided information about typical diet and were followed up for a mean period of about 10 years. By this time, 465 individuals had developed dementia. Total fish intake, type of fish intake (non or fatty), omega-3 PUFA intake, EPA + DHA intake, and ratio of omega-3 to omega-6 were all unrelated to dementia and specifically AD risk. Interestingly, however, when the observation periods were split into shorter (0–8 years) and longer (9–14 years) follow-ups, higher fish and omega-3 intakes were modestly related to lower dementia risk in the shorter follow-up period, but no relationship existed in the longer follow-up period. For example, specifically related to AD risk, those in the highest tertile of omega-3 PUFA intake had a 24 per cent reduced risk compared with those in the lowest tertile of intake. This finding may suggest that fish oil intake is associated with a decreased shorter-term dementia risk, but not over the longer term. One possible confounding issue, however, is that dietary assessment at baseline may be a less valid measure of fish and omega-3 PUFA intake in studies with longer follow-up periods. Furthermore, the participants were relatively low consumers of specifically fatty fish (Devore et al. 2009). However, Engelhart et al. (2002) also reported no association between omega-3 PUFA intake and dementia risk over the six-year follow-up period in the same cohort.

Specifically related to the risk of developing MCI, Roberts et al. (2010) assessed cognitive function and dietary intake of fatty acids in a cross-section of over 1200 adults with a mean age of 80 years. A dose-response relationship was observed between omega-3 PUFA intake and reduced odds of MCI. By contrast, Kalmijn, Feskens et al. (1997) measured fish consumption and omega-3 intake via dietary questionnaires, and cognitive function via the MMSE, in a cohort of older men (aged 69–89 years). No relationship was observed between omega-3 PUFA intake and cognitive impairment (MMSE ≤ 25) and an apparent association between fish consumption and cognitive impairment was not significant after confounding variables were taken into account. Furthermore, no reliable association existed between fish consumption and cognitive decline over three years in the sample.

However, it could be argued that the MMSE (the short screening instrument described earlier) lacks the sensitivity required for detecting small associations.

While ApoE is important for neuronal maintenance and repair in response to insult/stressors, the efficacy of these processes is variable with ApoE genotype; the ε4 allele has poor efficacy and is a risk factor for cognitive decline and AD (see Mahley et al. 2006 for a review). Barberger-Gateau et al. (2007) showed that frequent (at least weekly) consumption of fish was associated with reduced risk of dementia in over 8000 French participants aged 65 years or over at baseline, but the association was more pronounced for non-carriers of the ApoE ε4 allele. Furthermore, specifically among ApoE ε4 non-carriers, a detrimental effect of a diet rich in omega-6 oils was observed when consumption was not counterbalanced by omega-3 intake. In support of the importance of ApoE genotype when considering the potential effects of omega-3 PUFAs on older adult cognition, Huang et al. (2005) observed that consumption of fatty fish more than twice a week was associated with a 28 per cent decrease in dementia risk, as well as a 41 per cent reduction in risk of specifically AD, compared with those who consumed fish less than once a month. However, they noted that the association was selective to ApoE ε4 non-carriers. While these associations were attenuated below significance once education and income were taken into account, the role of ApoE genotype is noteworthy. Clearly, then, ApoE should be taken into account when possible. Nevertheless, other research that has done so has failed to observe roles for fish and omega-3 PUFAs. Kröger et al. (2009) observed no associations between total omega-3 PUFAs, DHA or EPA (erythrocyte membrane concentrations) and risk of dementia or AD in a subsample of the Canadian Study of Health and Ageing, and this remained the case even when ApoE ε4 status was taken into account (see also Laurin et al. 2003).

The protective effect of omega-3 PUFAs in cognitive decline or, more specifically, risk of dementia may be related to the physiological interaction of both omega-3 and omega-6, as implicated elsewhere in this chapter. Samieri et al. (2008) noted that increased plasma EPA was related to lower incidence of dementia, while DHA and total omega-3 PUFAs exhibited no association after adjustments for confounding variables. However, increased AA to DHA ratio was related to increased dementia risk. While excessive AA may result in pro-inflammatory metabolites, sufficient supplies of omega-3 PUFAs can help to counter their effects (Youdim et al. 2000).

The research regarding fish and omega-3 PUFA intake in cognitive impairment and dementia has again produced mixed findings, but the evidence is largely indicative of the presence of associations, particularly as more compelling dose-response relationships have been observed. It has been shown that complex interactions are likely to exist (such that ApoE genotype and relative levels of omega-6 fatty acids such as AA should be taken into account), and benefits may only be observed in the shorter term. It is interesting to note that Barberger-Gateau et al. (2007), who did observe an association between fish consumption and dementia risk, also demonstrated a protective effect of a diet rich in fruit and vegetables, and these authors argued that a diet rich in fish as well as fruit and vegetables is likely to be most beneficial for protection against dementia.

Omega-3 Supplementation

While the above evidence is not universally convincing in terms of the direct relationship between omega-3 fatty acids and more successful cognitive ageing, there is clearly some promising evidence that fish and omega-3 PUFA consumption is beneficial. Another approach to investigating the issue is to supplement the diet with omega-3 and observe whether any reliable cognitive benefits result.

Whalley et al. (2004) investigated the effect of fish oil supplementation on cognitive function (non-verbal reasoning, verbal memory, executive function, information-processing speed, and visuospatial ability) in a sample of 350 older adults aged 64 years for whom there was a valid measure of intelligence at age 11. The early intelligence measure was taken as part of a national survey of childhood mental ability by the Scottish Council for Research in Education, in 1936. This represents a great strength of the investigation because, in most studies, researchers have to estimate early cognitive ability relatively crudely via variables such as educational level achieved. Fish oil, vitamin, and 'other' supplement users exhibited higher age-64 IQ and processing speed scores than non-users of food supplements; however, once childhood IQ was taken into account, only the processing speed measure retained the difference. This finding demonstrates the importance of taking pre-existing ability into account. However, the study highlights the potential of fish oil supplementation, particularly as the cognitive function demonstrating the benefit (processing speed) is age-sensitive (Brown et al. 2012; Salthouse, 1996). Processing speed is also believed to impact on a range of higher-order cognitive functions. Another issue is that the participants may not have been of sufficiently old age to exhibit more extensive cognitive benefits of the fish oil supplement. Notably, however, use of vitamin and 'other' supplement types was also beneficial. In terms of the efficacy of fish oil observed in this study, it is crucial that future research helps to understand the specific benefits that may result from supplementation with this particular substance.

Another strength of the study by Whalley et al. (2004) was that blood sampling was carried out, and this allowed measurement of the omega-3 PUFA concentration of erythrocyte membranes of participants who were or were not using fish oil supplements. Measurement of erythrocyte membrane concentration likely reflects shorter-term intake than measurement from adipose tissue, which reflects intake over the longest term. However, erythrocyte membrane content is believed to be less affected by short-term variations in fatty acid intake than blood plasma phospholipids, and is a useful biomarker of fatty acid intake at least in the medium term (i.e. periods of months; Lattka et al. 2010; Sun et al. 2007). In the case-controlled comparison by Whalley et al. (2004), each group comprised 60 matched participants. With age-11 IQ taken into account, there were no differences in cognitive test scores between the two groups. The blood sampling demonstrated that omega-3 content was higher in the fish oil users than non-users, while the opposite was true of omega-6 content. Specifically, DHA content was higher in fish oil users than non-users, while AA was lower for users than non-users. Only one of the six cognitive test scores (visuospatial executive function, as measured by block design) exhibited better performance in the fish oil users. When the two groups were combined, and childhood IQ was controlled for, the ratio of DHA to AA was related both to IQ at age 64 and block design performance. This points to a benefit of optimal omega-3 PUFA intake to cognitive function in older age, even in a sample of older adults performing within the healthy range. Further work by the same group, which assessed the effects of total omega-3 PUFA and DHA erythrocyte concentration on cognitive function at later follow-ups, showed that both measures were associated with benefits for cognitive performance at age 64 and from age 64 to 68 years. However, once childhood IQ, sex, and ApoE ε4 allele status were taken into account, only the association between total omega-3 PUFAs and cognitive function remained. Importantly, the benefit of omega-3 PUFAs was restricted to those not in possession of ApoE ε4, again suggesting that the effects are limited by genotype (Whalley et al. 2008).While ApoE has received particular attention in recent years, there are other genes with potential to interact with omega-3 PUFA intake and action in older age. For example, the fatty acid desaturase (FADS)1 and FADS2 genes have recently been implicated in infant mental development as well as cardiovascular and metabolic disorders (Lattka et al. 2010). Although the level of long-chain PUFAs in the body is mainly dependent on dietary intake, they may also be formed via the elongation and desaturation of essential fatty acid precursors, and FADS1 and FADS2 encode the desaturase important for this process. Thus, the genetic contributions to omega-3 PUFA regulation are complex and numerous, and will require continued investigation with respect to the impact on cognitive functioning in older age.

The observational evidence discussed thus far is important for identifying the potential for fish and omega-3 PUFA consumption to exert beneficial effects on cognitive function in later life. However, randomised, placebo-controlled trials, in which both the experimenter and participant are blind to the experimental group to which the participant belongs (i.e. double-blind design), are the 'gold standard' for establishing causal relationships. Regarding cognitive benefits of fish oil in relation to adult ageing, there exists only a limited number of sufficient RCTs (Lim et al. 2009).

Van de Rest et al. (2008) conducted an RCT on the effect of EPA and DHA supplementation (1800 mg/d, 400 mg/d, or placebo over 26 weeks) on cognitive function in 302 healthy Dutch adults aged over 65 years (mean age of 70). Domains of attention, sensorimotor speed, memory, and executive function were assessed using a number of measures for each. Although blood plasma concentrations were increased by 236 per cent in the high-dose group and by 51 per cent in the low-dose group, there were no differences in cognitive performance between the three groups. It is possible that the study was not of sufficient duration to detect any effects of supplementation, or that little or no age-related cognitive decline had yet occurred in these individuals, rendering the study insensitive to the beneficial effects of omega-3 PUFAs. However, presence of the ApoE ε4 allele resulted in an improvement in attention performance with both low- and high-dose supplementation, suggesting an effect of ApoE genotype in the direction that supplementation is protective only in those most at risk of cognitive impairment. However, this would contradict the findings discussed earlier that suggest only the non-carriers of ApoE ε4 may benefit from fish and omega-3 PUFAs. Indeed, the authors cautioned that this could have been a spurious finding (van de Rest et al. 2008).

The OPAL study (Dangour et al. 2010) is a recent RCT that was longer in duration, larger in sample size and involved older participants than the trial conducted by van de Rest et al. (2008). Dangour et al. investigated the effect of 2-year supplementation with omega-3 (200 mg EPA + 500 mg DHA) on cognitive performance in 867 healthy older adults aged 70–79 years. Despite increased serum concentrations of EPA and DHA in the treatment group (in comparison with the control group, who received olive oil capsules), no differences in the cognitive outcome measures of memory, processing speed, and executive function were observed. A longer intervention period may have increased the sensitivity of the study. Additionally, despite a higher level in the treatment group compared with the control group overall, DHA levels appeared to be at sufficiently high levels in the control group at the end of the study. The study may have been more sensitive if more individuals with lower amounts of fish oil consumption were included.

While these two RCTs have produced disappointing results, three other RCTs have yielded more promising findings. Yurko-Mauro et al. (2010) conducted an RCT involving 485 healthy participants aged 55 years or over. The participants had self-reported memory complaints and exhibited age-related cognitive decline, but had passed the MMSE (score ≥ 26). The participants were administered either 900 mg/day of DHA or a placebo for 24 weeks. Those who had a high consumption of DHA in the two months prior to the study baseline were excluded, to avoid possible confounding of previously high consumption. Blood plasma levels of DHA evidenced compliance in the treatment group. Some cognitive measures exhibited improvement after DHA treatment in the domain of learning and episodic memory, but working memory and executive function did not improve. Freund-Levi et al. (2006) investigated the efficacy of treatment of mild to moderate AD with 6 months of 1.7 g/day of DHA and 0.6 g/day of EHA. Blood serum concentrations exhibited a 2.4-fold increase in DHA and a 3.6-fold increase in EPA in the treatment group, while the placebo group remained stable. Overall, rate of cognitive decline, as measured by the MMSE and the cognitive portion of the Alzheimer's Disease Assessment Scale (ADAS-cog; Rosen et al. 1984), did not differ between treatment and placebo. However, based on the premise that omega-3 PUFAs act on early pathological processes, subgroups of the sample were assessed individually. Indeed, a subsample of the patients with very mild impairment showed a slower rate of decline after treatment. This may suggest potential for omega-3 action at early stages of the disease process but, with progression, anti-inflammatory treatments such as fish oil may no longer be beneficial (Freund-Levi et al. 2006). It has therefore been suggested that omega-3 PUFAs offer no benefit once the AD pathophysiology has advanced (Panza et al. 2010).

Quinn et al. (2010) conducted an RCT to test the efficacy of 2 g/day of DHA over 18 months to slow cognitive decline in 295 patients with mild to moderate AD. The mean age of participants was 76 years. Global cognitive function, as measured by the ADAS-cog, was no different between treatment and placebo. In a subsample (n = 102), magnetic resonance imaging (MRI) failed to show any difference in rate of brain atrophy between the two groups after the treatment period. Interestingly, however, in exploratory analyses relating to ApoE ε4 status, those non-carriers receiving treatment showed a benefit in the ADAS-cog and MMSE measures. While the authors cautioned that they did not statistically adjust for multiple comparisons in their analyses, increasing the chance of spurious results, this is an intriguing finding in relation to the mounting evidence regarding the role of ApoE genotype.

Clearly, more RCTs are required (Lim et al. 2009) and we therefore await the results of further trials. For example, in the year 2013 results are expected from an RCT being conducted in frail older adults, which is investigating omega-3 supplementation, a multidomain intervention (physical, cognitive, and nutrition training as well as regular medical consultations) and the combination of the omega-3 supplementation and multidomain interventions (Gillette-Guyonnet et al. 2009).

Conclusions

It is clear from the above review that research addressing the possible role of omega-3 PUFAs in cognitive function in older age and dementia is increasing. While the evidence is by no means unanimous in supporting a protective effect of omega-3 PUFAs, each of the scientific approaches described above has resulted in some promising evidence. More RCTs are now required in order to clarify whether or not supplementation with omega-3 fatty acids can protect against age-related cognitive decline and dementia (Lim et al. 2009). Specifically, RCTs of very long duration would be beneficial. While it is likely that some complex interactions exist (e.g. with ApoE genotype and omega-6 to omega-3 ratio), it is important that we understand the specific cognitive benefits that result from omega-3 PUFA intake. We also need to continue to address the mechanisms of action, in whom the benefits may be expected, and at which stages of the ageing process. It is also important that we are clear on whether omega-3 PUFAs offer only a protective effect, or if they are also efficacious as a treatment option for pathological decline. Research will also be required to demonstrate when supplementation should be started, and at what dose (Cunnane et al. 2009).

References

Albanese, E., Dangour, A. D., Uauy, R., Acosta, D., Guerra, M., Gallardo Guerra, S. S., et al. (2009). Dietary fish and meat intake and dementia in Latin America, China, and India: A 10/66 Dementia Research Group population-based study. *American Journal of Clinical Nutrition, 90*, 392–400.

Barberger-Gateau, P., Letenncur, L., Deschamps, V., Pérès, K., Dartigues, J., & Renaud, S. (2002). Fish, meat, and risk of dementia: A cohort study. *British Medical Journal, 325*, 932–933.

Barberger-Gateau, P., Raffaitin, C., Letenneur, L., Berr, C., Tzourio, C., Dartigues, J. F., et al. (2007). Dietary patterns and risk of dementia. The Three-City cohort study. *Neurology, 69*, 1921–1930.

Beydoun, M. A., Kaufman, J. S., Satia, J. A., Rosamond, W., & Folsom, A. R. (2007). Plasma n-3 fatty acids and the risk of cognitive decline in older adults: The Atherosclerosis Risk in Communities Study. *American Journal of Clinical Nutrition, 85*, 1103–1111.

Boudrault, C., Bazinet, R. P., & Ma, D. W. L. (2009). Experimental models and mechanisms underlying the protective effects of n-3 polyunsaturated fatty acids in Alzheimer's disease. *Journal of Nutritional Biochemistry, 20*, 1–10.

Brown, L. A., Brockmole, J. R., Gow, A. J., & Deary, I. J. (2012). Processing speed and visuospatial executive function predict visual working memory ability in older adults. *Experimental Aging Research, 38*, 1–19.

Brown, L. A., Brockmole, J. R., Gow, A. J., & Deary, I. J. (in press). Processing speed and visuospatial executive function predict visual working memory ability in older adults. *Experimental Aging Research*.

Brown, L. A., Riby, L. M., & Reay, J. L. (2010). Supplementing cognitive aging: A selective review of the effects of ginkgo biloba and a number of everyday nutritional substances. *Experimental Aging Research, 36*, 105–122.

Cherubini, A., Andres-Lacueva, C., Martin, A., Lauretani, F., Di Iorio, A., Bartali, B., et al. (2007). Low plasma n-3 fatty acids and dementia in older persons: The InCHIANTI study. *Journal of Gerontology: Medical Sciences, 62A*, 1120–1126.

Connor, W. E. (2000). Importance of n-3 fatty acids in health and disease. *American Journal of Clinical Nutrition, 71*, 171S–175S.

Conquer, J. A., Tierney, M. C., Zecevic, J., Bettger, W. J., & Fisher, R. H. (2000). Fatty acid analysis of blood plasma of patients with Alzheimer's disease, other types of dementia, and cognitive impairment. *Lipids, 35*, 1305–1312.

Cunnane, S. C., Plourde, M., Pifferi, F., Bégin, M., Féart, C., & Barberger-Gateau, P. (2009). Fish, docosahexaenoic acid and Alzheimer's disease. *Progress in Lipid Research, 48*, 239–256.

Dangour, A. D., Allen, E., Elbourne, D., Fasey, N., Fletcher, A. E., Hardy, P., et al. (2010). Effect of 2-y n − 3 long-chain polyunsaturated fatty acid supplementation on cognitive function in older people: A randomized, double-blind, controlled trial. *American Journal of Clinical Nutrition, 91*, 1725–1732.

Dangour, A. D., Allen, E., Elbourne, D., Fletcher, A., Richards, M., & Uauy, R. (2009). Fish consumption and cognitive function among older people in the UK: Baseline data from the OPAL study. *Journal of Nutrition, Health & Aging, 13*, 198–202.

Devore, E. E., Grodstein, F., van Rooij, F. J. A., Hofman, A., Rosner, B., Stampfer, M. J., et al. (2009). Dietary intake of fish and omega-3 fatty acids in relation to long-term dementia risk. *American Journal of Clinical Nutrition, 90*, 170–170.

Dullemeijer, C., Durga, J., Brouwer, I. A., van de Rest, O., Kok, F. J., Brummer, R. M., et al. (2007). n-3 fatty acid proportions in plasma and cognitive performance in older adults. *American Journal of Clinical Nutrition, 86*, 1479–1485.

Engelhart, M. J., Geerlings, M. I., Ruitenberg, A., van Swieten, J. C., Hofman, A., Witteman, J. C. M., et al. (2002). Diet and risk of dementia: Does fat matter? The Rotterdam Study. *Neurology, 59*, 1915–1921.

Folstein, M. F., Folstein, S. E., & McHugh, P. R. (1975). "Mini-MentalState": A practical method for grading the cognitive state of patients for the clinician. *Journal of Psychiatric Research, 12*, 189–198.

Freund-Levi, Y., Eriksdotter- Jönhagen, M., Cederholm, T., Basun, H., Faxén-Irving, G., Garlind, A., et al. (2006). ω-3 fatty acid treatment in 174 patients with mild to moderate Alzheimer disease: OmegAD study. *Archives of Neurology, 63*, 1402–1408.

Gillette-Guyonnet, S., Andrieu, S., Dantoine, T., Dartigues, J., Touchon, J., Vellas, B., et al. (2009). Commentary on "A roadmap for the prevention of dementia II. Leon Thal Symposium 2008." The Multidomain Alzheimer Preventive Trial (MAPT): A new approach to the prevention of Alzheimer's disease. *Alzheimer's & Dementia, 5*, 114–121.

Hedden, T., & Gabrieli, J. D. E. (2004). Insights into the ageing mind: A view from cognitive neuroscience. *Nature Reviews Neuroscience, 5*, 87–96.

Heude, B., Ducimetière, P., & Berr, C. (2003). Cognitive decline and fatty acid composition of erythrocyte membranes – The EVA study. *American Journal of Clinical Nutrition, 77*, 803–808.

Huang, T. L., Zandi, P. P., Tucker, K. L., Fitzpatrick, A. L., Kuller, L. H., Fried, L. P., et al. (2005). Benefits of fatty fish on dementia risk are stronger for those without APOE ε4. *Neurology, 65*, 1409–1414.

Kalmijn, S., Feskens, E. J. M., Launer, L. J., & Kromhout, D. (1997). Polyunsaturated fatty acids, antioxidants, and cognitive function in very old men. *American Journal of Epidemiology, 145*, 33–41.

Kalmijn, S., Launer, L. J., Ott, A., Witteman, J. C. M., Hofman, A., & Breteler, M. M. B. (1997). Dietary fat intake and the risk of incident dementia in the Rotterdam study. *Annals of Neurology, 42*, 776–782.

Kalmijn, S., van Boxtel, M. P. J., Ocké, M., Verschuren, W. M. M., Kromhout, D., & Launer, L. J. (2004). Dietary intake of fatty acids and fish in relation to cognitive performance at middle age. *Neurology, 62*, 275–280.

Kesse-Guyot, E., Péneau, S., Ferry, M., Jeandel, C., Hercberg, S., Galan, P., et al. (2011). Thirteen-year prospective study between fish consumption, long-chain n-3 fatty acids intakes and cognitive function. *Journal of Nutrition, Health & Aging, 15*, 115–120.

Kröger, E., Verreault, R., Carmichael, P., Lindsay, J., Julien, P., Dewailly, É., et al. (2009). Omega-3 fatty acids and risk of dementia: The Canadian Study of Health and Aging. *American Journal of Clinical Nutrition, 90*, 184–192.

Lattka, E., Illig, T., Heinrich, J., & Koletzko, B. (2010). Do *FADS* genotypes enhance our knowledge about fatty acid related phenotypes? *Clinical Nutrition, 29*, 277–287.

Laurin, D., Verreault, R., Lindsay, J., Dewailly, É., & Holub, B. J. (2003). Omega-3 fatty acids and risk of cognitive impairment and dementia. *Journal of Alzheimer's Disease, 5*, 315–322.

Lim, W. S., Gammack, J. K., Van Niekerk, J. K., & Dangour, A. (2009). Omega 3 fatty acid for the prevention of dementia. *Cochrane Database of Systematic Reviews, 1*, CD005379.

Lopez, L. B., Kritz-Silverstein, D., & Barrett-Connor, E. (2011). High dietary and plasma levels of the omega-3 fatty acid docosahexaenoic acid are associated with decreased dementia risk: The Rancho Bernardo study. *Journal of Nutrition, Health & Aging, 15*, 25–31.

Lukiw, W. J., & Bazan, N. G. (2008). Docosahexaenoic acid and the aging brain. *Journal of Nutrition, 138*, 2510–2514.

Lukiw, W. J., Cui, J. G., Marcheselli, V. L., Bodker, M., Botkjaer, A., Gotlinger, K., et al. (2005). A role for docosahexaenoic acid-derived neuroprotein D1 in neural cell survival and Alzheimer disease. *Journal of Clinical Investigation, 115*, 2774–2783.

Mahley, R. W., Weisgraber, K. H., & Huang, Y. (2006). Apolipoprotein E4: A causative factor and therapeutic target in neuropathology, including Alzheimer's disease. *Proceedings of the National Academy of Sciences of the United States of America, 103*, 5644–5651.

Mazza, M., Pomponi, M., Janiri, L., Bria, P., & Mazza, S. (2007). Omega-3 fatty acids and antioxidants in neurological and psychiatric diseases: An overview. *Progress in Neuro-Psychopharmacology and Biological Psychiatry, 31*, 12–26.

Meyer, B. J., Mann, N. J., Lewis, J. L., Milligan, G. C., Sinclair, A. J., & Howe, P. R. C. (2003). Dietary intakes and food sources of omega-6 and omega-3 polyunsaturated fatty acids. *Lipids, 38*, 391–398.

Morris, M. C., Evans, D. A., Bienias, J. L., Tangney, C. C., Bennett, D. A., Wilson, R. S., et al. (2003). Consumption of fish and n-3 fatty acids and risk of incident Alzheimer disease. *Archives of Neurology, 60*, 940–946.

Morris, M. C., Evans, D. A., Tangney, C. C., Bienias, J. L., & Wilson, R. S. (2005). Fish consumption and cognitive decline with age in a large community study. *Archives of Neurology, 62*, 1849–1853.

Mozaffarian, D., & Rimm, E. B. (2006). Fish intake, contaminants, and human health. Evaluating the risks and the benefits. *Journal of the American Medical Association, 296*, 1885–1899.

Nurk, E., Drevon, C. A., Refsum, H., Solvoll, K., Vollset, S. E., Nygård, O., et al. (2007). Cognitive performance among the elderly and dietary fish intake: The Hordaland Health Study. *American Journal of Clinical Nutrition, 86*, 1470–1478.

Panza, F., Frisardi, V., Seripa, D., & Solfrizzi, V. (2010). Plasma levels of n-3 polyunsaturated fatty acids and cognitive decline: Possible role of depressive symptoms and Apolipoprotein E genotyping. *Journal of the American Geriatrics Society, 58,* 2249–2251.

Quinn, J. F., Raman, R., Thomas, R. G., Yurko-Mauro, K., Nelson, E. B., Van Dyck, C., et al. (2010). Docosahexaenoic acid supplementation and cognitive decline in Alzheimer disease. A randomized trial. *Journal of the American Medical Association, 304,* 1903–1911.

Roberts, R. O., Cerhan, J. R., Geda, Y. E., Knopman, D. S., Cha, R. H., Christianson, T. J. H., et al. (2010). Polyunsaturated fatty acids and reduced odds of MCI: The Mayo Clinic Study of Aging. *Journal of Alzheimer's Disease, 21,* 853–865.

Rosen, W. G., Mohs, R. C., & Davis, K. L. (1984). A new rating scale for Alzheimer's disease. *American Journal of Psychiatry, 141,* 1356–1364.

Salthouse, T. A. (1996). The processing-speed theory of adult age differences in cognition. *Psychological Review, 103,* 403–428.

Samieri, C., Féart, C., Letenneur, L., Dartigues, J., Pérès, K., Auriacombe, S., et al. (2008). Low plasma eicosapentaenoic acid and depressive symptomatology are independent predictors of dementia risk. *American Journal of Clinical Nutrition, 88,* 714–721.

Sanders, T. A. B. (2000). Polyunsaturated fatty acids in the food chain in Europe. *American Journal of Clinical Nutrition, 71,* 176S–178S.

Scientific Advisory Committee on Nutrition (2004). *Advice on fish consumption: Benefits & risks.* Norwich: TSO.

Sun, Q., Ma, J., Campos, H., Hankinson, S. E., & Hu, F. B. (2007). Comparison between plasma and erythrocyte fatty acid content as biomarkers of fatty acid intake in US women. *American Journal of Clinical Nutrition, 86,* 74–81.

Uauy, R., & Dangour, A. D. (2006). Nutrition in brain development and aging: Role of essential fatty acids. *Nutrition Reviews, 64,* S24–S33.

van de Rest, O., Geleijnse, J. M., Kok, F. K., van Staveren, W. A., Dullemeijer, C., Olderikkert, M. G. M., et al. (2008). Effect of fish oil on cognitive performance in older subjects. *Neurology, 71,* 430–438.

van de Rest, O., Spiro, A. III, Krall-Kaye, E., Geleijnse, J. M., de Groot, L. C. P. G. M., & Tucker, K. L. (2009). Intakes of (n-3) fatty acids and fatty fish are not associated with cognitive performance and 6-year cognitive change in men participating in the Veterans Affairs Normative Aging Study. *Journal of Nutrition, 139,* 2329–2336.

van Gelder, B. M., Tijhuis, M., Kalmijn, S., & Kromhout, D. (2007). Fish consumption, n-3 fatty acids, and subsequent 5-y cognitive decline in elderly men: The Zutphen Elderly Study. *American Journal of Clinical Nutrition, 85,* 1142–1147.

Vercambre, M., Boutron-Ruault, M., Ritchie, K., Clavel-Chapelon, F., & Berr, C. (2009). Long-term association of food and nutrient intakes with cognitive and functional decline: A 13-year follow-up study of elderly French women. *British Journal of Nutrition, 102,* 419–427.

Whalley, L. J., Deary, I. J., Starr, J. M., Wahle, K. W., Rance, K. A., Bourne, V. J., et al. (2008). n-3 fatty acid erythrocyte membrane content, APOE ε4, and cognitive variation: An observational follow-up study in late adulthood. *American Journal of Clinical Nutrition, 87,* 449–454.

Whalley, L. J., Fox, H. C., Wahle, K. W., Starr, J. M., & Deary, I. J. (2004). Cognitive aging, childhood intelligence, and the use of food supplements: Possible involvement of n-3 fatty acids. *American Journal of Clinical Nutrition, 80,* 1650–1657.

Youdim, K. A., Martin, A., & Joseph, J. A. (2000). Essential fatty acids and the brain: Possible health implications. *International Journal of Developmental Neuroscience, 18,* 383–399.

Yurko-Mauro, K., McCarthy, D., Rom, D., Nelson, E. B., Ryan, A. S., Blackwell, A., et al. (2010). Beneficial effects of docosahexaenoic acid on cognition in age-related cognitive decline. *Alzheimer's & Dementia, 6,* 456–464.

PART III

Phytochemicals and Mild Stimulants

CHAPTER 13

Energy Drink Consumption and the Effects on Stress and Cognitive Performance

Michele L. Pettit and Kathy DeBarr

Chapter Overview

- Energy drinks are beverages that commonly include sugar, caffeine, guarana, taurine, glucuronolactone, L-carnitine, ginseng and B vitamins.
- Energy drinks stimulate the central nervous system, thus resulting in a deceptive boost of 'energy' and a transitory improvement in mood, cognition and psychomotor activity.
- Underlying risks associated with energy drinks may include sleep deprivation, exhaustion, increased blood pressure, rapid heart rate and related effects.
- While individuals may consume energy drinks to allay fatigue resulting from physical and emotional stress, the physiological effects of caffeine, the leading ingredient in energy drinks, parallel activities of the stress response.
- Future research is needed to evaluate the safety of energy drinks and their potential for addiction, relationship to consumption of other substances (e.g. alcohol) and impact on perceived stress and cognitive performance.

Introduction

This chapter will explore energy drinks, their effects on stress and cognitive performance, and their potential for adverse health effects. Energy drink consumption is ubiquitous among adolescents and younger adults, with approximately onethird of American 12–24 year olds regularly consuming energy drinks (Simon & Mosher 2007). Red Bull, the leading energy drink, was introduced in the USA in 1997 (Higgins et al. 2010; Reissig et al. 2009). Today, energy drinks comprise a billion dollar plus industry. In 2002, $200 million were spent on energy drinks; by 2007, sales had grown to

$1 billion dollars, though some estimates of market expenditures were as high as $3 billion dollars (Agriculture and Agri-Food Canada 2008).

Because energy drinks impact on the way the body functions, they are part of a class labelled functional foods (Finnegan 2003; Reyner & Horne 2002). According to Ferguson (2009), 'functional foods benefit human health beyond the effect of nutrients alone' (p. 452). Energy drinks are also considered dietary supplements and thus are not regulated (Heckman et al. 2010) and are easy to obtain regardless of age. Seifert et al. (2011) estimate that between '30 to 50% of adolescents and young adults' drink energy drinks (p. 511). According to Agriculture and Agri-Food Canada (2009), energy drinks are widely available: 'From the local corner store, to grocery stores, supermarkets, retail outlets, vending machines, restaurants, cafeterias and bars, energy drinks are gaining as much, if not more, popularity than soft drinks.' This widespread availability makes further examination a necessity.

Parents, teachers, paediatricians and others responsible for the health and safety of children and adolescents need to be knowledgeable of energy drinks because of commonly held misconceptions. For example, adolescents and younger adults may erroneously believe that consumption of energy drinks ameliorates alcohol intoxication. Moreover, people sometimes confuse energy drinks with sports drinks, but there is a distinction. Most notably, sports drinks typically do not contain caffeine and have fewer calories than energy drinks. Moreover, sports drinks contain vital minerals (e.g. sodium, potassium and magnesium) for restoring electrolyte and fluid balances after vigorous exercise (Meadows-Oliver & Ryan-Krause 2007).

Babu et al. (2008) refer to energy drinks as the 'new eye opener for adolescents' (p. 35). Students often use energy drinks to remain alert while 'cramming' for exams, to enhance athletic performance and to socialise (Braganza & Larkin 2007). One concern, in particular, is that energy drinks may serve as a 'gateway' substance, leading to use of tobacco, alcohol and other drugs (Reissig et al. 2009; Temple 2009). 'Gateway' substances serve as precursors to the use of 'drugs in a developmental sequence'. Specifically:

> Individuals who use illicit drugs typically do so after first using one or more gateway substances.

> (Botvin et al. 2000, p. 769)

Interestingly, a marketing association exists between sports and energy drinks. Sports drinks are primarily used for rehydration and replacement of electrolytes and glucose ('Drinking to win' 1991). Caffeine and taurine, both present in energy drinks, independently act as diuretics (Riesenhuber et al. 2006). According to Riesenhuber et al. (2006), the combination of caffeine and taurine in energy drinks yields an effect that is no more dehydrating than any other caffeinated beverage. However, Braganza & Larkin (2007) indicate that because caffeine is a diuretic, greater caution must be taken for children who especially are prone to dehydration. Energy drinks are also known to be high in carbohydrates which, in combination with caffeine, may result in diarrhoea – another mechanism for dehydration (National Federation of State High School Associations 2011). For these reasons, the National Federation of State High School Associations does not support consumption of energy drinks.

What Are Energy Drinks?

Common constituents of energy drinks are sugar, caffeine, guarana, taurine, glucuronolactone, L-carnitine, ginseng and B vitamins (Curry & Stasio 2009). While some effects of these ingredients are known (see Table 13.1), little is known about the long-term impact of energy drinks on humans or levels of consumption that would be considered toxic.

Sugar

Findings related to sugar consumption are mixed. A 1995 meta-analysis of 23 studies indicated that sugar had no impact on children's cognition or behaviour (Wolraich et al. 1995). An analysis of more recent literature (1996–2006) found no relationship between sugar consumption and obesity, body mass index (BMI), attention deficit or mental health issues (Ruxton et al. 2010). Contrary to these findings, Bremer et al. (2010) indicate that consumption of sugar-sweetened beverages is associated with a number of adverse metabolic indicators, including reduced levels of high density lipoprotein and increased insulin resistance, systolic blood pressure, waist circumference and BMI. Insulin resistance, a precursor to type 2 diabetes, also increases cardiovascular risk (Steinberger & Daniels 2003).

Among those with diabetes, a decrease in cognitive function has been linked to increased blood sugar levels, as demonstrated through the Memory in Diabetes Study (MIND) (Wake Forest University Baptist Medical Center 2009). Yet another study indicated that hippocampal atrophy with related memory deficits could be found among younger people with well-controlled diabetes (Gold et al. 2007).

One concern that has recently been recognised is the potential for caffeine cross-sensitisation with sugar; in other words, the two substances are mutually reinforcing. This cross-sensitisation could lead adolescents and younger adults to develop strong preferences for sugar-laden foods and beverages, thus contributing to obesity (Oddy & O'Sullivan 2010; Temple 2009). In response to these concerns, sugar-free versions of energy drinks have emerged (Agriculture and Agri-Food Canada 2010).

Caffeine

Caffeine, the most prominent constituent of energy drinks (Ferreira et al. 2004; Simon & Mosher 2007), is a naturally occurring substance found in

Table 13.1 Common amounts and effects of substances present in energy drinks

Substance	Common amounts present per 8 ounce serving	Effects
Sugar, glucose, carbohydrates	35 g (Clauson et al. 2008)	Associated with mixed research findings; stimulates pleasure centres in the brain that relate to increased consumption of sugary foods, and subsequent contribution to obesity and type 2 diabetes (Oddy & O'Sullivan 2010); contributes to hippocampal atrophy with related memory deficits among younger persons with well-controlled diabetes (Gold et al. 2007); provides no impact on cognition or behaviour in children (Wolraich et al. 1995); possesses no relationship to obesity, body mass index, attention deficit or mental health issues (Ruxton et al. 2010)
Caffeine	80–300 mg (Clauson et al. 2008)	Acts as a stimulant with cardiac and haematological effects (Babu et al. 2008; Clauson et al. 2008; Seifert et al. 2011); enhances attentiveness, reaction time, short-term recall, mood and physical performance; reduces perceived tiredness (Ruxton 2009); improves high-intensity exercise performance (Astorino & Roberson 2010); suppresses appetite (Burgalassi et al. 2009); triggers insomnia, headaches, nervousness, tachycardia, frequent urination, nausea, vomiting, shakiness, anxiety, depression, dependence, increased blood pressure, and in large concentrations (i.e. 5–10 g), death (Clauson et al. 2008; Dugdale 2009; Greenberg et al. 2006; Kerrigan & Lindsey 2005); implicated in sleep impairment (Pollak & Bright 2003) and bone deterioration in children and adolescents when used in excess (Libuda et al. 2008)
Guarana	Negligible/insufficient to cause harm (Clauson et al. 2008)	Acts as a stimulant with cardiac and haematological effects (Babu et al. 2008; Clauson et al. 2008; Seifert et al. 2011); acts as an aphrodisiac and appetite suppressant; contains theobromine and theophylene which act as bronchodilators and diuretics (Smith & Atroch 2010); in dosages present, provides no therapeutic or other effects (Clauson et al. 2008)

Taurine	Negligible/insufficient to cause harm (Clauson et al. 2008)	Acts as a stimulant with cardiac and haematological effects (Babu et al. 2008; Clauson et al. 2008; Seifert et al. 2011); in dosages present, provides no therapeutic or other effects (Clauson et al. 2008)
Glucuronolactone	Unknown	Not determined (Higgins et al. 2010)
L-carnitine	Unknown	Positively affects blood lipid levels; provides benefits for individuals with selected heart conditions, including angina, arrhythmias, heart failure and heart attack; prevents nerve cell deterioration; assists with kidney disease, male infertility, chronic fatigue syndrome and hyperthyroidism (University of Maryland Medical Center 2011; Yonei et al. 2008)
Ginseng	Negligible/insufficient to cause harm (Clauson et al. 2008)	Purportedly improves cognition, concentration and memory; provides no therapeutic or other effects at less than two cans (Clauson et al. 2008)
B vitamins	Unknown	Promote energy conversion and generation of blood cells (Medline Plus 2011); assist with breakdown of fat, carbohydrates and protein; contribute to nervous system health (Dollemore et al. 1995); promote heart health; promote mental health; maintain healthy blood cells and prevent birth defects (Pressman & Buff 1997)

coffee and guarana (Smith & Atroch 2010). As a central nervous system stimulant (Heckman et al. 2010; Nehlig et al. 1992; Smith & Atroch 2010), caffeine counters fatigue by acting as an antagonist to adenosine receptors (Seifert et al. 2011). Mentally, caffeine enhances alertness and focus, improves mood and lessens fatigue (Brunye et al. 2010; Heckman et al. 2010; Ruxton 2009; see Chapter 14 for a comprehensive account of caffeine psychopharmacology), while physically, it improves psychomotor activity (Alford et al. 2001; Astorino & Roberson 2010; Seidl et al. 2000; Snyder & Sklar 1984) and acts as an appetite suppressant (Burgalassi et al. 2009).

Caffeine plays a prominent role in our society. According to Snyder & Sklar (1984), 'Caffeine is the most widely used psychoactive substance on earth. Besides being present in coffee, tea, and cocoa, caffeine is contained in a large number of soft drinks as well as in over-the-counter medications for headache and dieting' (p. 91). It is also found in dairy products, sweetened grains and other sweets (Frary et al. 2005). Caffeine is consumed as a 'daily ritual for 80 per cent of North Americans' (Agriculture and Agri-Food Canada 2010).

Caffeine consumers range from younger children to older adults. While children break down caffeine quicker than adults, they consume less of the substance. Using data from the 1999 Share of Intake Panel (SIP) survey, Knight et al. (2004) analysed caffeine consumption among American individuals throughout the lifespan. They discovered that, on average, younger children aged 1–5 and 6–9 consumed 14 mg and 22 mg of caffeine a day, respectively. Moreover, younger adults consumed an average of 106 mg of caffeine a day, while middle-aged adults (35–49) consumed an average of 170 mg of caffeine a day.

The addictive nature of caffeine is of particular concern to public health professionals. Johns Hopkins Bayview Medical Center (n.d.) notes that, 'The World Health Organization (ICD-10) recognizes a diagnosis of substance dependence due to caffeine. Despite the fact that *Diagnostic and Statistical Manual of Mental Disorders-IV* (DSM-IV) uses very similar criteria for making a diagnosis of substance dependence, caffeine dependence is not presently included in DSM-IV.' Caffeine dependence particularly may be problematic for adolescents as it has been related to increased anxiety and depression (Bernstein et al. 2002).

Substantial amounts of caffeine in selected energy drinks are sufficient to lead to caffeine toxicity for those who do not regularly consume the substance (Reissig et al. 2009). The caffeine content in energy drinks varies substantially, on average from 80 mg to 300 mg (Clauson et al. 2008). MacDonald et al. (2010) indicate that energy drinks may contain up to ten times the amount of caffeine in one can of cola. While death resulting from caffeine consumption is extremely rare (Kerrigan & Lindsey 2005; Mrvos et al. 1989; Nawrot et al. 2003), consumption at extreme doses (i.e. 5 g or more) can cause death (Clauson et al. 2008; Dugdale 2009; Greenberg et al. 2006; Kerrigan & Lindsey 2005). Table 13.2 presents caffeine concentrations of best-selling energy drinks.

Table 13.2 Caffeine concentrations of best-selling energy drinks

Top-selling energy drinks	US market share (%)	Ounces	Caffeine (mg/oz)	Sugar (g/oz)
Red Bull©	40.00	8.46	9.50	3.19
Monster©	23.00	16.00	10.00	3.38
Rockstar©	12.30	16.00	10.00	3.75
Amp©	8.00	16.00	8.90	3.12–3.62
Full Throttle©	4.00	16.00	9.00	3.62
Starbucks Doubleshot©	2.00	6.50	20.00	2.62
NOS©	1.50	16.00	16.20	3.38
No Fear©	1.40	16.00	10.90	4.12
No Fear Bloodshot				
Private label (a variety of store brands)	1.00	Not available	Not available	Not available
Sobe Adrenaline Rush©	0.70	8.30	9.50	4.22
Vitamin Energy©	0.50	16.00	9.40	1.65
SOBE Lean©	0.50	Not available	Not available	Not available
Venom©	0.44	16.00	10.00	0.38–3.50
Jolt©	0.40	23.50	11.90	4.00
Go Girl©	0.40	12.00	8.30	0.00

Source: Energy Fiend (http://www.energyfiend.com/) (Reproduced with permission).

Guarana

Guarana, a common constituent of energy drinks, is derived from an Amazonian rainforest vine (Smith & Atroch 2010). Guarana contains guaranine, a substance that is very similar to caffeine (Finnegan 2003). It also contains theobromine and theophylene, which act as bronchodilators and diuretics. Guarana was recently associated with cognitive improvements and decreased mental fatigue among 18–24-year-olds who completed subtraction tasks, the Rapid Visual Information Processing task and a scale measuring mental fatigue (Kennedy et al. 2008). Additionally, guarana is used as a tonic, an aphrodisiac and an appetite suppressant.

Because of its bitterness, guarana is usually combined with large quantities of sugar which, in itself, poses a problem as noted previously (Smith & Atroch 2010). According to Smith & Atroch (2010), the major medical concern associated with guarana is caffeine toxicity, which has resulted in adolescents and younger adults being seen in the emergency room. While Smith & Atroch (2010) contend that guarana has four times the caffeine found in coffee, Clauson et al. (2008) purport that the amount of guarana in energy drinks is insufficient to have either a positive or negative effect.

Taurine

Taurine, a 'semi-essential amino acid', is implicated in retinal, cardiovascular (Finnegan 2003; Franconi et al. 2004; Schuller-Levis & Park 2003), reproductive and central nervous system health (Schuller-Levis & Park 2003). With respect to the central nervous system, taurine has been found to have anti-anxiety properties (Zhang & Kim 2007). Taurine does not appear to impact memory (Warburton et al. 2001; Zhang & Kim 2007), but has been found to enhance attention, verbal reasoning, reaction time and accuracy (Warburton et al. 2001).

Taurine may have application in the treatment of type 1 diabetes and prevention of insulin resistance (Franconi et al. 2004). Schuller-Levis & Park (2004) further indicate that taurine plays a role in anti-inflammatory processes and may be effective in the treatment of infections and tumours. Taurine, when added to energy drinks, has been demonstrated to increase the heart's stroke volume, making it more efficient by pumping more blood with each beat during exercise (Baum & Weiss 2001). Nevertheless, Clauson et al. (2008) contend that the amount of taurine found in energy drinks has no therapeutic or other effects. Interestingly, a study in which rats were chronically exposed to alcohol demonstrated protective effects of taurine in relationship to reduction of fatty liver disease and prevention of cell damage from free radicals (Kerai et al. 1998). Whether these findings extend to humans remains to be seen.

Glucuronolactone

Glucuronolactone is a natural by-product of the liver's glucose metabolism. It is present in some plant-based foods, most prominently in wine, as well as in small quantities in meats (Finnegan 2003). However, according to Finnegan (2003), glucuronolactone found in plant sources 'is not readily bioavailable' (p. 152). In other words, the body cannot readily absorb it for use. While animal toxicity studies exist, the impact of glucuronolactone on humans has yet to be determined (Finnegan 2003).

L-Carnitine

Higgins et al. (2010) indicate that the liver and kidneys produce L-carnitine, which prevents injury to cells and aids in recuperation from strenuous exercise. However, the University of Maryland Medical Center (2011) states that there is 'no evidence that it enhances "exercise performance".' L-carnitine assists in converting fat to energy and, in most instances, there is no need for supplementation (University of Maryland Medical Center 2011).

Various forms of carnitine exist, including L-carnitine, Acetyl-L-carnitine and Propionyl-L-carnitine. Conditions for which L-carnitine is used include heart conditions (i.e. angina, arrhythmias, heart failure and heart attack), high cholesterol, kidney disease, male infertility, chronic fatigue syndrome, short-term memory loss and hyperthyroidism. Persons with these conditions should consult their physicians, as L-carnitine may interact with other drugs. L-carnitine has not proven effective in weight loss, though this is how it is marketed (University of Maryland Medical Center 2011; Yonei et al. 2008).

Ginseng

Ginseng is used as a dietary supplement and has been marketed as a 'healing panacea', purportedly improving memory, relieving stress, increasing energy and resulting in a host of other health benefits. It is widely believed to improve cognition, concentration and memory (Clauson et al. 2008). However, as noted by Lieberman (2001), 'there is clearly a lack of definitive evidence demonstrating any effects of ginseng on human physical and mental performance or any parameter related to perceived energy' (p. 94). Similarly, Clauson et al. (2008) contend that ginseng provides no therapeutic effects in the quantities present in a single energy drink serving.

B Vitamins

The B vitamins include B1 thiamine, B2 riboflavin, B3 niacin, B5 pantothenic acid, B6, B7 biotin, B12 and folic acid. B vitamins come from protein and dairy sources as well as green leafy vegetables. According to Pressman & Buff (1997), B vitamins perform the important task of cell communication between the brain and nerves. Beck & Rosenbach (2003) indicate that 'Even marginal deficiencies of the B vitamins have been associated with irritability, depression, and mood changes' (p. 220).

B vitamins are important in promoting heart health, promoting mental health, maintaining healthy blood cells and preventing birth defects (Pressman & Buff 1997). Additionally, B vitamins are vital for energy conversion, the generation of blood cells (Medline Plus 2011), nervous system health and the breakdown of fat, carbohydrates and protein (Dollemore et al. 1995).

Depletion of any one B vitamin impairs the ability of the others to fulfil their essential roles (Pressman & Buff 1997). For example, B1 (thiamine) is essential for nerve growth, memory and other mental functions (Pressman & Buff 1997), while B12 and folate are critical for childhood brain development (Black 2008).

B vitamins appear to provide some protection against elevated levels of homocysteine, an amino acid associated with cardiovascular disease and cognitive decline (Tucker et al. 2005). A study of British children aged 4–18 provides evidence that as children age, decreases in folate, B6, B12 and riboflavin levels correlate with increased homocysteine levels. Health status in adulthood mirrors health status in childhood and thus, early preventive behaviours are key (Kerr et al. 2009).

Benefits of Energy Drinks

Benefits of energy drinks are few and far between. College students have found energy drinks beneficial in overcoming exhaustion (Malinauskas et al. 2007). Similarly, Reyner & Horne (2002) indicate that energy drinks are efficacious for countering sleepiness while driving.

While energy drinks provide a quick burst of 'energy' through stimulation of the central nervous system, it is short-lived. Nonetheless, energy drinks have been demonstrated to improve mood, cognition and psychomotor activity (Alford et al. 2001; Seidl et al. 2000). One study indicated that the combination of sugar and caffeine resulted in improved secondary memory and speed of attention (Scholey & Kennedy 2004). Despite these findings, risks associated with energy drinks arguably outweigh benefits of the beverages (Seifert et al. 2011).

Risks Associated with Energy Drinks

Energy drinks are more than the sum of their constituents. The claim that they provide 'energy' is a fallacy. Many of the ingredients in energy drinks are stimulants that act synergistically (Bestervelt n.d.). Bestervelt (n.d.) indicates that this interaction could result in 'serious cardiovascular issues', as well as adverse reactions when combined with medications (p. 2). While the safety of energy drinks is uncertain, use of the beverages over time may result in cardiac irregularities (Higgins et al. 2010). Moreover, the health status of those with underlying medical conditions may be exacerbated, particularly for those who are unaware that they have a medical condition (Bestervelt n.d.).

Surprisingly, the amount of caffeine in energy drinks is not regulated (Babu et al. 2008). More troubling is that the impact of caffeine on children is still largely unknown (Temple 2009). In a recent study of 100 adolescents, 85 per cent had consumed caffeine; nearly half of this caffeine consumption occurred after school, and less than 20 per cent of the adolescents received the recommended amount of sleep (Calamaro et al. 2009). While many people consume caffeine to become more alert, it actually can have the opposite effect, with resultant sleep deprivation leading to greater exhaustion, and consequently, a self-perpetuating cycle of caffeine consumption and exhaustion (Millman 2005).

In addition to producing exhaustion, lack of sleep resulting from caffeine-containing energy drinks poses threats to cognitive performance. Memory represents an element of cognitive performance that potentially is influenced by caffeine-containing energy drinks. Research regarding the impact of caffeine-containing energy drinks on memory is conflicting. For example, Bichler et al. (2006) conducted a study in which college students were administered either a placebo or a pill containing caffeine and taurine. Students exposed to pills containing caffeine and taurine did not experience significant changes in short-term memory. To the contrary, Scholey & Kennedy (2004) examined effects of caffeine-containing energy drinks on cognitive performance among college students and discovered significant improvements in immediate word recall, delayed word recall and reaction time (i.e. 'speed of attention').

Not surprisingly, the physiological mechanisms through which energy drinks influence cognitive performance and more specifically, sleep patterns, parallel those which influence the stress response. Essentially, the quality and quantity of sleep are disrupted when caffeine binds to adenosine receptors in the brain. Not only does this process result in wakefulness, but it also inhibits blood vessels in the brain from dilating. Under normal circumstances, adenosine binds to receptors allowing blood vessels to dilate and enhance the flow of oxygen-rich blood to the brain. The flow of oxygen-rich blood to the brain creates an environment conducive to sleep (Fredholm et al. 1999; Seifert et al. 2011).

One health habit that is critical to student success and is compromised from energy drink consumption is attainment of sufficient sleep. Many external factors influence sleep changes in adolescents. Early school start times, employment and extracurricular activities frequently result in sleep deprivation among adolescents (Millman 2005). Sleep deprivation can have serious consequences for health and educational success (Calamaro et al. 2009). Roberts et al. (2010) note that 'adolescents with disturbed sleep report more depression, anxiety, anger, inattention and conduct problems, drug and alcohol use, impaired academic performance, and suicidal thoughts and behaviors. They also have been reported to have more fatigue, less energy, worse perceived health and symptoms such as headaches, stomachaches and backaches' (p. 1046).

While Riesenhuber et al. (2006) found that energy drinks have proven beneficial in maintaining alertness while driving, the caffeinated sleep deprivation cycle may lead to chronic exhaustion. Studies have indicated that sleep deprivation among adolescents puts them at increased risk for automobile accidents (Calamaro et al. 2009; Millman 2005; Pizza et al. 2010). Sleep deprivation has also been linked to weight gain and metabolic syndrome (Bass & Turek 2005; Lumeng et al. 2007; Spiegel et al. 1999; Vorona et al. 2005).

Combining caffeine-driven sleep deprivation with sugar found in energy drinks may contribute to the development of obesity. Even sugar-free varieties of energy drinks may predispose adolescents and younger adults to consume sugary food and drinks. While caffeinated sleep deprivation may pose a risk for obesity, recent studies indicate that those who consume caffeine regularly have a reduced risk of developing type 2 diabetes (Heckman et al. 2010; Seifert et al. 2011).

In addition to obesity, energy drinks may pose a risk for individuals with eating disorders. Our culture places high value on a slim figure; adolescents and younger adults are very conscious of this ideal and sometimes take measures such as foregoing meals. In a recent study, Attila & Cakir (2011) found that college students who consumed energy drinks were more likely to skip breakfast than students who did not consume energy drinks. Similarly, persons with anorexia or bulimia consume caffeine to control appetite (Burgalassi et al. 2009; Krahn et al. 1991; Seifert et al. 2011). Persons with eating disorders are particularly at risk of the adverse effects of caffeine given their greater risk for cardiac abnormalities and electrolyte imbalances (Seifert et al. 2011).

Recent studies have found a strong relationship between energy drink and alcohol consumption and non-medical use of prescription drugs among college students (Arria et al. 2010, 2011). In one study, college students with high frequency energy drink consumption were nearly two and a half times as likely to be alcohol dependent, while those who imbibed energy drinks less frequently were nearly twice as likely to be alcohol dependent; these findings held true while controlling for membership in a fraternity/sorority, parental alcoholism, depression and normal alcohol consumption (Arria et al. 2011).

Energy drinks may pose additional risks for those with mental health issues (Chelben et al. 2008). Specifically, those seeking relief from mental illness frequently self-medicate, and several of the ingredients in energy drinks are 'euphoria-inducing'. For example, the combination of ginseng, taurine and caffeine purportedly mitigate 'physical and mental fatigue' for individuals afflicted with mental health issues, such as depression (Chelben et al. 2008, p. 187). The 'euphoria-inducing' nature of energy drinks can be detrimental to individuals suffering from bipolar disorder who are prone to manic episodes (Clauson et al. 2008).

While some individuals afflicted with mental illness self-medicate, others are prescribed medications for their conditions. For example, many adolescents and younger adults are prescribed medication for attention deficit hyperactivity disorder (ADHD). ADHD drugs are stimulants, as are ingredients in energy drinks, thus placing adolescents and younger adults who consume energy drinks at risk of increased blood pressure, a more rapid heart rate (Seifert et al. 2011), and ultimately, cardiac pathology (Elia & Vetter 2010).

In extreme instances, death can result from the consumption of energy drinks. Clauson et al. (2008) reported four deaths and four cases of seizures associated with caffeine in energy drinks. Additionally, energy drinks were suspected of causing the death of a teenage Irish basketball player (Finnegan 2003) and three Swedish youths, two of whom combined energy drinks with alcohol (Finnegan 2003). In Australia, a young man suffered cardiac infarct after consuming large amounts of an energy drink throughout the day, while involved in motocross. The infarct is hypothesised to have resulted from the combination of energy drinks with extreme exertion (Berger & Alford 2009). The bottom line is that energy drinks have not been determined to be safe, and the effects of energy drink consumption over time are unknown (Higgins et al. 2010).

The Stress Response

Throughout history, humans have experienced and responded to stress on a daily basis. Stress can be acute (i.e. it emerges, is intense and promptly dissipates) or chronic (i.e. it emerges, is less intense and remains for an extended time frame) and can manifest from many different sources (Seaward 2009). Sources of stress for adolescents and younger adults include, but are not limited to, relationship difficulties, school and/or work activities, body image, family issues and peer pressure. The human body is masterfully designed to respond not only to sources like these, but also to immediate threats, or stressors, through a systematic chain of events commonly referred to as 'fight or flight'.

The 'fight or flight' response occurs when individuals are confronted with stimuli that signal the brain to take action or, in the case of a non-threatening event, resume normal activity (Cannon 1932). When a stressor, or perceived threat, signals the brain to take action, the central nervous system (brain and spinal cord) coordinates a host of physiological responses. The peripheral nervous system, a complex network of sensory and motor neurons, operates in conjunction with the central nervous system (i.e. the brain and spinal cord). Sensory neurons transmit messages from the five senses (in response to stimuli in the physical environment) to the central nervous system, and motor neurons transmit messages from the central nervous system to muscle fibres, organs and other tissues. The peripheral nervous system includes the autonomic nervous system, which is divided into two branches – parasympathetic and sympathetic. These branches represent key centres of activity for the stress response. Specifically, the parasympathetic branch enables the body to return to homeostasis after activities of the sympathetic branch have run their course in response to a stressful situation. Activities of the sympathetic branch include decreased digestive activity (salivation, intestinal secretions, etc.) and blood flow to inactive muscles, and increased perspiration, sodium retention, heart rate, blood pressure, circulation of oxygenated blood to skeletal muscles, glucose production, bronchiole dilation and pupil dilation (Blonna 2005; Seaward 2009).

The aforementioned activities occur in accordance with the release of stress hormones including epinephrine (adrenaline), norepinephrine (noradrenaline), aldosterone and cortisol. Not only are epinephrine and norepinephrine released immediately (2–3 seconds) through direct actions of the hypothalamus and sympathetic nervous system, but they are also released intermediately (20–30 seconds) from the adrenal medulla through the bloodstream upon activation of the posterior hypothalamus. Aldosterone and cortisol, a hormone known for its prolonged effects on the body, are released through the adrenal cortex, another component of the adrenal gland (Allen 1983). Specifically, aldosterone and cortisol are released through the hyphothalamic-pituitary-adrenal (HPA) axis, a less rapid hormonal response to stress (minutes, hours, days or weeks) than the sympathetically mediated pathway. The HPA axis is initiated when the anterior hypothalamus releases corticotrophin-releasing factor (CRF), which signals the anterior pituitary gland to emit adrenocorticotropic hormone (ACTH) through the bloodstream. ACTH, in turn, signals the adrenal cortex to emit aldosterone and cortisol, resulting in the complex array of stress-related activities described above (Seaward 2009).

Stress and Energy Drink Consumption

Stress represents a common and unavoidable experience of the human race. While humans share selected physiological and emotional responses to stress, their behavioural responses to stress vary. Some people respond to stress in ways that are health enhancing (e.g. exercise), while others respond to stress in ways that are health compromising (e.g. consumption of food, alcohol, drugs, etc.). Energy drink consumption represents an emerging behavioural response to perceived stress (Pettit & DeBarr 2011).

The precise relationship between stress and energy drink consumption is complex and, consequently, subject to scientific inquiry. From a metaphorical perspective, the relationship between stress and energy drink consumption poses a conundrum. Specifically, the notion of which phenomenon precedes the other is debatable (Pettit & DeBarr 2011). On one hand, individuals may consume energy drinks to alleviate fatigue resulting from physical and emotional stress. On the other hand, the physiological effects of caffeine, the leading ingredient in energy drinks, mimic activities of the stress response (e.g. increased heart rate, blood pressure and blood glucose) (Seaward 2009). According to Seaward (2009), stimulation of the sympathetic nervous system and production of stress-related hormones that result from caffeine consumption lead to 'a heightened state of alertness, which makes [an] individual more susceptible to perceived stress' (p. 497).

The physiological effects of energy drink consumption stem from the interplay of caffeine and the central nervous system. Not only does caffeine purportedly stimulate dopamine levels in the brain to evoke a sense of pleasure, but it also stimulates the pituitary gland, an instrumental catalyst of the stress response. Adenosine, a compound abundantly present in the basal forebrain, binds to receptors, inhibits nerve cell action and prepares the body for sleep. Caffeine interferes with this process by binding to adenosine receptors and signalling the pituitary gland to initiate activities of the stress response. Specifically, caffeine stimulates the pituitary gland to emit ACTH and, ultimately, cortisol (Al'Absi et al. 1998; Fredholm et al. 1999; Seifert et al. 2011).

From a scientific perspective, evidence suggests that the relationship between stress and energy drink consumption potentially is bi-directional. In a recent study, Pettit & DeBarr (2011) found a positive relationship between perceived stress and energy drink consumption. They also found a negative relationship between academic performance (GPA) and energy drink consumption. Specifically, an inverse relationship existed between academic performance and the largest number of energy drinks consumed on any occasion during the past 30 days. This finding suggests that energy drink consumption may interfere with cognitive performance. The researchers attributed the latter finding to 'students' propensity to procrastinate and consume more energy drinks when preparing for stressful events such as exams or deadlines for major projects' (p. 339). They also indicated that students are more likely to succeed academically when they practise regular study (Willingham 2002) and health (Trockel et al. 2000) habits.

Marketing

Despite the adverse health effects of energy drinks, the energy drink industry has made a concerted effort to market its products to the public. Vendors often place energy drinks next to sports drinks, which may mislead people who already are confused about differences between the two types of beverages (Higgins et al. 2010). The lack of key minerals for balancing fluid and electrolytes, and the presence of caffeine and greater calories in energy drinks, differentiate them from sports drinks (Meadows-Oliver & Ryan-Krause 2007). One of the most troubling aspects of the energy drink market is the appeal to adolescents and younger adults, particularly their sense of adventure and daring. Energy drinks primarily are marketed to teens and younger males (Agriculture and Agri-Food Canada 2008, 2009; Reissig et al. 2009) as performance enhancing. Highly aggressive promotions connect energy drinks to celebrities and sports (Babu et al. 2008).

Some countries have responded to aggressive marketing campaigns on behalf of the energy drink industry. For example, Agriculture and Agri-Food Canada (2010) indicates that there has been an 'attitude shift' and 'backlash' due to health concerns posed by energy drinks. In Canada, energy drinks are beginning to attain the stigma normally reserved for tobacco products due to a perception that adolescents and younger adults are being targeted by marketing campaigns (Agriculture and Agri-Food Canada 2010).

Regulation

In light of the adverse health consequences and widespread marketing of energy drinks, regulation of the beverages has emerged as an important public health priority for many countries. Coffee, a beverage seen as a legal social drug (Topik 2009), is increasingly available on high school campuses. According to Agriculture and Agri-Food Canada (2010), 'There has been talk of an evolving new generation of consumers that will replace coffee with energy drinks as their primary caffeine delivery method'. The question becomes will we see energy drinks enter the schools, at a time when soft drink vending machines are either removed from the schools or access to them is restricted? In Canada, some are calling for stricter regulation of energy drinks, including age requirements for purchase, label requirements, marketing regulations and limiting sales to particular venues (Agriculture and Agri-Food Canada 2010).

The use of energy drinks is reminiscent of the concern evoked by ephedra, a controversial stimulant. Ephedra also was once readily available in over-the-counter products and, similar to energy drinks, promised and delivered increased energy, endurance and weight loss (Lieberman 2001). Ephedra was suspect in several deaths (Lieberman 2001), and was banned by the US Food and Drug Administration (FDA) in 2004 due to its adverse outcomes (Seamon & Clauson 2005). Although the ban was briefly overturned, the FDA pressed the issue. The Nutraceuticals Corporation argued that ephedra is a 'natural' product present in the Chinese herb ma huang and, therefore, should be considered a dietary supplement (Seamon & Clauson 2005). Ultimately, ephedra was banned once again in 2006 (US Food and Drug Administration 2006). Like ephedra, energy drinks lay claim to being comprised of 'natural' ingredients.

The USA has one of the least restrictive regulatory environments with respect to energy drinks. According to Heckman et al. (2010), 'In the United States, energy drink companies have no limitations over the caffeine content of their beverages because the FDA has placed no restrictions on an upper caffeine limit in these types of beverages' (p. 315). In 2008, a group of scientists and physicians petitioned the FDA to regulate energy drinks by limiting and labelling their amounts of caffeine and requiring warning labels for them (Weise 2008). That same year, Kentucky, Maine and Michigan sought to ban sales of products with large doses of caffeine to minors ('Pressure on energy' 2009).

To date, the only successful regulation of energy drinks in the USA has targeted alcoholic energy drinks. In November 2010, following investigation of research regarding harmful and adverse effects of caffeinated alcoholic beverages, the FDA targeted selected beverage companies to address safety violations and banned alcoholic energy drinks (Leinwand 2010; US Food and Drug Administration 2010). Interestingly, alcoholic energy drinks are legal in Canada, where it was determined that as long as caffeine present in alcoholic energy drinks is from natural sources, it is acceptable (Howland et al. 2010; Schmidt 2010).

Several countries have instituted regulations to address energy drink consumption. For example, in Norway energy drinks only can be purchased in pharmacies, while the Argentineans attempted to prevent the sale of energy drinks in clubs (Oddy & O'Sullivan 2010). Additionally, energy drinks were banned in Turkey, Uruguay, Demark and France, with France later adhering to the European Union and its regulations. Sweden warns consumers not to combine energy drinks with exercise or alcohol, and prohibits children under 15 from purchasing energy drinks (Oddy & O'Sullivan 2010; Seifert et al. 2010). Similarly, as noted by Oddy & O'Sullivan (2010), 'In the United Kingdom, the stimulant drinks committee recommended that labels on energy drinks should state that they are unsuitable for children (< 16 years), pregnant or lactating women, and people who are sensitive to caffeine' (p. b5268).

In response to growing concerns among public health professionals and consumers, the energy drink industry has found loopholes to circumvent regulatory authority over its products.

In Australia and New Zealand, energy drink companies were able to avoid regulation by claiming that energy drinks are 'dietary supplements' and, therefore, not subject to regulation (Oddy & O'Sullivan 2010, p. b5268). These examples arguably are indicative of considerable disagreement with respect to the safety of energy drinks.

Future Implications

As noted by Scholey & Kennedy (2004), the combination of caffeine and sugar found in energy drinks has the capacity to improve cognition (see chapters 8 and 4). Specifically, enhanced reaction time, attention and memory have been associated with consumption of caffeine and sugar-containing energy drinks. Despite these documented benefits, caffeine, when paired with sugar, may make children and adolescents more susceptible to conditioned preferences for sugar-laden drinks and foods (Temple 2009).

The advertising of energy drinks has emphasised their purported claims 'to rapidly increase energy, endurance, and performance' (Seifert et al. 2011, p. 515). Currently, lack of regulation of energy drinks presents an opportunity for marketers to target adolescents and younger adults, particularly younger males interested in performance enhancement and stimulant-related effects (Reissig et al. 2009). Lack of regulation of energy drinks also presents challenges for public health professionals. Most notably, concentrations of selected ingredients vary among energy drink brands, thus potentially contributing to confusion and deception among consumers. Moreover, selected ingredients (e.g. caffeine) in energy drinks are addictive in nature and, thereby, comparable to alcohol, tobacco and other gateway substances that appeal to at-risk adolescents and younger adults (Bernstein et al. 2002; Temple 2009). Cross-sensitisation can occur where sensitivity to one drug increases sensitivity 'to other drugs acting on the same neurobiological site' (Temple 2009, p. 798). According to Temple (2009), this is one of the reasons why caffeine could be considered a 'gateway' drug, possibly leading to use of other drugs with potential for abuse. Not only do gateway substances individually elicit adverse physiological effects, but they also potentially encourage experimentation with equally or more powerful substances that individually or synergistically compromise health and well-being (Degenhardt et al. 2010).

The combination of energy drinks with other addictive substances (e.g. alcohol) represents an emerging health risk behaviour among adolescents and younger adults. This trend especially is noteworthy because research has indicated that individuals who engage in one health risk behaviour (e.g. energy drink consumption) are more likely to engage in multiple health risk behaviours (e.g. binge drinking). For example, in a sample of college students, Miller (2008a) found an association between energy drink consumption and selected detrimental health behaviours, including failure to wear a safety belt, risky sexual practices, physical violence, and use of alcohol, tobacco, marijuana and illicit prescription drugs. She posits that these behaviours may be indicative of a desire for 'sensation seeking' (p. 495).

Individuals who combine energy drinks with alcohol are more likely to consume alcohol in greater quantities than those who only consume alcohol. It has been established that those who habitually consume caffeinated alcoholic beverages are at greater risk of developing alcoholism (National Council on Alcohol and Drug Dependence 2010; Oteri et al. 2007). Like energy drink consumption, alcoholic energy drink consumption is most common among males, particularly college males (O'Brien et al. 2008). Not surprisingly, this trend coincides with binge drinking, a behaviour that disproportionately occurs among males (Harrell & Karim 2008).

A National Coalition on Alcohol and Drug Dependence (2010) report indicates that caffeinated alcoholic beverages effectively result in 'wide awake drunk' consumers who perceive themselves to be less intoxicated than they in fact are. These findings are supported by a naturalistic study conducted in seven college bars to determine the impact of combining energy drinks with alcohol. Patrons who combined energy drinks with alcohol perceived themselves to be less impaired than they actually were, drank more and were four times more likely to express an intention to drive impaired than those who did not consume energy drinks (Thombs et al. 2009). It is this perception that may lead to driving while intoxicated and resultant accidents (Oteri et al. 2007).

To date, the role of perceived stress in relation to alcoholic energy drink consumption is tenuous. Similarly, a dearth of literature exists to explain gender disparities in energy drink consumption (including alcoholic energy drink consumption). At least one researcher has speculated reasons for gender disparities regarding energy drink consumption. Miller (2008b) identified 'jock identity, conformity to masculine norms, and risk-taking behaviour' as potential reasons for increased energy drink consumption among college males (p. 485). As energy drink consumption continues to develop as a trend among adolescents and younger adults, additional reasons for gender disparities are likely to surface.

To gain insight into the precise relationships among perceived stress, cognitive performance and energy drink consumption, an examination of the dose-response relationship is necessary. Specifically, research is needed to examine which doses of selected energy drinks correspond to optimal cognitive performance and which doses counteract proposed cognitive benefits of the beverages by eliciting symptoms of the stress response. Research is also needed to determine appropriate doses of energy drink constituents and which constituent(s) work most effectively in combination to render optimal cognitive performance.

According to Seifert et al (2011), 'Adults who consume low-to-moderate amounts of caffeine (1–3 mg/kg or 12.5–100 mg/day) have improved exercise endurance, cognition, reaction time, and mood with sleep deprivation' (p. 518). Unfortunately, these cognitive benefits are counteracted by the exorbitant caffeine and sugar concentrations in many energy drinks (Malinauskas et al. 2007). For example, an inverse relationship exists between reaction time and caffeine consumption in children (Seifert et al. 2011). This relationship is noteworthy to parents, teachers, coaches and other parties who have a vested interest in the health and academic success of youth.

Conclusions

This chapter has explored energy drink consumption as it relates to perceived stress, cognitive performance and potential for adverse health effects. Energy drink consumption represents an emerging trend, particularly among male adolescents and younger adults (Miller 2008b). The ubiquitous presence of energy drinks in convenience stores, grocery stores and other public venues has created a culture conducive to their usage. Increased access to and availability of energy drinks has sparked the attention of the media, the public and researchers interested in the public's health. Potential effects and regulation of energy drinks are of particular concern to the latter group because of common ingredients in the beverages.

Primary ingredients in energy drinks include, but are not limited to, carbohydrates, taurine, glucoronolactone and caffeine (Ivy et al. 2009). While caffeine concentrations in leading energy drink brands vary, the caffeine content in many energy drinks overwhelmingly exceeds that which comprises coffee and popular soft drinks (McCusker et al. 2006). Caffeine, a well-known stimulant, has been associated with a myriad of physical, mental and emotional effects, many of which relate to stress.

When combined with taurine and other ingredients commonly found in energy drinks, caffeine reportedly has resulted in favourable cognitive effects, including improved reaction time, enhanced attention in stressful situations and enhanced secondary memory (i.e. immediate and delayed word recall) (Scholey & Kennedy 2004; Seidl et al. 2000). However, the latter finding was contradicted by Bichler et al. (2006), who found no effect of caffeine/taurine consumption on short-term memory. Positive effects of caffeine-containing energy drinks are further contradicted by research linking them to increased blood glucose (Scholey & Kennedy 2004), blood pressure and heart rate (American Heart Association 2007) – distinctive characteristics of the stress response (Seaward 2009).

The precise relationship between stress and energy drink consumption presents a conundrum. Pettit & DeBarr (2011) found a positive relationship between perceived stress and energy drink consumption, however, their findings were inconclusive. While individuals may consume energy drinks to combat fatigue from physical (e.g. athletic performance) or emotional (e.g. exam preparation) stress, ingredients in energy drinks may induce perceived stress.

References

Agriculture and Agri-Food Canada. (2008). The energy drink segment in North America. Retrieved from http://www.ats.agr.gc.ca/info/4387-eng.htm

Agriculture and Agri-Food Canada. (2009). Market update: Energy drinks in North America. Retrieved from http://www.ats.agr.gc.ca/info/5234-eng.htm

Agriculture and Agri-Food Canada. (2010). Energy and relaxation drinks in North America: A changing landscape. Retrieved from http://www.ats.agr.gc.ca/info/5579-eng.htm

Al'Absi, M., Lovallo, W. R., McKey, B., Sung, B. H., Whitsett, T. L., & Wilson, M. F. (1998). Hypothalamic-pituitary-adrenocortical responses to psychological stress and caffeine in men at high and low risk for hypertension. *Psychosomatic Medicine, 60,* 521–527.

Alford, C., Cox, H., & Wescott, R. (2001). The effects of Red Bull energy drink on human performance and mood. *Amino Acids, 21,* 139–150.

Allen, R. (1983). *Human stress: Its nature and control.* Minneapolis, MN: Burgess.

American Heart Association. (2007). Energy drinks may pose risks for people with high blood pressure, heart disease. Retrieved from http://americanheart.mediaroom.com/index.php?s=43&item=206

Arria, A. M., Caldeira, K. M., Kasperski, S. J., O'Grady, K., Vincent, K. B., Griffiths, R. R., & Wish, E. (2010). Increased alcohol consumption, nonmedical prescription drug use, and illicit drug use are associated with energy drink consumption among college students. *Journal of Addiction Medicine, 4*(2), 74–80.

Arria, A. M., Caldeira, K. M., Kasperski, S. J., Vincent, K. B., Griffiths, R. R., & O'Grady, K. E. (2011). Energy drink consumption and increased risk for alcohol dependence. *Alcoholism: Clinical and Experimental Research, 35*(2), 365–375.

Attila, S., & Cakir, B. (2011). Energy drink consumption in college students and associated factors. *Nutrition, 27,* 316–322.

Astorino, T. A., & Roberson, D. W. (2010). Efficacy of acute caffeine ingestion for short-term high-intensity exercise performance: A systematic review. *Journal of Strength and Conditioning Research, 24*(1), 257–265.

Babu, K. M., Church, R. J., & Lewander, W. (2008). Energy drinks: The new eye-opener for adolescents. *Clinical Pediatric Emergency Medicine, 9*(1), 35–42.

Bass, J., & Turek, F. (2005). Sleepless in America: A pathway to obesity and the metabolic syndrome. *Archives of Internal Medicine, 165,* 15–16.

Baum, M., & Weiss, M. (2001). The influence of a taurine containing drink on cardiac parameters before and after exercise measured by echocardiography. *Amino Acids, 20*(1), 75–82.

Beck, L., & Rosenbach, A. V. (2003). *The ultimate nutrition guide for women: How to stay healthy with diet, vitamins, minerals, and herbs.* Hoboken, NJ: John Wiley and Sons, Inc.

Berger, A. J., & Alford, K. (2009). Cardiac arrest in a young man following excess consumption of caffeinated "energy drinks." *Medical Journal of Australia, 190*(1), 41–43.

Bernstein, G. A., Carroll, M. E., Thuras, P. D., Cosgrove, K. P., & Roth, M. E. (2002). Caffeine dependence in teenagers. *Drug and Alcohol Dependence, 66,* 1–6.

Bestervelt, L. (n.d.). Raising the red flag on some energy drinks. Retrieved from http://www.nsf.org/business/athletic_banned_substances/energy_drinks.pdf

Bichler, A., Swenson, A., & Harris, M. A. (2006). A combination of caffeine and taurine has no effect on short-term memory but induces changes in heart rate and mean arterial blood pressure. *Amino Acids, 31*(4), 471–476.

Black, M. M. (2008). Effects of vitamin B12 and folate deficiency on brain development in children. *Food and Nutrition Bulletin, 29*(2 Suppl.), S125–S131.

Blonna, R. (2005). *Coping with stress in a changing world* (3rd ed.). New York: McGraw-Hill.

Botvin, G. J., Griffen, K. W., Diaz, T., Scheier, L. M., Williams, C., & Epstein, J. A. (2000). Preventing illicit drug use in adolescents: Long-term follow-up data from a randomized control trial of a school population. *Addictive Behaviors, 25*(5), 769–774.

Braganza, S., & Larkin, M. (2007). Riding high on energy drinks. *Contemporary Pediatrics, 24*(5), 61–73.

Bremer, A. A., Auinger, P., & Byrd, R. S. (2010). Sugar-sweetened beverage intake trends in US adolescents and their association with insulin resistance-related parameters. *Journal of Nutrition and Metabolism, 2010.* Retrieved from http://downloads. hindawi.com/journals/jnume/2010/196476.pdf. doi:10.1155/2010/196476

Brunye, T. T., Mahoney, C. R., Lieberman, H. R., & Taylor, H. A. (2010). Caffeine modulates attention network function. *Brain and Cognition, 72*, 181–188.

Burgalassi, A., Ramacciotti, C. E., Bianchi, M., Coli, E., Polese, L., Bondi, E. & Dellosso, L. (2009). Caffeine consumption among eating disorder patients: Epidemiology, motivations, and potential of abuse. *Eating and Weight Disorders, 14*(4), e212–e218.

Calamaro, C. J., Mason, T. B. A., & Ratcliffe, S. J. (2009). Adolescents living the 24/7 lifestyle: Effects of caffeine and technology on sleep duration and daytime functioning. *Pediatrics, 123*(6), e1005–e1010.

Cannon, W. (1932). *The wisdom of the body.* New York: W. W. Norton.

Chelben, J., Piccone-Sapir, A., Ianco, I., Shoenfeld, N., Kotler, M., & Strous, R. (2008). Effects of amino acid energy drinks leading to hospitalization. *General Hospital Psychiatry, 30*(2), 187–189.

Clauson, K. A., Shields, K. M., McQueen, C. E., & Persad, N. (2008). Safety issues associated with commercially available energy drinks. *Journal of the American Pharmacists Association, 48*(3), e55–e67.

Curry, K., & Stasio, M. J. (2009). The effects of energy drinks alone and with alcohol on neuropsychological functioning. *Human Psychopharmacology: Clinical and Experimental, 24*, 473–481.

Degenhardt, L., Dierker, L., Chiu, W. T., Medina-Mora, M. E., Neumark, Y., Sampson, N., & Kessler, R. C. (2010). Evaluating the drug use "gateway" theory using cross-national data: Consistency and associations of the order of initiation of drug use among participants in the WHO World Mental Health Surveys. *Drug and Alcohol Dependence, 108*(1–2), 84–97.

Dollemore, D., Giuliucci, M., Haigh, J., Kirchheimer, S., & Callahan, J. (1995). *New choices in natural healing.* Emmaus, PA: Rodale Press, Inc.

Drinking to win. (1991). *The Lancet, 338*(8772), 940–941.

Dugdale, D. C. (2 May 2009). Caffeine in the diet. Retrieved from http://www.nlm. nih.gov/medlineplus/ency/article/002445.htm

Elia, J. & Vetter, V. L. (2010) Cardiovascular effects of medication for the treatment of attention-deficit hyperactivity disorder: What is known and how should it influence prescribing in children? *Pediatric Drugs, 12*(3), 165–175.

Energy Fiend. (n.d.) Energy drink ingredients. Retrieved from http://www. energyfiend.com/energy-drink-ingredients

Ferguson, L. R. (2009). Nutrigenomics approaches to functional foods. *Journal of the American Dietetic Association, 109*(3), 452–458.

Ferreira, S. E., de Mello, M. T., Rossi, M. V., & Souza-Formigoni, M. L. (2004). Does an energy drink modify the effects of alcohol in a maximal effort test? *Alcoholism: Clinical and Experimental Research, 28*(9), 1408–1412.

Finnegan, D. (2003). The health effects of stimulant drinks. *Nutrition Bulletin, 28*, 147–155.

Franconi, F., Di Leo, M. A. S., Bennardini, F., & Ghirlanda, G. (2004). Is taurine beneficial in reducing risk factors for diabetes mellitus? *Neurochemical Research, 29*(1), 143–150.

Frary, C. D., Johnson, R. K., & Wang, M.Q. (2005). Food sources and intakes of caffeine in the diets. *Journal of the American Dietetic Association, 105,* 110–113.

Fredholm, B. B., Battig, K., Holmen, J., Nehlig, A., & Zvartau, E. E. (1999). Actions of caffeine in the brain with special reference to factors that contribute to its widespread use. *Pharmacological Reviews, 51*(1), 83–133.

Gold, S. M., Dziobek, I., Sweat, V., Tirsi, A., Rogers, K., Bruehl, H., et al. (2007). Hippocampal damage and memory impairments as possible early brain complications of type 2 diabetes. *Diabetologia, 50*(4), 711–719.

Greenberg, J. A., Boozer, C. N., & Geliebter, A. (2006). Coffee, diabetes, and weight control. *American Journal of Clinical Nutrition, 84,* 682–693.

Harrell, Z. A., & Karim, N. M. (2008). Is gender relevant only for problem alcohol behaviors? An examination of correlates of alcohol use among college students. *Addictive Behaviors, 33*(2), 359–365.

Heckman, M. A., Weil, J., & de Meija, E. G. (2010). Caffeine (1, 3, 7-trimethylxanthine) in foods: A comprehensive review on consumption, functionality, safety, and regulatory matters. *Journal of Food Service, 75*(3), R77–R87.

Higgins, J. P., Tuttle, T. D., & Higgins, C. L. (2010). Energy beverages: Content and safety. *Mayo Clinic Proceedings, 85*(11), 1033–1041.

Howland, J., Rohsenow, D. J., Arnedt, J. T., Bliss, C. A., Hunt, S. K., Calise, T. V., et al. (2010). The acute effects of caffeinated versus non-caffeinated alcoholic beverages on driving performance and attention/reaction time. *Addiction, 106,* 335–341.

Ivy, J. L., Kammer, L., Ding, Z., Wang, B., Bernard, J. R., & Liao, Y. (2009). Improved cycling time-trial performance after consumption of a caffeine energy drink. *International Journal of Sport Nutrition and Exercise Metabolism, 19*(1), 61–78.

Johns Hopkins Bayview Medical Center. (n.d.). Information about caffeine dependence. Retrieved from http://www.caffeinedependence.org/caffeine_dependence.html#addiction

Kennedy, D. O., Haskell, C. F., Robertson, B., Reay, J., Brewster-Maund, C., Luedemann, J., Scholey, A. B. (2008). Improved cognitive performance and mental fatigue following a multi-vitamin and mineral supplement with added guaraná (Paullinia cupana). *Appetite, 50*(2–3), 506–513.

Kerai, M. D., Waterfield, C. J., Kenyon, S. H., Asker, D. S., & Timbrell, J. A. (1998). Taurine: Protective properties against ethanol-induced hepatic steatosis and lipid peroxidation during chronic ethanol consumption in rats. *Amino Acids, 15*(1–2), 53–76.

Kerrigan, S., & Lindsey, T. (2005). Fatal caffeine overdose: Two case reports. *Forensic Science International, 153*(1), 67–69.

Knight, C. A., Knight, I., Mitchell, D. C., & Zepp, J. E. (2004). Beverage caffeine intake in US consumers and subpopulations of interest: Estimates from the Share of Intake Panel survey. *Food and Chemical Toxicology, 42,* 1923–1930.

Kerr, M. A., Livingstone, B., Bate, C. J., Bradbury, I., Scott, J. M., Ward, M., et al. (2009). Folate, related B vitamins, and homocysteine in childhood and adolescence: Potential implications for disease risk in later life. *Pediatrics, 123*(2), 627–635.

Krahn, D. D., Haase, S., Ray, A., Gosnell, B., & Drewnowski, A. (1991). Caffeine consumption in patients with eating disorders. *Hospital and Community Psychiatry, 42,* 313–315.

Leinwand, D. (17 November 2010). Alcoholic energy drinks targeted. *USA Today,* p. A1.

Libuda, L., Lexy, U., Remer, T., Stehle, P., Shoenau, E., & Kirsting, M. (2008). Association between long-term consumption of soft drinks and variables of bone modeling and remodeling in a sample of healthy German children and adolescents. *American Journal of Clinical Nutrition, 8*(6), 1670–1677.

Lieberman, H. R. (2001). The effects of ginseng, ephedrine, and caffeine on cognitive performance, mood, and energy. *Nutrition Reviews, 59*(4), 91–102.

Lumeng, J. C., Somashekar, D., Appugliese, D., Kaciroti, N., Corwyn, R., & Bradley, R. H. (2007). Shorter sleep duration is associated with increased risk for being overweight at ages 9 to 12 years. *Pediatrics, 120*(5), 1020–1029.

MacDonald, N., Stanbrook, M., & Hebert, P. C. (2010). "Caffeinating" children and youth. *CMAJ, 182*(15), 1597.

Malinauskas, B. M., Aeby, V. G., Overton, R. F., Carpenter-Aeby, T., & Barber-Heidal, K. (2007). A survey of energy drink consumption patterns among college students. *Nutrition Journal, 6*(35). Retrieved from http://www.nutritionj.com/content/6/1/35

McCusker, R. R., Goldberger, B. A., & Cone, E. J. (2006). Caffeine content of energy drinks, carbonated sodas, and other beverages. *Journal of Analytical Toxicology, 30*(2), 112–114.

Meadows-Oliver, M., & Ryan-Krause, P. (2007). Powering up with sports and energy drinks. *Journal of Pediatric Health Care, 21*(6), 413–416.

Medline Plus. (14 February 2011). B vitamins. Retrieved from http://www.nlm.nih.gov/medlineplus/bvitamins.html

Miller, K. E. (2008a). Energy drinks, race, and problem behaviors among college students. *Journal of Adolescent Health, 43*, 490–497.

Miller, K. E. (2008b). Wired: Energy drinks, jock identity, masculine norms, and risk taking. *Journal of American College Health, 56*(5), 481–489.

Millman, R. P. (2005). Excessive sleepiness in adolescents and young adults: Causes, consequences, and treatment strategies. *Pediatrics, 115*, 1774–1786.

Mrvos, R. M., Reilly, P. E., Dean, B. S., & Krenzelok, R. P. (1989). Massive caffeine ingestion resulting in death. *Veterinary and Human Toxicology, 31*(6), 571–572.

National Commission on Alcohol and Drug Dependence (NCADD). (2010). *Washington report, 13*(11/12), p. 1.

National Federation of State High School Associations. (2011). Position statement and recommendations for the use of energy drinks. Retrieved from http://www.nfhs.org/search.aspx?searchtext=Position statement and recommendations for the use of energy drinks

Nawrot, P., Jordan, S., Eastwood, J., Rostein, J., Hugenholtz, A., & Feely, M. (2003). Effects of caffeine on human health. *Food Additives and Contaminants, 20*, 1–30.

Nehlig, A., Daval, J. L., & Debry, G. (1992). Caffeine and the central nervous system: Mechanisms of action, biochemical, metabolic and psychostimulant effects. *Brain Research Reviews, 17*, 139–170.

O'Brien, M. C., McCoy, T. P., Rhodes, S. D., Wagoner, A., & Wolfson, M. (2008). Caffeinated cocktails: Energy drink consumption, high-risk drinking, and alcohol-related consequences among college students. *Academic Emergency Medicine, 15*(5), 453–460.

Oteri, A., Salvo, F., Caputi, A. P., & Calapai, G. (2007). Intake of energy drinks in association with alcoholic beverages in a cohort of students of the school of medicine of the University of Messina. *Alcoholism: Clinical and Experimental Research, 31*(10), 1677–1680.

Oddy, W. H., & O'Sullivan, T. A. (2010). Energy drinks for children and adolescents. *British Medical Journal, 339,* b5268.

Pettit, M. L., & DeBarr, K. A. (2011). Perceived stress, energy drink consumption, and academic performance among college students. *Journal of American College Health, 59*(5), 335–341.

Pizza, F., Contardi, S., Antognini, A. B., Zagoraiou, M., Borrotti, M., Mostacci, B., et al. (2010). Sleep quality and motor vehicle crashes in adolescents. *Journal of Clinical Sleep Medicine, 6*(1), 41–45.

Pollak, C. P., & Bright, D. (2003). Caffeine consumption and weekly sleep patterns in U.S. seventh-, eighth-, and ninth-graders. *Pediatrics, 111*(1), 42–46.

Pressman, A. H., & Buff, S. (1997). *The complete idiot's guide to vitamins and minerals.* New York: Alpha Books.

Pressure on energy drink manufacturers. (2009). *Nutrition Business Journal.* Retrieved from http://newhope360.com/beverage/pressure-energy-drink-manufacturers

Reissig, C. J., Strain, E. C., & Griffiths, R. R. (2009). Caffeinated energy drinks – a growing problem. *Drug and Alcohol Dependence, 99*(1–3), 1–10.

Reyner, L. A., & Horne, J. A. (2002). Efficacy of a "functional energy drink" in counteracting driver sleepiness. *Physiology and Behavior, 101*(3), 331–335.

Riesenhuber, A., Boehm, M., Posch, M., & Aufricht, C. (2006). Diuretic potential of energy drinks. *Amino Acids, 31,* 81–83.

Roberts, R. E., Roberts, C. R., & Duong, H. T. (2010). Sleepless in adolescence: Prospective data on sleep deprivation, health, and functioning. *Journal of Adolescence, 32*(5), 1045–1057.

Ruxton, C. (2009). Health aspects of caffeine: Benefits and risks. *Nursing Standard, 24*(9), 41–48.

Ruxton, C. H. S., Gardner, E. J., & McNulty, H. M. (2010). Is sugar consumption detrimental to health? A review of the evidence. *Critical Review in Food Science and Nutrition, 50,* 1–19.

Schmidt, S. (17 May 2010). Health Canada strikes an uneasy regulatory balance on alcoholic energy drinks. *The Vancouver Sun.* Retrieved from http://www.vancouversun.com/health/Health+Canada+strikes+uneasy+regulatory+balance+alcoholic+energy+drinks/3039397/story.html

Scholey, A. B., & Kennedy, D. O. (2004). Cognitive and physiological effects of an "energy drink": An evaluation of the whole drink and of glucose, caffeine and herbal flavoring. *Psychopharmacology, 176*(3), 320–330.

Schuller-Levis, G. B., & Park, E. (2003). Taurine: New implications for an old amino acid. *FEMS Microbiology Letters, 226,* 195–202.

Seamon, M. J., & Clauson, K. A. (2005). Ephedra: Yesterday, DSHEA, and tomorrow – A ten year perspective on the Dietary Supplement Health and Education Act of 1994. *Journal of Herbal Pharmacotherapy, 5*(3), 67–86.

Seaward, B. L. (2009). *Managing stress: Principles and strategies for health and well-being* (6th ed.). Sudbury, MA: Jones and Bartlett Publishers.

Seidl, R., Peyrl, R., Nicham, R., & Hauser, E. (2000). A taurine and caffeine-containing drink stimulates cognitive performance and well-being. *Amino Acids, 19*(3), 635–642.

Seifert, S. M., Schaecter, S. L., Hershorin, E. R., & Lipshultz, S. E. (2011). Health effects of energy drinks on children, adolescents, and young adults. *Pediatrics, 127*(3), 511–528.

Simon, M., & Mosher, J. (2007). Alcohol, energy drinks, and youth: A dangerous mix. Retrieved from http://www.marininstitute.org/alcopops/resources/EnergyDrinkReport.pdf

Smith, N., & Atroch, A. L. (2010). Guarana's journey from regional tonic to aphrodisiac and global energy drink. *Evidence-Based Complementary and Alternative Medicine, 7*(3), 279–282.

Spiegel, K., Leproult, R., & Van Cauter, E. (1999). Impact of sleep debt on metabolic and endocrine function. *Lancet, 354,* 1435–1439.

Snyder, S. H., & Sklar, P. (1984). Behavioral and molecular actions on caffeine: Focus on adenosine. *Journal of Psychiatric Research, 18*(2), 91–106.

Steinberger, J., & Daniels, S, R. (2003). Obesity, insulin resistance, diabetes, and cardiovascular risk in children: An American Heart Association scientific statement from the Atherosclerosis, Hypertension, and Obesity in the Young Committee (Council on Cardiovascular Disease in the Young) and the Diabetes Committee (Council on Nutrition, Physical Activity, and Metabolism). *Circulation, 107,* 1448.

Temple, J. L. (2009). Caffeine use in children: What we know, what we have left to learn, and why we should worry. *Neuroscience and Biobehavioral Reviews, 33*(6), 793–806.

Thombs, D. L., O'Mara, R. J., Tsukamoto, M., Rossheim, M. E., Weiler, R. M., Merves, M. L., & Goldberger, B. A. (2009). Event-level analyses of energy drink consumption and alcohol intoxication in bar patrons. *Addictive Behaviors, 35*(4), 325–330.

Topik, S. (2009). Coffee as a social drug. *Cultural Critique, 71,* 81–106.

Trockel, M. T., Barnes, M. D., & Egget, D. L. (2000). Health-related variables and academic performance among first-year college students: Implications for sleep and other behaviors. *Journal of American College Health, 49*(3), 125–131.

Tucker, K. L., Qiao, N., Scott, T., Rosenberg, I., & Spiro, A. (2005). High homocysteine and low B vitamins predict cognitive decline in aging men: The Veterans Affairs Normative Aging Study. *American Journal of Clinical Nutrition, 82*(3), 627–635.

University of Maryland Medical Center. (2011). Carnitine (L-carnitine). Retrieved from http://www.umm.edu/altmed/articles/carnitine-l-000291.htm

U.S. Food and Drug Administration. (21 August 2006). FDA statement on tenth circuit's ruling to uphold FDA decision banning dietary supplements containing ephedrine alkaloids. Retrieved from http://www.fda.gov/NewsEvents/Newsroom/PressAnnouncements/2006/ ucm108715.htm

U.S. Food and Drug Administration. (17November 2010). FDA warning letters issued to four makers of caffeinated alcoholic beverages. Retrieved from http://www.fda.gov/NewsEvents/Newsroom/PressAnnouncements/2010/ucm234109.htm

Vorona, R., Winn, M. P., Babineau, T. W., Eng, B. P., Feldman, H. R., & Ware, J. C. (2005). Overweight and obese patients in a primary care population report less sleep than patients with a normal body mass index. *Archives of Internal Medicine, 165,* 25–30.

Wake Forest University Baptist Medical Center. (11 February 2009). Higher blood sugar levels linked to lower brain function in diabetics. Retrieved from http://www.wakehealth.edu/News-Releases/2009/Higher_Blood_Sugar_Levels_Linked_to_Lower_Brain_Function_in_Diabetics,_Study_Shows.htm

Warburton, D. M., Bersellini, E., & Sweeney, E. (2001). An evaluation of a caffeinated taurine drink on mood, memory, and information processing in healthy volunteers without caffeine abstinence. *Psychopharmacology, 158*(3), 322–328.

Weise, E. (22 October 2008). Petition calls for FDA to regulate energy drinks. *USA Today*, p. D6. Retrieved from http://www.usatoday.com/news/health/2008-10-21-energy-drinks_N.htm?csp=34&loc=interstitialskip

Willingham, D. T. (2002). How we learn – Ask the cognitive scientist: Allocating student study time, "massed" versus "distributed" practice. *American Educator, 26*(2), 37–39.

Wolraich, M. L., Wilson, D. B., & White, J. W. (1995). The effect of sugar on behavior or cognition in children. *Journal of the American Medical Association, 274*(20), 1617–1621.

Yonei, Y., Takahashi, Y., Hibino, S., Watanabe, M., & Yoshioka, T. (2008). Effects on the human body of a dietary supplement containing L-carnitine *Garcinia cambogia* extract: A study using double-blind tests. *Journal of Clinical Biochemistry and Nutrition, 42*(2), 89–103.

Zhang, C. G., & Kim, S. J. (2007). Taurine induces anti-anxiety by activating strychnine-sensitive glycine receptor in-vivo. *Annals of Nutrition and Metabolism, 51*(4), 379–386.

CHAPTER 14

Caffeine Psychopharmacology and Effects on Cognitive Performance and Mood

Jack E. James

Chapter Overview

- This chapter provides an overview of the main sources of caffeine and a description of prevailing patterns of usage.
- The putative neuroprotective actions of caffeine are considered, with particular reference to age-related cognitive degeneration.
- The effects of caffeine on cognitive performance and mood, until recently, have produced relatively few findings that can be interpreted with confidence due to methodological difficulties.
- After controlling for caffeine withdrawal and withdrawal reversal research has consistently shown that caffeine has little or no acute net beneficial effects for either cognitive performance or mood.
- Regarding cognitive ageing, the question of caffeine's putative neuroprotective potential requires large-scale intervention studies involving random allocation of participants and long-term manipulation of caffeine consumption.

Introduction

Caffeine is the most widely consumed psychoactive compound in history (James 1991, 2010). None of the drugs that humans ingest has ever been more popular than caffeine. With more than 80 per cent of people worldwide consuming caffeine daily, current usage transcends almost every social barrier, including age, gender, geography and culture. The popularity of caffeine far exceeds that of any other psychoactive substance, including nicotine, alcohol, prescribed medications and illicit drugs. Indeed, caffeine is unusual among psychoactive compounds in being part of the daily diet of most people on Earth. This possibly accounts for a common, but erroneous, belief that caffeine has always been widely present in the human diet. In fact, it was not until after European colonisation in the 17th and 18th centuries that caffeine products, previously unavailable to most people, became widely accessible. The ubiquitous presence of caffeine in the human diet and culture is a phenomenon of fairly recent origin.

Main Sources of Caffeine and Patterns of Consumption

Caffeine is a naturally occurring toxin. Its botanical function is explained by two main theories (Ashihara & Crozier 2001) that describe non-mutually exclusive processes. The chemical-defence theory posits that caffeine in young leaves, fruits and flower buds acts to protect these delicate structures from predators, such as insect larvae and beetles. The allelopathic theory posits that caffeine in seed coats is released into the soil in order to inhibit the germination of other seeds. This action not only serves to control the population density of other plant species, but also serves as a form of autotoxic self-regulation that enables caffeine-bearing species to control their own population density in the immediate environment.

The main dietary sources of caffeine are tea and coffee beverages, and increasingly, carbonated soft drinks (e.g. colas) and energy drinks (James 2011). The tea plant is indigenous to regions of China, South Asia and India. Written accounts in China of tea leaves being used to brew a beverage date to as early as AD 350 and, by about AD 600, tea had been introduced to Japan from China. It is unclear, however, to what extent tea was consumed by the general population of either country during these early periods. In the 17th century, Dutch traders introduced tea to Europe and America, and today tea is cultivated commercially in about 30 countries. Coffee is indigenous to Ethiopia, from where it was transported for cultivation to Arabia in the 15th century. By the early 16th century, the practice of extracting caffeine by infusing ground roasted coffee beans had been established in the Islamic world. Dutch traders brought coffee plants to Europe in the early 17th century, and established plantations in the Dutch East Indies. Subsequent colonisation by other European powers led to new and extensive plantations being established in the West Indies, Latin America, Africa and India.

By the late 18th century, coffee replaced tea in popularity in the USA, and today coffee is the main source of caffeine globally. Tea continues to be consumed more widely, but qualifies as the second main caffeine source because its caffeine content is generally lower than that of coffee (Gilbert 1984). Other common sources of caffeine include cocoa and chocolate (in both solid and beverage form), but the caffeine content of these is generally low and represents a negligible fraction of the total amount of caffeine consumed. In addition, although the daily intake of caffeine from sources specific to particular regions (e.g. maté in parts of South America) may be substantial for individual consumers, the overall intake from such sources is small relative to total global consumption of the drug. Similarly, some medications, both prescribed and over-the-counter, contain as much as 200 mg caffeine (approximately 2–4 cups of coffee or tea) per tablet or capsule, and could be an important (even the main) source of caffeine for some individuals. For the general population, however, caffeine-containing medications are typically taken intermittently, or not at all, thereby contributing little to total population caffeine intake. Notwithstanding variations in per capita consumption within and between geographic regions, intake throughout much of the world ranges from about 200 to 400 mg of caffeine per day (the approximate equivalent of 2–6 cups of coffee or tea per day) (James 1991).

Caffeine soft drinks are an important (and frequently the main) source of the drug for children (James et al. 2011; Kristjansson et al. 2011; National Sleep Foundation 2006). Additionally, concerns have been expressed in relation to the consumption by young people of more recently developed so-called 'energy' drinks, some of which have particularly high caffeine content (Reissig et al. 2009; Temple 2009). Whereas the caffeine in sodas and energy drinks sometimes partly derives from plant products involved in manufacture (e.g. cacao, cola nut, guarana), most of the caffeine content of such drinks is added in refined form. That is, it is not the case, as is sometimes implied, that the caffeine concentration of such drinks is explained by the presence of natural plant products during drink manufacture. One obvious implication of the deliberate addition of refined caffeine is that the drinks in question are expressly designed to have particular effects (namely, psychoactive effects) on consumers, who in this instance are mostly children. Among the concerns that have been expressed is the view that caffeine in the form of sodas and energy drinks could serve as an early 'gateway' in the experience of many children to later use of both licit and illicit drugs (James et al. 2011; Reissig et al. 2009; Temple 2009). Indeed, evidence from a variety of sources shows that higher caffeine consumption by young people is associated with increased usage of licit and illicit drugs (e.g. Arria et al. 2010; James et al. 2011). Moreover, in a recent study of choice behaviour in healthy adults, stronger preference for caffeine was found to be associated with more positive subjective effects and fewer negative effects of *d*-amphetamine (Sigmon & Griffiths 2011).

Although consumption patterns relating to the various main sources of caffeine may change during the lifespan of consumers (e.g. an individual may switch from soft drinks during childhood to coffee in adulthood), exposure to caffeine is essentially lifelong for the majority of people. Indeed, people are sometimes surprised to learn that the first exposure generally precedes birth (James 1997a). This is because most women consume caffeine while pregnant, and caffeine crosses the placenta (Bonita et al. 2007; Brazier & Salle 1981; Van't Hoff 1982). Consequently, the majority of newborns show pharmacologically active levels of caffeine in the blood (Dumas et al. 1982). Exposure typically continues during childhood, with patterns of use tending to consolidate during adolescence and early adulthood. Thereafter, usage tends to stabilise, generally undergoing little change for the remainder of life

(James 1997a). The unparalleled prevalence of caffeine consumption and the fact that its use is essentially lifelong raise major questions concerning possible implications of dietary caffeine for population health (e.g. James 2004).

Pharmacology of Caffeine

Caffeine is a white, odourless bitter-tasting powder belonging to a family of purine derivative methylated xanthines often referred to as methylxanthines or simply xanthines. First isolated from green coffee beans in 1820 by Ferdinand Runge in Germany, caffeine was later found to be present in a variety of other species (e.g. tea, maté and cacao). Figure 14.1 shows the structure of caffeine (1, 3, 7-trimethylxanthine) and its three dimethylxanthine primary metabolic products (paraxanthine, theobromine and theophylline) in humans. Following oral ingestion, caffeine is rapidly absorbed into the bloodstream from the gastrointestinal tract (Arnaud 1987). Approximately 90 per cent of the caffeine contained in a cup of coffee is cleared from the stomach within 20 minutes (Chvasta & Cooke 1971), and peak plasma concentration is typically reached within about 40–60 minutes (Rall 1990).

Once ingested, caffeine is readily distributed throughout the body, and the concentrations attained in blood are highly correlated with those found in the brain, saliva, breast milk, semen, amniotic fluid and foetal tissue (James 1991). The drug has an elimination half-life of about 5 hours in adults (Pfeifer

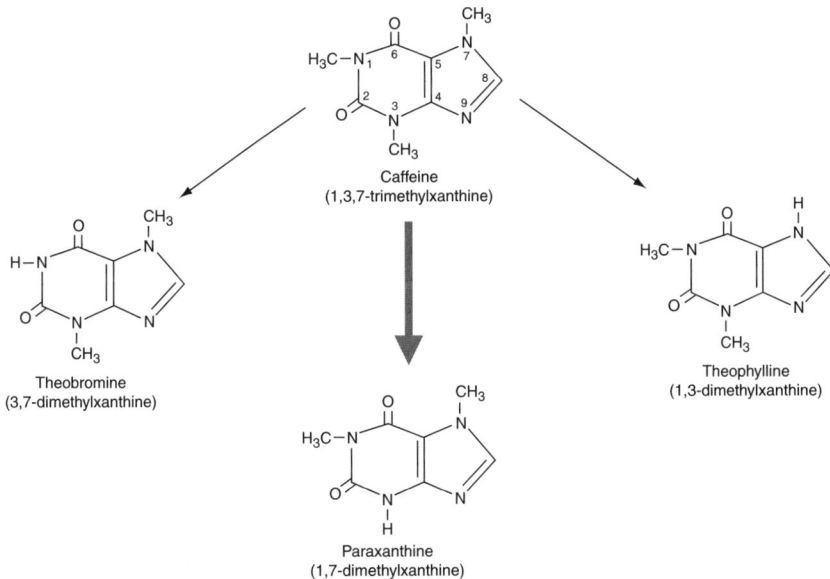

Figure 14.1   Caffeine and its dimethylated metabolites in humans (arrow widths indicate relative proportions of the metabolites in plasma)

& Notari 1988), and typical consumption patterns in the order of 3–4 cups of coffee per day result in plasma concentrations that remain at pharmacologically active levels for most waking hours for the majority of people. In adults, caffeine is virtually completely transformed by the liver, with less than 2 per cent of the ingested compound being recoverable in urine (Somani & Gupta 1988). Although the beverages and foods that contain caffeine may have other constituents (e.g. sugar, milk) that possess nutritional value, caffeine itself has no nutritional value.

Main Mechanism of Action

Caffeine exerts a variety of pharmacological actions at diverse sites, both centrally and peripherally, that are generally believed to be due mostly to competitive blockade of endogenous adenosine (Dunwiddie & Masino 2001). Adenosine is a neuromodulator that has mostly inhibitory effects on the central nervous system, but also acts on specific cell-surface receptors distributed throughout the body (Bush et al. 1989; Marangos and Boulenger 1985; Schiffman et al. 1989; Watt et al. 1989). Due to similarities in the molecular structure of caffeine and adenosine, caffeine occupies adenosine receptor sites, with A_1 and A_{2A} receptors appearing to be the primary targets (Biaggioni et al. 1991; Carter et al. 1995; Franchetti et al. 1994). It also appears that A_1 and A_{2A} receptors may interact in functionally important ways with dopamine receptors (Ferré et al. 1994; Garrett & Holtzman 1994). In particular, A_{2A} receptors may be involved in the control of the dopaminergic signalling system essential to motor control (Cunha et al. 2008).

In addition, caffeine has been reported to stimulate neuroendocrine activity, especially the catecholamine stress hormones of epinephrine and norepinephrine (e.g. Lane & Williams 1987). Increases in serum cortisol and/or urinary cortisol metabolites have also been reported (Lane et al. 1990; Lovallo et al. 1989, 2006; Pincomb et al. 1987, 1988). However, results have not been entirely consistent in that some investigators have found cortisol levels to be unresponsive to caffeine (Lane 1994; Oberman et al. 1975). It may be that the inconsistencies indicate that the typical challenge of about 250 mg (2–3 cups of coffee) represents a 'borderline dose' with respect to activation of cortisol release. For example, in one study, 250 mg of caffeine had no effect on cortisol levels, whereas 500 mg increased plasma cortisol (Spindel et al. 1984).

Physical Dependence

Repeated use of caffeine, such as occurs in the context of dietary use, generally leads to the development of physical dependence, evidenced by the appearance of behavioural, physiological and subjective 'withdrawal' effects provoked by abrupt cessation of use (Juliano & Griffiths2004). Although incompletely understood, the mechanism responsible for caffeine physical dependence is believed to involve adenosine. Repeated exposure to caffeine, including dietary use, is thought to lead to adenosine up-regulation (increased receptor number and/or increased affinity), resulting in adenosine hypersensitivity during caffeine abstinence (Biaggioni et al. 1991; Paul et al. 1993; von Borstal & Workman 1982). Sleepiness, lethargy and headache are common symptoms of caffeine withdrawal in humans (Evans & Griffiths 1991; Hughes et al. 1991; James 1998; Lane 1997; Lane & Phillips-Bute 1998; Phillips-Bute & Lane 1998; Stafford & Yeoman's 2005; Streford et al. 1995; van Dusseldorf & Karan 1990), and cessation of as little as 100 mg (e.g. one cup of coffee) per day, and possibly considerably less, can produce symptoms (e.g. Lieberman et al. 1987; Smite & Rogers 2000). These may be felt within about 12–16 hours of abstinence, with a peak at around 24–48 hours, generally abating within 3–5 days, and only infrequently extending for up to 1 week (Griffiths et al. 1990; Hughes et al. 1992, 1993). Notably, studies show that decreases in psychomotor performance (not necessarily discernible to the individual) are detectable after as little as 6–8 hours since caffeine was last ingested (Heatherly et al. 2005).

Tolerance

Drug tolerance refers to the progressive reduction in responsiveness that sometimes accompanies repeated exposure to a drug. It is evidenced by a decline in efficacy, whereby the same drug dose has less effect following repeated use or an increased dose is required to produce effects previously experienced. Although caffeine tolerance has been shown in relation to the locomotor stimulant effects of the drug in rats (Finn & Holtzman 1987; Holtzman & Finn 1988), there have been relatively few empirical demonstrations of caffeine tolerance in humans. One focus of attention in relation to caffeine tolerance in humans has been the drug's cardiovascular effects (Denary et al. 1991; James 1994a, b), which it has been widely believed undergo tolerance. The most frequently reported source (and in many articles, the only source) cited in support of the claim for hemodynamic tolerance is a study by Robertson and colleagues (Robertson et al. 1981), which is widely misquoted as having demonstrated complete hemodynamic tolerance to dietary caffeine. James (1991, pp. 111–113) has shown that the Robertson et al. (1981) study did not demonstrate complete tolerance to caffeine, nor was it capable (due to methodological limitations) of demonstrating complete tolerance. On the contrary, empirical evidence from diverse sources converges to show that blood pressure remains reactive to the pressor effects of caffeine despite repeated exposure, such as occurs when caffeine is part of the daily diet (e.g. Bak & Grobbee 1990; James 1994b, 2004; Jeep et al. 1999).

It is notable that overnight abstinence, which characterises usual patterns of consumption, results in almost complete depletion of systemic caffeine by early morning (Lela et al. 1986; Pfeifer & Notari 1988; Shi et al. 1993). Several lines of inquiry suggest that any tendency towards tolerance, if such tendency exists, is likely to be expressed partially, if at all, because of the pattern of diurnal depletion of caffeine as experienced by most consumers. Indeed, the very fact that many hundreds of published experiments have reported significant caffeine-induced behavioural, physiological and subjective effects provides strong evidence that usual patterns of consumption do not in general produce complete tolerance. Most participants in such experiments have been typical caffeine consumers who arrive at the experimental laboratory following a brief period of abstinence. Notwithstanding the brevity of the typical abstinence period (e.g. overnight) employed in experimental studies of the acute effects of caffeine, participants generally remain responsive. That is, under typical conditions of dietary caffeine consumption, there is a general absence of complete tolerance to the bio behavioural effects of the drug. The one exception to this general observation appears in relation to self-reported subjective effects, which Sigmon et al. (2009) found were subject to 'complete' tolerance, whereas the physiological responses of cerebral blood flow velocity and EEG were not.

Psychopharmacology of Caffeine: Caffeine Withdrawal and Withdrawal Reversal

The earliest systematic examinations of the psychopharmacology of caffeine were conducted a century ago (Hollingsworth 1912a, b). The strong consensus for most of the intervening period has been that caffeine is a stimulant capable of enhancing aspects of human psychomotor performance and mood. However, inconsistencies between studies have long provided sound empirical reasons for doubting that consensus. When studies of the effects of caffeine on performance and mood are compared, a substantial lack of consistency is revealed in choice of performance tasks and method for measuring changes in mood. With specific reference to performance, tasks have ranged from relatively 'simple' psychomotor activities, such as hand steadiness and manual dexterity, to more 'complex' cognitive activities, such as mental arithmetic and memory. One trend in the literature is that complex higher-level cognitive processes have generally been reported to be less responsive to caffeine than more simple, repetitive and prolonged psychomotor activities (James 1997a).

To the extent that comparisons can be made between studies, a high level of inconsistency in findings is revealed. Although many studies report caffeine-induced positive performance effects, many also report null results and not uncommonly there have been reports of caffeine-induced impairment in performance. Some reviewers have interpreted the available evidence as indicating that any enhancement of performance by caffeine is, at best, likely to be small and unstable (e.g. Dews 1984; James 1991; Satiric 1988). It is surprising, therefore, that caffeine has been lauded as having properties that are 'extremely beneficial' for cognitive performance (e.g. Smith 2009). However, claims of marked benefits, which sometimes also appear in marketing campaigns for individual caffeine products, have been disproved. As explained below, advances in knowledge about the dynamics of caffeine withdrawal and withdrawal reversal have radically transformed our understanding of caffeine psychopharmacology, showing that caffeine has little or no net beneficial effects for cognitive performance and mood.

Confounding Due to Reversal of Withdrawal Effects

Drawing upon the time-honoured practice of placebo-controlled studies of therapeutic drugs, experimental studies of caffeine (as mentioned above) typically involve caffeine being withheld for a period prior to testing for effects. The reason for withholding caffeine is to ensure all participants are essentially equivalent (i.e. free of caffeine) at the time of testing. Such efforts to achieve experimental control appear especially warranted in the case of caffeine, because the drug is so readily available and so widely consumed. When employing the placebo-controlled paradigm, caffeine researchers frequently make use of the naturally occurring period of overnight abstinence by simply asking participants to forego their usual morning caffeine beverage prior to laboratory testing. Figure 14.2 gives a schematic representation of results from a typical experiment, whereby caffeine has traditionally been interpreted as having enhancing effects for performance and mood. However, as Figure 14.2 illustrates, a critical appraisal of the typical study design shows that the findings yielded by such studies are, at best, ambiguous (James 1994c, 1995; James & Rogers 2005). In short, in a typical study, there is no way of knowing whether participants at baseline were performing 'normally' or whether they were experiencing the negative effects of overnight caffeine withdrawal. Similarly, following drug challenge, there is no way of knowing whether performance was improved by caffeine or whether improved performance was due to reversal of negative withdrawal effects experienced at baseline. If the latter, the performance 'improvement' following caffeine would not represent a net benefit and would not have occurred at all had the person not been a caffeine consumer in the first place.

Until relatively recently, it has not been adequately appreciated that study participants, having avoided caffeine since the evening before, are generally entering the early stages of caffeine withdrawal by the time they are tested in the laboratory (typically, at least 12–14 hours since caffeine was last ingested). As mentioned above, habitual use of caffeine produces physical dependence, evidenced by the appearance of readily measurable withdrawal symptoms following periods of abstinence (e.g. Juliano & Griffiths 2004). Accordingly, although generally presented as investigations of the effects of caffeine compared with caffeine-free controls, traditional studies of caffeine effects on performance and mood can equally be conceptualised as studies of

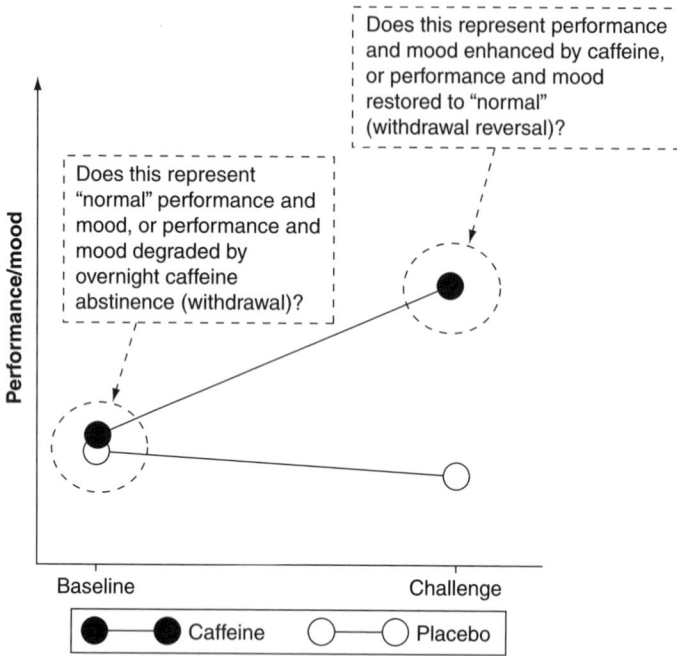

Figure 14.2   A schematic representation showing the ambiguous nature of the results of a typical double-blind, placebo-controlled experiment to test the acute effects of caffeine on cognitive performance (see text)

the effects of caffeine withdrawal, and sometimes have been conceptualised that way (e.g. Mitchell et al. 1995; Phillips-Bute & Lane 1998; Streford et al. 1995; Watson et al. 2000). Thus, the crucial question is: To what extent do effects (e.g. improvements in performance and mood), traditionally regarded as net benefits, represent genuine net effects of caffeine or reversal of withdrawal effects induced by brief periods of abstinence (James 1994c)? Indeed, a third possibility exists, namely, that observed 'benefits' are a combination of net effects and withdrawal reversal. Overall, studies support the conclusion that caffeine-induced 'improvements' in cognitive performance and mood are primarily due to reversal of withdrawal effects. Symptoms of caffeine withdrawal in habitual consumers include 'headache, tiredness/fatigue, decreased energy/activeness, decreased contentedness/well-being, difficulty concentrating, irritability, and foggy/not clearheaded' (Juliano & Griffiths 2004, p. 25), as well as degraded performance and mood (e.g. James 1998; Juliano & Griffiths (2004); Richardson et al. 1995; Rogers et al. 2003; Yeoman's et al. 2002). In turn, it has been established beyond doubt that withdrawal-induced dysphoric effects are relieved when caffeine is re-ingested (Griffiths et al. 2003; Juliano & Griffiths 2004).

It has been speculated that conclusions based on studies of caffeine withdrawal effects may suffer from a confound due to participant expectancies (Dews et al. 2002; Smith 2002). Although it is impossible to rule out some influence of participant expectancies in some studies, control of participant expectancies has been a strong feature of much of the relevant empirical research (e.g. Garrett & Griffiths 1998; James 1998; James & Gregg 2004a; James et al. 2005; Richardson et al. 1995; Rogers et al. 2005; Tinley et al. 2003). For example, in addition to employing standard double-blind controls, the context of experiments conducted by Griffiths and his colleagues (see Juliano & Griffiths 2004) was such that participants were often not even aware that caffeine and their reactions to it were being studied. In double-blind experiments conducted by James and his colleagues (e.g. James 1998; James et al. 2005), participants, who were invited to guess whether they had ingested caffeine or a placebo, usually performed no better than chance. Since participants generally did not know which compound (whether caffeine or placebo) had been ingested, it is hard to imagine that expectancies could have had much systematic influence on objective performance and subjective mood. Based on a comprehensive review of caffeine withdrawal, Juliano & Griffiths (2004) concluded that the available evidence 'overwhelmingly' supports withdrawal effects being pharmacological rather than being due to participant expectancies.

Referring to the functional role of adenosine in the brain, Dunwiddie & Masino (2001) posed what they labelled a 'challenging' question: Why is it that caffeine antagonism of adenosine produces what are generally considered to be improvements in cognitive function, whereas antagonism of most other neurotransmitter receptors produces either deficits or pathological effects? The answer would appear to be that, in reality, caffeine is not too dissimilar to other neurotransmitter antagonists. As explained below, withdrawal from habitual caffeine consumption has negative effects on cognitive function that are reversed when caffeine is re-ingested. The net result appears to be largely nil, or even an overall net negative effect as has been found in recent research outlined below.

Attempts to Control for Confounding Due to Caffeine Withdrawal and Withdrawal Reversal

Considering the limitations of the typical drug-challenge protocol for clarifying the effects of caffeine on human performance and mood, three main alternative empirical paradigms have been proposed (James 1997a, James & Rogers 2005). The three approaches, each of which employs a different method for dealing with the problem of confounding due to caffeine withdrawal, may be succinctly described as: (a) studies that compare consumers and low/non-consumers, (b) pre-treatment and ad lib consumption studies, and (c) long-term withdrawal studies.

Studies Comparing Consumers and Low/Non-Consumers

In this approach, caffeine is administered to low or non-consumer 'naïve' participants for whom the likelihood of caffeine withdrawal would be lessened or removed. However, since more than 80 per cent of the global population consumes one or more caffeine beverages daily (James 1997a), low use is atypical. Importantly, then, since persons who consume little or no caffeine represent a small self-selected minority, the generality of any effects they might show is open to question. Their low use of caffeine might not only make them unrepresentative with regard to general characteristics, but their reaction to caffeine might also be atypical. For example, a proportion of infrequent consumers may have adverse reactions to caffeine (e.g. Alsense et al. 2003), and this may explain their low use of the drug. In addition, the existence of polymorphisms of the cytochrome P450 oxidase enzyme system in the liver that is responsible for caffeine metabolism (Kot & Daniel 2008) may contribute to individual differences in caffeine-consuming habits.

There are good reasons for undertaking scientific study of persons who consume little or no caffeine, including the aim of better understanding the influence of genotype and gene–environment interactions on individual differences in behaviour. However, studying low and non-consumers as a way of understanding caffeine effects in habitual consumers violates a central tenet of experimental science. People self-select whether or not to consume caffeine. To generalise from one group (those who do not consume caffeine) to the other (those who do consume caffeine) violates the axiom of random assignment that unpins the logic of experimental science and the statistics that are used to interpret experimental results. Unfortunately, conclusions purporting to reflect generalised effects of caffeine continue to be reported on the basis of studies of low/non-consumers without due regard for the profound interpretational problems that such studies pose (e.g. Adan & Serra-Grabulos 2010; Haskell et al. 2005; Serra-Grabulosa et al. 2010).

Pre-Treatment and Ad Lib Consumption Studies

In pre-treatment and ad lib consumption studies, participants are 'pre-treated' with caffeine so that they might be 'minimally', or not at all, caffeine deprived when tested for performance and mood after a subsequent (experimentally delivered) dose of caffeine. As a test of withdrawal reversal versus net benefit, ad lib consumption studies are also inherently flawed. Because caffeine consumption patterns and rate of caffeine metabolism vary between individuals, it is problematic to try to estimate the precise amount of pre-treatment caffeine needed to ensure uniform and complete removal of caffeine withdrawal effects from one individual to the next. Perhaps reflecting that problem, the approach has produced inconsistent findings. Several studies have reported enhanced performance and mood after a second or subsequent caffeine dose following pre-treatment (Christopher et al. 2005; van Duinen et al. 2005; Warburton 1995; Warburton et al. 2001), whereas others have failed to observe any enhancement of either performance or mood after caffeine was ingested within less than 6–8 hours following pre-treatment (Heatherly et al. 2005; Robelin & Rogers 1998; Yeoman's et al. 2002).

Interestingly, in the studies reporting positive results, participants were relied upon to self-administer the pre-treatment dose(s) while unsupervised (Christopher et al. 2005; Warburton 1995; Warburton et al. 2001; van Duinen et al. 2005). The problem of unsupervised 'pre-treatment' has been compounded by inconsistency in instructions to participants. For example, Warburton et al. (2001) told participants 'they could consume their normal quantities [of caffeine] during the day', whereas van Duinen et al. (2005) informed participants they 'were allowed to consume one cup of coffee' prior to testing. In one study, saliva samples were collected with the aim of verifying systemic caffeine levels in participants who received unsupervised pre-treatment (Christopher et al. 2005), but James & Rogers (2005) have shown that the saliva results were not interpreted correctly. In contrast to unsupervised pre-treatment, supervised pre-treatment studies have reported no caffeine effects following a second dose unless the interval between the two doses exceeded 6 hours (Heatherly et al. 2005; Robelin & Rogers 1998; Yeoman's et al. 2002). The implication arising from systematic comparison of the findings from studies in which pre-treatment was unsupervised or supervised is that a proportion of participants in the unsupervised studies arrived at the experimental laboratory in a greater state of relative caffeine deprivation than the experimenters had intended.

Indeed, because pre-treatment studies involve the administration of one or more doses of caffeine followed by caffeine or placebo, they may reasonably be conceptualised as dose-response studies (James & Rogers 2005). When viewed as such, findings indicate a rather flat dose-response relationship. In other words, in physically dependent short-term abstinent individuals, withdrawal effects for performance and mood are reversed in an essentially all-or-nothing fashion when caffeine, in typical dietary amounts, is ingested. As such, although caffeine reverses negative withdrawal effects, it does not enhance performance and mood to levels above those that are 'normal' for the individual.

Long-Term Withdrawal Studies

In recognition of the inherent limitations of low/non-consumer studies and pre-treatment/ad lib consumption studies, a third strategy has been advocated as the preferred approach for establishing the net effects of caffeine on performance and mood uncontaminated by withdrawal reversal (James 1997a, 1998; James & Rogers 2005). This approach recognises the need for study designs that permit direct within-subject comparisons of the effects of sustained periods of caffeine use and abstinence in persons who are habitual consumers. Studies of this kind are particularly demanding on participants and research resources, because of the requirement that participants be tested repeatedly over extended time periods, during which participants alternate between being with and without caffeine.

Taking the core features of the traditional drug-challenge paradigm, with its attendant strengths of double blinding and placebo control, James (1998) extended that protocol to include four consecutive one-week periods, with a strictly prescribed and biologically verified regimen of caffeine intake for every day of each week (see Table 14.1). For six consecutive days ('run-in' phase) of each of the four weeks of the protocol, participants ingest a placebo ('washout') or caffeine ('dietary exposure') to ensure stability of responding when 'challenged'. During caffeine phases, participants ingest the approximate equivalent of one cup of coffee three times daily (gelatine capsules containing caffeine and starch filler), thereby simulating the typical population pattern of caffeine consumption. During placebo washout, participants ingest capsules containing starch alone. On the seventh ('challenge') day of each alternating one-week period, participants are challenged with placebo or caffeine. The one-week time frame was chosen on the basis of findings from previous research. Although tolerance has been confirmed infrequently in human studies, when it has been observed (e.g. partial tolerance of caffeine-induced increases in blood pressure), it has generally been found to plateau within 3–5 days of continuous use (Denaro et al. 1991; James 1994b; Robertson et al. 1981). Withdrawal effects have been examined more extensively, and have been found to abate within a similar time frame (i.e. 3–5 days and not more a week) (e.g. Griffiths et al. 1986; Hughes et al. 1993). The use of alternating periods of caffeine exposure and placebo abstinence means that the separate acute and chronic effects of caffeine can be examined and compared in the one experiment.

As had been speculated (James 1994c), James (1998) found that overnight caffeine abstinence had negative effects on performance and

Table 14.1 Summary of a long-term withdrawal protocol, involving double-blind, placebo-controlled crossover incorporating alternating periods of caffeine exposure and abstinence[a]

Week	'Long-term' run-in/ washout (Days 1–6)	'Challenge' (Day 7)	Effect revealed by challenge
1	Placebo	Placebo	Caffeine non-consumption
2	Placebo	Caffeine	Acute caffeine challenge
3	Caffeine	Placebo	Acute caffeine withdrawal
4	Caffeine	Caffeine	Dietary exposure (habitual caffeine consumption)

[a] Design originally described by James (1997, 1998), versions of which have been employed in subsequent studies (e.g. James & Gregg 2004a, b; James et al. 2005; Keane et al. 2007).

mood (withdrawal), and that these effects were reversed when caffeine was re-ingested (withdrawal reversal). There was no evidence of caffeine having any beneficial effects on performance or mood under conditions of 'long-term' caffeine use compared with long-term abstinence ('caffeine non-consumption' versus 'dietary exposure' in Table 14.1). The finding that caffeine-induced improvements in performance and mood are due to reversal of withdrawal effects associated with brief (typically, overnight) abstinence has been confirmed in independent studies that employed the same or similar controls (James & Gregg 2004a; James et al. 2005; Lane & Phillips-Bute 1998; Phillips-Bute & Lane 1998; Rogers et al. 2003, 2005; Sigmon et al. 2009; Yeomans et al. 2002). Notwithstanding the evidence for withdrawal reversal, claims of beneficial effects for caffeine continue to be reported on the basis of results obtained from studies in which inadequate or no account has been taken of the processes of withdrawal and withdrawal reversal (Adan & Serra-Grabulos 2010; Brunyé et al. 2010a, b; Childs & de Wit 2005; Haskell et al. 2008; Hogervorst et al. 2008; Nehlig 2010; Seng et al. 2010; Smillie & Gökcen 2010).

Age-Related Cognitive and Neuromotor Degeneration

In recent years, attention has been given to the possibility that lifelong caffeine consumption may improve cognitive function in later life and provide partial protection against age-related cognitive and neuromotor degeneration. These possibilities have sometimes been linked to caffeine's putative acute enhancement of cognitive performance, in the apparent belief that acute advantage might confer long-term biobehavioural benefits. However, because caffeine does not possess significant acute cognitive-enhancing potential (as seen above), any long-term advantages it may provide would have to be due to mechanisms other than acute enhancement. Moreover, it should be noted that much of the current interest in long-term cognitive enhancement and neuroprotective potential derives from the findings of epidemiological studies of caffeine consumption. While such studies are useful for identifying the presence of associations (i.e. correlations) between variables, they are unable to establish cause-and-effect relationships. Moreover, particular care is needed with epidemiological data when putative etiological variables are 'distant' from relevant outcome variables (Turkheimer 1998), and by any analysis the etiological distance between dietary caffeine and age-related neurological functioning could not be considered other than considerable. That is, the acute biobehavioural effects of caffeine are generally small, and lifelong consumption of caffeine is but one of an essentially limitless number of biobehavioural variables that might conceivably influence the course of human cognitive and neuromotor competence.

In an early cross-sectional population study, Jarvis (1993) reported that higher caffeine intake was associated with better performance on certain psychomotor and cognitive tasks, and the relationship was reported to be more pronounced in older than younger participants, possibly implying that the effects of caffeine are more pronounced when risk of age-related degeneration is greater. However, a subsequent prospective study involving a larger population sample found little evidence of improved cognitive performance or reduced age-related cognitive decline associated with dietary caffeine consumption (van Boxtel et al. 2003). Nevertheless, continued interest in caffeine and age-related cognitive function has been encouraged by a substantial body of epidemiological research reporting an inverse ('protective') association between caffeine consumption and the incidence of Alzheimer's disease (e.g. Eskelinen & Kivipelto 2010; Lindsay et al. 2002; Maia & de Mendonca 2002; Quintana et al. 2007). However, recent findings from the Lothian Birth Cohort 1936 Study in Scotland (Corley et al. 2010) suggest that the widely reported epidemiological link between caffeine and Alzheimer's disease is likely to be spurious, being the result of uncontrolled confounding, as is next discussed.

Corley et al. (2010) examined almost 1000 healthy adults aged 70 years for whom data were available from intelligence quotient (IQ) tests conducted at age 11 years. Individuals with higher childhood IQ performed better in adulthood than those with lower childhood IQ, irrespective of caffeine intake. In addition, higher-IQ children consumed more caffeine/coffee in adulthood than lower-IQ children (a lifestyle-related choice by individuals possessing higher IQ and higher social status). Analyses showed that coffee-related superior cognitive ability in adulthood was the result of lifelong cognitive advantage stemming from superior cognitive ability in childhood. Caffeine had no protective effect for cognitive function.

Akin to findings concerning caffeine and Alzheimer's disease, several recent epidemiological studies have reported an inverse association between caffeine consumption and the development of Parkinson's disease (e.g. Ascherio et al. 2001; Costa et al. 2010; Hernan et al. 2002; Ross et al. 2000). This finding has been widely assumed to be causal and has contributed to speculation about possible caffeine-related neuroprotective mechanisms. In particular, attention has focused on interactions between the dopaminergic and adenosinergic systems and caffeine's putative ability to forestall dopaminergic neuron degeneration through its action on the A_{2A} adenosine receptor (e.g. Chen et al. 2001; Kalda et al. 2006; Menza 2000; Schwarzschild et al. 2003; Todes & Lees 1985). Nevertheless, the fact remains that the evidence linking caffeine consumption to Parkinson's disease is largely correlational. Despite routinely controlling for multiple potential confounders, no epidemiological study can guarantee having taken account of all relevant confounders, as is amply demonstrated by the findings outlined above concerning childhood IQ and coffee-related cognitive ability in adulthood. Further paralleling the findings with Alzheimer's disease, the causal nature of epidemiological observations linking caffeine to decreased risk of Parkinson's disease has been called into question on the basis of failure to take account of variation in individual differences.

Evans et al. (2006) have suggested that the inverse association between caffeine consumption and Parkinson's disease, as well as the similar relationship that exists with cigarette smoking (which has fostered the belief that nicotine is neuroprotective), may be 'epiphenomena' rather than causal. Broadly, Evans et al. (2006) argued that confounding due to individual differences in the personality disposition of impulsive sensation seeking may have led to misunderstanding of the epidemiological findings. The authors cited evidence that sensation seeking is inversely associated with Parkinson's disease, with higher sensation seeking also being associated with higher caffeine consumption and smoking. Evans et al. (2006) hypothesised that there are biological features characteristic of low-sensation-seeking individuals that also predispose to Parkinson's disease. Thus, rather than indicating any neuroprotective capability, higher caffeine and nicotine intake may simply be behavioural manifestations of a generalised disposition towards impulsive sensation seeking, which itself is the expression of a biological substrate that confers a level of protection against the development of Parkinson's disease. According to this logic, caffeine and nicotine do not protect against Parkinson's disease. Rather, increased caffeine consumption, cigarette smoking and lower risk of Parkinson's disease are all expressions of biological correlates of an underlying personality difference.

Sleep and Wakefulness

The new understanding about withdrawal reversal described above has shown that long-held beliefs about the effects of caffeine on performance and mood are no longer tenable. This new understanding also challenges equally strong beliefs about how caffeine might moderate the effects of sleep and wakefulness on performance and mood. Even a cursory examination of the relevant literature shows that the key methodological limitations described above in relation to studies of caffeine and performance/mood also typify studies of caffeine and sleep/wakefulness. As an illustrative example, Shilo et al. (2002) employed a double-blind crossover design with caffeine-consuming healthy volunteers given either decaffeinated or regular coffee during each of two 24-hour periods separated by one week. Actigraphic measurements indicated that caffeine increased sleep latency and negatively affected a variety of other sleep variables. In keeping with a long tradition of similar research, the results were interpreted as confirming the widely held belief that coffee/caffeine consumption counteracts sleep and promotes wakefulness. However, these conclusions are unwarranted when considered in light of understanding of caffeine withdrawal and withdrawal reversal.

Specifically, by the time participants in the Shilo et al. (2002) study retired for the night in the decaffeinated coffee condition, they had been without caffeine for 24 hours and remained caffeine-free for an additional 7 hours (approximate) while in bed. This timing coincides with the period of peak caffeine withdrawal effects. It can be seen that the study design employed by Shilo et al. (2002), as well as that employed in many similar studies, was confounded and that the results are entirely ambiguous. Rather than demonstrating caffeine-induced wakefulness, such studies can equally be seen as having demonstrated the sleep-inducing effects of caffeine withdrawal in the control condition, with the latter effects being reversed when participants ingested caffeine. Indeed, considering the strength of evidence showing sleepiness to be an effect of caffeine withdrawal (e.g. Jones et al. 2000; Juliano & Griffiths 2004), withdrawal reversal can justifiably be seen as the more plausible of the two alternative explanations. As with the large traditional body of research into the effects of caffeine on performance and mood, a third possibility is that both of the aforementioned interpretations are partially correct, such that caffeine interferes with sleep (producing wakefulness) and caffeine withdrawal induces sleepiness which is reversed when caffeine is re-ingested.

Topographic Quantitative EEG

EEG studies have been widely interpreted as confirming caffeine as a psychostimulant capable of producing wakefulness. However, there is little cause for confidence in a substantial proportion of the findings, due to researchers having ignored the problems of caffeine withdrawal and withdrawal reversal (James & Keane 2007). The crucial relevance of withdrawal is confirmed by evidence that EEG is affected by caffeine abstinence, and that effects on brain activity are reversed when caffeine is re-ingested (Jones et al. 2000; Reeves et al. 1995). Specifically, Jones et al. (2000) employed a double-blind crossover design in which ten healthy moderate caffeine consumers were observed during a baseline period while maintaining their normal diet (including their usual caffeine intake), and during two one-day periods when they consumed caffeine-free diets. During the caffeine-free periods, they received capsules containing placebo or caffeine in amounts equal to their baseline daily consumption. Compared with the caffeine condition, after 21 hours of caffeine withdrawal in the placebo condition, EEG theta power was significantly increased (indicating increased sleepiness). Importantly, the EEG pattern was found to be essentially the same in the baseline and caffeine conditions, indicating reversal of the withdrawal effects observed in the placebo condition. As such, the findings further confirm the potential for confounding in studies that fail to control for withdrawal and withdrawal reversal.

In a more recent study of the effects of caffeine on EEG, which explicitly controlled for withdrawal and withdrawal reversal, Keane et al. (2007) employed the study design summarised in Table 14.1. Participants alternated weekly between ingesting placebo and caffeine (1.75 mg/kg) three times daily for four consecutive weeks, and EEG activity was measured at 32 sites during eyes closed, eyes open and performance of a vigilance task. After controlling for withdrawal and withdrawal reversal, caffeine was found to have few and modest affects on EEG in the theta and alpha bandwidths, and no effects in the delta and beta bandwidths. In contrast, caffeine-induced changes in EEG pattern were observed for withdrawal, withdrawal reversal and tolerance during all three behavioural conditions (eyes closed, eyes open and task performance). Interestingly, similar effects were found for caffeine challenge and acute caffeine withdrawal. Thus, while demonstrating that caffeine affects electrophysiological activity in the brain, the findings cast doubt on whether the drug can be viewed as having stimulant effects. The Keane et al. (2007) study suggests that it would be more apposite to say that caffeine 'interferes with' or 'disrupts', rather than 'stimulates', brain activity.

Caffeine, Sleep Loss and Cognitive Performance

A large research effort has been aimed at assessing the potential that caffeine may have for allaying sleepiness caused by sleep loss and atypical sleep-wake cycles (e.g. shift work), with particular interest being paid to the potential of caffeine to ameliorate performance decrements associated with sleep loss. Findings have generally been interpreted as showing that caffeine promotes wakefulness, and researchers have been quick to recommend use of the drug for that purpose (e.g. Rosenthal et al. 1991; Walsh et al. 1990). Indeed, it has become commonplace for caffeine to be recommended as an antidote for shift work-induced sleepiness, especially in jobs where sustained optimal performance is required for reasons of safety (e.g. air pilots, air traffic controllers, nurses, nuclear power plant operators). Such recommendations, however, derive from a misunderstanding of results from studies of caffeine, sleep and performance.

In a manner not unlike caffeine's effects on electrophysiological brain activity (mentioned above), caffeine is disruptive to normal sleep functions in both humans and animals, producing increases in latency-to-sleep and increases in sleep fragmentation (Basheer 2004; Dunwiddie & Masino 2001). Labelling such disruptions as 'stimulating' suggests benefits to performance and mood degraded by sleep loss. However, empirical studies that have controlled for caffeine withdrawal and withdrawal reversal show that the drug does not counter sleepiness-induced decrements in performance and mood, showing instead that it has the potential to exacerbate negative effects of sleepiness (Keane & James 2008; Rogers et al. 2005). The extent of the potential harm due to caffeine's disruptive effects on sleep are suggested in a recent cross-sectional population study by James et al. (2011), which examined caffeine use and academic achievement in 7377 adolescents. In addition to caffeine consumption and academic performance, participants were surveyed for cigarette smoking, alcohol use, daytime sleepiness and a host of potential confounders. Higher caffeine consumption was found to be independently associated with poorer academic performance. Importantly, daytime sleepiness, a possible effect of loss of overnight sleep (due to the direct disruptive effects of caffeine on sleep) and increased daytime sleepiness due to caffeine withdrawal, was found to be an important mediator of the apparent negative effect of caffeine on academic achievement.

Sleepiness and Driving

In a series of experiments, Reyner & Horne (1997, 2000, 2002; Horne & Reyner 1996, 2001a) claimed to have shown beneficial effects of caffeine beverages, especially a well-known brand of 'functional energy' drink. However, none of the studies controlled for confounding due to caffeine withdrawal and withdrawal reversal. An illustrative example is provided by Reyner & Horne (2002) who administered the drink in question, with and without caffeine, to 12 young adults who experienced 5 hours' sleep restriction on 2 nights. In the late afternoon following each night, participants drank the assigned beverage 30 minutes before being tested for 2 hours on a driving simulator. On the day caffeine was ingested, simulated driving performance was reported to have improved, and this was interpreted as showing that the drink is beneficial in promoting wakefulness and reducing sleep-related driving incidents.

However, all participants were moderate consumers (2–4 cups of coffee daily), and all were tested between 14.00 and 17.00 hours after having consumed no caffeine, other than that administered by the researchers, since 18.00 hours the day before. As such, participants in the no-caffeine condition were assessed after being caffeine deprived for a minimum of 22 hours, well beyond the point in time when withdrawal-induced sleepiness and decrements in performance begin to appear. After ingesting caffeine, increased sleepiness and decreased performance were reversed, as would be expected, considering the processes of caffeine withdrawal and withdrawal reversal. Thus, rather than demonstrating net increases in wakefulness and performance due to caffeine ingestion, all of the studies by Reyner & Horne (Horne & Reyner 1996, 2001a; Reyner & Horne 1997, 2000, 2002), though ambiguous in a strict sense, are consistent with the interpretation that participants experienced reversal of withdrawal effects characteristic of physically dependent moderate caffeine consumers during periods of caffeine withdrawal.

In another study of driving performance, Philip et al. (2006) took the unusual step (ethically) of experimentally examining the effects of caffeine on sleepiness and driver performance in situ. That is, whereas others have examined simulated driving, Philip et al. examined actual driving performance on public roads at night in sleep-deprived participants. Participants were 'not allowed to sleep' before driving 200 km over 90 minutes (i.e. at average speed of 133 km per hour) during the early-morning hours of 02.00–03.30. The experimental vehicle was equipped with dual controls, and a 'professional driving instructor' remained at the ready to take control 'if needed' and to take control altogether if 'a participant could no longer drive' (p. 787). 'Extended time awake and sleepiness at the wheel' (p. 789) resulted in a significant increase in line crossings, which were found to be less frequent following 200 mg caffeine compared with placebo. Philip et al. (2006) attributed the results to the net effect of caffeine, but no such conclusion is warranted considering their failure to take account of the processes of caffeine withdrawal and withdrawal reversal.

It should be of concern that in study after study, and in advice intended to influence road safety policy (e.g. Horne & Reyner 2001b), the evidence cited in support of the use of caffeine drinks while driving is ambiguous at best and does not provide a factual basis for recommendations that caffeine be used as a prophylactic strategy for addressing fatigue-related safety and performance concerns. The actual facts are that most people consume caffeine daily, that sleepiness is caused and exacerbated by caffeine withdrawal, and that susceptibility to these effects occurs within the time frame of the several hours that frequently separate beverages as typically consumed. Consequently, the cumulative facts would suggest that caffeine withdrawal effects are likely to increase the rate of accidents and harm in situations where safety depends on wakefulness.

Increased risk of withdrawal-induced sleepiness can, in theory, be avoided by maintaining systemic caffeine concentrations at the pharmacological levels needed to allay or reverse withdrawal effects. In practice, such states of perfect pharmacological equilibrium are likely to be achieved inconsistently, if at all, and only at a cost to health and wellbeing. Short-term attempts at maintaining such equilibrium are likely to create unwanted side-effects, including feelings of agitation and acute anxiety (cf. James and Rogers 2005), whereas long-term attempts are likely to contribute to chronic health problems, such as elevated blood pressure and increased risk of cardiovascular disease (e.g. James 1997b, 2004; James & Gregg 2004b). The only certain way of avoiding caffeine withdrawal-induced sleepiness is not to be a caffeine consumer in the first place. Thus, caffeine may have implications for road safety, though not in its frequently proposed role as a countermeasure to sleepiness. On the contrary, caffeine may pose a risk to road safety as a drug of habitual use, wherein users experience intermittent withdrawal-induced sleepiness and associated decrements in driving competence.

Military Operations

The potentially serious consequences of not taking adequate account of confounding due to reversal of caffeine withdrawal effects are possibly most dramatically exemplified in the burgeoning literature concerned with caffeine use under the extreme conditions of military combat. Although the details of the methodologies employed have varied, the central design feature of these studies has generally involved healthy young participants in simulated combat experiencing a period of caffeine abstinence (i.e. withdrawal) accompanied by sleep restriction, followed by administration of caffeine (withdrawal reversal) or placebo (continued abstinence). Based on results involving outcome variables having high face validity (e.g. marksmanship), it has generally been concluded that caffeine enhances operational capabilities (McLellan et al. 2005a, b; Tikuisis et al. 2004). This conclusion has prompted researchers to develop new methods, deemed more suitable than beverages, for delivering caffeine to personnel during active operations, and chewing gum has emerged as the preferred delivery vehicle (Kamimori et al. 2002; LaJambe et al. 2005; Syed et al. 2005).

One group of researchers, responsible for a series of studies involving defence forces from Canada, New Zealand and the USA (McLellan et al. 2005a, b), has claimed that the success of a military operation could depend on the use of caffeine to induce wakefulness in operational personnel suffering sleep deprivation (McLellan et al. 2005a, p. 43). In reality, the studies concerned do not justify such conclusions, because of a failure to control for caffeine withdrawal and withdrawal reversal. For example, McLellan et al. (2005a) measured performance 38–46 hours after caffeine was last consumed, at a time when withdrawal effects were likely to have been particularly pronounced. However, no account was taken of withdrawal-induced performance decrements in the placebo group, or of withdrawal reversal effects in the caffeine group. Thus, the study was not capable of showing net effects of caffeine. As with driver safety, caffeine may threaten, rather than aid, the safety of military personnel, due to disruptive effects of the drug on sleep-wake cycles, not to mention the potential for long-term harm on other aspects of health (James 1997a).

Putative Restorative Effects of Caffeine

Several studies have sought to compare the efficacy of caffeine with other strategies promoting wakefulness, such as naps (Bonnet & Arand 1994; Bonnet et al. 1995; Schweitzer et al. 2006) and bright light (Hayashi et al. 2003; Jay et al. 2006; Wright et al. 1997, 2000). Again, researchers have either been unaware of the potentially confounding influence of caffeine withdrawal and withdrawal reversal, or have sought to address those concerns by recruiting only 'moderate' or 'mild' consumers as participants (e.g. Jay et al. 2006; Wright et al. 1997, 2000). However, as mentioned above, caffeine withdrawal effects have been reliably demonstrated following cessation of as little as 100 mg (one moderate-strength cup of coffee) per day and less (e.g. Evans et al. 1994; Griffiths et al. 1990; Lieberman et al. 1987; Smit & Rogers 2000). Therefore, recruitment of 'moderate' or 'mild' consumers as participants for such studies does not address the core methodological problem it is intended to address. Considering the overall evidence, little confidence can be placed in the sometimes seemingly authoritative recommendations for using caffeine as a stimulant to enhance performance. One such authority was the task force established by the American Academy of Sleep Medicine to examine the use of caffeine (and other stimulants) for countering the effects of sleep loss (Bonnet et al. 2005). The available evidence does not support the task force's claim that caffeine is an effective sleep-loss prophylactic.

Conclusions

A major conclusion of this review is that the substantial research effort to elucidate the effects of caffeine on cognitive performance and mood, undertaken over many decades, until recently, had produced relatively few findings that can be interpreted with confidence (James & Rogers 2005). A major problem has been the widespread failure of researchers to recognise that habitual use of caffeine, even at moderate levels, leads to physical dependence, evidenced by measurable behavioural and subjective effects in response to abstinence for periods as short as 6–8 hours (Heatherley et al. 2005). Because researchers have generally not controlled for withdrawal and reversal of withdrawal effects, results have been ambiguous and conclusions open to challenge. More recent studies that have controlled for reversal of withdrawal effects do not support long-held beliefs about caffeine being a general enhancer of performance and mood. On the contrary, studies that included effective experimental controls against confounding due to withdrawal and withdrawal reversal have consistently shown that caffeine has little or no acute net beneficial effects for either cognitive performance or mood.

The empirical foundations of recent interest in putative neuroprotective actions of caffeine do not elicit any greater level of confidence than traditional beliefs about caffeine enhancement of cognitive performance and mood. Considering the correlational nature of much of that evidence and the fact that findings are open to alternative interpretations, it is certainly premature to recommend caffeine/coffee as a prophylactic against age-related cognitive and neuromotor decline. The question of caffeine's putative neuroprotective potential requires large-scale intervention studies involving random allocation of participants and long-term manipulation of caffeine consumption.

Similarly, in the absence of controls for withdrawal and withdrawal reversal, relatively few findings concerning the effects of caffeine on sleep and wakefulness can be interpreted with confidence. The few adequately controlled studies that have been conducted were mostly concerned with the adverse effects of sleep loss on performance and mood, and the potential of caffeine to counteract such effects. More studies are needed, especially in relation to the implications of caffeine as a countermeasure to sleepiness in situations where safety is a concern, including long-distance driving, shift work and military operations, in which caffeine has been claimed to offer substantial, even life-saving, benefits. However, taking account of withdrawal and withdrawal reversal effects, findings indicate that caffeine does not reverse negative effects of sleep loss on performance and mood (James et al. 1998; Keane & James 2008; Rogers et al. 2005). Indeed, current evidence indicates that rather than decreasing risk of harm, there is potential for dietary caffeine to increase risk of mishap, arising from the direct disruptive effects of caffeine on sleep (contributing to sleep debt) and sleepiness being exacerbated when consumers experience periods of relative caffeine deprivation.

References

Adan, A., & Serra-Grabulos, J. M. (2010). Effects of caffeine and glucose, alone and combined, on cognitive performance. *Human Psychopharmacology: Clinical and Experimental*, 25, 310–317.

Alsense, K., Deckert, J., Sand, P., & de Wit, H. (2003). Association between A_{2A} receptor gene polymorphisms and caffeine-induced anxiety. *Neuropsychopharmacology*, 28, 1694–1702.

Arnaud, M. J. (1987). The pharmacology of caffeine. *Progress in Drug Research*, 31, 273–313.

Arria, A. M., Caldeira, K. M., Kasperski, S. J., O'Grady, K. E., Vincent, K. B., Griffiths, R. R., & Wish, E. D. (2010). Increased alcohol consumption, nonmedical prescription drug use, and illicit drug use are associated with energy drink consumption among college students. *Journal of Addiction Medicine*, 4, 74–80.

Ascherio, A., Zhang, S. M., Hernan, M. A., Kawachi, I., Colditz, G. A., Speizer, F. E. et al. (2001). Prospective study of caffeine consumption and risk of Parkinson's disease in men and women. *Annals of Neurology*, 50, 56–63.

Ashihara, H. & Crozier, A. (2001). Caffeine: A well known but little mentioned compound in plant science. *Trends in Plant Science*, 6, 407–413.

Back, A. A., & Grobbee, D. E. (1990). A randomized study on coffee and blood pressure. *Journal of Human Hypertension*, 4, 259–264.

Basheer, R., Strecke, R. E., Thakkar, M. M., & McCarley, R. W. (2004). Adenosine and sleep–wake regulation. *Progress in Neurobiology*, 73, 379–396.

Biaggioni, I., Paul, S., Puckett, A., & Arzubiaga, C. (1991). Caffeine and theophylline as adenosine receptor antagonists in humans. *Journal of Pharmacology and Experimental Therapeutics*, 258, 588–593.

Bonita, J. S., Mandarano, M., Shuta, D., & Vinson, J. (2007). Coffee and cardiovascular disease: *In vitro*, cellular, animal, and human studies. *Pharmacological Research*, 55, 187–198.

Bonnet, M. H. & Arand, D. L. (1994). Impact of naps and caffeine on extended nocturnal performance. *Physiology and Behavior*, 56, 103–109.

Bonnet, M. H., Balkin, T. J., Dinges, D. F., Roehrs, T., Rogers, N. L, & Wesensten, N. J. (2005). The use of stimulants to modify performance during sleep loss: A review by the sleep deprivation and stimulant task force of the American Academy of Sleep Medicine. *Sleep*, 28: 1163–1187.

Bonnet, M. H., Gomez, S., Wirt, O., & Arand, D. (1995). The use of caffeine versus prophylactic naps in sustained performance. *Sleep*, 18: 97–104.

Brazier, J. L., & Salle, B. (1981). Conversion of theophylline to caffeine by the human fetus. *Seminars in Perinatology*, 5, 315–320.

Brunyé, T. T., Mahoney, C. R., Lieberman. H. R., & Taylor, H. A. (2010a). Caffeine modulates attention network function. *Brain and Cognition*, 72, 181–188.

Brunyé, T. T., Mahoney, C. R., Lieberman. H. R., Giles, G. E., & Taylor, H. A. (2010b). Acute caffeine consumption enhances the executive control of visual attention in habitual consumers. *Brain and Cognition*, 74, 186–192.

Bush, A., Busst, C. M., Clarke, B., & Barnes, P. J. (1989). Effect of infused adenosine on cardiac output and systemic resistance in normal subjects. *British Journal of Clinical Pharmacology*, 27, 165–171.

Carter, A. J., O'Connor, W. T., Carter, M. J., & Ungerstedt, U. (1995). Caffeine enhances acetylcholine release in the hippocampus in vivo by a selective interaction with adenosine A1 receptors. *Journal of Pharmacology and Experimental Therapeutics*, 273, 637–642.

Chen, J. F., Xu, K., Petzer, J. P., Staal, R., Xu, Y. H., Beilstein, M. et al. (2001). Neuroprotection by caffeine and A(2A) adenosine receptor inactivation in a model of Parkinson's disease. *Journal of Neuroscience*, 21, RC143.

Childs, E., & de Wit, H. (2005). Enhanced mood and psychomotor performance by a caffeine-containing energy capsule in fatigued individuals. *Experimental and Clinical Psychopharmacology*, 16, 13–21.

Christopher, G., Sutherland, D., &Smith, A. (2005). Effects of caffeine in non-withdrawn volunteers. *Human Psychopharmacology: Clinical and Experimental*, 20, 47–53.

Chvasta, T. E., & Cooke, A. R. (1971). Absorption and emptying of caffeine from the human stomach. *Gastroenterology*, 61, 838–843.

Corley, J., Jia, X., Kyle, J. A. M., Gow, A. J., Brett, C. E., Starr, J. M., McNeill, G., & Deary, I. J. (2010). Caffeine consumption and cognitive function at age 70: The Lothian Birth Cohort 1936 Study. *Psychosomatic Medicine*, 72, 206–214.

Costa, J., Lunet, N., Santos, C., Santos, J., & Vaz-Carneiro, A. (2010). Caffeine exposure and the risk of Parkinson's disease: A systematic review and meta-analysis of observational studies. *Journal of Alzheimer's Disease*, 20, S221–S238.

Cunha, R. A., Ferré, S., Vaugeois, J-M., & Chen, J-F. (2008). Potential therapeutic interest of adenosine A_{2A} receptors in psychiatric disorders. *Current Pharmaceutical Design*, 14, 1512–1524.

Denaro, C. P., Brown, C. R., Jacob, P. I., & Benowitz, N. L. (1991). Effects of caffeine with repeated dosing. *European Journal of Clinical Pharmacology*, 40, 273–278.

Dews, P. B. (1984). *Behavioral effects of caffeine.* Berlin: Springer-Verlag.

Dews, P. B., O'Brien, C. P., Bergman, J. (2002). Caffeine: Behavioural effects of withdrawal and related issues. *Food and Chemical Toxicology*, 40, 1257–1261.

Dumas, M., Gouyon, J. B., Tenenbaum, D., Michiels, Y., Escousse, A., & Alison, M. (1982). Systematic determination of caffeine plasma concentrations at birth in preterm and full-term infants. *Developmental Pharmacology and Therapeutics*, 4, 182–186.

Dunwiddie, T. V., & Masino, S. A. (2001). The role and regulation of adenosine in the central nervous system. *Annual Review of Neuroscience*, 24, 31–55.

Eskelinen, M. H., & Kivipelto, M. (2010). Caffeine as a protective factor in dementia and Alzheimer's disease. *Journal of Alzheimer's Disease*, 20, S167–S174.

Evans, A. H., Lawrence, A. D., Potts, J., MacGregor, L., Katzenschlager, R., Shaw, K. et al. (2006). Relationship between impulsive sensation seeking traits, smoking, alcohol and caffeine intake, and Parkinson's disease. *Journal of Neurology, Neurosurgery, and Psychiatry*, 77, 317–321.

Evans, S. M., Critchfield, T. S., & Griffiths, R. R. (1994). Caffeine reinforcement demonstrated in a majority of moderate caffeine users. *Behavioural Pharmacology*, 5, 231–238.

Evans, S. M., & Griffiths, R. R. (1991). Dose-related caffeine discrimination in normal volunteers: Individual differences in subjective effects and self-reported cues. *Behavioural Pharmacology*, 2, 345–356.

Ferré, S., Schwarcz, R., Li, X. M., Snaprud, P., Ögren, S. O., & Fuxe, K. (1994). Chronic haloperidol treatment leads to an increase in the intramembrane interaction between adenosine A2 and dopamine D2 receptors in the neostriatum. *Psychopharmacology*, 116, 279–284.

Finn, I. B., & Holtzman, S. G. (1987). Pharmacologic specificity of tolerance to caffeine-induced stimulation of locomotor activity. *Psychopharmacology*, 93, 428–434.

Franchetti, P., Messini, L., Cappellacci, L., Grifantini, M., Lucacchini, A., Martini, C., & Senatore, G. (1994). 8-Azaxanthine derivatives as antagonists of adenosine receptors. *Journal of Medicinal Chemistry*, 37, 2970–2975.

Garrett, B. E., & Griffiths, R. R. (1998). Physical dependence increases the relative reinforcing effects of caffeine versus placebo. *Psychopharmacology*, 139, 195–202.

Garrett, B. E., & Holtzman, S. G. (1994). Caffeine cross-tolerance to selective dopamine D1 and D2 receptor agonists but not to their synergistic interaction. *European Journal of Pharmacology*, 262, 65–75.

Gilbert, R. M. (1984). Caffeine consumption. In: Spiller, G. A. (ed.) *The Methylxanthine Beverages and Foods: Chemistry, Consumption, and Health Effects.* New York: Alan R. Liss, Inc.

Gilbert, S. G., & Rice, D. C. (1994). In utero caffeine exposure affects feeding pattern and variable ratio performance in infant monkeys. *Fundamental and Applied Toxicology*, 22, 41–50.

Griffiths, R. R., Bigelow, G. E., & Liebson, I. A. (1986). Human coffee drinking: Reinforcing and physical dependence producing effects of caffeine. *Journal of Pharmacology and Experimental Therapeutics*, 239, 416–425.

Griffiths, R. R., Evans, S. M., Heishman, S. J., Preston, K. L., Sannerud, C. A., Wolf, B., & Woodson, P. P. (1990). Low-dose caffeine physical dependence in humans. *Journal of Pharmacology and Experimental Therapeutics*, 255, 1123–1132.

Griffiths, R. R., Juliano, L. M., & Chausmer, A. L. (2003). Caffeine: Pharmacology and clinical effects. In: Graham, A. W., Schultz, T. K, Mayo-Smith. M. F., Ries, R. K., & Wilford, B. B. (eds) *Principles of Addiction Medicine* (3rd ed.). Chevy Chase, MD: American Society of Addiction Medicine, pp. 193–134.

Haskell, C. F., Kennedy, D. O., Wesnes, K. A., & Scholey, A. B. (2005). Cognitive and mood improvements of caffeine in habitual consumers and habitual non-consumers of caffeine. *Psychopharmacology*, 179, 813–825.

Haskell, C. F., Kennedy, D. O., Milne, A. L., Wesnes, K. A., & Scholey, A. B. (2008). The effects of L-theanine, caffeine and their combination on cognition and mood. *Biological Psychology*, 77, 113–122.

Hayashi. M., Masuda, A., &Hori, T. (2003). The alerting effects of caffeine, bright light and face washing after a short daytime nap. *Clinical Neurophsyiology*, 114, 2268–2278.

Heatherley, S. V., Hayward, R. C., Seers, H. E., & Rogers, P. J. (2005). Cognitive and psychomotor performance, mood, and pressor effects of caffeine after 4, 6 and 8 h caffeine abstinence. *Psychopharmacology*, 178, 461–470.

Heatherley, S. V., Hancock, K. M. F., & Rogers, P. J. (2006). Psychostimulant and other effects of caffeine in 9- to 11-year-old children. *Journal of Child Psychology and Psychiatry*, 47, 135–142.

Hernan, M. A., Takkouche, B., Caamano-Isorna, F., & Gestal-Otero, J. J. (2002). A meta-analysis of coffee drinking, cigarette smoking, and the risk of Parkinson's disease. *Annals of Neurology*, 52, 276–284.

Hogervorst, E., Bandelow, S., Schmitt, J., Jentjens, R., OliveirA, M., Allgrove, J., Carter, T., & Gleeson, M. (2008). Caffeine improves physical and cognitive performance during exhaustive exercise. *Medicine and Science in Sports and Exercise*, 40, 1841–1851.

Hollingworth, H. L. (1912a). The influence of caffeine on mental and motor efficiency. *Archives of Psychology*, 22, 1–166.

Hollingworth, H. L. (1912b). The influence of caffeine on the speed and quality of performance in typewriting. *Psychological Review*, 19, 66–73.

Holtzman, S. G., & Finn, I. B. (1988). Tolerance to behavioral effects of caffeine in rats. *Pharmacology, Biochemistry, and Behavior*, 29, 411–418.

Horne, J. A., & Reyner, L. A. (1996). Counteracting driver sleepiness: Effects of napping, caffeine, and placebo. *Psychophysiology*, 33, 306–309.

Horne, J. A., & Reyner, L. A. (2001a). Beneficial effects of an "energy drink" given to sleepy drivers. *Amino Acids*, 20: 83–89.

Horne, J. A., & Reyner, L. A. (2001b). Sleep-related vehicle accidents: Some guidelines for road safety policies. *Transportation Research Part F: Traffic Psychology and Behaviour*, 4, 63–74.

Hughes, J. R., Higgins, S. T., Bickel, W. K., Hunt, W. K., Fenwick, J. W., Gulliver, S. B., & Mireault, G. C. (1991). Caffeine self-administration, withdrawal, and adverse effects among coffee drinkers. *Archives of General Psychiatry*, 48, 611–617.

Hughes, J. R., Oliveto, A. H., Bickel, W. K., Higgins, S. T., & Badger, G. J. (1993). Caffeine self-administration and withdrawal: Incidence, individual differences and interrelationships. *Drug and Alcohol Dependence*, 32, 239–246.

Hughes, J. R., Oliveto, A. H., Helzer, J. E., Higgins, S. T., & Bickel, W. K. (1992). Should caffeine abuse, dependence or withdrawal be added to DSM-IV and ICD-10? *American Journal of Psychiatry*, 149, 33–40.

James, J. E. (1991). *Caffeine and health* (pp. 430). London: Academic Press.

James, J. E. (1994a). Psychophysiological effects of habitual caffeine consumption. *International Journal of Behavioral Medicine*, 1, 247–263.

James, J. E. (1994b). Chronic effects of habitual caffeine consumption on laboratory and ambulatory blood pressure levels. *Journal of Cardiovascular Research*, 1, 159–164.

James, J. E. (1994c). Does caffeine enhance or merely restore degraded psychomotor performance? *Neuropsychobiology*, 30, 124–125.

James, J. E. (1995). Caffeine and psychomotor performance revisited. *Neuropsychobiology*, 31, 202–203.

James, J. E. (1997a). *Understanding Caffeine: A Biobehavioral Analysis*. Thousand Oaks, CA: Sage Publications.

James, J. E. (1997b). Caffeine and blood pressure: Habitual use is a preventable cardiovascular risk factor. *The Lancet*, 349, 279–281.

James, J. E. (1998). Acute and chronic effects of caffeine on performance, mood, headache, and sleep. *Neuropsychobiology*, 38, 32–41.

James, J. E. (2004). A critical review of dietary caffeine and blood pressure: A relationship that should be taken more seriously. *Psychosomatic Medicine*, 66, 63–71.

James, J. E. (2010). Caffeine. In B. Johnson (ed.), *Addiction Medicine: Science and Practice*. New York, NY: Springer, pp. 551–583.

James, J. E. (2011). Editorial: A new journal to advance caffeine research. *Journal of Caffeine Research*, 1, 1–3.

James, J. E., & Gregg, M. E. (2004a). Effects of dietary caffeine on mood when rested and sleep restricted. *Human Psychopharmacology: Clinical and Experimental*, 19, 333–341.

James, J. E., & Gregg, M. E. (2004b). Hemodynamic effects of dietary caffeine, sleep restriction, and laboratory stress. *Psychophysiology*, 41, 914–923.

James, J. E., Gregg, M. E., Kane, M., & Harte, F. (2005). Dietary caffeine, performance and mood: Enhancing and restorative effects after controlling for withdrawal relief. *Neuropsychobiology*, 52, 1–10.

James, J. E., & Keane, M. A. (2007). Caffeine, sleep and wakefulness: Implications of new understanding about withdrawal reversal. *Human Psychopharmacology: Clinical & Experimental*, 22, 549–558.

James, J. E., Kristjansson, A. L., & Sigfusdottir, I. D. (2011). Adolescent substance use, sleep, and academic achievement: Evidence of harm due to caffeine. *Journal of Adolescence*, 34, 665–673.

James, J. E., & Rogers, P. J. (2005). Effects of caffeine on performance and mood: Withdrawal reversal is the most plausible explanation. *Psychopharmacology*, 182, 1–8.

Jarvis, M. J. (1993). Does caffeine intake enhance absolute levels of cognitive performance? *Psychopharmacology*, 110, 45–52.

Jay, S. M., Petrilli, R. M., Ferguson, S. A., Dawson, D., & Lamond, N. (2006). The suitability of a caffeinated energy drink for night-shift workers. *Physiology and Behaviour*, 87, 925–931.

Jee, S. H., He, J., Whelton, P. K., Suh, I., & King J. (1999). The effect of chronic coffee drinking on blood pressure: A meta-analysis of controlled clinical trials. *Hypertension*, 33, 647–652.

Jones, H. E., Herning, R. I., Cadet, J. L., & Griffiths, R. R. (2000). Caffeine withdrawal increases cerebral blood flow and alters quantitative electroencephalography (EEG) activity. *Psychopharmacology*, 147, 371–377.

Juliano, L. M., and Griffiths, R. R. (2004). A critical review of caffeine withdrawal: Empirical validation of symptoms and signs, incidence, severity, and associated features. *Psychopharmacology*, 176, 1–29.

Kalda, A., Yu, L., Oztas, E., & Chen, J. F. (2006). Novel neuroprotection by caffeine and adenosine A(2A) receptor antagonists in animal models of Parkinson's disease. *Journal of the Neurological Sciences*, 248, 9–15.

Kamimori, G H, Karyekar, C. S., Otterstetter, R., Cox, D. S., Balkin, T. J., Belenky, G. L., et al. (2002). The rate of absorption and relative bioavailability of caffeine administered in chewing gum versus capsules to normal healthy volunteers. *International Journal of Pharmaceutics*, 234, 159–167.

Keane, M. A., & James, J. E. (2008). Effects of dietary caffeine on EEG, performance, and mood when rested and sleep restricted. *Human Psychopharmacology: Clinical and Experimental*, 23, 669–680.

Keane, M. A., James, J. E., & Hogan, M. J. (2007). Effects of dietary caffeine on topographic EEG after controlling for withdrawal and withdrawal reversal. *Neuropsychobiology*, 56, 197–207.

Kot, M., & Daniel, W. A. (2008). Caffeine as a marker substrate for testing cytochrome P450 activity in human and rat. *Pharmacological Reports*, 60, 789–797.

Kristjansson, A. L., Sigfusdottir, I. D., Allegrante, J. P., & James, J. E. (2011). Adolescent caffeine consumption, daytime sleepiness, and anger. *Journal of Caffeine Research*, 1, 75–82.

LaJambe, C. M., Kamimori, G. H., Belenky, G., Balkin, T. J. (2005). Caffeine effects of recovery sleep following 27 h total sleep deprivation. *Aviation, Space, and Environmental Medicine*, 76, 108–113.

Lane, J. D. (1994). Neuroendocrine responses to caffeine in the work environment. *Psychosomatic Medicine*, 546, 267–270.

Lane, J. D. (1997). Effects of brief caffeine deprivation on mood, symptoms, and psychomotor performance. *Pharmacology, Biochemistry, and Behavior*, 58, 203–208.

Lane, J. D., & Phillips-Bute, B. G. (1998). Caffeine deprivation affects vigilance performance and mood. *Physiology and Behavior*, 65, 171–175.

Lane, J. D., & Williams, R. B. (1987). Cardiovascular effects of caffeine and stress in regular coffee drinkers. *Psychophysiology*, 24, 157–164.

Lane, J. D., Adcock, R. A., Williams, R. B., & Kuhn, C. M. (1990). Caffeine effects on cardiovascular and neuroendocrine responses to acute psychosocial stress and their relationship to level of habitual caffeine consumption. *Psychosomatic Medicine*, 52, 320–336.

Lelo, A., Miners, J. O., Robson, R., & Birkett, D. J. (1986). Assessment of caffeine exposure: Caffeine content of beverages, caffeine intake, and plasma concentrations of methylxanthines. *Clinical Pharmacology and Therapeutics*, 39, 54–59.

Lieberman, H. R., Wurtman, R. J., Emde, G. G., Roberts, C., & Coviella, I. L. G. (1987). The effects of low doses of caffeine on human performance and mood. *Psychopharmacology*, 92, 308–312.

Lindsay, J., Laurin, D., Verreault, R., Hébert, R., Helliwell, B., Hill, G. B., & McDowell, I. (2002). Risk factors for Alzheimer's disease: A prospective analysis from the Canadian study of aging. *American Journal of Epidemiology*, 156, 445–453.

Lovallo, W. R., Farag, N. H., Vincent, A. S., Thomas, T. L., & Wilson, M. F. (2006). Cortisol responses to mental stress, exercise, and meals following caffeine intake in men and women. *Pharmacology, Biochemistry, and Behavior*, 83, 441–447.

Lovallo, W. R., Pincomb, G. A., Sung, B. H., Passey, R. B., Suasen, K. P., & Wilson, M. F. (1989). Caffeine may potentiate adrenocortical stress responses in hypertension-prone men. *Hypertension*, 14, 170–176.

Maia, L., & de Mendonca, A. (2002). Does caffeine intake protect from Alzheimer's disease? *European Journal of Neurology*, 9, 377–382.

Marangos, P. J., & Boulenger, J. P. (1985). Basic and clinical aspects of adenosinergic neuromodulation. *Neuroscience and Biobehavioral Reviews*, 9, 421–430.

McLellan, T. M., Kamimori, G. H., Bell, D. G, Smith, I. G, Johnson, D., & Belenky, G. (2005)a. Caffeine maintains vigilance and marksmanship in simulated urban operations with sleep deprivation. *Aviation, Space, and Environmental Medicine*, 76, 39–45.

McLellan, T. M., Kamimori, G. H., Voss, D. M, Bell, D. G, Cole, K. G, & Johnson, D. (2005b). Caffeine maintains vigilance and improves run times during night operations for special forces. *Aviation, Space, and Environmental Medicine*, 76, 647–654.

Menza, M. (2000). The personality associated with Parkinson's disease. *Current Psychiatry Reports*, 2, 421–426.

Mitchell, S. H., de Wit, H., & Zacny, J. P. (1995). Caffeine withdrawal symptoms and self-administration following caffeine deprivation. *Pharmacology, Biochemistry, and Behavior*, 51, 941–945.

National Sleep Foundation. (2006). *2006 Sleep in America Poll: Summary of Findings*. Washington, DC: National Sleep Foundation.

Nehlig, A. (2010). Is caffeine a cognitive enhancer? *Journal of Alzheimer's Disease*, 20, S85–S94.

Oberman, Z., Harell, A., Herzberg, M., Hoerer, E., Jaskolka, H., & Laurian, L. (1975). Changes in plasma cortisol, glucose and free fatty acids after caffeine ingestion in obese women. *Israel Journal of Medical Sciences*, 11, 33–36.

Paul, S., Kurunwune, B., & Biaggioni, I. (1993). Caffeine withdrawal: Apparent heterologous sensitization to adenosine and prostacyclin actions in human platelets. *Journal of Pharmacology and Experimental Therapeutics*, 267, 838–843.

Pfeifer, R. W., & Notari, R. E. (1988). Predicting caffeine plasma concentrations resulting from consumption of food or beverages: A simple method and its origin. *Drug Intelligence and Clinical Pharmacy*, 22, 953–959.

Philip, P., Taillard, J., Moore, N., Delord, S., Valtat, C., Sagaspe, P., et al. (2006). The effects of coffee and napping on nighttime highway driving. *Annals of Internal Medicine*, 144, 785–791.

Phillips-Bute, B. G., & Lane, J. D. (1998). Caffeine withdrawal symptoms following brief caffeine deprivation. *Physiology and Behavior*, 63, 35–39.

Pincomb, G. A., Lovallo, W. R., Passey, R. B., & Wilson, M. F. (1988). Effect of behavior state on caffeine's ability to alter blood pressure. *American Journal of Cardiology*, 61, 798–802.

Pincomb, G. A., Lovallo, W. R., Passey, R. B., Brackett, D. J., & Wilson, M. F. (1987). Caffeine enhances the physiological response to occupational stress in medical students. *Health Psychology*, 6, 101–112.

Quintana, J. L. B., Allam, M. F., Del Castillo, A. S., & Navajas, R. F-C. (2007). Alzheimer's disease and coffee: A quantitative review. *Neurological Research*, 29, 91–95.

Rall, T. W. (1990). Drugs used in the treatment of asthma. The methylxanthines, cromolyn sodium, and other agents. In A. G. Gilman, T. W. Rall, A. S. Nies, & P. Taylor (Eds.), *Goodman and Gilman's The Pharmacological Basis of Therapeutics* (pp. 618–637). New York: Pergamon Press.

Reeves, R. R, Struve, F. A., Patrick, G., (Bullen, J. A. (1995). Topographic quantitative EEG measures of alpha and theta power changes during caffeine withdrawal: Preliminary findings from normal subjects. *Clinical Electroencephalography*, 26, 154–162.

Reissig, C. J., Strain, E. C., & Griffiths, R. R. (2009). Caffeinated energy drinks: A growing problem. *Drug and Alcohol Dependence*, 99, 1–10.

Reyner, L. A., & Horne, J. A. (1997). Suppression of sleepiness in drivers: Combination of caffeine with a short nap. *Psychophysiology*, 34: 721–725.

Reyner, L. A., & Horne, J. A. (2000). Early morning driver sleepiness: Effectiveness of 200 mg caffeine. *Psychophysiology*, 37: 251–256.

Reyner, L. A., & Horne, J. A. (2002). Efficacy of a 'functional energy drink' in counteracting driver sleepiness. *Physiology and Behavior*, 75, 331–335.

Richardson, N. J., Rogers, P. J., Elliman, N. A., O'Dell, R. J. (1995). Mood and performance effects of caffeine in relation to acute and chronic caffeine deprivation. *Pharmacology, Biochemistry, and Behavior*, 52, 313–320.

Robelin, M., & Rogers, P. J. (1998). Mood and psychomotor performance effects of the first, but not of subsequent, cup-of-coffee equivalent doses of caffeine consumed after overnight caffeine abstinence. *Behavioural Pharmacology*, 9, 611–618.

Robertson, D., Wade, D., Workman, R., Woosley, R. L., & Oates, J. A. (1981). Tolerance to the humoral and hemodynamic effects of caffeine in man. *Journal of Clinical Investigation*, 67, 1111–1117.

Rogers, P. J., Heatherley, S. V., Hayward, R. C., Sears, H. E., Hill, J.,& Kane, M. (2005). Effects of caffeine and caffeine withdrawal on mood and cognitive performance degraded by sleep restriction. *Psychopharmacology*, 179, 742–752.

Rogers, P. J., Martin, J., Smith, C., Heatherley, S. V., Smit, H. J. (2003). Absence of reinforcing, mood and psychomotor performance effects of caffeine in habitual non-consumers of caffeine. *Psychopharmacology*, 167, 54–62.

Rosenthal, L., Roehrs, T., Zwyghuizen-Doorenbos, A., Plath, D., & Roth, T. (1991). Alerting effects of caffeine after normal and restricted sleep. *Neuropsychpharmacology*, 4, 103–108.

Ross, G. W., Abbott, R. D., Petrovitch, H., Morens, D. M., Grandinetti, A., Tung, K. H. et al. (2000). Association of coffee and caffeine intake with the risk of Parkinson disease. *Journal of the American Medical Association*, 283, 2674–2679

Schiffman, S. S., & Warwick, Z. S. (1989). *Use of Flavor-amplified Foods to Improve Nutritional Status in Elderly Persons*. New York: The New York Academy of Sciences.

Schwarzschild, M. A., Xu, K., Oztas, E., Petzer, J. P., Castagnoli, K., Castagnoli, N. Jr. et al. (2003). Neuroprotection by caffeine and more specific A2A receptor antagonists in animal models of Parkinson's disease. *Neurology*, 61(Suppl. 6), S55–S61.

Schweitzer, P. K., Randazzo, A. C., Stone, K., Erman, M., & Walsh, J. K. (2006). Laboratory and field studies of naps and caffeine as practical countermeasures for sleep-wake problems associated with night work. *Sleep*, 29, 39–50.

Seng, K-Y., Teo. W-L. G., Fun, C-Y. D., Law, Y-L. L., & Lim, C-L. (2010).Inter-relations between plasma caffeine concentrations and neurobehavioural effects in healthy volunteers: Model analysis using NONMEM. *Biopharmaceutics and Drug Disposition*, 31: 316–330.

Serra-Grabulosa1, J. M., Adan. A., Falcón, C., & Bargallo, N. (2010). Glucose and caffeine effects on sustained attention: An exploratory fMRI study. *Human Psychopharmacology: Clinical and Experimental*, 25, 543–552

Shi, J., Benowitz, N. L., Denaro, C. P., & Sheiner, L. B. (1993). Pharmacokinetic-pharmacodynamic modeling of caffeine: Tolerance to pressor effects. *Clinical Pharmacology and Therapeutics*, 53, 6–14.

Shilo, L., Sabbah, H., Hadari, R., Kovatz, S., Weinberg, U., Dolev, S., et al. (2002). The effects of coffee consumption on sleep and melatonin secretion. *Sleep Medicine*, 3, 271–273.

Sigmon, S. S., & Griffiths, R. R. (2011). Caffeine choice prospectively predicts positive subjective effects of caffeine and *d*-amphetamine. *Drug and Alcohol Dependence*, 118, 341– 348.

Smillie, L. D., & Gökcen, E. (2010). Caffeine enhances working memory for extraverts. *Biological Psychology*, 85, 496–498.

Sigmon, S. C., Herning, R. I., Better, W., & Cadet, J. L., & Griffiths, R. R. (2009). Caffeine withdrawal, acute effects, tolerance, and absence of net beneficial effects of chronic administration: Cerebral blood flow velocity, quantitative EEG, and subjective effects. *Psychopharmacology*, 204, 573–585.

Smit, H. J., & Rogers, P. J. (2000). Effects of caffeine on cognitive performance, mood and thirst in lower and higher caffeine consumers. *Psychopharmacology*, 152, 167–173.

Smith, A. (2002). Effects of caffeine on human behaviour. *Food and Chemical Toxicology*, 40,1243–1255.

Smith, A. (2009). Effects of caffeine in chewing gum on mood and attention. *Human Psychopharmacology: Clinical and Experimental*, 24, 239–247.

Somani, S. M., & Gupta, P. (1988). Caffeine: A new look at an age-old drug. *International Journal of Clinical Pharmacology, Therapy, and Toxicology*, 26, 521–533.

Spindel, E. R., Wurtman, R. J., McCall, A., Carr, D., Conlay, L., Griffith, L., & Arnold, M. A. (1984). Neuroendocrine effects of caffeine in normal subjects. *Clinical Pharmacology and Therapeutics*, 36, 402–407.

Stafford, L. D., & Yeomans, M. R. (2005). Caffeine deprivation state modulates coffee consumption but not attentional bias for caffeine-related stimuli. *Behavioral Pharmacology*, 16(7), 559–571.

Stavric, B. (1988). Methylxanthines: Toxicity to humans. 2. Caffeine. *Food and Chemical Toxicology*, 26, 645–662.

Streufert, S., Pogash, R., Miller, J., Gingrich, D., Landis, R., Lonardi, L., Severs, W., Roache, J. D. (1995). Effects of caffeine deprivation on complex human functioning. *Psychopharmacology*, 118, 377–384.

Syed, S. A., Kamimori, G. H., Kelly, W., & Eddington, N. D. (2005). Multiple dose pharmacokinetics of caffeine administered in chewing gum to normal healthy volunteers. *Biopharmaceutics and Drug Disposition*, 26, 403–409.

Temple, J. L. (2009). Caffeine use in children: What we know, what we have left to learn, and why we should worry. *Neuroscience and Biobehavioral Reviews*, 33, 793–806.

Tikuisis, P., Keefe, A. A., McLellan, T. M., & Kamimori, G. (2004). Caffeine restores engagement speed but not shooting precision following 22 h of active wakefulness. *Aviation, Space, and Environmental Medicine*, 75, 771–776.

Tinley, E. M., Yeomans, M. R., & Durlach, P. J. (2003). Caffeine reinforces flavour preference in caffeine-dependent, but not long-term withdrawn, caffeine consumers. *Psychopharmacology*, 166, 416–423.

Todes, C. J., & Lees, A. J. (1985). The pre-morbid personality of patients with Parkinson's disease. *Journal of Neurology, Neurosurgery, and Psychiatry*, 48, 97–100.

Turkheimer, E. (1998). Heritability and biological explanation. *Psychological Review*, 105, 782–791.

van Boxtel, M. P., Schmitt, J. A., Bosma, H., Jolles, J. (2003). The effects of habitual caffeine use on cognitive change: A longitudinal perspective. *Pharmacology, Biochemistry, and Behavior*, 75, 921–927.

van Duinen, H., Lorist, M. M., & Zijdewind, I. (2005). The effect of caffeine on cognitive task performance and motor fatigue. *Psychopharmacology*, 180, 539–547.

van Dusseldorp, M., & Katan, M. B. (1990). Headache caused by caffeine withdrawal among moderate coffee drinkers switched from ordinary to decaffeinated coffee: A 12 week double blind trial. *British Medical Journal*, 300, 1558–1559.

Van't Hoff, W. (1982). Caffeine in pregnancy. *Lancet*, 2, 1020.

von Borstel, R. W., & Wurtman, R. J. (1982). Caffeine withdrawal enhances sensitivity to physiologic level of adenosine in vivo. *Federation Proceedings*, 41, 1669.

Walsh, J. K., Muelbach, M. J., Humm, T. M., Dickens, Q. S., & Sugerman, J. L. (1990). Effect of caffeine on physiological sleep tendency and ability to sustain wakefulness at night. *Psychopharmacology*, 101, 271–273.

Warburton, D. M. (1995). Effects of caffeine on cognition and mood without caffeine abstinence. *Psychopharmacology*, 119, 66–70.

Warburton, D. M., Bersellini, E., & Sweeney, E. (2001). An evaluation of a caffeinated taurine drink on mood, memory and information processing in healthy volunteers without caffeine abstinence. *Psychopharmacology*, 158, 322–328.

Watson, J. M., Lunt, M. J., Morris, S., Weiss, M. J., Hussey, D., & Kerr, D. (2000). Reversal of caffeine withdrawal by ingestion of a soft beverage. *Pharmacology, Biochemistry, and Behavior*, 66, 15–18.

Watt, A. H., Bayer, A., Routledge, P. A., & Swift, C. G. (1989). Adenosine-induced respiratory and heart rate changes in young and elderly adults. *British Journal of Clinical Pharmacology*, 27, 265–267.

Wright, K. P., Badia, P., Myers, B. L., & Plenzler, S. C. (1997). Combination of bright light and caffeine as a countermeasure for impaired alertness and performance during extended sleep deprivation. *Journal of Sleep Research*, 6: 26–35.

Wright, K. P., Myers, B. L., Plenzler, S. C., Drake, C. L., & Badia, P. (2000). Acute effects of bright light and caffeine on nighttime melatonin and temperature levels in women taking and not taking oral contraceptives. *Brain Research*, 873, 310–317.

Yeomans, M. R., Ripley, T., Davies, L. H., Rusted, J. M., & Rogers, P. J. (2002). Effects of caffeine on performance and mood depend on the level of caffeine abstinence. *Psychopharmacology*, 164, 241–249.

CHAPTER 15

Herbal Extracts and Cognition in Adulthood and Ageing

David Camfield, Lauren Owen, Andrew Pipingas, Con Stough and Andrew Scholey

Chapter Overview

- Cognitive ageing involves multiple interacting systems which may be influenced by components of plants. Thus, certain herbal extracts may be effective in maintaining psychological health or 'wellbeing', particularly in ageing.
- Certain plants have evolved with components that can modulate behaviour, including cognitive performance.
- There is good evidence that certain extracts have properties that enhance cognition. These include extracts of ginkgo, ginseng, salvia, guarana, lemon balm, bacopa and polyphenols.
- In the domain of mood, lemon balm has consistently been shown to have a calming effect; evidence for other herbals is less clear, although cocoa polyphenols may have anti-fatigue effects.
- One constant challenge for the psychopharmacology of herbal extracts is the use of standardised extracts and the use of multiple extracts in some medicinal systems.

Introduction

The use of herbal products has become widespread in recent years, with surveys reporting that around 19 per cent of the population in the USA (38 million people) are now using herbal medicines (Kennedy 2005; Tindle et al. 2005). Similar figures have also been reported in Australia, with an estimated 16.3 per cent of respondents reporting the use of western herbal medicine and 7.0 per cent reporting the use of Chinese herbal medicine (Xue et al. 2007). The total expenditure on complementary and alternative medicine (CAM) products in Australia is estimated to be around AU$1.86 billion per year (Xue et al. 2007), with spending on

302

alternative therapies estimated at nearly four times that on pharmaceuticals (MacLennan et al. 2002). While specific data as to the numbers of people using herbal extracts specifically for cognition are yet to be published, a large number of products with known cognitive-enhancing effects are among the most popular; for example, 23 per cent of those using herbal products report using ginseng and 20.1 per cent report using ginkgo biloba (Kennedy 2005).

There are numerous reasons as to why herbal extracts represent an attractive alternative for individuals seeking to enhance their cognitive function. First, it may be perceived that these 'natural' treatments have fewer side effects and of less severity compared with conventional pharmaceutical treatments. Second, it may be perceived that these types of treatments have mechanisms of action that are more suitable for achieving cumulative benefits with long-term (chronic) use. Third, the majority of herbal extracts are currently available over the counter at a significantly cheaper price than prescription-only pharmaceuticals.

Herbal treatments may be particularly attractive for adults approaching late middle age who are concerned with maintaining their cognitive abilities into old age. The world's population is ageing rapidly, with the proportion of the population over 60 years growing at a rate of around 2 per cent per annum in the developed world (United Nations 2009). Average life expectancies in Australia are among the highest in the world; 79.3 years for males and 83.9 years for females (Australian Bureau of Statistics 2009). These figures are 79 years for the USA and 80 years for the UK (World Health Organization). In 2007, individuals aged 65 years and over made up 13 per cent of the Australian population, with this proportion projected to increase to between 23 per cent and 25 per cent by 2056 (Australian Bureau of Statistics 2011). More recent data for the UK reveal that individuals aged over 65 years made up 17 per cent of the population in 2010, with this figure projected to reach 23 per cent by 2035. Considering that up to 50 per cent of adults aged 64 and over have reported difficulties with their memory (Reid & MacLullich 2006), it is apparent that effective interventions to enhance cognitive abilities in the elderly will become increasingly popular in an ageing western world.

The Cellular Basis of Age-Related Cognitive Decline

Declines in a range of cognitive abilities, including processing speed, episodic memory, spatial ability and reasoning, have all been found to be associated with normal ageing (Hedden & Gabrieli 2004). In the Seattle longitudinal study of ageing, where over 5000 participants were tested over a 35-year period, declines across all domains were evident after the age of 55 (Schaie 1994). These cognitive declines are correlated with age-related volumetric decreases in a number of brain regions; with the prefrontal cortex (PFC) being one of the most heavily affected brain regions, losing 5 per cent volume per decade after the age of 20 (compared with around 2–3 per cent per decade in other

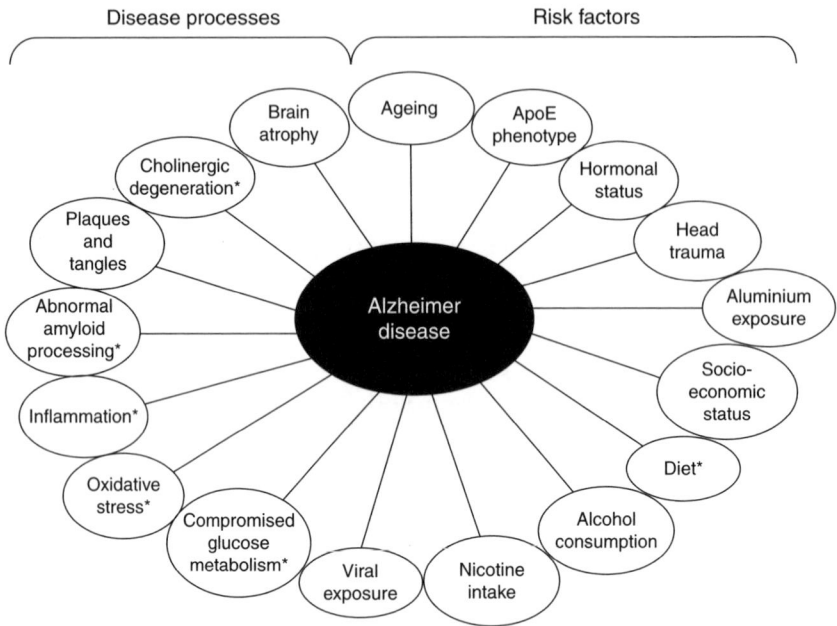

Figure 15.1   Schematic diagram of influences on the progression of Alzheimer's disease, illustrating the multi-faceted nature of the disorder, including disease processes towards the left and risk factors on the right (of which age is the single most important factor)

Processes marked with * indicate known influence of herbal extracts. Note that many of these processes also contribute to non-pathological cognitive ageing.

areas (Hedden & Gabrieli 2004)). These decreases in cortical volume can be attributed to a number of detrimental cellular processes that occur during ageing (Figure 15.1).

According to the 'free radical theory of ageing' (Harman 1957), organisms age because of the accumulation of damage caused by highly reactive (free radical) molecules with unpaired electrons in their outer shell (Riley 1994). Reactive oxygen species (ROS), such as superoxide, hydrogen peroxide and the hydroxyl radical, are found in the environment and are also produced in the body as a by-product of normal metabolism. While these molecules are necessary for certain biological processes, excessive levels do considerable cellular damage, especially to brain tissue (Halliwell 1992). The human brain, which consumes 20 per cent of the body's total oxygen and contains a large proportion of polyunsaturated fatty acids (PUFAs) in its neurons, is particularly vulnerable to damage by ROS (Halliwell 1992). In a process known as lipid peroxidation, ROS sequester electrons from cell membranes, causing oxidative damage to neurons. Mitochondrial deoxyribonucleic acid (DNA), ribonucleic acid (RNA) and proteins are common targets of this damage (Cadenas & Davies 2000).

A number of endogenous antioxidants, such as glutathione (GSH) and superoxide dismutase (SOD), are produced by the body in order to counteract the effects of excess ROS (Rossi et al. 2008). However, compared with other tissue, the brain produces relatively low levels of antioxidant enzymes (Floyd 1999), and as the brain ages it becomes increasingly vulnerable to oxidative damage (Gracy et al. 1999). Reasons for this increased susceptibility include an accumulation of transition metals, such as iron and copper, which catalyse hydroxyl radical production (Halliwell & Gutteridge 1990; Zecca et al. 2004), as well as a decreased ability to modify genes involved in the production of antioxidant enzymes (Lu et al. 2004).

In addition to cumulative damage from free radicals, a number of other cellular processes are also implicated in brain ageing. The brain's inflammatory response to oxidative stress is another important factor, with advanced age being found to be associated with a chronic low-grade inflammatory response (Sarkar & Fisher 2006). A decrease in mitochondrial efficiency also plays a role in cognitive decline. The high energy requirements of the mammalian brain mean that it has a relatively high number of mitochondria per cell (Veltri et al. 1990). While mitochondria are essential for meeting the energy requirements of the cell, they are also major contributors to ROS production. A large turnover of molecular oxygen is required by mitochondria in order to provide energy in the form of adenosine triphosphate (ATP), the cellular energy 'currency'. However, as the mitochondria process oxygen, a certain proportion of ROS are generated due to electron leakage (Kidd 2005). Over time, oxidative damage to the mitochondria accumulates, rendering them less efficient at producing energy. Further, oxidative damage to mitochondrial and cellular DNA eventually leads to cell death (Kidd 2005). Finally, in addition to the direct cellular damage to the brain that occurs during ageing, insults to microvascular structures also impair brain function, resulting in further cognitive declines (Pase et al. 2010).

Herbal Extracts and Mechanisms of Cognitive Enhancement

In consideration of the diverse aetiology of age-related cognitive decline, herbal treatments hold great potential for ameliorating the impact of brain ageing. This is because plant extracts have evolved in such a way as to typically contain multiple active constituents, with each constituent capable of targeting a different mechanism of brain ageing. As will be detailed below, when taken as a chronic dietary supplement, numerous herbal extracts have been found to have powerful antioxidant and anti-inflammatory properties in the brain, as well as aiding in metal-chelation, enhancing mitochondrial efficiency and bringing improvement to cerebrovascular function.

In addition to the neuroprotective effects associated with the chronic use of herbal extracts by the elderly, there are a number of acute (single dose) effects associated with certain herbs that may benefit adults of any age. These effects can be attributed to modulation of certain neurotransmitter systems in the brain, most notably (but not limited to) acetylcholine (ACh). The cortical cholinergic system is a diffuse, modulatory input system that innervates the entire cortex (Edeline 2003). While the primary purpose of the cholinergic system was originally thought to be the enhancement of sensory input processing, more recent evidence has extended this view to implicate it as playing a central role in attentional processes and memory encoding via region-specific regulation of cortical functions (Hasselmo & McGaughy 2004; Sarter & Bruno 1997; Weinberger 2004; Zaborszky 2002). Herbal extracts can enhance cholinergic neurotransmission in a number of ways, the most notable being direct agonism of ACh receptors (nicotinic or muscarinic) or inhibition of the acetylcholinesterase (ChE) enzyme, which is responsible for removing excess ACh in the extracellular space.

Evidence of Efficacy for Cognitive Enhancement Associated with Herbal Extracts

The following section will briefly summarise the current evidence for the acute and chronic cognitive-enhancing effects associated with a selection of herbal extracts.

Ginkgo Biloba

Ginkgo biloba extract (GBE) is taken from a single type of tree, known to be one of the world's oldest living species – a living fossil that has been in existence for over 150 million years (McKenna et al. 2001). Recorded uses of GBE in traditional Chinese medicine date back over 5000 years, with the seeds and leaves being used to treat a range of ailments, including pulmonary disorders, alcohol abuse, bladder inflammation, heart and lung dysfunctions, as well as skin infection (Mahady 2002; Smith & Luo 2004). Since the 1960s, when Dr Schwabe first introduced GBE to Germany, this extract has become increasingly popular in the western world as a herbal extract with purported cognitive-enhancing effects (Birks & Grimley Evans 2009; Smith & Luo 2004).

A standardised extract of *ginkgo biloba* (EGb 761) has been available in Europe since the early 1990s (Frank & Gupta 2005). There are two active groups of compounds in this extract: flavonoid glycosides (24 per cent) and terpene lactones (6 per cent) (Smith & Luo 2004). The flavonoids consist of quercetin, kaempferol and isorhamnetin, which are antioxidants that can trap ROS, modify the expression of endogenous antioxidants and chelate transition metal ions. The terpene lactones consist of the ginkgolides A, B, C, J and M, as well as bilobalide. The ginkgolides improve blood circulation through their action as platelet-activating factor antagonists (DeFeudis &

Drieu 2000; Ramassamy et al. 2007). GBE has also been found to be anti-inflammatory, enhance cholinergic transmission and preserve mitochondrial function (Mahadevan & Park 2008; Ramassamy et al. 2007; Smith & Luo 2004).

In studies investigating the cognitive effects of GBE in healthy adults, the findings have been mixed, with a number of extensive reviews failing to find consistent benefits associated with either acute or chronic use (Canter & Ernst 2007; Crews Jr et al. 2005). However, in relation to acute (single dose) studies in healthy adults, there is a certain amount of evidence to suggest that reliable cognitive benefits emerge when participants are tested under conditions of fatigue, or when cognitive tests of a higher degree of difficulty are used. Kennedy et al. (2000) administered 120, 240 and 360 mg of ginkgo (GK501, Pharmaton SA, standardised to contain 24 per cent flavone glycosides and 6 per cent terpene lactones) or placebo to 20 young adult participants in a single-dose study. Using the cognitive research test battery (comprising brief computerised neuropsychological assessments, including measures of attention, reaction time, working memory and episodic memory) to measure cognitive performance following dosing, a significant dose-dependent improvement in the 'speed of attention' factor was noted from 2.5 to 6 hours post-dose using both the 240 and 360 mg dose. In another acute study using young adults, Scholey et al. (2002) administered 120, 240 or 360 mg ginkgo (GK501 Pharmaton SA) to 18 participants as part of a multi-arm design. A dose-dependent improvement in the speed of response in the Serial Threes subtraction task was observed following ginkgo consumption. Improvements were observed for all doses of ginkgo at 4 hours, while for the 240 mg dose, improvement in response speed was also observed at 6 hours post-dose. In a re-analysis of three separate single-dose studies, Kennedy et al. (2007) investigated the effects of 120, 240 and 350 mg ginkgo (Indena SpA, Milan standardised to contain 24 per cent flavone glycosides and 6 per cent terpene lactones) in a combined sample of 78 healthy young participants. A significant improvement in the cognitive drug research (CDR) 'quality of memory' factor was reported at both 1 and 4 hours following 120 mg ginkgo. However, a negative effect on 'speed of attention' with this dose was also reported – an effect that was most evident at one and 6 hours post-dose. The authors noted that the effect on speed of attention at this lower dose is in the opposite direction to what has been previously observed using larger doses (Kennedy et al. 2000).

A number of studies investigating the chronic effects of GBE supplementation in healthy adult participants have also been undertaken. Mix & Crews (2000) administered 180 mg/day of ginkgo (EGb 761) to 48 healthy elderly participants (55–86 years) for 6 weeks and reported significantly greater improvements in tasks assessing speed of processing abilities for the EGb 761 group. In a larger follow-up study involving 262 healthy elderly adults (60 years and older), Mix & Crews (2002) again administered 180 mg/day EGb761 over a 6-week period. Significantly greater improvements were reported for selective reminding tasks involving delayed free recall and recognition of non-contextual, auditory-verbal material for the EGb 761 group compared with the placebo group. Using a different GBE extract, Blackmore's Ginkgo Biloba Forte (standardised to contain 24 per cent flavanol glycosides and 6 per cent terpene lactones), Stough and colleagues (2001a) investigated the cognitive effects associated with 30-day supplementation in 61 healthy younger adults. Significantly greater improvements in speed of information processing, working memory and executive processing were reported for those receiving ginkgo compared with the placebo group (Stough et al. 2001a). Since this time, a number of other studies have also been published. While a detailed discussion of the individual trials is beyond the scope of this chapter, the reader is referred to an excellent review by Kaschel (2009), which summarises data from 29 randomised controlled studies. In this review it was concluded that there is consistent evidence to suggest that chronic GBE administration improves selective attention, some executive processes, as well as long-term memory in healthy adults (Kaschel 2009).

A number of studies have also investigated the cognitive effects associated with both acute and chronic use of ginkgo in combination with a variety of other substances, including ginseng (Kennedy et al. 2001; Scholey & Kennedy 2002), vinpocetine (Polich & Gloria 2001), phosphatidylserine (Kennedy et al. 2007) and bacopa monniera (Nathan et al. 2004). However, the individual contribution of ginkgo to the cognitive effects in these studies is more difficult to interpret due to the effects of the other substances administered.

A large number of clinical trials have also been conducted investigating GBE for the treatment of dementia and mild cognitive impairment (MCI), with the findings being generally less consistent than for the research in healthy populations. A recent meta-analysis by Cochrane reviews (Birks & Grimley Evans 2009) concluded that there is currently inconsistent and unreliable evidence to suggest a clinically significant benefit from the use of GBE in the treatment of dementia or cognitive impairment. In contrast, the recently completed GuidAge study (Andrieu et al. 2008), in which over 2000 elderly individuals were followed over a 5-year period, revealed more positive findings. In the GuidAge study, individuals treated for at least four years with GBE were found to have a significantly lower rate of conversion to Alzheimer's disease (AD) compared with the placebo group (1.6 per cent of individuals developed AD in the GBE group compared with 3.0 per cent of individuals in the placebo group) (Ipsen press release, June 2010).

In summary, there is currently stronger evidence for cognitive-enhancing effects associated with GBE supplementation in healthy adult populations compared with clinical populations with existing dementia and MCI. For healthy adults, both young and elderly, there is now considerable evidence to suggest that both acute and chronic cognitive effects (including working memory, long-term memory, attention and executive processes) can be associated with GBE supplantation. There also appears to be a complex non-linear relationship between GBE dosage and the cognitive effects across different domains that warrants further investigation.

Panax Ginseng

Panax (Asian) *ginseng* has been used for millennia in traditional Chinese medicine as a 'tonic' and adaptogen to provide energy and aid convalescence in the ill and elderly (Fulder 1990). The constituents of the *Panax* genus which are thought to contribute to its bioactivity are the ginsenoside saponins. Ginsenosides can be classified into three groups on the basis of their chemical structure: the panaxadiol group (Rb1, Rb2, Rb3, Rc, etc.), the panaxatriol group (Re, Rf, Rg1, Rg2, Rh1) and the oleanolic acid group (e.g. Ro) (Tachikawa et al. 1999). Only two standardised extracts of ginseng exist: G115, marketed by Pharmaton SA, contains an invariant 4 per cent ginsenosides; Naturex manufactures three ginseng products under the brand name Ginsenipure™, two of these are standardised to 4 per cent and 15 per cent ginsenosides, respectively. The safety profile of G115 has been well established both from clinical studies in healthy volunteers and patients and from its use for over 30 years as a marketed medicinal product in many countries worldwide. A recent review (Scaglione et al. 2005) examined the safety profile of G115 and found that few clinical studies were to report adverse effects.

In preclinical studies, ginsenosides have been found to exert effects on the cholinergic system, with Rb1 stimulating acetylcholine (Ach) release (Benishin et al. 1991) and increasing expression of choline acetyltransferase (Salim et al. 1997). Ginsenosides Rg1 and Rb1 have also been found to elicit marked alterations in other systems which are relevant to mood and cognitive function, such as serotonin (Zhang et al. 1990) and nerve growth factor (Salim et al. 1997), while the ginsenosides Rd and Re have been found to increase levels of norepinephrine, dopamine, serotonin and gamma-Aminobutyric acid (GABA) in the rat brain (Tsang et al. 1985).

A series of double-blind, placebo-controlled studies have assessed the cognitive and mood effects associated with acute administration of ginseng in healthy young adults. In the first study (Kennedy et al. 2001b), doses of 200, 400 and 600 mg of ginseng (G115) were administered. Enhancement of 'secondary episodic memory' was found following administration of 400 mg at four post-dose testing sessions, while the lower and higher dosage reduced performance for 'speed of attention'. In a further study assessing combinations of ginseng and ginkgo (ratio 100 : 60) at dosages of 320, 640 and 960 mg, a similar pattern was observed (Kennedy et al. 2001a), with performance of secondary memory being improved by 960 mg and performance on speed of attention being reduced for the other doses (320 and 640 mg). A later study (Kennedy et al. 2002) replicated the finding that a 400 mg dosage improved secondary memory. A further study assessed the effect of 200, 400 and 600 mg ginseng on mental arithmetic performance, where cognitive demand was manipulated. Again, performance on this task was improved by a 400 mg dosage but only for the most demanding task (Serial Sevens) (Reay et al. 2006).

There are few behavioural studies addressing the potential cognitive-enhancing effects associated with chronic ginseng administration. One study reported the effects in healthy young volunteers of 12 weeks' administration of *Panax ginseng* (200 mg G115 per day) on cognitive and psychomotor performance (D'angelo et al. 1986). Within (change from baseline) and between (treatment versus placebo) groups, comparisons revealed that ginseng administration led to significantly more correct responses on a mental arithmetic task. Another study investigated the effects of 8–9 weeks of ginseng ingestion (400 mg standardised Gerimax ginseng extract) on cognitive performance in 112 healthy middle-aged participants (Sørensen & Sonne 1996). Participants completed a cognitive test battery at a pre-treatment baseline and then again following 8–9 weeks of ginseng ingestion. The results revealed that ginseng led to significantly faster performance in only the most rapid auditory reaction time test (10th percentile). Accuracy in performing the Wisconsin card sort test was also reported to be superior following ginseng administration; however, there was no baseline completion of this task, as there are no parallel versions, therefore this result is confounded.

A third empirical investigation (Labadorf 2004) examined the mood, memory and attentional effects following 14 days of ginseng treatment (200 mg G115) in 18 healthy young volunteers. An independent measure (eight volunteers receiving ginseng and ten volunteers receiving placebo), placebo-controlled, double-blind, randomised design was utilised. Volunteers were required to ingest 200 mg (G115) or placebo for a 3-week period. Cognitive assessment was conducted using CDR computerised assessment battery at a pre-dose baseline and thereafter at 1, 3 and 6 hours post-treatment on days 1, 7 and 14. Results revealed that ginseng led to better performance on all four factors: speed of attention; quality of attention; speed of memory; quality of memory – derived from the CDR research battery. Few conclusive interpretations can be made from these chronic studies as they have investigated different cohorts (i.e. healthy young and middle aged), administered different ginseng extracts (e.g. G115 or Gerimax) and doses (200 and 400 mg), for different durations (2–12 weeks) and implemented different assessment tools designed to assess different aspects/domains of cognitive functioning. Further research is required to examine the chronic use of *Panax ginseng* on neurocognitive outcome measures.

Recently, another species of ginseng – *Panax quinquefolius* – has also been found to acutely improve aspects of working memory in healthy young adults (Scholey et al. 2010). *Panax quinquefolius* has a slightly different ginsenoside profile to *Panax ginseng*. It remains to be seen which ginsenosides are responsible for the cognition-enhancing effects of ginseng, but studies directly comparing standardised extracts with different profiles may offer some insight.

Bacopa Monnieri

Bacopa monnieri is a herb that has been used for centuries in ayurvedic medicine as a memory enhancer, sedative, analgesic, anti-inflammatory and anti-epileptic treatment (Jain 1994; Stough et al. 2001b). Saponins (bacosides, bacopasides or bacopasaponins) are the active ingredients that have been attributed the memory-enhancing effects. Suggested mechanisms of action include cholinergic upregulation, γ-aminobutyric acid (GABA)-ergic modulation, antioxidant effects, brain protein synthesis, serotonin agonism, modulation of brain stress hormones, as well as reduction of β-amyloid (Calabrese et al. 2008).

A high-quality standardised extract of *bacopa monnieri*, known as CDRI08, is manufactured by the Central Drug Research Institute in Lucknow, India. This extract has been standardised for bacosides A and B, with no less than 55 per cent of combined bacosides. Stough et al. (2001b) conducted a clinical trial in 46 healthy volunteers aged 18–60 using 300 mg CDRI08 per day to examine the effects on cognition. A battery of cognitive tests was administered after 5 weeks and 12 weeks, with CDRI08 found to significantly improve performance on the Rey Auditory Verbal Learning Test (RAVLT) as well as State Anxiety at 12 weeks. Further three-month clinical trials of *bacopa monnieri* in elderly adults have reported similar improvements in a number of measures, including the retention of new information in delayed recall of word pairs (Roodenrys et al. 2002), improvements in subsets of the Wechsler Memory Scale (Raghav et al. 2006) and improvements on the Stroop task assessing the ability to ignore irrelevant information (Calabrese et al. 2008).

In a follow-up randomised trial by Stough et al. (2008), 300 mg CDRI08 bacopa extract was administered to 62 healthy volunteers aged 18–60 years over a 90-day period. Significant improvements in the working memory factor from the CDR battery were observed in the group receiving *bacopa monnieri*. Greater accuracy on the rapid visual information processing (RVIP) test was also observed in this group following treatment. These findings corroborated the results of the previous study, providing further evidence for the cognitive-enhancing effects of *bacopa monnieri*.

Melissa Officinalis

Melissa officinalis (lemon balm) is a perennial herb from the Lamiaceae family that is native to southern Europe and the Mediterranean and has a history of use dating back over 2000 years. Traditionally it was used as a mild sedative and anxiolytic; however, early records from the middle ages indicate that it has also been long recognised for its positive effects on memory (Kennedy et al. 2002). The putative biologically active compounds in *M. officinalis* include monoterpenoid aldehydes (including citronellal, neral and geranial), flavonoids and polyphenolic compounds, such as rosmarinic acid and monoterpene glycosides (Carnat et al. 1998; Mulkens et al. 1985; Sadraei et al. 2003).

There is evidence to suggest that *M. officinalis* enhances cholinergic transmission, based on the fact that it binds to both nicotinic and muscarinic acetylcholine receptors within the central nervous system (Perry et al. 1996; Wake et al. 2000). However, large variations in receptor-binding affinities have been noted between varying strains and preparations of *M. officinalis*, with the more reliable action of the plant across samples being its calming effects (Kennedy et al. 2003). The anxiolytic effects of *M. officinalis* are most likely due to a non-cholinergic mechanism that is yet to be identified. There is also evidence to suggest that *M. officinalis* acts as a moderately effective free radical scavenger, which can be attributed to its flavonoid content (Hohmann et al. 1999; Mantle et al. 2000).

A number of studies have investigated the anxiolytic effects of *M. officinalis*; however, a discussion of these findings is beyond the scope of this chapter. In relation to cognitive effects, two recent studies by Kennedy et al. (2002, 2003) have investigated the effects of acute administration. In the first of these studies, 20 healthy young participants received single doses of 200, 600 and 900 mg *M. officinalis* ethanolic extract versus placebo in a randomised, double-blind, crossover design. At 1, 1.5, 4 and 6 hours following dosing their cognitive performance was assessed using the CDR test battery, together with serial subtraction tasks. Accuracy of attention was found to be significantly improved following the middle dose of 600 mg of *M. officinalis*; however, at the highest dose (900 mg), decrements in memory performance together with reduced alertness were observed (Kennedy et al. 2002).

In vitro analysis of the extract revealed low binding affinity for nicotinic and muscarinic receptors, prompting Kennedy and colleagues (2003) to conduct a further study using an extract with greater cholinergic activity. In the second study, eight samples of *M. officinalis* were screened using in vitro analysis, and the extract with the highest binding affinity for both muscarinic and nicotinic receptors was used. Following a similar crossover design to the previous study, 600, 1000 and 1600 mg *M. officinalis* extract versus placebo was then administered to 20 healthy young participants. At the highest dosage level of 1600 mg, performance on the quality of memory factor of the CDR was found to be significantly improved at both 3 and 6 hours post-dose. At the lowest dose (600 mg), performance decrements were noted for the same timed memory tasks used in the previous study, together with a newly introduced task, the RVIP. These effects were found to decrease as the dose was increased, with the authors speculating that two distinct mechanisms of action could explain the results. Presumably the cholinergic effect associated with the extract was responsible for the improvements to cognition, an effect which only counteracted a sedative effect at the higher dosage levels (>1000 mg).

In relation to the chronic effects of *M. officinalis*, a study by Akhondzadeh et al. (2003a) investigated the efficacy of 60 drops/day tincture versus placebo over a 4-month period in 35 patients with mild to moderate AD, aged 65 and 80 years. Cognitive function, as measured by the ADAS-cog and clinical dementia rating scale, was found to be significantly improved in comparison to placebo at four months. Regarding *M. officinalis*, it appears that there are other nutraceutical compounds with more well-established cholinergic effects. The primary effect for this substance appears to be one of mood modulation, rather than cognitive enhancement. While careful selection of an extract which demonstrates higher binding affinity for nicotinic and muscarinic receptors may improve its efficacy, it appears that there are more potent nootropic agents to choose from.

Salvia Officinalis

Salvia officinalis is part of the *Salvia* genus in the Labiatae family, containing over 700 species of plants. It has been used over several millennia across a number of different cultures, including ayurvedic medicine, as well as early Greek and Chinese civilisations, as a treatment for the amelioration of age-related memory loss. The proposed mechanisms of action for *S. officinalis* include acetylcholinesterase inhibition (ChEI), inhibition of glia-derived butyrylcholinesterase (BuChe), antioxidant, anti-inflammatory and oestrogenic effects (Kennedy et al. 2006; Perry et al. 1999). To date, two randomised controlled trials have been conducted to assess the acute memory-enhancing effects of *S. officinalis*. A study by Kennedy et al. (2006) examined the acute effects of *S. officinalis* on cognition in 30 healthy participants, who completed a test battery at baseline as well as 1 and 4 hours post-dose on three separate testing occasions. On each occasion they received a different treatment: a placebo, 300 or 600 mg of dried sage leaf. The higher dose was found to be associated with improved performance on the Stroop test as well as an aggregate score obtained from a battery of tests, including tasks of mathematical processing and memory search tasks, at both post-dose time points.

In a more recent study by Scholey et al. (2008), the acute effects of *S. Officinalis* on memory were examined using 20 elderly volunteers (over the age of 65) administered 167, 333, 666 and 1332 mg of dried sage and tested 1, 2.5, 4 and 6 hours post-dose. Significant improvements in secondary memory performance (aggregate percentage accuracy in word recognition, picture recognition, immediate word recall and delayed word call from the CDR battery) were noted for the 333 mg dose in comparison to placebo at all post-dose time points. The extracts used in the study were subjected to in vitro analysis, confirming cholinesterase-inhibiting properties in comparison to an ethanol control sample. These findings have been corroborated by investigations into the acute effects of *Salvia lavandulaefolia* essential oil, another ChEI of the Sage family containing similar components to *S. officinalis*. A study by Tildesley et al. (2003) reported a significant improvement in immediate and delayed word recall post-dose using a 50 μl dose of the oil in 20 young healthy volunteers. A second study by Tildesley et al. (2005), using the CDR battery, reported an improvement in secondary memory performance at 1 hour post-dose and speed of memory at 2.5 hours post-dose using 25 μl of *S. lavandulaefolia*. Improvements in speed of memory at 4 and 6 hours post-dose were also reported with the higher dose of 50 μl. More recently, a different extract of *S. lavandulaefolia* has been shown to improve memory function (Kennedy et al. 2010), suggesting that these effects are robust across a range of extracts of sage. With regards to studies of the chronic effects of *S. officinalis* among the clinical population, a study by Akhondzadeh et al. (2003b) examined the effects of *S. officinalis* on memory in 39 AD patients. Significantly improved scores on the ADAS-cog were reported for those in the *S. officinalis* group compared with the placebo at 16 weeks. However, this study was not without criticism, with reviewers drawing attention to the unexpectedly large effect size, an ill-defined herb extract and no description of the placebo (Kennedy & Scholey 2006). However, an open-label trial using the *S. lavandulaefolia* essential oil from the Sage family by Perry et al. (2003) also reported a significant improvement in the accuracy of performing a vigilance task at the 6-week end-point among 11 AD patients. To date, the findings from the relatively few studies that have been conducted using *S. officinalis* and *S. lavandulaefolia* are promising, suggesting efficacy associated with both acute and chronic supplementation. Further randomised controlled trials, as well as longitudinal studies, using larger samples from both the non-clinical population as well as MCI and AD patients, are warranted in order to properly establish the efficacy of these nutraceuticals as nootropics.

Guarana

The plant species guarana originates from the central Amazonian Basin and has a long history of local usage, initially as a stimulant by indigenous tribes people (Henman 1982) and more latterly as a ubiquitous ingredient in Brazilian soft drinks. An extensive range of products that include guarana (*Paullinia cupana*) seed extracts as ingredients are commercially available. Examples include confectionery (e.g. chocolate products), fruit juice-based drinks, 'energy' drinks (see Chapter 13), dietary and herbal supplements, and, most controversially, natural weight loss products. The putative stimulant properties are generally taken to reflect the presence of caffeine, which comprises 2.5–5 per cent of the extract's dry weight, although other purine alkaloids (theophylline and theobromine) are present in smaller quantities (Weckerle et al. 2003). The psychoactive properties of guarana have also been attributed to a high content of both saponins and tannins (Espinola et al. 1997), the latter of which may well underlie the demonstrated antioxidant properties of the plant (Mattei et al. 1998).

An early investigation into the potential effects of guarana in normal young volunteers failed to find any effects of guarana using tests of digit span, free recall, digit symbol, cancellation tests and the mosaic test (Galduróz & Carlini 1994). The same study also evaluated sleep interference and anxiety and, again, found no effects. The authors present possible explanations for their lack of positive results, such as task insensitivity. They also failed to find effects of 25 mg caffeine in the same study, a dose twice that of the lowest known psychoactive dose (Smit & Rogers 2000). In this first investigation in humans, 1000 mg guarana was tested, containing only 2.1 per cent caffeine. Given the lack of data in this area, it is quite possible that any effects could have been missed simply as a result of inappropriate dose selection. Finally, the time course of testing may not have been sufficient, with acute testing only being carried out at 1 hour post-treatment and chronic testing following three days of treatment administration. In a follow-up study (Galduróz & Carlini 1994), the same doses and tasks were used to assess chronic (five months) effects in an elderly population. They found only one improvement – a significant effect of guarana on mosaic performance at five months.

In one randomised, double-blind, placebo-controlled, counterbalanced study, 75 g of a proprietary extract of guarana (Pharmaton extract PC-102), ginseng and their combination were compared with placebo over the course of 6 hours, using a battery of computerised assessments (Kennedy et al. 2004). Improvements in the speed of task performance were observed following 75 mg of guarana during the tasks making up a 'speed of attention' factor – a heavily loaded serial subtraction task (Serial Sevens) – and during a sentence verification task. In the case of the former, the effects were seen at 1, 4 and 6 hours post-dose, and in the case of the latter, improvements were seen at 2.5, 4 and 6 hours post-dose. For Serial Sevens, significant improvements were found at all four post-dose testing sessions, albeit with increased errors at 4 hours, suggesting that there may have been a speed–accuracy trade-off at this time point. Although less pronounced, performance on the 'secondary memory' factor was again improved, at 2.5 hours post-dose, with improvements seen at 1, 4 and 6 hours post-dose on a (contributing) picture recognition task.

In another similarly controlled study from the same group, the cognitive and mood effects of different doses of guarana were assessed. The doses used were 37.5, 75, 150 and 300 mg, and their effects were assessed using the same outcomes in 30 healthy participants. Testing took place pre-dose and at 1, 3 and 6 hours thereafter, with a seven-day 'wash out'. The data confirm the positive effects on secondary memory, which, in this case, were evident following 37.5 and 75 mg of extract. There was a significant positive effect on 'alert' following the highest dose only and significant improvements of 'content' ratings associated with all doses (Haskell et al. 2007).

The data suggest that guarana can positively modulate cognitive performance and mood. An unpublished study directly compared the neurocognitive effects of guarana with caffeine administered at the same dose contained within guarana treatment. The data showed that the cognitive and mood profile of guarana is distinct from that of its caffeine content.

Guarana extracts appear to show positive effects on cognitive performance which may be robust in the secondary memory domain. These effects are probably not underpinned by caffeine alone and may be attributable to modulation of caffeine by other guarana components or by direct effects of non-caffeine constituents. Examination of the behavioural effects of decaffeinated guarana may resolve this issue. The effects of chronic guarana administration are not known.

Polyphenols

Polyphenols, substances with one or more phenol units in their chemical structure, are ubiquitous in plant materials, with over 8000 phenolic structures currently known (Bravo 1998). The main groupings of polyphenols are tannins, lignins and flavonoids, with the latter being the most common. Polyphenols are produced by plants in greater quantities when in the presence of environmental stress; having the effect of enhancing a plant's chances of survival (Morris 2008). The xenohormesis hypothesis (Lamming et al. 2004) (combining the prefix *xeno* for stranger and *hormesis* for a protective response induced by mild stress) suggests that other animals have evolved to pick up on these chemical cues in plant-based foods in order to anticipate a deteriorating environment and stimulate their own adaptive survival responses (Baur & Sinclair 2008). A number of polyphenols have been found to exhibit potent antioxidant and anti-inflammatory effects in the central nervous system (Bravo 1998). There is also evidence to suggest that certain polyphenols may promote longevity when consumed in sufficient quantities. For example, Howitz & Sinclair (2003) demonstrated that through activation of the sirtuin (SIRT1) gene, the polyphenol resveratrol can extend the lifespan of yeast cerevisiae by 70 per cent; with similar effects on lifespan extension also observed in a variety of other species (Tissenbaum & Guarente 2001; Valenzano & Cellerino 2006; Valenzano et al. 2006). In recent years there has been growing epidemiological and clinical evidence to suggest that cognitive-enhancing effects are associated with a wide range of polyphenols.

Curcumin is found in rhizomes of tropical ginger and turmerics, and is the spice which gives yellow curry its vibrant colour (Frank & Gupta 2005). For several hundred years curcumin has been used in ayurvedic medicine to treat inflammation and pain (Mishra & Palanivelu 2008). Epidemiological studies have shown that increased consumption of curry by Asians is associated with better performance on the Mini Mental State Exam (MMSE) later in life. Further, the prevalence of AD among adults aged 70–79 in India is 4.4 times less than that of adults aged 70–79 years in the USA (Ganguli et al. 2000; Ng et al. 2006).

Resveratrol is a phytoalexin polyphenolic compound occurring in grapes and wine, with higher concentration in red wine, that is a powerful antioxidant contributing to the cardio-protective, anti-inflammatory and neuroprotective properties of red wine intake (Pervaiz 2003). The resveratrol content of wine has been used as an explanation for the 'French paradox', whereby a lower incidence of coronary mortality is observed in France even though the French consume foods that are high in saturated fats (de Leiris & Boucher 2008). There is also evidence to suggest that wine intake is associated with a lower risk of developing neurodegenerative disease. A number of epidemiological studies have associated red wine intake, but not other alcoholic drinks, with a lower risk of developing AD (Vingtdeux et al. 2008). A study in the Canadian population determined that wine consumption was the most protective variable against AD, reducing the risk by 50 per cent, and more protective than the use of non-steroidal anti-inflammatory drugs (Lindsay et al. 2002). The protective effects of resveratrol on neurological disorders associated with ageing are considered in detail in Chapter 16.

Green tea catechins account for around 30–40 per cent of the dry weight of green tea leaves, which is four times that found in black tea (Khokhar & Magnusdottir 2002; Wang et al. 1994; Yang & Wang 1993). There are various catechins found in green tea, the major constituent being (–)-epigallocatechin-3-gallate (EGCG) that accounts for more than 10 per cent of the dry weight, while other catechins are found in lesser amounts: (–)-epigallocatechin (EGC) > (–)-epicatechin (EC) > = (–)-epicatchin-3-gallate (ECG) (Mandel et al. 2008). All four catechins have been found to be powerful antioxidants and radical scavengers, although EGCG has been found to be the most potent (Nanjo et al. 1996; Salah et al. 1995). The catechins have been shown to strongly inhibit lipid peroxidation as well as display an ability to induce endogenous antioxidant defences and chelate transitional metals, such as iron and copper (Mandel et al. 2004).

Kuriyama et al. (2006) conducted a cross-sectional study of 1002 elderly Japanese individuals over the age of 70 years. They administered the MMSE, together with a questionnaire containing questions regarding the individuals' consumption of tea and coffee. Higher consumption of green tea was found to be associated with a lower prevalence of cognitive impairment. Using <26 on the MMSE as a cut-off, those drinking > = two cups of green tea a day were found to have an odds ratio for cognitive impairment of 0.46 (compared with an odds ratio of 1.00 for < = three cups per week). Green tea was found to be associated with a greater reduction in risk of cognitive impairment in comparison to black tea and oolong tea, as well as coffee. Similarly, a large-scale 12-year prospective study by Hu and colleagues (2007) of 29,335 Finnish participants aged 25–74 years found greater tea consumption to also be associated with a reduced risk of developing Parkinson's disease.

Pycnogenol® (PYC; Horphag Research, Geneva, Switzerland) is the trade name given to a specific blend of procyanidins, extracted from the bark of French maritime pine. Use of the pine bark dates back over 2000 years, with documented usage for the treatment of inflammation, skin disorders, wound healing and scurvy in Europe, as well as among native American Indians.

Today, it is used as a nutritional supplement and phytochemical remedy for a variety of diseases, ranging from chronic inflammation to circulatory dysfunction (Packer et al. 1999). The procyanidins found in PYC belong to the same family of flavonoid polyphenols as green tea catechins.

Extensive reviews by Packer (1999) and Rohdewald (2002) have established the powerful antioxidant capacity of PYC. In a double-blind trial examining the effects of PYC on cognitive performance, Ryan et al. (2008) administered 150 mg/day PYC versus placebo to 101 elderly participants (60–85 years) over a 3-month period. The group receiving PYC, in comparison to controls, were found to display significantly better working/spatial memory function at three months as measured by the CDR test battery. Interestingly, concentrations of F2-isoprostanes in blood plasma, which are biomarkers of lipid peroxidation, were also found to be reduced at three months in comparison to baseline for the PYC group.

An extract related to PYC which is rich in flavonoid proanthocyanidins is the *Pinus radiata* bark extract branded Enzogenol. Pipingas et al. (2008) examined the effects of Enzogenol in combination with vitamin C on cognitive performance in a double-blind trial using 42 males aged 42–65 years over a 5-week period. The speed of response for the spatial working memory and immediate recognition tasks was found to be significantly improved for the group receiving Enzogenol plus vitamin C after five weeks, whereas vitamin C alone showed no improvements.

Cocoa polyphenols are found at high levels in cocoa which contains simple monomeric flavanols, primarily (–)-epicatechin and structurally related procyanidin flavanol dimers and oligomers (Lazarus et al. 1999). The consumption of flavanol-containing cocoa products has been found to lead to a range of cognition-relevant health benefits, including improved insulin sensitivity (Grassi et al. 2005), endothelial function (Heiss et al. 2003), reduced platelet aggregation (Holt et al. 2002) and blood pressure (Taubert et al. 2007). Epidemiological studies have also reported cardio-protective effects associated with cocoa consumption, due to the procyanidin component, the monomeric component, or both (Williamson & Manach 2005), that inhibit the oxidation of low density lipoproteins (LDL) (Kondo et al. 1996; Serafini et al. 2003; Waterhouse et al. 1996).

Clinical evidence for the efficacy of cocoa flavanols (CF) for enhancing cognition has also recently begun to emerge. Francis et al. (2006) reported that 150 mg/day CF was associated with greater activation of task-relevant brain loci (dorsolateral prefrontal cortex, anterior cingulate and parietal cortex) using functional magnetic resonance imaging (fMRI). Recently, the possibility of acute cognitive effects of CF administration was assessed using a test procedure involving a high cognitive load (Scholey et al. 2009). Volunteers consumed drinks containing 520 mg, 994 mg CF and a matched (low flavanol) control. Assessments included repeated cycles of two serial subtraction tasks (Serial Threes and Serial Sevens), an (RVIP) task and a 'mental fatigue' scale over the course of 1 hour. Both 520 and 994 mg CF significantly improved

Serial Threes performance while self-reported 'mental fatigue' was significantly attenuated by the consumption of the 520 mg CF beverage only (Scholey et al. 2009).

The Importance of Phyto-Equivalence in Clinical Studies

An issue of critical importance in clinical psychopharmacological studies of herbal extracts is that a high-quality standardised extract is used. Two different herbal extracts are said to be phyto-equivalent when they have the same psychopharmacological effects. However, in practice, many of the health claims that are attributed to plant species have been established in relation to specific standardised extracts rather than the species as a whole; and for this reason phyto-equivalence cannot be assumed unless the same chemical fingerprint associated with each extract is demonstrated using chromatographic or electrophoretic techniques (Liang et al. 2004).

There are a number of factors involved in the production chain of a herbal product that can impact on its chemical properties. These factors include seeding, cultivation, harvesting, drying, extraction, formulation of the dry extract and quality control monitoring (Groot & Van Der Roest 2006). In order to ensure a consistently high-quality extract, a detailed crop production protocol needs to implemented, as well as testing the finished product to ensure that it contains active ingredients that comply with the qualitative and quantitative composition of the marketing authorisation (Groot & Van Der Roest 2006). As an example of how widely herbal products can vary in the quantities of their active constituents, an analysis of 25 different ginseng products in the USA revealed that concentrations of ginsenosides varied by 15-fold in capsules and 36-fold in liquids (Harkey et al. 2001). For several years now, the burgeoning herbal medicine industry has been poorly regulated compared with the strict regulatory standards of the pharmaceutical industry. However, recent legislation in Europe indicates that the tide is turning. As of April 2011, only products that have been assessed by the Medicine and Healthcare Regulatory Agency (MHRA) are permitted to go on sale. Part of the MHRA assessment process includes evidence of strict manufacturing standards and consistent quantities of active ingredients.

Future Directions

In this chapter we have outlined a number of herbal extracts for which there exists evidence to suggest that they have cognitive-enhancing properties, but the list is by no means exhaustive. Research in this field is still in its infancy and, as such, there is a need for continued research focus regarding the cognitive effects of herbal extracts. Future research in this field would benefit from several processes. As already mentioned above, the use of a high-quality standardised extract is paramount. Further, the use of precise and highly accurate cognitive test batteries, including computerised cognitive assessments, are likely to be more sensitive to cognitive change. A further recommendation is that in addition to the use of cognitive outcome measures, researchers also co-monitor relevant biomarkers in order to further elucidate the complex relationships between cognitive and physiological parameters that occur during ageing. Such markers should include oxidative stress, inflammation, genetic and cardiovascular measures, as well as increasingly sophisticated brain imaging techniques, including fMRI, electroencephalography and magnetoencephalography. Additionally, there are several psychological processes which can impinge on cognitive function. These include sleep, stress and anxiety, as well as overall mood. Future research which takes into account these important factors will help to provide more accurate understanding as to the necessary conditions under which nutraceutical interventions may provide cognitive benefits, as well as the mechanisms of action by which they occur.

References

Akhondzadeh, S., Noroozian, M., Mohammadi, M., Ohadinia, S., Jamshidi, A. H., & Khani, M. (2003a). Melissa officinalis extract in the treatment of patients with mild to moderate Alzheimer's disease: a double blind, randomised, placebo controlled trial. *Journal of Neurology Neurosurgery and Psychiatry, 74*(7), 863–866.

Akhondzadeh, S., Noroozian, M., Mohammadi, M., Ohadinia, S., Jamshidi, A. H., & Khani, M. (2003b). Salvia officinalis extract in the treatment of patients with mild to moderate Alzheimer's disease: a double blind, randomized and placebo-controlled trial. *Journal of Clinical Pharmacy and Therapeutics, 28*(1), 53–59.

Andrieu, S., Ousset, P. J., Coley, N., Ouzid, M., Mathiex-Fortunet, H., & Vellas, B. (2008). GuidAge study: a 5-year double blind, randomised trial of EGb 761 for the prevention of Alzheimer's disease in elderly subjects with memory complaints. I. Rationale, design and baseline data. *Current Alzheimer Research, 5*(4), 406–415.

Australian Bureau of Statistics (ABS). (2009). *Deaths, Australia, 2009.* cat. no. 3302.0.

Australian Bureau of Statistics (ABS). (2011). *Population Projections Australia 2006 to 2101.* cat. no. 3222.0.

Baur, J. A., & Sinclair, D. A. (2008). What is xenohormesis? *American Journal of Pharmacology and Toxicology, 3*(1), 149–156.

Benishin, C., Lee, R., Wang, L., & Liu, H. (1991). Effects of Ginsenoside Rbi on Central Cholinergic Metabolism. *Pharmacology, 42*, 223–229.

Birks, J., & Grimley Evans, J. (2009). *Ginkgo biloba* for cognitive impairment and dementia. *Cochrane Database of Systematic Reviews,* (1), CD003120.

Bravo, L. (1998). Polyphenols: chemistry, dietary sources, metabolism, and nutritional significance. *Nutrition Reviews, 56*(11), 317–333.

Cadenas, E., & Davies, K. J. A. (2000). Mitochondrial free radical generation, oxidative stress, and aging. *Free Radical Biology and Medicine, 29*(3–4), 222–230.

Calabrese, C., Gregory, W. L., Leo, M., Kraemer, D., Bone, K., & Oken, B. (2008). Effects of a standardized Bacopa monnieri extract on cognitive performance, anxiety, and depression in the elderly: a randomized, double-blind, placebo-controlled trial. *Journal of Alternative and Complementary Medicine, 14*(6), 707–713.

Canter, P. H., & Ernst, E. (2007). *Ginkgo biloba* is not a smart drug: an updated systematic review of randomised clinical trials testing the nootropic effects of G. biloba extracts in healthy people. *Human Psychopharmacology, 22*(5), 265–278.

Carnat, A. P., Carnat, A., Fraisse, D., & Lamaison, J. L. (1998). The aromatic and polyphenolic composition of lemon balm (*Melissa officinalis* L. subsp. *officinalis*) tea. *Pharmaceutica Acta Helvetiae, 72*(5), 301–305.

Crews Jr, W. D., Harrison, D. W., Griffin, M. L., Falwell, K. D., Crist, T., Lomgest, L., et al. (2005). The neuropsychological efficacy of Ginkgo preparations in healthy and cognitively intact adults: a comprehensive review. *HerbalGram, 67,* 43–62.

D'angelo, L., Grimaldi, R., Caravaggi, M., Marcoli, M., Perucca, E., Lecchini, S., et al. (1986). A double-blind, placebo-controlled clinical study on the effect of a standardized ginseng extract on psychomotor performance in healthy volunteers. *Journal of Ethnopharmacology, 16*(1), 15–22.

de Leiris, J., & Boucher, F. (2008). Does wine consumption explain the French paradox? *Dialogues in Cardiovascular Medicine, 13*(3), 183–192.

DeFeudis, F. V., & Drieu, K. (2000). *Ginkgo biloba* extract (EGb 761) and CNS functions: basic studies and clinical applications. *Current Drug Targets, 1*(1), 25–58.

Edeline, J. M. (2003). The thalamo-cortical auditory receptive fields: regulation by the states of vigilance, learning and the neuromodulatory systems. *Experimental Brain Research, 153*(4), 554–572.

Espinola, E., Dias, R., Mattei, R., & Carlini, E. (1997). Pharmacological activity of Guarana (Paullinia cupana Mart.) in laboratory animals. *Journal of Ethnopharmacology, 55*(3), 223–229.

Floyd, R. A. (1999). Antioxidants, oxidative stress, and degenerative neurological disorders. *Proceedings of the Society for Experimental Biology and Medicine, 222*(3), 236–245.

Francis, S., Head, K., Morris, P. G., & Macdonald, I. A. (2006). The effect of flavanol-rich cocoa on the fMRI response to a cognitive task in healthy young people. *Journal of Cardiovascular Pharmacology, 47*(suppl. 2), S215–S220.

Frank, B., & Gupta, S. (2005). A review of antioxidants and Alzheimer's disease. *Annals of Clinical Psychiatry, 17*(4), 269–286.

Fulder, S. (Ed.). (1990). *The book of ginseng*. Rochester, VT: Healing Arts Press.

Galduróz, J., & Carlini, E. (1994). Acute efects of the Paulinia cupana, "Guaraná" on the cognition of normal volunteers. *Sao Paulo Medical Journal, 112,* 607–611.

Ganguli, M., Chandra, V., Kamboh, M. I., Johnston, J. M., Dodge, H. H., Thelma, B. K., et al. (2000). Apolipoprotein E polymorphism and Alzheimer disease: the Indo-US cross-national dementia study. *Archives of Neurology, 57*(6), 824–830.

Gracy, R. W., Talent, J. M., Kong, Y., & Conrad, C. C. (1999). Reactive oxygen species: the unavoidable environmental insult? *Mutation Research – Fundamental and Molecular Mechanisms of Mutagenesis, 428*(1–2), 17–22.

Grassi, D., Necozione, S., Lippi, C., Croce, G., Valeri, L., Pasqualetti, P., et al. (2005). Cocoa reduces blood pressure and insulin resistance and improves endothelium-dependent vasodilation in hypertensives. *Hypertension, 46*(2), 398–405.

Groot, M. J., & Van Der Roest, J. (2006). Quality control in the production chain of herbal products. In R. J. Bogers, L. E. Craker & D. Lange (Eds.), *Medicinal and aromatic plants* (pp. 253–260). The Netherlands: Springer.

Halliwell, B. (1992). Reactive oxygen species and the central nervous system. *Journal of Neurochemistry, 59*(5), 1609–1623.

Halliwell, B., & Gutteridge, J. M. C. (1990). Role of free radicals and catalytic metal ions in human disease: an overview. *Methods in Enzymology, 186,* 1–85.

Harkey, M. R., Henderson, G. L., Gershwin, M. E., Stern, J. S., & Hackman, R. M. (2001). Variability in commercial ginseng products: An analysis of 25 preparations. *American Journal of Clinical Nutrition, 73*(6), 1101–1106.

Harman, D. (1957). Aging: A theory based on free radical and radiation chemistry. *Journal of Gerentology, 2,* 298–300.

Haskell, C., Kennedy, D., Wesnes, K., Milne, A., & Scholey, A. (2007). JPsychopharm. *Journal of Psychopharmacology, 21*(1), 65–70.

Hasselmo, M. E., & McGaughy, J. (2004). High acetylcholine levels set circuit dynamics for attention and encoding and low acetylcholine levels set dynamics for consolidation, *Progress in Brain Research,* 145, 207–231.

Hedden, T., & Gabrieli, J. D. E. (2004). Insights into the ageing mind: A view from cognitive neuroscience. *Nature Reviews Neuroscience, 5*(2), 87–96.

Heiss, C., Dejam, A., Kleinbongard, P., Schewe, T., Sies, H., & Kelm, M. (2003). Vascular effects of cocoa rich in flavan-3-ols. *Research Letters, 290*(8), 1030–1031.

Henman, A. (1982). Guarana (Paullinia cupana var. sorbilis): Ecological and social perspectives on an economic plant of the central Amazon basin. *Journal of Ethnopharmacology, 6*(3), 311–338.

Hohmann, J., Zupkó, I., Rédei, D., Csányi, M., Falkay, G., Máthé, I., et al. (1999). Protective effects of the aerial parts of Salvia officinalis, Melissa officinalis and Lavandula angustifolia and their constituents against enzyme- dependent and enzyme-independent lipid peroxidation. *Planta Medica, 65*(6), 576–578.

Holt, R., Schramm, D., Keen, C., Lazarus, S., & Schmitz, H. (2002). Chocolate consumption and platelet function. *Journal of the American Medical Association, 287,* 2212–2213.

Howitz, K. T., Bitterman, K. J., Cohen, H. Y., Lamming, D. W., Lavu, S., Wood, J. G., et al. (2003). Small molecule activators of sirtuins extend. *Saccharomyces cerevisiae* lifespan. *Nature, 425*(6954), 191–196.

Hu, G., Bidel, S., Jousilahti, P., Antikainen, R., & Tuomilehto, J. (2007). Coffee and tea consumption and the risk of Parkinson's disease. *Movement Disorders, 22*(15), 2242–2248.

Jain, S. K. (1994). Ethnobotany and research on medicinal plants in India. *Ciba Foundation symposium, 185,* 153–164; discussion 164.

Kaschel, R. (2009). *Ginkgo biloba*: specificity of neuropsychological improvement – a selective review in search of differential effects. *Human Psychopharmacology, 24*(5), 345–370.

Kennedy, D., Haskell, C., Wesnes, K., & Scholey, A. (2004). Improved cognitive performance in human volunteers following administration of guarana (Paullinia cupana) extract: comparison and interaction with Panax ginseng. *Pharmacology Biochemistry and Behavior, 79*(3), 401–411.

Kennedy, D., Scholey, A., & Wesnes, K. (2001a). Differential, dose dependent changes in cognitive performance following acute administration of a *Ginkgo biloba*/Panax ginseng combination to healthy young volunteers. *Nutritional Neuroscience, 4*(5), 399–412.

Kennedy, D., Scholey, A., & Wesnes, K. (2001b). Dose dependent changes in cognitive performance and mood following acute administration of Ginseng to healthy young volunteers. *Nutritional Neuroscience, 4*(4), 295–310.

Kennedy, D., Scholey, A., & Wesnes, K. (2002). Modulation of cognition and mood following administration of single doses of *Ginkgo biloba*, ginseng, and a ginkgo/ginseng combination to healthy young adults. *Physiology and Behavior, 75*(5), 739–752.

Kennedy, D. O., Dodd, F. L., Robertson, B. C., Okello, E. J., Reay, J. L., Scholey, A. B., et al. (2011). Monoterpenoid extract of sage (Salvia lavandulaefolia) with cholinesterase inhibiting properties improves cognitive performance and mood in healthy adults. *Journal of Psychopharmacology, 25*, 1088–1100.

Kennedy, D. O., Haskell, C. F., Mauri, P. L., & Scholey, A. B. (2007). Acute cognitive effects of standardised *Ginkgo biloba* extract complexed with phosphatidylserine. *Human Psychopharmacology, 22*(4), 199–210.

Kennedy, D. O., Jackson, P. A., Haskell, C. F., & Scholey, A. B. (2007). Modulation of cognitive performance following single doses of 120mg *Ginkgo biloba* extract administered to healthy young volunteers. *Human Psychopharmacology, 22*(8), 559–566.

Kennedy, D. O., Pace, S., Haskell, C., Okello, E. J., Milne, A., & Scholey, A. B. (2006). Effects of cholinesterase inhibiting sage (*Salvia officinalis*) on mood, anxiety and performance on a psychological stressor battery. *Neuropsychopharmacology, 31*(4), 845–852.

Kennedy, D. O., & Scholey, A. B. (2006). The psychopharmacology of European herbs with cognition-enhancing properties. *Current Pharmaceutical Design, 12*(35), 4613–4623.

Kennedy, D. O., Scholey, A. B., Tildesley, N. T. J., Perry, E. K., & Wesnes, K. A. (2002). Modulation of mood and cognitive performance following acute administration of *Melissa officinalis* (lemon balm). *Pharmacology Biochemistry and Behavior, 72*(4), 953–964.

Kennedy, D. O., Scholey, A. B., & Wesnes, K. A. (2000). The dose-dependent cognitive effects of acute administration of *Ginkgo biloba* to healthy young volunteers. *Psychopharmacology, 151*(4), 416–423.

Kennedy, D. O., Scholey, A. B., & Wesnes, K. A. (2001). Differential, dose dependent changes in cognitive performance following acute administration of a *Ginkgo biloba*/Panax ginseng combination to healthy young volunteers. *Nutritional Neuroscience, 4*(5), 399–412.

Kennedy, D. O., Wake, G., Savelev, S., Tildesley, N. T. J., Perry, E. K., Wesnes, K. A., et al. (2003). Modulation of mood and cognitive performance following acute administration of single doses of Melissa officinalis (Lemon balm) with human CNS nicotinic and muscarinic receptor-binding properties. *Neuropsychopharmacology, 28*(10), 1871–1881.

Kennedy, J. (2005). Herb and supplement use in the US adult population. *Clinical Therapeutics, 27*(11), 1847–1858.

Khokhar, S., & Magnusdottir, S. G. M. (2002). Total phenol, catechin, and caffeine contents of teas commonly consumed in the United Kingdom. *Journal of Agricultural and Food Chemistry, 50*(3), 565–570.

Kidd, P. M. (2005). Neurodegeneration from mitochondrial insufficiency: Nutrients, stem cells, growth factors, and prospects for brain rebuilding using integrative management. *Alternative Medicine Review, 10*(4), 268–293.

Kondo, K., Hirano, R., Matsumoto, A., Igarashi, O., & Itakura, H. (1996). Inhibition of LDL oxidation by cocoa. *Lancet, 348*(9040), 1514.

Kuriyama, S., Hozawa, A., Ohmori, K., Shimazu, T., Matsui, T., Ebihara, S., et al. (2006). Green tea consumption and cognitive function: A cross-sectional study from the Tsurugaya Project. *American Journal of Clinical Nutrition, 83*(2), 355–361.

Labadorf, C., Manktelow, T., Labadorf, S., et al. (2004). The global cognitive effects of ginseng taken by healthy volunteers over a 21 day period (*abstract*). *Journal of Psychopharmacology,* 18(suppl. 1), A46.

Lamming, D. W., Wood, J. G., & Sinclair, D. A. (2004). Small molecules that regulate lifespan: Evidence for xenohormesis. *Molecular Microbiology, 53*(4), 1003–1009.

Lazarus, S. A., Hammerstone, J. F., & Schmitz, H. H. (1999). Chocolate contains additional flavonoids not found in tea. *Lancet, 354*(9192), 1825.

Liang, Y. Z., Xie, P., & Chan, K. (2004). Quality control of herbal medicines. *Journal of Chromatography B: Analytical Technologies in the Biomedical and Life Sciences, 812*(1–2 special issue), 53–70.

Lindsay, J., Laurin, D., Verreault, R., Hébert, R., Helliwell, B., Hill, G. B., et al. (2002). Risk factors for Alzheimer's disease: A prospective analysis from the Canadian Study of Health and Aging. *American Journal of Epidemiology, 156*(5), 445–453.

Lu, T., Pan, Y., Kao, S. Y., Li, C., Kohane, I., Chan, J., et al. (2004). Gene regulation and DNA damage in the ageing human brain. *Nature, 429*(6994), 883–891.

MacLennan, A. H., Wilson, D. H., & Taylor, A. W. (2002). The escalating cost and prevalence of alternative medicine. *Preventive Medicine, 35*(2), 166–173.

Mahadevan, S., & Park, Y. (2008). Multifaceted therapeutic benefits of *Ginkgo biloba* L.: Chemistry, efficacy, safety, and uses. *Journal of Food Science, 73*(1), R14–R19.

Mahady, G. B. (2002). *Ginkgo biloba* for the prevention and treatment of cardiovascular disease: A review of the literature. *The Journal of Cardiovascular Nursing, 16*(4), 21–32.

Mandel, S., Weinreb, O., Amit, T., & Youdim, M. B. H. (2004). Cell signaling pathways in the neuroprotective actions of the green tea polyphenol (–)-epigallocatechin-3-gallate: Implications for neurodegenerative diseases. *Journal of Neurochemistry, 88*(6), 1555–1569.

Mandel, S. A., Amit, T., Weinreb, O., Reznichenko, L., & Youdim, M. B. H. (2008). Simultaneous manipulation of multiple brain targets by green tea catechins: A potential neuroprotective strategy for Alzheimer and Parkinson diseases. *CNS Neuroscience and Therapeutics, 14*(4), 352–365.

Mantle, D., Pickering, A. T., & Perry, E. K. (2000). Medicinal plant extracts for the treatment of dementia: A review of their pharmacology, efficacy and tolerability. *CNS Drugs, 13*(3), 201–213.

Mattei, R., Dias, R., Espínola, E., Carlini, E., & Barros, S. (1998). Guaraná (Paullinia cupana): Toxic behavioral effects in laboratory animals and antioxidant activity in vitro. *Journal of Ethnopharmacology, 60*(2), 111–116.

McKenna, D. J., Jones, K., & Hughes, K. (2001). Efficacy, safety, and use of ginkgo biloga in clinical and preclinical applications. *Alternative Therapies in Health and Medicine, 7*(5), 70–90.

Mishra, S., & Palanivelu, K. (2008). The effect of curcumin (turmeric) on Alzheimer's disease: An overview. *Annals of Indian Academy of Neurology, 11*(1), 13–19.

Mix, J. A., & Crews Jr, W. D. (2000). An examination of the efficacy of *Ginkgo biloba* extract EGb 761 on the neuropsychologic functioning of cognitively intact older adults. *Journal of Alternative and Complementary Medicine, 6*(3), 219–229.

Mix, J. A., & Crews Jr, W. D. (2002). A double-blind, placebo-controlled, randomized trial of *Ginkgo biloba* extract EGb 761® in a sample of cognitively intact older adults: Neuropsychological findings. *Human Psychopharmacology, 17*(6), 267–277.

Morris, B. J. (2008). How xenohormetic compounds confer health benefits. In E. Le Bourg & S. I. S. Rattan (Eds.), *Mild stress and healthy aging: applying hormesis in aging research and interventions.* Springer: The Netherlands.

Mulkens, A., Stephanou, E., & Kapetanidis, I. (1985). Glycosides with volatile genins in leaves of Melissa officinalis. *Heterosides a Genines Volatiles dans les Feuilles de Melissa officinalis L. (Lamiaceae), 60*(9–10), 276–278.

Nanjo, F., Goto, K., Seto, R., Suzuki, M., Sakai, M., & Hara, Y. (1996). Scavenging effects of tea catechins and their derivatives on 1,1-diphenyl-2-picrylhydrazyl radical. *Free Radical Biology and Medicine, 21*(6), 895–902.

Nathan, P. J., Tanner, S., Lloyd, J., Harrison, B., Curran, L., Oliver, C., et al. (2004). Effects of a combined extract of *Ginkgo biloba* and Bacopa monniera on cognitive function in healthy humans. *Human Psychopharmacology, 19*(2), 91–96.

Ng, T. P., Chiam, P. C., Lee, T., Chua, H. C., Lim, L., & Kua, E. H. (2006). Curry consumption and cognitive function in the elderly. *American Journal of Epidemiology, 164*(9), 898–906.

Packer, L., Rimbach, G., & Virgili, F. (1999). Antioxidant activity and biologic properties of a procyanidin-rich extract from pine (pinus maritima) bark, pycnogenol. *Free Radical Biology and Medicine, 27*(5–6), 704–724.

Pase, M. P., Pipingas, A., Kras, M., Nolidin, K., Gibbs, A. L., Wesnes, K. A., et al. (2010). Healthy middle-aged individuals are vulnerable to cognitive deficits as a result of increased arterial stiffness. *Journal of Hypertension, 28*(8), 1724–1729.

Perry, E. K., Pickering, A. T., Wang, W. W., Houghton, P. J., & Perry, N. S. L. (1999). Medicinal plants and Alzheimer's disease: From ethnobotany to phytotherapy. *Journal of Pharmacy and Pharmacology, 51*(5), 527–534.

Perry, N., Court, G., Bidet, N., Court, J., & Perry, E. (1996). European herbs with cholinergic activities: Potential in dementia therapy. *International Journal of Geriatric Psychiatry, 11*(12), 1063–1069.

Perry, N. S. L., Bollen, C., Perry, E. K., & Ballard, C. (2003). Salvia for dementia therapy: Review of pharmacological activity and pilot tolerability clinical trial. *Pharmacology Biochemistry and Behavior, 75*(3), 651–659.

Pervaiz, S. (2003). Resveratrol: From grapevines to mammalian biology. *FASEB Journal, 17*(14), 1975–1985.

Pipingas, A., Silberstein, R. B., Vitetta, L., Van Rooy, C., Harris, E. V., Young, J. M., et al. (2008). Improved cognitive performance after dietary supplementation with a Pinus radiata bark extract formulation. *Phytotherapy Research, 22*(9), 1168–1174.

Polich, J., & Gloria, R. (2001). Cognitive effects of a *Ginkgo biloba*/vinpocetine compound in normal adults: Systematic assessment of perception, attention and memory. *Human Psychopharmacology, 16*(5), 409–416.

Raghav, S., Singh, H., & Dalal, P. (2006). Randomized controlled trial of standardized bacopa monniera in age-associated memory impairment. *Indian Journal of Psychiatry, 48*, 238–242.

Ramassamy, C., Longpre, F., & Christen, Y. (2007). *Ginkgo biloba* extract (EGb 761) in Alzheimer's disease: Is there any evidence? *Current Alzheimer Research, 4*(3), 253–262.

Reay, J., Kennedy, D., & Scholey, A. (2006). Effects of Panax ginseng, consumed with and without glucose, on blood glucose levels and cognitive performance during sustained "mentally demanding" tasks. *Journal of Psychopharmacology, 20*(6), 771.

Reid, L. M., & MacLullich, A. M. J. (2006). Subjective memory complaints and cognitive impairment in older people. *Dementia and Geriatric Cognitive Disorders, 22*(5–6), 471–485.

Riley, P. A. (1994). Free radicals in biology: Oxidative stress and the effects of ionizing radiation. *International Journal of Radiation Biology, 65*(1), 27–33.

Rohdewald, P. (2002). A review of the French maritime pine bark extract (Pycnogenol®), a herbal medication with a diverse clinical pharmacology. *International Journal of Clinical Pharmacology and Therapeutics, 40*(4), 158–168.

Roodenrys, S., Booth, D., Bulzomi, S., Phipps, A., Micallef, C., & Smoker, J. (2002). Chronic effects of Brahmi (Bacopa monnieri) on human memory. *Neuropsychopharmacology, 27*(2), 279–281.

Rossi, L., Mazzitelli, S., Arciello, M., Capo, C. R., & Rotilio, G. (2008). Benefits from dietary polyphenols for brain aging and Alzheimer's disease. *Neurochemical Research, 33*(12), 2390–2400.

Ryan, J., Croft, K., Mori, T., Wesnes, K., Spong, J., Downey, L., et al. (2008). An examination of the effects of the antioxidant Pycnogenol® on cognitive performance, serum lipid profile, endocrinological and oxidative stress biomarkers in an elderly population. *Journal of Psychopharmacology, 22*(5), 553–562.

Sadraei, H., Ghannadi, A., & Malekshahi, K. (2003). Relaxant effect of essential oil of Melissa officinalis and citral on rat ileum contractions. *Fitoterapia, 74*(5), 445–452.

Salah, N., Miller, N. J., Paganga, G., Tijburg, L., Bolwell, G. P., & Rice-Evans, C. (1995). Polyphenolic flavanols as scavengers of aqueous phase radicals and as chain-breaking antioxidants. *Archives of Biochemistry and Biophysics, 322*(2), 339–346.

Salim, K., McEwen, B., & Chao, H. (1997). Ginsenoside Rb1 regulates ChAT, NGF and trkA mRNA expression in the rat brain. *Molecular Brain Research, 47*(1–2), 177–182.

Sarkar, D., & Fisher, P. B. (2006). Molecular mechanisms of aging-associated inflammation. *Cancer Letters, 236*(1), 13–23.

Sarter, M., & Bruno, J. P. (1997). Cognitive functions of cortical acetylcholine: Toward a unifying hypothesis. *Brain Research Reviews, 23*(1–2), 28–46.

Scaglione, F., Pannacci, M., & Petrini, O. (2005). The Standardised G115 (R) Panax ginseng CA Meyer Extract: A Review of its Properties and Usage. *Evidence-Based Integrative Medicine, 2*(4), 195.

Schaie, K. W. (1994). The course of adult intellectual development. *American Psychologist, 49*(4), 304–313.

Scholey, A., French, S., Morris, P., Kennedy, D., Milne, A., & Haskell, C. (2010). Consumption of cocoa flavanols results in acute improvements in mood and cognitive performance during sustained mental effort. *Journal of Psychopharmacology, 24*, 1505–1514.

Scholey, A., & Kennedy, D. (2002). Acute, dose-dependent cognitive effects of *Ginkgo biloba*, Panax ginseng and their combination in healthy young volunteers: Differential interactions with cognitive demand. *Human Psychopharmacology: Clinical and Experimental, 17*(1), 35–44.

Scholey, A., Ossoukhova, A., Owen, L., Ibarra, A., Pipingas, A., He, K., et al. (2010). Effects of American ginseng (Panax quinquefolius) on neurocognitive function: An acute, randomised, double-blind, placebo-controlled, crossover study. *Psychopharmacology, 212*(3), 345–356.

Scholey, A. B., Tildesley, N. T. J., Ballard, C. G., Wesnes, K. A., Tasker, A., Perry, E. K., et al. (2008). An extract of Salvia (sage) with anticholinesterase properties improves memory and attention in healthy older volunteers. *Psychopharmacology, 198*(1), 127–139.

Serafini, M., Bugianesi, R., Maiani, G., Valtuena, S., De Santis, S., & Crozier, A. (2003). Plasma antioxidants from chocolate. *Nature, 424*(6952), 1013.

Smit, H., & Rogers, P. (2000). Effects of low doses of caffeine on cognitive performance, mood and thirst in low and higher caffeine consumers. *Psychopharmacology (Berl), 152*(2), 167–173.

Smith, J. V., & Luo, Y. (2004). Studies on molecular mechanisms of *Ginkgo biloba* extract. *Applied Microbiology and Biotechnology, 64*(4), 465–472.

Sørensen, H., & Sonne, J. (1996). A double-masked study of the effects of ginseng on cognitive functions. *Current Therapeutic Research, 57*(12), 959–968.

Stough, C., Clarke, J., Lloyd, J., & Nathan, P. J. (2001a). Neuropsychological changes after 30-day *Ginkgo biloba* administration in healthy participants. *International Journal of Neuropsychopharmacology, 4*(2), 131–134.

Stough, C., Downey, L. A., Lloyd, J., Silber, B., Redman, S., Hutchison, C., et al. (2008). Examining the nootropic effects of a special extract of Bacopa monniera on human cognitive functioning: 90 day double-blind placebo-controlled randomized trial. *Phytotherapy Research, 22*(12), 1629–1634.

Stough, C., Lloyd, J., Clarke, J., Downey, L. A., Hutchison, C. W., Rodgers, T., et al. (2001b). The chronic effects of an extract of Bacopa monniera (Brahmi) on cognitive function in healthy human subjects. *Psychopharmacology, 156*(4), 481–484.

Tachikawa, E., Kudo, K., Harada, K., Kashimoto, T., Miyate, Y., Kakizaki, A., et al. (1999). Effects of ginseng saponins on responses induced by various receptor stimuli. *European journal of pharmacology, 369*(1), 23–32.

Taubert, D., Roesen, R., Lehmann, C., Jung, N., & Schomig, E. (2007). Effects of low habitual cocoa intake on blood pressure and bioactive nitric oxide: A randomized controlled trial. *Journal of the American Medical Association, 298*(1), 49–60.

Tildesley, N. T. J., Kennedy, D. O., Perry, E. K., Ballard, C. G., Savelev, S., Wesnes, K. A., et al. (2003). Salvia lavandulaefolia (Spanish Sage) enhances memory in healthy young volunteers. *Pharmacology Biochemistry and Behavior, 75*(3), 669–674.

Tildesley, N. T. J., Kennedy, D. O., Perry, E. K., Ballard, C. G., Wesnes, K. A., & Scholey, A. B. (2005). Positive modulation of mood and cognitive performance following administration of acute doses of Salvia lavandulaefolia essential oil to healthy young volunteers. *Physiology and Behavior, 83*(5), 699–709.

Tindle, H. A., Davis, R. B., Phillips, R. S., & Eisenberg, D. M. (2005). Trends in use of complementary and alternative medicine by us adults: 1997–2002. *Alternative Therapies in Health and Medicine, 11*(1), 42–49.

Tissenbaum, H. A., & Guarente, L. (2001). Increased dosage of a sir-2 gene extends lifespan in Caenorhabditis elegans. *Nature, 410*(6825), 227–230.

Tsang, D., Yeung, H., Tso, W., & Peck, H. (1985). Ginseng saponins: Influence on neurotransmitter uptake in rat brain synaptosomes. *Planta medica, 51*(3), 221.

United Nations. (2009). *World Population Prospects: The 2008 Revision, Highlights, Working Paper No. ESA/P/WP.210.* Department of Economic and Social Affairs Population Division.

Valenzano, D. R., & Cellerino, A. (2006). Resveratrol and the pharmacology of aging: A new vertebrate model to validate an old molecule. *Cell Cycle, 5*(10), 1027–1032.

Valenzano, D. R., Terzibasi, E., Genade, T., Cattaneo, A., Domenici, L., & Cellerino, A. (2006). Resveratrol prolongs lifespan and retards the onset of age-related markers in a short-lived vertebrate. *Current Biology, 16*(3), 296–300.

Veltri, K. L., Espiritu, M., & Singh, G. (1990). Distinct genomic copy number in mitochondria of different mammalian organs. *Journal of Cellular Physiology, 143*(1), 160–164.

Vingtdeux, V., Dreses-Werringloer, U., Zhao, H., Davies, P., & Marambaud, P. (2008). Therapeutic potential of resveratrol in Alzheimer's disease. *BMC Neuroscience, 9*(suppl. 2), S6.

Wake, G., Court, J., Pickering, A., Lewis, R., Wilkins, R., & Perry, E. (2000). CNS acetylcholine receptor activity in European medicinal plants traditionally used to improve failing memory. *Journal of Ethnopharmacology, 69*(2), 105–114.

Wang, Z. Y., Huang, M. T., Lou, Y. R., Xie, J. G., Reuhl, K. R., Newmark, H. L., et al. (1994). Inhibitory effects of black tea, green tea, decaffeinated black tea, and decaffeinated green tea on ultraviolet B light-induced skin carcinogenesis in 7,12-dimethylbenz[a]anthracene-initiated SKH-1 mice. *Cancer Research, 54*(13), 3428–3435.

Waterhouse, A. L., Shirley, J. R., & Donovan, J. L. (1996). Antioxidants in chocolate. *Lancet, 348*(9030), 834.

Weckerle, C., Stutz, M., & Baumann, T. (2003). Purine alkaloids in Paullinia. *Phytochemistry, 64*(3), 735–742.

Weinberger, N. M. (2004). Specific long-term memory traces in primary auditory cortex. *Nature Reviews Neuroscience, 5*(4), 279–290.

Williamson, G., & Manach, C. (2005). Bioavailability and bioefficacy of polyphenols in humans. II. Review of 93 intervention studies. *The American Journal of Clinical Nutrition, 81*(Suppl. 1), S230–S242.

Xue, C. C. L., Zhang, A. L., Lin, V., Da Costa, C., & Story, D. F. (2007). Complementary and alternative medicine use in Australia: A national population-based survey. *Journal of Alternative and Complementary Medicine, 13*(6), 643–650.

Yang, C. S., & Wang, Z. Y. (1993). Tea and cancer. *Journal of the National Cancer Institute, 85*(13), 1038–1049.

Zaborszky, L. (2002). The modular organization of brain systems. Basal forebrain: The last frontier. *Progress in Brain Research, 136*, 359–372.

Zecca, L., Youdim, M. B. H., Riederer, P., Connor, J. R., & Crichton, R. R. (2004). Iron, brain ageing and neurodegenerative disorders. *Nature Reviews Neuroscience, 5*(11), 863–873.

Zhang, J. T., Qu, Z. W., Liu, Y., & Deng, H. L. (1990). Preliminary study on anti-amnestic mechanism of ginsenoside Rg1 and Rb1. *Chinese Medical Journal, 103*(11), 932–938.

CHAPTER 16

Preventive Effects of Resveratrol on Age-Associated Neurological Disorders

Stephane Bastianetto and Remi Quirion

Chapter Overview

- A growing number of epidemiological studies indicate that older adults who consume red wine (in moderation), green tea, fruits and vegetables have a lower risk of developing age-related neurological disorders.
- In vitro studies and animal models have demonstrated that polyphenols, and particularly the stilbene called resveratrol, display a neuroprotective action in various models of toxicity, suggesting that they mediate, at least in part, the prophylactic effects of food and beverages in reducing the risk of developing neurological disorders.
- In this chapter we review the most recent findings about mechanisms of action possibly involved in the beneficial effect of resveratrol and high-light the possible role of stilbenes in the prevention of various age-related neurological disorders.

Introduction

There is strong evidence suggesting that a healthy diet may delay the incidence of neurodegenerative disorders related to ageing, such as Alzheimer's disease (AD), Parkinson's disease (PD) and stroke (Luchsinger et al. 2007; Morris 2009). For example, it has been reported that older adults who regularly consume fruits, vegetables and fish have a lower risk of cognitive decline related or not to dementia (Luchsinger et al. 2007). Moreover, people who consume red wine (one to three drinks per day) in moderation and on a regular basis display a reduced incidence (up to 50 per cent) of dementia, such as AD and vascular dementia, as well as macular degeneration (Luchsinger et al. 2007; Obisesan et al. 1998; Orgogozo et al. 1997).

These findings from epidemiological studies are of great interest for three main reasons. First, with an upsurge in life expectancy, an increased number of people will be affected by or susceptible to age-related neuropsychiatric disorders. Second, to date, there is no cure for the treatment of dementia. Third, the economic burden of care and treatment of patients with age-related neurological disorders will increase. Considerable effort has been made to explain the protective role of diet, and particularly red wine, in the prevention of age-related neurodegenerative diseases. Although it is likely that the combination of the numerous polyphenols present in red wine accounts for its beneficial effect, resveratrol has drawn particular attention because its presence in wine has been suggested to explain the cardioprotective effects of wine (for a review, see Baur & Sinclair 2006). Numerous studies have shown that resveratrol displays anti-inflammatory and antioxidant properties, and modulates protein aggregation (e.g. beta-amyloid, Aß) and intracellular effectors (e.g. haem oxygenase 1) involved in neuronal cell survival/death. Here, we will review possible mechanisms underlying the purported neuroprotective action of resveratrol in in vitro as well as rodent models of diseases. Table 16.1 summarises the protective effects of resveratrol.

Dietary Source and Bioavailability of Resveratrol

Resveratrol (3,5,4′-trihydroxystilbene) was first isolated in 1940 from the roots of white hellebore (Veratrum grandiflorum O. Loes) and later from the medicinal plant named Polygonum cuspidatum (Japanese 'Ko-jo-kon'). Forty years later, it was characterised as a phytoalexin produced after microbial infections (Baur & Sinclair 2006). Resveratrol has been identified from a number of dietary sources, including red grapes, muscadine grapes, berries (e.g. cranberries, bilberries, blueberries) and peanuts (Baur & Sinclair 2006), and trans-resveratrol is the preferred steric form. It is also present in red wine and to a much lesser extent in white wine. The concentration of trans-resveratrol in red wine varies widely, from 0.1 to 14.3 mg/l (i.e. 0.4–57μM), and depends on the type of red wine and its geographic origin (Baur & Sinclair 2006). The range of concentration could be doubled if the cis steric and the β-glucoside (i.e. piceid) forms of resveratrol are included (Baur & Sinclair 2006).

The bioavailability of resveratrol is poor and oral administration of resveratrol undergoes extensive metabolism in the intestine and liver (Walle et al. 2004; Walle 2011). Hence, only trace amounts of free resveratrol (approximately 7 μg/l; 30 nM) could be detected in plasma 30 minutes after a 25 mg oral dose, whereas the total (free and conjugated) estimated resveratrol quantity was much higher (around 450 μg/l; 2 μM) (Goldberg et al 2003). It is therefore unlikely that resveratrol is the sole polyphenol that may explain the purported beneficial effects of red wine in the elderly population, since resveratrol intake of 25 mg corresponds to the consumption of more than

Table 16.1 Summary of the protective effects of resveratrol

Effects	Proposed underlying mechanisms	References
Increases cerebral blood flow (250 and 500 mg) during cognitive task performance in healthy adults	Vasodilatory activities	Kennedy et al. (2010)
Protects and rescues (5–25 µM) hippocampal neuronal cells against toxicity induced by nitric oxide	Free radical and ion metal scavenging activities	Bastianetto et al. (2000)
Protects (25 µM) primary neuronal cells against excitotoxicity	Increase in HO1 protein levels	Sakata et al. (2010)
Protects (10–60 µM) mitochondria against oxidative stress	Phosphorylation of GSK3β	Shin et al. (2009)
Protects (10–40 mg/kg/day for 10 weeks) dopaminergic neuronal cells exposed to 6-hydroxydopamine (6-OHDA)	Reduction of the expression of COX-2 protein and tumour necrosis factor α (TNFα) in the substantia nigra	Jin et al. (2008)
Inhibits (up to 50 µM) the activation of microglia exposed to lipopolysaccharide (LPS)	Reduction of the production/expression of prostaglandins, NO, TNFα, COX-1 and NF-κB	Bi et al. (2005) Candelario-Jalil et al. (2007) Kim et al. (2007) Meng et al., (2008)
Resveratrol (15–60 µM) protects against LPS-induced dopaminergic neuronal death	Attenuating of the activation of MAPK and NF-κB in microglia	Zhang et al. (2010)

Table 16.1 (Continued)

Protects (15–40 µM) hippocampal cells exposed to Aβ peptides	– Induction of the phosphorylation of PKCδ isoform – Inhibition and destabilisation of $Aβ_{1-42}$ fibrils formation	Han et al. (2004) Feng et al. (2009) Richard et al. (2011)
Protects (10–20 µM) HT22 hippocampal cells against Aβ-induced toxicity	Inhibition of the $Aβ_{1-42}$-induced activation of GSK3β and AMPK activity	Kwon et al. (2010)
Reduces (diet supplemented with 3.5 g/l of resveratrol for 2 weeks) in brain Aβ levels and amyloid deposition in mice	– Activation of the degradation of Aβ via a mechanism that involves proteasome – Activation of the AMPK signalling pathway	Vingtdeux et al. (2010)
Reduces (75 and 100 µM) neuronal death in the CA1 region of the hippocampus exposed to ischaemia	Activation of SIRT1	Raval et al. (2006)
Protects (7.5 µM) neuroblastoma cell line exposed to 6-OHDA	Activation of SIRT1	Albani et al. (2009)

4 litres of red wine, while a daily intake of two glasses of red wine contains a mean content of 5 mg/l of resveratrol.

Human Studies

A few studies have been performed to assess the effects of oral resveratrol on cognitive performance in both healthy and cognitively impaired patients. Based on the vasodilatory action of resveratrol, Kennedy et al. (2010) aimed to investigate whether the polyphenol aimed to improve blood flow, and thereby promote cognitive function. In a randomised, double-blind, placebo-controlled trial involving 22 healthy adults, it was found that administration of resveratrol (250 and 500 mg) resulted in dose-dependent increases in cerebral blood flow during cognitive task performance that activates the frontal cortex. However, cognitive function was not affected. The presence of free resveratrol was confirmed by high-performance liquid chromatography (HPLC), with concentrations peaking at 5.65 and 14.4 $\mu g/l$ after 250 and 500 mg, respectively, 90 minutes after dosing (Kennedy et al. 2010). In contrast, the glucuronidated and sulfated conjugates were present in much higher concentrations 45 minutes after the start of the cognitive tasks. Concentrations, particularly of the sulfated conjugate, continued to rise to reach about 300 and 700 $\mu g/l$ after 250 and 500 mg of resveratrol, respectively, 90 minutes post-dose, which corresponded to the end of the cognitive tasks (Kennedy et al. 2010).

Clinical studies are underway to evaluate resveratrol as a dietary ingredient with beneficial effects in mild to moderate AD. At present, three randomised, placebo-controlled pilot studies are being conducted to determine whether resveratrol supplements (250–1000 mg/day for 12–52 weeks) improve memory and physical performance in older adults (ClinicalTrials.gov Identifier: NCT01126229), ameliorate cognitive and global functioning in patients with mild to moderate AD on standard therapy (ClinicalTrials.gov Identifier: NCT00743743) or slow the progression of AD (ClinicalTrials.gov Identifier: NCT00678431). These trials will confirm if resveratrol can be used as a potential therapeutic agent alone or in combination with other polyphenols in neurodegenerative disorders associated with ageing.

Antioxidant and Anti-Inflammatory Effects of Resveratrol

The mechanisms underlying the pathological processes occurring in dementia remain to be elucidated. However, it has been postulated that these behavioural deficits and neuronal death are caused by an increasing vulnerability to oxidative stress and inflammatory molecules (i.e. cytokines) (for a review, see Agostinho et al. 2010). Resveratrol is a free radical scavenger, a copper chelator and displays potent antioxidant activity by up-regulating endogenous antioxidant enzymes, including glutathione peroxidase (Belguendouz et al. 1997; de la Lastra et al. 2005). We have shown that resveratrol is able to protect and rescue hippocampal cells against toxicity induced by nitric oxide (NO) (Bastianetto et al. 2000). Both the purported antioxidant activities as well as the reactive oxygen species (ROS)-scavenging properties could likely explain, at least in part, the protective and rescuing abilities of resveratrol (Bastianetto et al. 2000). Resveratrol may also play a beneficial role in the cellular response by modulating enzymes involved in stress response, such as quinone reductase 2 (QR2) – a cytosolic enzyme which enhances the production of damaging activated quinone and ROS. Our group demonstrated that QR2 is overexpressed in the hippocampus (a brain region involved in learning and memory) in rodents that exhibit learning deficits, suggesting that this enzyme plays a deleterious role in cognition (Benoit et al. 2010). This hypothesis is confirmed by the fact that adult QR2 knock-out mice show improved learning abilities compared with wild-type mice. Interestingly, selective inhibitors of QR2 were able to block hippocampal neuronal cell death induced by menadione, a QR2 substrate. Resveratrol appeared to be a good inhibitor of QR2 in the presence of nicotinamide riboside (NRH), likely a natural QR2 co-substrate, with inhibitory concentration $(IC)_{50}$ of 2.9 µM (Ferry et al. 2010). Another study confirmed that resveratrol is a potent inhibitor of QR2 enzymatic activity with a dissociation constant of 35 nM (Buryanovskyy et al. 2004). This inhibition of QR2 may lead to an up-regulation of antioxidant enzymes, resulting in an increase in cellular resistance to oxidation-related neuronal damage (Buryanovskyy et al. 2004).

Besides its inhibitory action on QR2, resveratrol has been shown to modulate haem oxygenase (HO), an enzyme that metabolises the prooxidant haem to produce carbon monoxide, iron and biliverdin, which is immediately reduced to the antioxidant bilirubin. The inducible form of HO, known as HO-1, is induced in response to various noxious stimuli, such as hypoxia and oxidative stress, whereas it is inhibited by beta-amyloid ($A\beta$), suggesting that this isoform is a promising therapeutic target for the treatment of various diseases, including ischaemia/reperfusion and AD (Baranano and Snyder 2001; Ma et al. 2010; Panahian et al. 1999; Takahashi et al. 2000). Doré et al. have recently shown that resveratrol (25 µM) protected against excitotoxicity in primary neuronal cells, and this effect was accompanied by an increase in HO-1 protein levels (Sakata et al. 2010). The possible role of HO1 in the neuroprotective action of resveratrol was further confirmed by showing that a peripheral administration of HO1 (20 mg/kg) significantly attenuated infarct size in wild-type but not in HO1 knock-out (HO1 -/-) mice subjected to ischaemia and reperfusion.

Finally, using neuronal cell lines, Dasgupta and Milbrandt (2007) showed that resveratrol is a potent activator of AMP-activated protein kinase (AMPK), a key regulator of cell survival in response to oxidative stress insults (Culmsee et al. 2001; Shibata et al. 2005). Moreover, resveratrol protected against ROS production induced by a combination of arachidonic acid and iron, a mechanism that involves phosphorylation of glycogen synthase kinase-3β (GSK3β) and depends on AMPK (Shin et al. 2009).

Besides its antioxidant activities, resveratrol has been reported to display anti-inflammatory properties (for reviews, see De la Lastra et al. 2005; Zhang et al. 2010). It is well established that resveratrol is an effective inhibitor of cyclooxygenases, particularly COX-1.These enzymes are involved in the production of pro-inflammatory molecules, cytokines (Jang et al. 1997). Using a rat model of 6-hydroxydopamine (6-OHDA)-induced PD, Jin et al. (2008) showed that resveratrol exerted a neuroprotective effect, and this effect was related to its ability to reduce expression of COX-2 protein and tumour necrosis factor α (TNFα) in the substantia nigra.

Resveratrol is also able to reduce the release of pro-inflammatory mediators through the inhibition of the transcriptional factors such as nuclear factor-kappaB (NF-κB) and activator protein-1 (AP-1) (Das & Das 2007).

It has been hypothesised that microglial activation may contribute to neuronal death during brain damage by releasing neurotoxic pro-inflammatory molecules (McGeer & McGeer 2004; Perry et al. 2010). Resveratrol has been shown to inhibit the activation of microglia and reduce the production of pro-inflammatory factors through cellular cascade signalling pathways. Using rat primary microglia cultures exposed to lipopolysaccharide (LPS), it has been reported that resveratrol (up to 50 μM) reduced the production of prostaglandins (e.g. PGE2), NO, TNFα, as well as the expression of COX-1 and activation of NF-κB (Bi et al. 2005; Candelario-Jalil et al. 2007; Kim et al. 2007; Meng et al. 2008). Finally, it has recently been shown that resveratrol (15–60 μM) protects against LPS-induced dopaminergic neuronal death by attenuating the activation of mitogen-activated protein kinases (MAPK) and nuclear factor-kappaB signalling pathways in microglia, supporting the hypothesis that resveratrol-related neuroprotection involved inhibition of microglia-mediated neuroinflammation (Zhang et al. 2010).

Amyloidogenesis

Although inflammation and free radicals are likely to be involved in the progression of AD pathology, the dominant hypothesis is still that progressive accumulation of Aß deposits plays a pivotal role in neurodegeneration processes seen in AD (Hardy et al. 2002). It has been postulated that the reduction of Aß generation/accumulation may be an effective therapeutic approach to block neuronal damage in AD (Aisen et al. 2005; Hardy 2009). Our group reported that Aß peptides-induced hippocampal neuronal death was dose-dependently reduced in the presence of resveratrol (15–40 μM) (Han et al. 2004, 2006). Further, in vitro studies have shown the protective action of resveratrol and derivatives against Aß-associated neuronal death in PC12 cells (Jang et al. 2003) and murine HT22 hippocampal cell line

(Kwon et al. 2010). In the most recent study, the authors reported that resveratrol inhibited both the $A\beta_{1-42}$-induced activation of glycogen synthase kinase-3β (GSK3β) and AMP-activated protein kinase (AMPK) activity (Kwon et al. 2010). Other mechanisms have also been proposed to explain the neuroprotective action of resveratrol in $A\beta$-related neurotoxicity. For example, resveratrol and other derivatives (e.g. piceid, piceatannol and viniferin glucoside) were shown to directly interact with $A\beta$ peptides by inhibiting and destabilising the formation of $A\beta_{1-42}$ fibrils (Feng et al. 2009; Richard et al. 2011). Resveratrol has also been shown to display anti-amyloidogenic effect by promoting $A\beta$ clearance, rather than inhibiting its production (Riviere et al. 2007). Indeed, resveratrol did not alter the activities of both β-and γ-secretase, two enzymes recognised for their role in the production of amyloidogenic $A\beta$ peptides but instead promoted intracellular degradation of $A\beta$ via a mechanism that involves proteasome, a multicatalytic protease complex (Marambaud et al. 2005). However, Jeon et al. demonstrated that resveratrol displayed inhibitory activities of β-secretase with an IC_{50} of 15 μM (Jeon et al., 2007).

Although insoluble fibrils are hallmarks of AD pathology, a newer aspect receiving considerable attention is the build-up of soluble forms of $A\beta$ oligomers (known as $A\beta$-derived diffusible neurotoxic ligands or ADDLs) in the brains and cerebrospinal fluid of AD patients (Gongn et al. 2003; Klein 2002). Their accumulation in the brains of affected individuals has been suggested to explain the neuronal dysfunction and associated early cognitive decline characteristic of AD (Klein 2006). A recent study showed that resveratrol could dose-dependently inhibit $A\beta_{1-42}$ fibril formation and cytotoxicity. According to the study's authors, resveratrol may change $A\beta_{1-42}$ oligomer conformation but does not prevent its formation (Feng et al. 2009). Finally, it has been postulated that the activation of protein kinase C (PKC) may stimulate α-secretase and then non-amyloidogenic pathway in APP processing and resulting in a reduction in the production of $A\beta$ (Stefan et al. 2004). Our group reported that GF 109203X, a PKC inhibitor attenuated the neuroprotective effects of resveratrol against $A\beta$-induced toxicity in hippocampal cell cultures (Han et al. 2004). Moreover, western blot data suggested that resveratrol (20–30 μM) induced the phosphorylation of PKCδ isoform, whose inhibition is required for initiation of the apoptotic pathway (DeVries-Seimon et al. 2007). Taken together, these results support the idea that resveratrol protect neuronal cells against $A\beta$-induced toxicity through a mechanism that involves a stimulation of PKC isoforms (particularly the delta isoform), possibly leading to an inactivation of GSK3β. In support of this hypothesis, the PKC activator known as phorbol 12-myristate 13-acetate has been shown to block primary hippocampal neuronal cell death exposed to $A\beta$ and to inhibit GSK3β via a mechanism that involves serine 9 phosphorylation (Garrido et al. 2002). Vingtdeux et al. (2010) suggested that AMPK signalling pathway plays a role in the anti-amyloidogenic effect of resveratrol.

They showed that peripheral administration of resveratrol stimulated AMPK and reduced Aβ levels and amyloid deposition in the cerebral cortex in mice, whereas the inhibition of AMPK counteracted the effect of resveratrol on Aβ accumulation (Vingtdeux et al. 2010).

Sirtuins

It has been reported that activation of sirtuins leads to the activation of anti-apoptotic, anti-inflammatory and anti-stress processes, suggesting that these enzymes might be new targets for the treatment of neurological disorders such as stroke, AD and PD (Albani et al. 2009; Outeiro et al. 2008). Various groups have shown that resveratrol is a SIRT1 activator (Albani et al. 2010; Borra et al. 2005; Karuppagounde et al. 2009), an effect that might be indirect and via AMPK activation (Pacholec et al. 2010; Tang et al. 2010). Some evidence suggests that the neuroprotective action of resveratrol in models of ischaemia or PD involves SIRT1. For example, Raval et al. (2006) demonstrated that the inhibition of SIRT1 blocks the ability of resveratrol to mimic ischaemic preconditioning, a brief stressful episode that protects against subsequent ischaemic insults, whereas a down-regulation of SIRT1 expression abolished this protective action in neuroblastoma cell line exposed to 6-OHDA (Albani et al. 2009).

Conclusions

In previous studies, resveratrol displayed pleiotropic activities, including antioxidant and anti-inflammatory effects, as well as an inhibition action, suggesting that it is one of the most promising compounds for the development of AD therapies. Several intracellular targets of resveratrol have also been identified. However, the poor absorption of resveratrol makes the development of analogues very challenging. The drug company Sirtris Pharmaceuticals developed a formulation of resveratrol called SRT501, with about five times higher bioavailability than the chemical alone. However, SRT501 induced multiple side effects that caused the suspension of the clinical trial in patients with multiple myeloma. Different strategies need to be investigated to increase its solubility and bioavailability. Recently, nanotechnologies have been shown to improve the solubility and stability of different natural ingredients. Hence, various methods such us encapsulation in polymeric micelles and lipid based nanoparticles and hydrogels have been shown to improve the solubility of curcumin, a natural compound derived from the spice turmeric. These new technologies should be used to confirm if resveratrol is a potential therapeutic agent in age-related neurodegenerative disorders associated with ageing.

Acknowledgements

The authors wish to thank Mira Thakur for proofreading the manuscript.

References

Agostinho P, Cunha RA, Oliveira C. (2010) Neuroinflammation, oxidative stress and the pathogenesis of Alzheimer's disease. *Curr Pharm* 16:2766–78.

Aisen P. (2005) The development of anti-amyloid therapy for Alzheimer's disease: from secretase modulators to polymerisation inhibitors. *CNS Drugs* 19:989–96.

Albani D, Polito L, Batelli S, De Mauro S, Fracasso C, Martelli G, Colombo L, Manzoni C, Salmona M, Caccia S, Negro A, Forloni G. (2009) The SIRT1 activator resveratrol protects SK-N-BE cells from oxidative stress and against toxicity caused by alpha-synuclein or amyloid-beta (1–42) peptide. *J Neurochem* 110:1445–56.

Albani D, Polito L, Forloni G. (2010) Sirtuins as novel targets for Alzheimer's isease and other neurodegenerative disorders: experimental and genetic evidence. *J Alzheimers Dis* 19, September 11.

Albani D, Polito L, Signorini A, Forloni G. (2010) Neuroprotective properties of resveratrol in different neurodegenerative disorders. *Biofactors* 36:370–6.

Bastianetto S, Zheng WH, Quirion R. (2000) Neuroprotective abilities of resveratrol and other red wine constituents against nitric oxide-related toxicity in cultured hippocampal neurons. *Br J Pharmacol* 131:711–20.

Baur JA, Sinclair DA. (2006) Therapeutic potential of resveratrol: the in vivo evidence. *Nat Rev Drug Discov* 5:493–506.

Belguendouz L, Fremont L, Linard A. (1997) Resveratrol inhibits metal ion-dependent and independent peroxidation of porcine low-density lipoproteins. *Biochemi Pharmacol* 53:1347–55.

Benoit CE, Bastianetto S, Brouillette J, Tse Y, Boutin JA, Delagrange P, Wong T, Sarret P, Quirion R. (2010) Loss of quinone reductase 2 function selectively facilitates learning behaviors. *J Neurosci* 30:12690–700.

Bi XL, Yang JY, Dong YX, Wang JM, Cui YH, Ikeshima T, Zhao YQ, Wu CF. (2005) Resveratrol inhibits nitric oxide and TNF-alpha production by lipopolysaccharide-activated microglia. *Int Immunopharmacol* 5:185–93.

Buryanovskyy L, Fu Y, Boyd M, Ma Y, Hsieh TC, Wu JM, Zhang Z. (2004) Crystal structure of quinone reductase 2 in complex with resveratrol. *Biochemistry* 43:11417–26.

Candelario-Jalil E, de Oliveira AC, Gräf S, Bhatia HS, Hüll M, Muñoz E, Fiebich BL. (2007) Resveratrol potently reduces prostaglandin E2 production and free radical formation in lipopolysaccharide-activated primary rat microglia. *J Neuroinflammation* 4:25–36.

Culmsee C, Monnig J, Kemp BE, Mattson MP. (2001) AMP-activated protein kinase is highly expressed in neurons in the developing rat brain and promotes neuronal survival following glucose deprivation. *J Mol Neurosci* 17:45–58.

Das S, Das DK. (2007) Anti-inflammatory responses of resveratrol. *Inflamm Allergy Drug Targets* 6:168–73.

Dasgupta B, Milbrandt J. (2007) Resveratrol stimulates AMP kinase activity in neurons. *Proc Natl Acad Sci USA* 104:7217–22.

De la Lastra CA, Villegas I. (2005) Resveratrol as an anti-inflammatory and anti-aging agent: mechanisms and clinical implications. *Mol Nutr Food Res* 49:405–30.

DeVries-Seimon TA, Ohm AM, Humphries MJ, Reyland ME. (2007) Induction of apoptosis is driven by nuclear retention of protein kinase C delta. *J Biol Chem* 282:22307–14.

Feng Y, Wang XP, Yang SG, Wang YJ, Zhang X, Du XT, Sun XX, Zhao M, Huang L, Liu RT. (2009) Resveratrol inhibits beta-amyloid oligomeric cytotoxicity but does not prevent oligomer formation. *Neurotoxicology* 30:986–95.

Ferry G, Hecht S, Berger S, Moulharat N, Coge F, Guillaumet G, Leclerc V, Yous S, Delagrange P, Boutin JA. (2010) Old and new inhibitors of quinone reductase 2. *Chem Biol Interact* 186:103–9.

Goldberg DM, Yan J, Soleas GJ. (2003) Absorption of three wine-related polyphenols in three different matrices by healthy subjects. *Clin Biochem* 36:79–87.

Gong Y, Chang L, Viola KL, Lacor PN, Lambert MP, Finch CE, Krafft GA, Klein WL. (2003) *Proc Natl Acad Sci USA* 100:10417–22.

Han YS, Zheng WH, Bastianetto S, Chabot JG, Quirion R. (2004) Neuroprotective effects of resveratrol against beta-amyloid-induced neurotoxicity in rat hippocampal neurons: involvement of protein kinase C. *Br J Pharmacol* 141:997–1005.

Han YS, Bastianetto S, Dumont Y, Quirion R. (2006) Specific plasma membrane binding sites for polyphenols, including resveratrol, in the rat brain. *J Pharmacol Exp Ther* 318:238–45.

Hardy J (2009) The amyloid hypothesis for Alzheimer's disease: a critical reappraisal. *J Neurochem* 110:1129–34.

Hardy J, Selkoe DJ. (2002) The amyloid hypothesis of Alzheimer's disease: progress and problems on the road to therapeutics. *Science* 297:353–6.

Jang M, Cai L, Udeani GO, Slowing KV, Thomas CF, Beecher CW, Fong HH, Farnsworth NR, Kinghorn AD, Mehta RG, Moon RC, Pezzuto JM. (1997) Cancer chemopreventive activity of resveratrol, a natural product derived from grapes. *Science* 275:218–20.

Jin F, Wu Q, Lu YF, Gong QH, Shi JS. (2008) Neuroprotective effect of resveratrol on 6-OHDA-induced Parkinson's disease in rats. *Eur J Pharmacol* 600: 78–82.

Karuppagounder SS, Pinto JT, Xu H, Chen HL, Beal MF, Gibson GE. (2009) Dietary supplementation with resveratrol reduces plaque pathology in a transgenic model of Alzheimer's disease. *Neurochem Int* 54:111–8.

Kennedy DO, Wightman EL, Reay JL, Lietz G, Okello EJ, Wilde A, Haskell CF. (2010) Effects of resveratrol on cerebral blood flow variables and cognitive performance in humans: a double-blind, placebo-controlled, crossover investigation. *Am J Clin Nutr* 91:1590–7.

Kim YA, Kim GY, Park KY, Choi YH. (2007) Resveratrol inhibits nitric oxide and prostaglandin E2 production by lipopolysaccharide-activated C6 microglia. *J Med Food* 10:218–24.

Kirkby KA, Adin CA. (2006) Products of heme oxygenase and their potential therapeutic applications. *Am J Physiol Renal Physiol* 290:F563–F571.

Klein WL. (2002) ADDLs & protofibrils – the missing links? *Neurobiol Aging* 23:231–5.

Klein WL. (2006) Synaptic targeting by Abeta oligomers (ADDLS) as a basis for memory loss in early Alzheimer's disease. *Alzheimers Dement* 2:43–55.

Kwon KJ, Kim HJ, Shin CY, Han SH. (2010) Melatonin potentiates the neuroprotective properties of resveratrol against beta-amyloid-induced neurodegeneration by modulating AMP-activated protein kinase pathways. *J Clin Neurol* 6: 127–37.

Lichtenthaler SF, Haass C. (2004) Amyloid at the cutting edge: activation of alpha-secretase prevents amyloidogenesis in an Alzheimer disease mouse model. *J Clin Invest* 113:1384–7.

Luchsinger JA, Noble JM, Scarmeas N. (2007) Diet and Alzheimer's disease. *Curr Neurol Neurosci Rep* 7:366–72.

Marambaud P, Zhao H, Davies P. (2005) Resveratrol promotes clearance of Alzheimer's disease amyloid-beta peptides. *J Biol Chem* 280:37377–82.

Martinez J, Moreno JJ. (2000) Effect of resveratrol, a natural polyphenolic compound, on reactive oxygen species and prostaglandin production. *Biochemi Pharmacol* 59:865–70.

McGeer PL, McGeer EG. (2004) Inflammation and neurodegeneration in Parkinson's disease. *Parkinsonism Relat Disord* 10:S3–7.

McNaull BB, Todd S, McGuinness B, Passmore AP. (2010) Inflammation and anti-inflammatory strategies for Alzheimer's disease – a mini-review. *Gerontology* 56:3–14.

Meng XL, Yang JY, Chen GL, Wang LH, Zhang LJ, Wang S, Li J, Wu CF. (2008). Effects of resveratrol and its derivatives on lipopolysaccharide-induced microglial activation and their structure-activity relationships. *Chem Biol Interact* 174:51–9.

Morris MC. (2009)The role of nutrition in Alzheimer's disease: epidemiological evidence. *Eur J Neurol* 16(Suppl. 1):1–7.

Obisesan TO, Hirsh R, Kosoko O, Carlson LM, Parrott M. (1998) Moderate wine consumption is associated with decreased odds of developing age-related macular degeneration in NHANES-1. *J Am Ger Soc* 46:1–7.

Orgogozo JM, Dartigues JF, Lafont S, Letenneur L, Commenges D, Salomon R, Renaud S, Breteler MB. (1997) Wine consumption and dementia in the elderly: a prospective community study in the Bordeaux area. *Revue Neurologique* 153:185–92.

Outeiro TF, Marques O, Kazantsev A. (2008) Therapeutic role of sirtuins in neurodegenerative disease. *Biochim Biophys Acta* 1782:363–69.

Pallàs M, Casadesús G, Smith MA, Coto-Montes A, Pelegri C, Vilaplana J, Camins A. (2009) Resveratrol and neurodegenerative diseases: activation of SIRT1 as the potential pathway towards neuroprotection. *Curr Neurovasc Res* 6:70–81.

Panahian N, Yoshiura M, Maines MD. (1999) Overexpression of heme oxygenase-1 is neuroprotective in a model of permanent middle cerebral artery occlusion in transgenic mice. *J Neurochem* 72:1187–203.

Patel KR, Scott E, Brown VA, Gescher AJ, Steward WP, Brown K. (2011) Clinical trials of resveratrol. *Ann N Y Acad Sci* 1215:161–9.

Perry VH, Nicoll JA, Holmes C. (2010) Microglia in neurodegenerative disease. *Nat Rev Neurol* 6:193–201.

Raval AP, Dave KR, Pérez-Pinzón MA. (2006) Resveratrol mimics ischemic preconditioning in the brain. *J Cereb Blood Flow Metab* 26:1141–47.

Riviere C, Richard T, Quentin L, Krisa S, Merillon JM, Monti JP. (2007) Inhibitory activity of stilbenes on Alzheimer's beta-amyloid fibrils in vitro. *Bioorg Med Chem* 15:1160–67.

Shibata R, Sato K, Pimentel DR, Takemura Y, Kihara S, Ohashi K, Funahashi T, Ouchi N, Walsh K. (2005) Adiponectin protects against myocardial ischemia-reperfusion injury through AMPK- and COX-2-dependent mechanisms. *Nat Med* 11:1096–103.

Shin SM, Cho IJ, Kim SG. (2009) Resveratrol protects mitochondria against oxidative stress through AMP-activated protein kinase-mediated glycogen synthase kinase-3beta inhibition downstream of poly(ADP-ribose)polymerase-LKB1 pathway. *Mol Pharmacol* 76:884–95.

Siemann EH, Creasy LL. (1992) Concentration of the phytoalexin resveratrol in wine. *Am J Eno Vitic* 43:49–52.

Soleas GJ, Diamandis EP, Goldberg DM. (1997) Resveratrol: a molecule whose time has come? And gone? *Clin Biochem* 30:91–113.

Vingtdeux V, Giliberto L, Zhao H, Chandakkar P, Wu Q, Simon JE, Janle EM, Lobo J, Ferruzzi MG, Davies P, Marambaud P. (2010) AMP-activated protein kinase signalling activation by resveratrol modulates amyloid-beta peptide metabolism. *J Biol Chem* 285:9100–13.

Walle T, Hsieh F, DeLegge MH, Oatis JE, Walle UK. (2004) High absorption but very low bioavailability of oral resveratrol in humans. *Drug Metab Dispos* 32:1377–82.

Walle T. (2011) Bioavailability of resveratrol. *Ann N Y Acad Sci* 1215:9–15.

Zhang F, Liu J, Shi JS. (2010) Anti-inflammatory activities of resveratrol in the brain: role of resveratrol in microglial activation. *Eur J Pharmacol* 636:1–7.

Zhang F, Shi JS, Zhou H, Wilson B, Hong JS, Gao HM. (2010) Resveratrol protects dopamine neurons against lipopolysaccharide-induced neurotoxicity through its anti-inflammatory actions. *Mol Pharmacol* 78:466–77.

Zhuang H, Kim YS, Koehler RC, Doré S. (2003) Potential mechanism by which 281 resveratrol, a red wine constituent, protects neurons. *Ann NY Acad Sci* 993:276–86.

Index